Advance Praise for *Head First Algebra*

"*Head First Algebra* is a clear, easy-to-understand method to learn a subject that many people find intimidating. Because of its somewhat irreverent attitude in presenting mathematical topics for beginners, this book inspires students to learn algebra at a depth they might have otherwise thought unachievable."

— Ariana Anderson

"The way this book presents information is so conversational and intriguing it helps in the learning process. It truly feels like you're having a conversation with the author."

— Amanda Borcky

"What do punk bands need to know about algebra? How will quadratics make your listening experiences better? Crack the spine on this to find out in a fun and engaging way!"

— Cary Collett

"This has got to be the best book out there for learning basic algebra. It's genuinely entertaining."

— Dawn Griffiths, author of "Head First Statistics"

"I wish I had a book like *Head First Algebra* when I was in high school. I love how the authors relate math concepts to real-life situations. Not only does it make learning Algebra easy, but also fun!"

— Karen Shaner

"*Head First Algebra* is an engaging read. The book does a fantastic job of explaining concepts and taking the reader step-by-step through solving problems. The problems were challenging and applicable to everyday life."

— Shannon Stewart, Math Teacher

"The book is driven by excellent examples from the world in which students live. No trains leaving from the same station at the same time moving in opposite directions. The authors anticipate well the questions that arise in students' minds and answer them in a timely manner. A very readable look at the topics encountered in Algebra 1."

— Herbert Tracey, Instructor of Mathematical Sciences, Loyola University

Other related books from O'Reilly

Statistics in a Nutshell

Statistics Hacks™

Mind Hacks™

Mind Performance Hacks™

Your Brain: The Missing Manual

Other books in O'Reilly's *Head First* series

Head First Java™

Head First Object-Oriented Analysis and Design (OOA&D)

Head First HTML with CSS and XHTML

Head First Design Patterns

Head First Servlets and JSP

Head First EJB

Head First PMP

Head First SQL

Head First Software Development

Head First JavaScript

Head First Ajax

Head First Physics

Head First Statistics

Head First Rails

Head First Web Design

Head First PHP & MySQL

Head First Algebra

Tracey Pilone M.Ed.
Dan Pilone

O'REILLY®

Beijing • Cambridge • Köln • Sebastopol • Taipei • Tokyo

Head First Algebra

by Tracey Pilone M.Ed. and Dan Pilone

Printed in the United States of America.

Published by O'Reilly Media, Inc., 1005 Gravenstein Highway North, Sebastopol, CA 95472.

O'Reilly Media books may be purchased for educational, business, or sales promotional use. Online editions are also available for most titles (*safari.oreilly.com*). For more information, contact our corporate/institutional sales department: (800) 998-9938 or *corporate@oreilly.com*.

Series Creators: Kathy Sierra, Bert Bates

Series Editor: Brett D. McLaughlin

Editors: Brett D. McLaughlin, Louise Barr, Sanders Kleinfeld

Design Editor: Louise Barr

Cover Designers: Louise Barr, Steve Fehler

Production Editor: Brittany Smith

Indexer: Julie Hawks

Page Viewers: Vinny and Nick

Printing History:

December 2008: First Edition.

No variables were harmed in the making of this book.

RepKover™ This book uses RepKover™, a durable and flexible lay-flat binding.

ISBN: 978-0-596-51486-0

[M]

This book is dedicated to my parents and teachers for believing that I could be good at math, even when I didn't agree.

— Tracey

This book is dedicated to the amazing teachers I've had in life—starting with my parents who taught me that the most important is to keep learning.

— Dan

Authors of Algebra

Tracey Pilone

Dan Pilone

Tracey Pilone would first like to thank her co-author and husband for being unwavering in his support and open enough to share the Head First world with her.

She is a freelance technical writer who supported mission planning and RF analysis software for the Navy, right before she decided to write a math book.

She spent several years before becoming a writer working as a construction manager on large commercial construction sites around Washington DC. That's where she actually used Algebra on a somewhat regular basis and saw first hand that math is what makes buildings stay up.

She has a Civil Engineering degree from Virginia Tech, holds a Professional Engineer's License, and received a Masters of Education from the University of Virginia.

Dan Pilone is a Software Architect for Vangent, Inc. and has led software development teams for the Naval Research Laboratory and NASA. He's taught graduate and undergraduate Software Engineering at Catholic University in Washington, D.C.

This is Dan's second Head First Book, but it still comes with some firsts: his first book outside of Computer Science and his first book co-authored by his wife (who, incidentally, is much better looking than his last co-author. Sorry, Russ.) Working with Tracey on this book changed it from being work to being family fun time. Well, not entirely, but still an amazing experience.

Dan's degree is in Computer Science with a minor in Mathematics. For anyone who needs inspiration that Algebra can be fun, fire up a good game of Halo and think about all the x's, y's, and z's that make it all possible.

Table of Contents (Summary)

Table of Contents (the real thing)

Intro

Your brain on Algebra. Here *you* are trying to *learn* something, while here your *brain* is doing you a favor by making sure the learning doesn't *stick*. Your brain's thinking, "Better leave room for more important things, like which wild animals to avoid and whether naked snowboarding is a bad idea." So how *do* you trick your brain into thinking that your life depends on knowing Algebra?

1 what is algebra?

Solving for unknowns...

Do you ever wish you knew more than you know? Well, that's what Algebra's all about: ***making unknowns known***. By the time you're through this first chapter, you'll already have a handle on X being a lot more than a mark where treasure's buried.

SUPER Buy!
KillerX 2.0 Gaming System
The brand new KillerX 2.0 includes full circle entertainment value. One game controller included. (KILLX-112)
special value
$199
Gaming Headset
Headset with microphone, ideal for online gaming (HS-AL1-B67)
special value
$39
Big Bag o' Games
A variety of games for the KillerX system (HD-ISH-5309)
special value
$49

2 (more) complicated equations

Taking Algebra on the road

Imagine a world where there is more than ONE thing you don't know. Not only are there problems with **more than one unknown**, but sometimes you've got **one unknown** that appears ***multiple times in the same equation***! No worries, though... with the tools you'll learn in this chapter, you'll be solving more complicated expressions in no time at all.

ADMIT ONE
151332

3 rules for numeric operations

Follow the rules

Sometimes you just gotta follow the stinking rules.

But when it comes to Algebra, **rules are a good thing**. They'll keep you from getting the wrong answer. In fact, lots of times, rules will **help you solve for an unknown** without a lot of extra work. Leave your dunce cap behind for this chapter because we'll be following a few handy rules all the way to a perfect score.

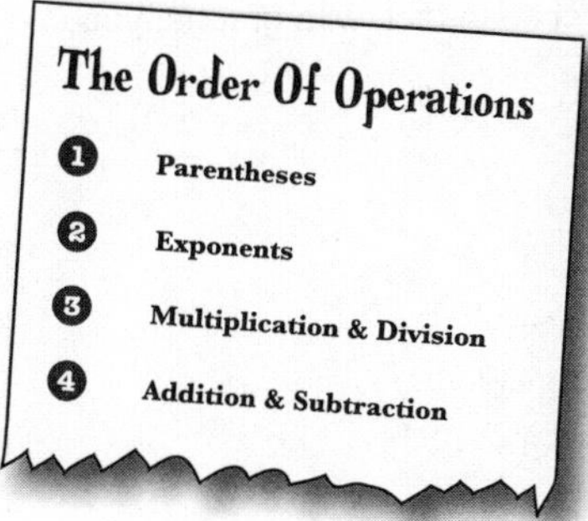
The Order Of Operations
1 Parentheses
2 Exponents
3 Multiplication & Division
4 Addition & Subtraction

4 exponent operations

Podcasts that spread like the plague

Could you multiply that again... and again?

There's another way to express multiplication that's repeated over and over and over again, without just repeating yourself. **Exponents** are a way of **repeating multiplication**. But there's more to exponents, including some smaller-than-usual numbers (and we don't just mean fractions). In this chapter, you'll brush up on **bases**, **roots**, and **radicals**.

5 graphing

A picture's worth 1,000 words

Sometimes an equation might be hiding things.

Ever looked at an equation and thought, "But what the heck does that ***mean***?" In times like that, you just might need a **visual representation** of your equation. That's where **graphs** come in. They let you ***look*** at an equation, instead of just reading it. You can see where **important points** are on the graph, like when you'll run out of money, or how long it will take you to save up for that new car. In fact, with graphs, you can make **smart decisions** with your equations.

Edward's Lawn Service

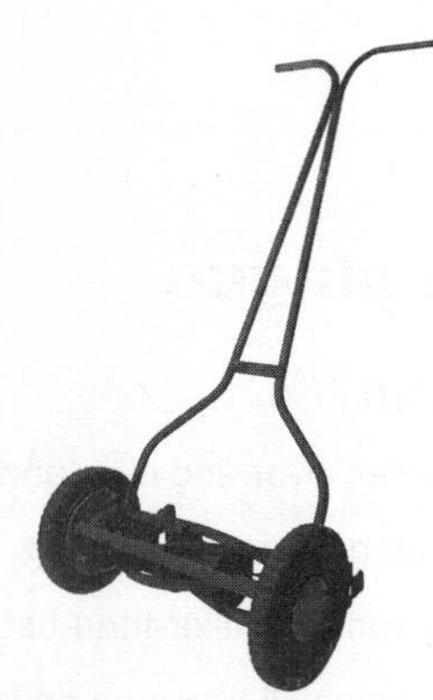

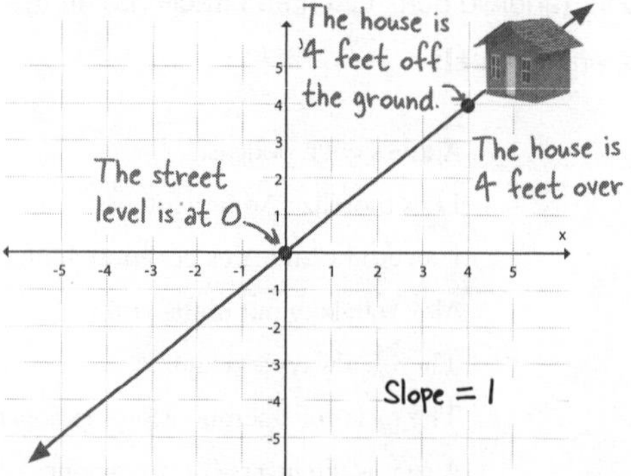

The house is 4 feet off the ground.
The house is 4 feet over
The street level is at 0.
x
Slope = 1

inequalities

6 Can't quite get enough?

Sometimes enough is enough... and sometimes it's not.

Have you ever thought, "I just need a little bit **more**"? But what if someone gave you ***more*** than just a bit more? Then you'd have **more than you need**... but life might still be pretty good. In this chapter, you'll see how Algebra lets you say, "Give me a little more... and then some!" With **inequalities**, you'll go ***beyond two values*** and allow yourself to get **more**, or **less**.

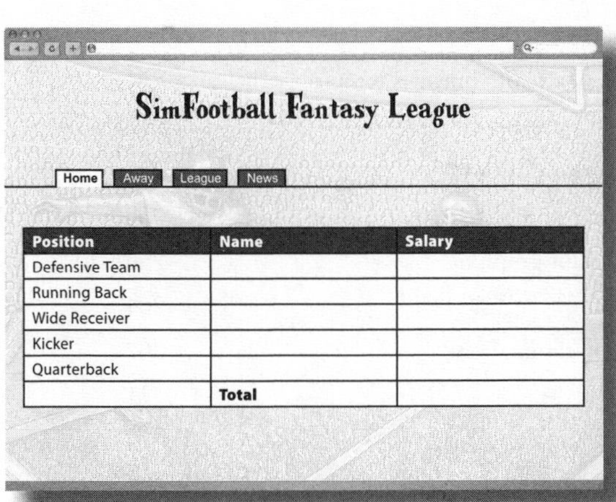

Position	Name	Salary
Defensive Team		
Running Back		
Wide Receiver		
Kicker		
Quarterback		
	Total	

Defensive Teams

Team	Cost
Broncos	$300,000
Eagles	$200,000
Steelers	$333,000
Ravens	$250,000

Running Backs

Name	Cost
Mike Anta	$197,000
Bobby Hull	$202,187
Rick Timmer	$185,200
Ed Babens	$209,115

Wide Receivers

Name	Cost
Ben Toppy	$195,289
Eric Freidr	$212,000
Ron Jupper	$185,200
Mark Marten	$165,950

Kickers

Team	Cost
Joe Amten	$183,500
Rick Vuber	$155,000
Pete Hock	$203,200
Matt Eatens	$209,100

Quarterbacks

Name	Cost
Tony Jaglen	$208,200
Eric Hemal	$175,000
Pat Brums	$199,950
Dan Dreter	$202,400

7 systems of equations

Know what you don't know

You can graph equations with two unknowns, but can you actually solve them? You've been graphing all kinds of expressions lately: ***C*** and ***t***, ***x***, and ***y***, and more. But what about actually ***solving*** equations with **two variables**? That's going to take more than one equation. In fact, you need an equation for every unknown you've got. But what then? Well, a little **substitution**, a few **lines**, and an **intersection** are all you need to solve two-variable equations.

8 expanding binomials & factoring

Breaking up is hard to do

Sometimes being square is enough to give you fits. So far, we've dealt with variables like x and y. But what happens when x is **squared** in your equations? It's time to find out—and you already have the tools to solve these problems! Remember the distributive rule? In this chapter, you're going to learn how to use **distribution** and a special technique called **FOIL** to solve a *new* kind of equation: **binomials**. Get ready—it's time to ***break down*** some really tough equations.

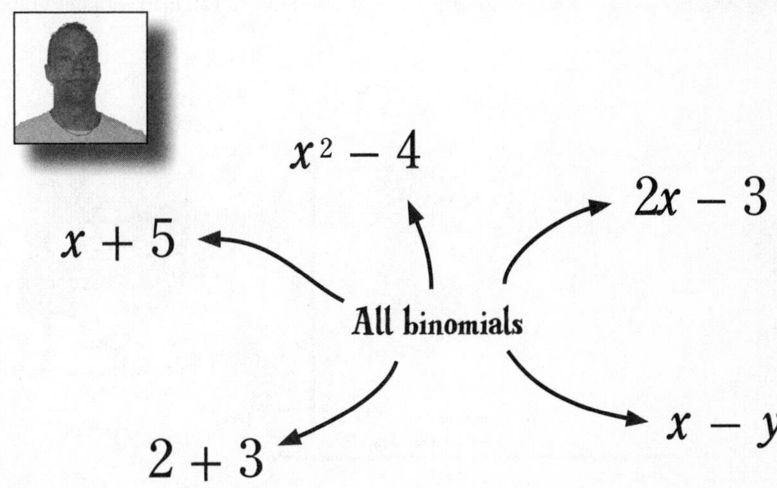

9 quadratic equations

Getting out of line

Not everything in life is linear. But just because an equation doesn't graph as a **straight line**, doesn't mean it's unimportant. In fact, some of the most important **unknowns** you'll have to work with in life end up being **non-linear**. Sometimes you've got to deal with terms that have **exponents greater than 1**. In fact, some equations with **squared terms** graph as **curves**! How's that work? Well, there's only one way to find out...

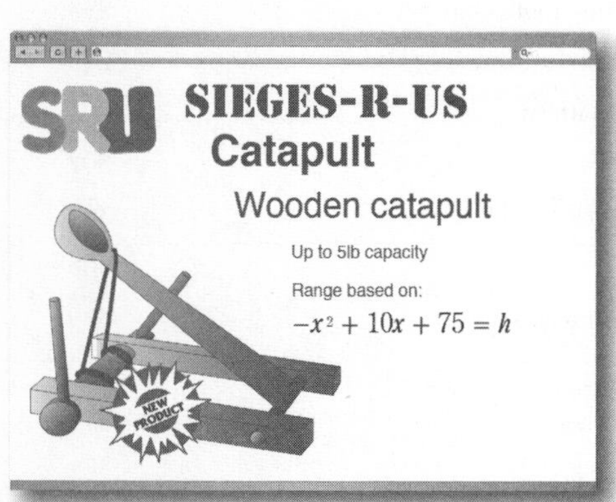

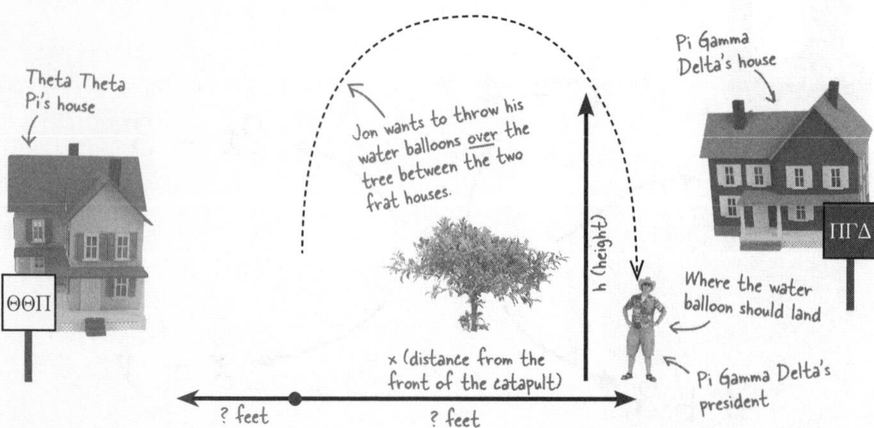

10 functions

Everyone has limits

Some equations are like suburban neighborhoods... ...they're fenced in.

You'll find that in the real world, many equations are **limited**. There are only certain values that an equation is good for. For instance, you can't drive a car -5 miles or dig a hole 13 feet up. In those cases, you need to set **boundaries** on your equations. And when it comes to putting some limits on your equations, there's nothing better than a **function**. A function? What the heck is that? Well, turn the page, and find out... through the lens of reality TV.

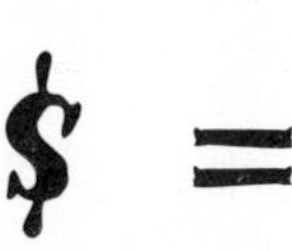

11 real-world algebra

Solve the world's problems

The world's got big problems... you've got big answers. Hundreds of pages of math, and what do you really have? A bunch of ***x***'s and ***y***'s, **a**'s and **b**'s? Nope... you've got **skills** to **solve for an unknown**, even in the most difficult situations. So what's that good for? Well, in this chapter, it's all about the **real world**: you're going to use your Algebra skills to ***solve some real problems***. By the time you're done, you'll have won friends, influenced people, and saved yourself a whole bucket full of cash. Interested? Let's get started.

i leftovers

The Top Five Things (we didn't cover)

You've learned a lot in this book, but Algebra has even more to offer. Don't worry, we're almost done! Before we go, there are a just few gaps we want to fill in. Then you'll be onto Algebra 2, and that's a whole additional book...

pre-Algebra review

Build on a solid foundation

Do you ever feel like you can't even get started?

Algebra is great, but if you want to learn it, you have to have a good understanding of number rules. Suppose you're rolling along and realize that you forgot how to multiply integers, add fractions, or divide a decimal? Well, you've come to the right place! Here we're going to cover all the pre-Algebra you need—*fast*.

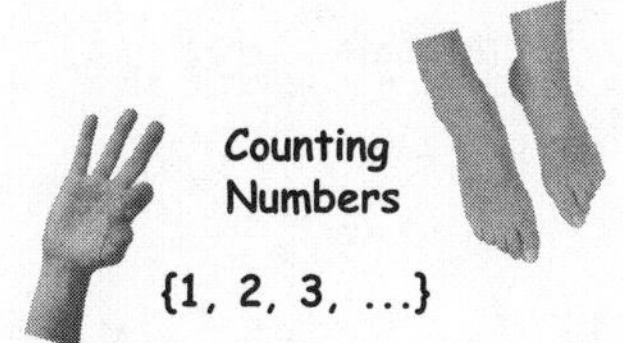

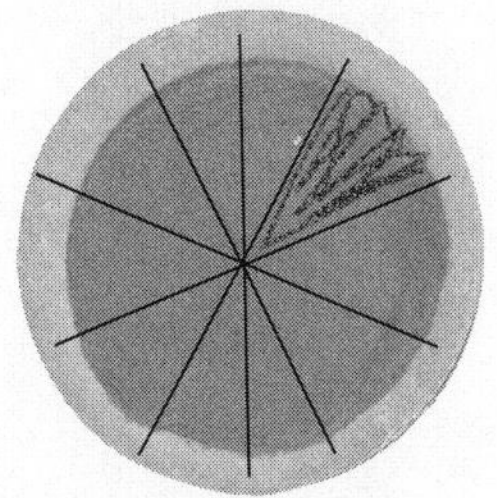

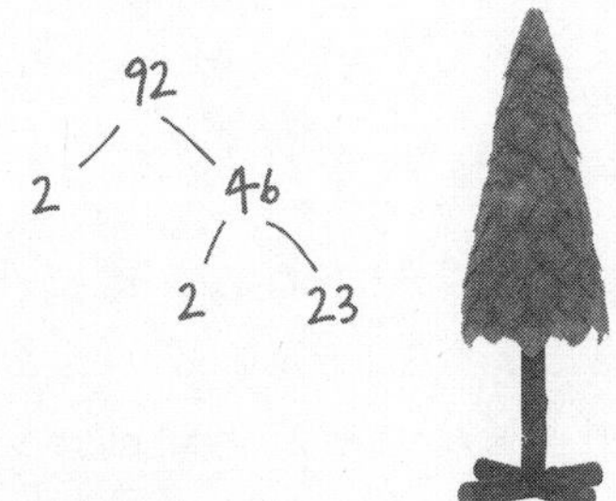

how to use this book

Intro

In this section we answer the burning question: "So why DID they put that in an Algebra book?"

Who is this book for?

If you can answer "yes" to all of these:

1. Are you comfortable with numbers and pre-algebra?
2. Do you want to learn Algebra by learning the concepts and not just looking for practice problems?
3. Are you familiar with integers and fractions and ready to move onto solving for unknowns?

this book is for you.

Who should probably back away from this book?

If you can answer "yes" to any of these:

1. Are you someone who is really uncomfortable with fractions and decimals?
2. Who is looking for Algebra 2 or Statistics information?
3. Are you someone who is obsessed with plugging things into a calculator?

If this is the case, pick up Head First Statistics.

this book is not for you.

[Note from marketing: this book is for anyone with a credit card.]

We know what you're thinking

"How can *this* be a serious Algebra book?"

"What's with all the graphics?"

"Can I actually *learn* it this way?"

We know what your *brain* is thinking

Your brain craves novelty. It's always searching, scanning, *waiting* for something unusual. It was built that way, and it helps you stay alive.

So what does your brain do with all the routine, ordinary, normal things you encounter? Everything it *can* to stop them from interfering with the brain's *real* job—recording things that *matter*. It doesn't bother saving the boring things; they never make it past the "this is obviously not important" filter.

How does your brain *know* what's important? Suppose you're out for a day hike and a tiger jumps in front of you, what happens inside your head and body?

Neurons fire. Emotions crank up. *Chemicals surge.*

And that's how your brain knows...

This must be important! Don't forget it!

But imagine you're at home, or in a library. It's a safe, warm, tiger-free zone. You're studying. Getting ready for an exam. Or trying to learn some tough technical topic your boss thinks will take a week, ten days at the most.

Just one problem. Your brain's trying to do you a big favor. It's trying to make sure that this *obviously* non-important content doesn't clutter up scarce resources. Resources that are better spent storing the really *big* things. Like tigers. Like the danger of fire. Like remembering where all of the warp zones are in Super Mario Brothers. And there's no simple way to tell your brain, "Hey brain, thank you very much, but no matter how dull this book is, and how little I'm registering on the emotional Richter scale right now, I really *do* want you to keep this stuff around."

Your brain thinks THIS is important.
Great. Only 520 more dull, dry, boring pages.
Your brain thinks THIS isn't worth saving.

We think of a "Head First" reader as a learner.

So what does it take to *learn* something? First, you have to *get* it, then make sure you don't *forget* it. It's not about pushing facts into your head. Based on the latest research in cognitive science, neurobiology, and educational psychology, *learning* takes a lot more than text on a page. We know what turns your brain on.

Some of the Head First learning principles:

100% + or 40% = 52%

Make it visual. Images are far more memorable than words alone, and make learning much more effective (up to 89% improvement in recall and transfer studies). It also makes things more understandable. **Put the words within or near the graphics** they relate to, rather than on the bottom on another page, and learners will be up to *twice* as likely to solve problems related to the content.

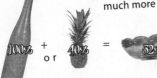

Use a conversational and personalized style. In recent studies, students performed up to 40% better on post-learning tests if the content spoke directly to the reader, using a first-person, conversational style rather than taking a formal tone. Tell stories instead of lecturing. Use casual language. Don't take yourself too seriously. Which would *you* pay more attention to: a stimulating dinner party companion, or a lecture?

Whaddup, girl? I can help you out... I've got tons of friends, you know. Have you seen my Facebook page?

Get the learner to think more deeply. In other words, unless you actively flex your neurons, nothing much happens in your head. A reader has to be motivated, engaged, curious, and inspired to solve problems, draw conclusions, and generate new knowledge. And for that, you need challenges, exercises, and thought-provoking questions, and activities that involve both sides of the brain and multiple senses.

Get—and keep—the reader's attention. We've all had the "I really want to learn this but I can't stay awake past page one" experience. Your brain pays attention to things that are out of the ordinary, interesting, strange, eye-catching, unexpected. Learning a new, tough, technical topic doesn't have to be boring. Your brain will learn much more quickly if it's not.

Touch their emotions. We now know that your ability to remember something is largely dependent on its emotional content. You remember what you care about. You remember when you *feel* something. No, we're not talking heart-wrenching stories about a boy and his dog. We're talking emotions like surprise, curiosity, fun, "what the...?", and the feeling of "I Rule!" that comes when you solve a puzzle, learn something everybody else thinks is hard, or realize you know something that "I'm more technical than thou" Bob from engineering *doesn't*.

Metacognition: thinking about thinking

If you really want to learn, and you want to learn more quickly and more deeply, pay attention to how you pay attention. Think about how you think. Learn how you learn.

Most of us did not take courses on metacognition or learning theory when we were growing up. We were *expected* to learn, but rarely *taught* to learn.

But we assume that if you're holding this book, you really want to master Algebra. And you probably don't want to spend a lot of time. If you want to use what you read in this book, you need to *remember* what you read. And for that, you've got to *understand* it. To get the most from this book, or *any* book or learning experience, take responsibility for your brain. Your brain on *this* content.

The trick is to get your brain to see the new material you're learning as Really Important. Crucial to your well-being. As important as a tiger. Otherwise, you're in for a constant battle, with your brain doing its best to keep the new content from sticking.

So just how *DO* you get your brain to treat Algebra like it was a hungry tiger?

There's the slow, tedious way, or the faster, more effective way. The slow way is about sheer repetition. You obviously know that you *are* able to learn and remember even the dullest of topics if you keep pounding the same thing into your brain. With enough repetition, your brain says, "This doesn't *feel* important to him, but he keeps looking at the same thing *over* and *over* and *over*, so I suppose it must be."

The faster way is to do ***anything that increases brain activity,*** especially different *types* of brain activity. The things on the previous page are a big part of the solution, and they're all things that have been proven to help your brain work in your favor. For example, studies show that putting words *within* the pictures they describe (as opposed to somewhere else in the page, like a caption or in the body text) causes your brain to try to makes sense of how the words and picture relate, and this causes more neurons to fire. More neurons firing = more chances for your brain to *get* that this is something worth paying attention to, and possibly recording.

A conversational style helps because people tend to pay more attention when they perceive that they're in a conversation, since they're expected to follow along and hold up their end. The amazing thing is, your brain doesn't necessarily *care* that the "conversation" is between you and a book! On the other hand, if the writing style is formal and dry, your brain perceives it the same way you experience being lectured to while sitting in a roomful of passive attendees. No need to stay awake.

But pictures and conversational style are just the beginning…

Here's what WE did:

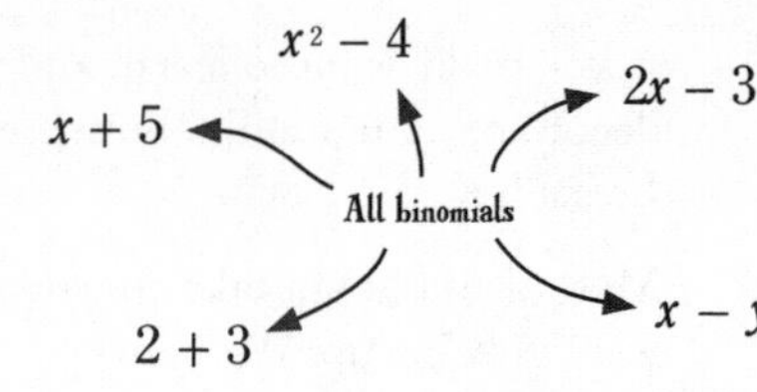

We used ***pictures***, because your brain is tuned for visuals, not text. As far as your brain's concerned, a picture really *is* worth a thousand words. And when text and pictures work together, we embedded the text *in* the pictures because your brain works more effectively when the text is *within* the thing the text refers to, as opposed to in a caption or buried in the text somewhere.

We used ***redundancy***, saying the same thing in *different* ways and with different media types, and *multiple senses*, to increase the chance that the content gets coded into more than one area of your brain.

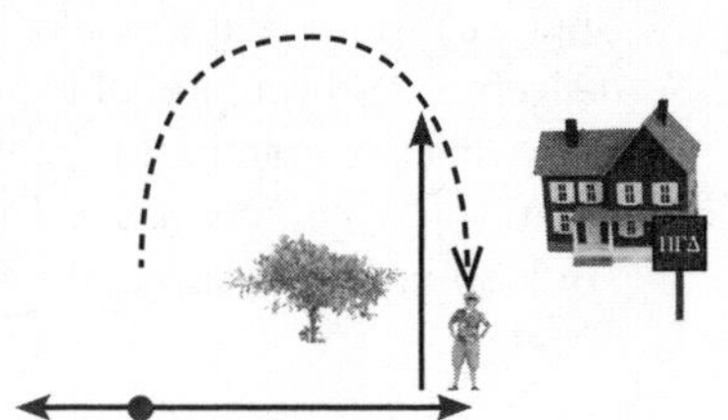

We used concepts and pictures in ***unexpected*** ways because your brain is tuned for novelty, and we used pictures and ideas with at least *some* ***emotional*** *content*, because your brain is tuned to pay attention to the biochemistry of emotions. That which causes you to *feel* something is more likely to be remembered, even if that feeling is nothing more than a little ***humor***, ***surprise***, or ***interest.***

We used a personalized, ***conversational style***, because your brain is tuned to pay more attention when it believes you're in a conversation than if it thinks you're passively listening to a presentation. Your brain does this even when you're *reading*.

We included more than 80 ***activities***, because your brain is tuned to learn and remember more when you ***do*** things than when you *read* about things. And we made the exercises challenging-yet-do-able, because that's what most people prefer.

We used ***multiple learning styles***, because *you* might prefer step-by-step procedures, while someone else wants to understand the big picture first, and someone else just wants to see an example. But regardless of your own learning preference, *everyone* benefits from seeing the same content represented in multiple ways.

We include content for ***both sides of your brain***, because the more of your brain you engage, the more likely you are to learn and remember, and the longer you can stay focused. Since working one side of the brain often means giving the other side a chance to rest, you can be more productive at learning for a longer period of time.

And we included ***stories*** and exercises that present ***more than one point of view,*** because your brain is tuned to learn more deeply when it's forced to make evaluations and judgments.

We included ***challenges***, with exercises, and by asking ***questions*** that don't always have a straight answer, because your brain is tuned to learn and remember when it has to *work* at something. Think about it—you can't get your *body* in shape just by *watching* people at the gym. But we did our best to make sure that when you're working hard, it's on the *right* things. That ***you're not spending one extra dendrite*** processing a hard-to-understand example, or parsing difficult, jargon-laden, or overly terse text.

We used ***people***. In stories, examples, pictures, etc., because, well, because *you're* a person. And your brain pays more attention to *people* than it does to *things*.

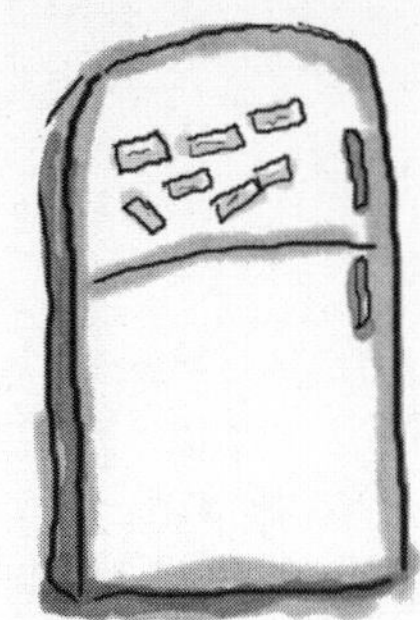

Here's what YOU can do to bend your brain into submission

So, we did our part. The rest is up to you. These tips are a starting point; listen to your brain and figure out what works for you and what doesn't. Try new things.

Cut this out and stick it on your refrigerator.

---

1 Slow down. The more you understand, the less you have to memorize.

Don't just *read*. Stop and think. When the book asks you a question, don't just skip to the answer. Imagine that someone really *is* asking the question. The more deeply you force your brain to think, the better chance you have of learning and remembering.

2 Do the exercises. Write your own notes.

We put them in, but if we did them for you, that would be like having someone else do your workouts for you. And don't just *look* at the exercises. **Use a pencil.** There's plenty of evidence that physical activity *while* learning can increase the learning.

3 Read the "There are No Dumb Questions"

That means all of them. They're not optional sidebars, ***they're part of the core content!*** Don't skip them.

4 Make this the last thing you read before bed. Or at least the last challenging thing.

Part of the learning (especially the transfer to long-term memory) happens *after* you put the book down. Your brain needs time on its own, to do more processing. If you put in something new during that processing time, some of what you just learned will be lost.

5 Talk about it. Out loud.

Speaking activates a different part of the brain. If you're trying to understand something, or increase your chance of remembering it later, say it out loud. Better still, try to explain it out loud to someone else. You'll learn more quickly, and you might uncover ideas you hadn't known were there when you were reading about it.

6 Drink water. Lots of it.

Your brain works best in a nice bath of fluid. Dehydration (which can happen before you ever feel thirsty) decreases cognitive function.

7 Listen to your brain.

Pay attention to whether your brain is getting overloaded. If you find yourself starting to skim the surface or forget what you just read, it's time for a break. Once you go past a certain point, you won't learn faster by trying to shove more in, and you might even hurt the process.

8 Feel something.

Your brain needs to know that this *matters*. Get involved with the stories. Make up your own captions for the photos. Groaning over a bad joke is *still* better than feeling nothing at all.

9 Use Algebra in the Real World.

There's only one way to get comfortable with Algebra: **do it a lot**. Now, that doesn't mean you need to lock yourself in a room with graph paper and pencils. But it does mean you should think about how Algebra fits in to the world around you. What problem are you trying to solve? What are the knowns and unknowns? How do they relate to each other? The point is that you won't really ***get*** Algebra if you just read about it—you need to ***do*** it. We're going to give you a lot of practice: every chapter is full of exercises and asks questions that you need to think about. Don't just skip over them—most of the learning actually happens when you work on the exercises. Don't be afraid to peek at the solutions if you get stuck, but at least give the problems a try first.

Read Me

This is a learning experience, not a reference book. We deliberately stripped out everything that might get in the way of learning whatever it is we're working on at that point in the book. And the first time through, you need to begin at the beginning because the book makes assumptions about what you've already seen and learned.

We start off by teaching how to solve algebraic equations.

Believe it or not, even if you've never taken Algebra, you can jump right in and start solving for unknowns. You'll also learn about the deeper motivations for the study of Algebra, and why you should learn it in the first place.

Calculators are only for arithmetic you can't solve easily, NOT for solving equations.

There are lots of calculators out there that can do lots of things, including solving equations and plotting graphs. Since the entire purpose of working through this book is to learn how to solve and graph equations yourself, using a calculator to do it would just cheat you out of your learning experience!

If you're rusty on some pre-Algebra topics, we can help.

You need to be able to work with fractions, decimals, integers, and exponents to get into Algebra and solve for unknowns. The good news is that if you have a decent understanding of these concepts, but you can't quite remember how to get a common denominator, there is a big appendix at the back to help you. It's quick and dirty, but it can bring you back up to speed on how to work with those tricky pre-Algebra topics.

Algebra is not just about getting the right "answer."

There's a lot in this book about the process: writing out the steps, understanding what's going on at each point, and really understanding what you're doing. We have taken a lot of time to make sure that each exercise is well explained, and there's a reason for it—you're trying to learn here, right? So don't just skip to the $x = 5$ and see if you're right, because that's only a piece of the answer.

The activities are NOT optional.

The exercises and activities are not add-ons; they're part of the core content of the book. Some of them are to help with memory, some are for understanding, and some will help you apply what you've learned. Don't skip the exercises.

The redundancy is intentional and important.

One distinct difference in a Head First book is that we want you to really get it. And once you finish the book, we want you to remember what you've learned. Most reference books don't have retention and recall as a goal, but this book is about learning, so you'll see some of the same concepts come up more than once.

Everyone can learn Algebra, even if you think you're not a "math person."

You need to leave all of this "I'm not a math person" stuff behind. Everybody is a "math person," you just might not know it yet. You actually do a lot of Algebra every day, it's just not labelled that way. If you haven't yet found your inner "math person," or you're rusty, you've come to the right place. You're going to finish the book knowing how to handle Algebra. Now get going and solve some equations!

The technical review team

Ariana Anderson

Amanda Borcky

Dawn Griffiths

Karen Shaner

Shannon Stewart

Herbert Tracey

Cary Collett

Technical Reviewers:

Ariana Anderson is a PhD Candidate in Statistics at UCLA and a member of the Collegium of University Teaching Fellows. Her research involves the integration of neuro-imaging and statistics to create "mind reading" machines.

Amanda Borcky is a student at Virginia Tech in Blacksburg, VA. She is studying Dietetics and hopes to practice Clinical Dieteics in the future. This is her first time technically reviewing a book.

Dawn Griffiths is the author of *Head First Statistics*. When Dawn's not working on Head First books, you'll find her honing her Tai Chi skills, making bobbin lace, or spending time with her lovely husband David.

Karen Shaner is a grad student at Emerson College in Boston pursuing a MA in Publishing and Writing in addition to working at O'Reilly. In the little free time she has, she enjoys contra dancing, spending time with friends, singing with the Praise Band, and enjoying all that Boston has to offer.

Shannon Stewart is a former fifth grade math teacher. During her five years in Mesquite, she was grade level chair as well as recognized in Who's Who of American Teachers. She graduated from Hardin Simmons with a BS in elementary education and then went on to graduate cum laude from A&M Commerce with a Masters in Education. She currently resides in Texas with her husband Les and her son Nathan.

Herbert Tracey received his BS from Towson University and a MS from The Johns Hopkins University. Currently, he is an instructor of Mathematical Sciences at Loyola University Maryland and served as Department Chair of Mathematics (retired) at Hereford High School.

Cary Collett majored in physics and astrophysics in college and grad school, respectively, so needless to say, he learned a great deal of mathematics and will tell anyone that algebra is the hardest subject in the field. He current works in IT and lives in central Ohio.

Acknowledgments

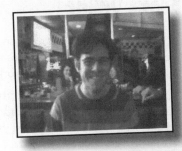

Sanders Kleinfeld

Our editors:

Thanks to ***Sanders Kleinfeld***, who took this book from the first outline through the first draft. He also put up with endless questions (mostly from Tracey), and let us wax philosophical about the math books that 80's TV stars write.

And to **Brett McLaughlin**, who in addition to running the whole series, got us from the first draft across the finish line. His feedback had a whole lot of "why didn't we think of that" in it, which was incredibly helpful. His understanding about the kids and dog in the background on conference calls was also a plus.

Brett McLaughlin

Lou Barr

Thanks also to ***Lou Barr***, who somehow managed to take notes that say things like "Lou, can you make this look cool?" and made things look cool. Since neither of us have any artistic talent at all, anything that looks fantastic is clearly her work.

The O'Reilly team:

To **Caitrin McCullough**, for the cool website and **Karen Shaner**, for being both a tech reviewer and keeping the review process running smoothly.

To **Brittany Smith**, the book's Production Editor, who always answered questions really fast, somehow made sense of all of the computer files that went into this thing, and always sent happy emails.

Last but not least, to **Laurie Petrycki**, who gave us a chance to write a math book that we're very excited about!

To the reviewers:

Thanks to all of you for reading the whole book with such enthusiasm. To ***Amanda Borcky,*** for being our sample audience and telling us when we were not being cool. To ***Herbert Tracey***, who, in addition to teaching Tracey Trig and Calculus, also gave us extremely detailed feedback that made this a much better **math** book. To ***Ariana Anderson*** and ***Shannon Stewart***, who, as math teachers, could point out gaps in our assumptions and good questions to ask. Finally, to ***Cary Collett*** and ***Dawn Griffiths***, who helped with the math and made sure that we were keeping true to the Head First way.

To our friends and family:

To all of the ***Pilones*** and all of the ***Chadwicks***, if it hadn't been for your love and support, we wouldn't have passed Algebra the first time! To Tracey's math teachers—***Mr. Tracey, Mrs. Vesley,*** and ***Mrs. Booth***—who turned her from a being math hater into an engineer; and to Dan's math teachers—***Br. Leahy, Mr. Cleary, Fr. Shea,*** and ***Mrs. Newell***—you saw past him getting his head stuck in the door and put the first draft of this book in motion so many years ago....

And last but not least, thanks to ***Vinny*** and ***Nick***—the first two projects that Dan and Tracey worked on together—who put up with a lot of "Daddy and Mommy have a call" and learned more Algebra than any preschooler or kindergartner should know.

Safari® Books Online

When you see a Safari® icon on the cover of your favorite technology book that means the book is available online through the O'Reilly Network Safari Bookshelf.

Safari offers a solution that's better than e-books. It's a virtual library that lets you easily search thousands of top tech books, cut and paste code samples, download chapters, and find quick answers when you need the most accurate, current information. Try it for free at `http://safari.oreilly.com`.

1 what is algebra?

Solving for unknowns...

This is so sweet, but how can we possibly afford it?
I don't know... but it can't be that much, right? The salesman said that twelve times its price was $22,400, and that didn't seem too much at the time.

Do you ever wish you knew more than know? Well, that's what Algebra's all about: ***making unknowns known***. By the time you're through this first chapter, you'll already have a handle on X being a lot more than a mark where treasure's buried. You'll get a handle on **equations**, keeping both sides of an equation **balanced**, and why **solving for unknowns** really isn't that big of a deal. What are you waiting for? Go on and get started!

It all started with a big gaming sale

Jo has been watching the game system battles for a while now and has finally decided on the one she wants. Her favorite system's on sale this week, and she's ready to buy. But can she afford it? That's where she needs a little help from you.

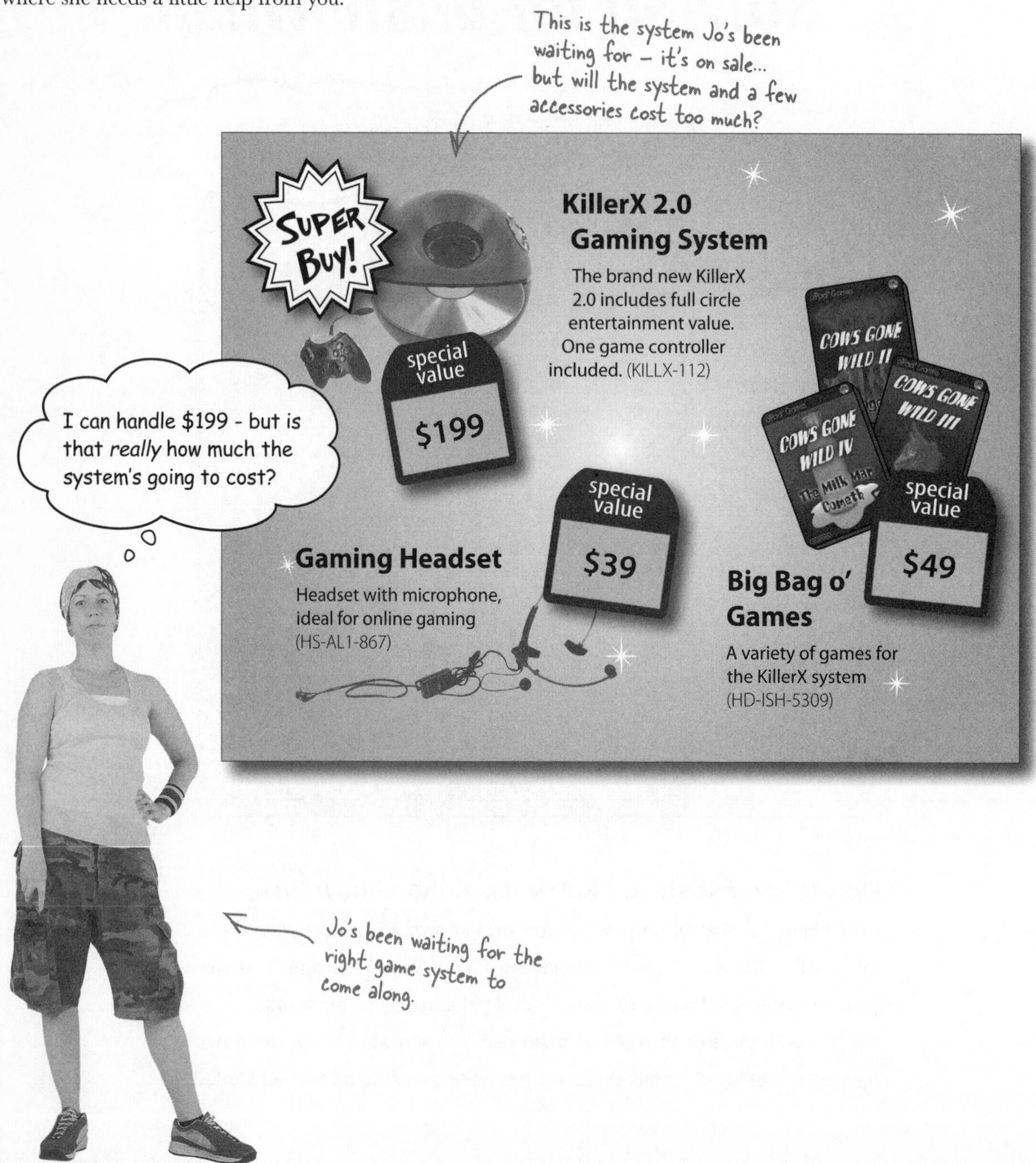
This is the system Jo's been waiting for – it's on sale... but will the system and a few accessories cost too much?
SUPER BUY!
KillerX 2.0 Gaming System
The brand new KillerX 2.0 includes full circle entertainment value. One game controller included. (KILLX-112)
special value
$199
I can handle $199 - but is that *really* how much the system's going to cost?
Gaming Headset
Headset with microphone, ideal for online gaming (HS-AL1-867)
special value
$39
Big Bag o' Games
A variety of games for the KillerX system (HD-ISH-5309)
special value
$49
COWS GONE WILD II
COWS GONE WILD III
COWS GONE WILD IV
The Milk Man Cometh
Jo's been waiting for the right game system to come along.

What does a system really cost?

When you buy things—especially expensive electronic things—there are lots of pieces that add into the price besides just the number on a sales flyer: sales tax, an extended warranty, shipping and handling, etc. So what will a KillerX system really cost?

There's tax on the system...

The base price of the system is \$199. After that, we need to think about taxes, which are 5%. Let's figure out how much Jo'll have to pay in taxes:

KillerX 2.0
Gaming Syst
The brand new Killer
2.0 includes full circle
entertainment value.
game controller inclu
(KILLX-112)
special value
$199

Here's the original \$199 we pulled from the ad.

5% sales tax means multiply by 0.05...

...figure it out. Remember this, you'll need it again in a minute.

$$\$199 \times 0.05 = \$ \boxed{}$$

If you're rusty on your decimal math, just turn to the appendix and brush up!

...and the extended warranty, too.

Jo's about to spend \$199 on a game machine, and she wants to purchase an extended warranty plan for an additional \$20. Let's put that in the price, too. What price will Jo need to pay?

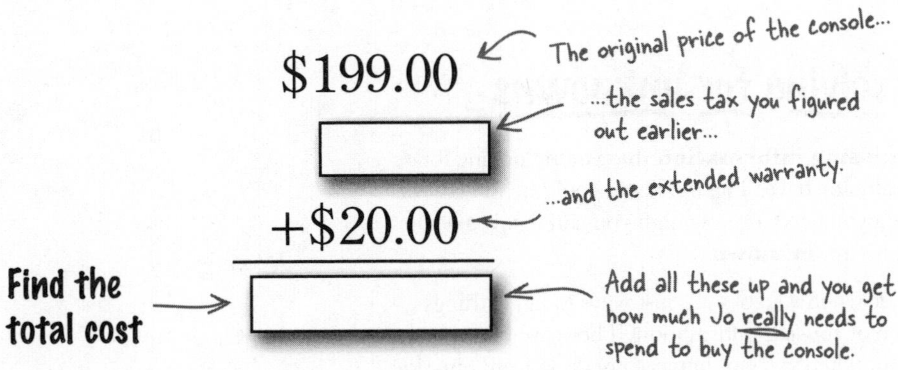
$199.00
The original price of the console...
...the sales tax you figured out earlier...
+$20.00
...and the extended warranty.
Find the total cost
Add all these up and you get how much Jo really needs to spend to buy the console.

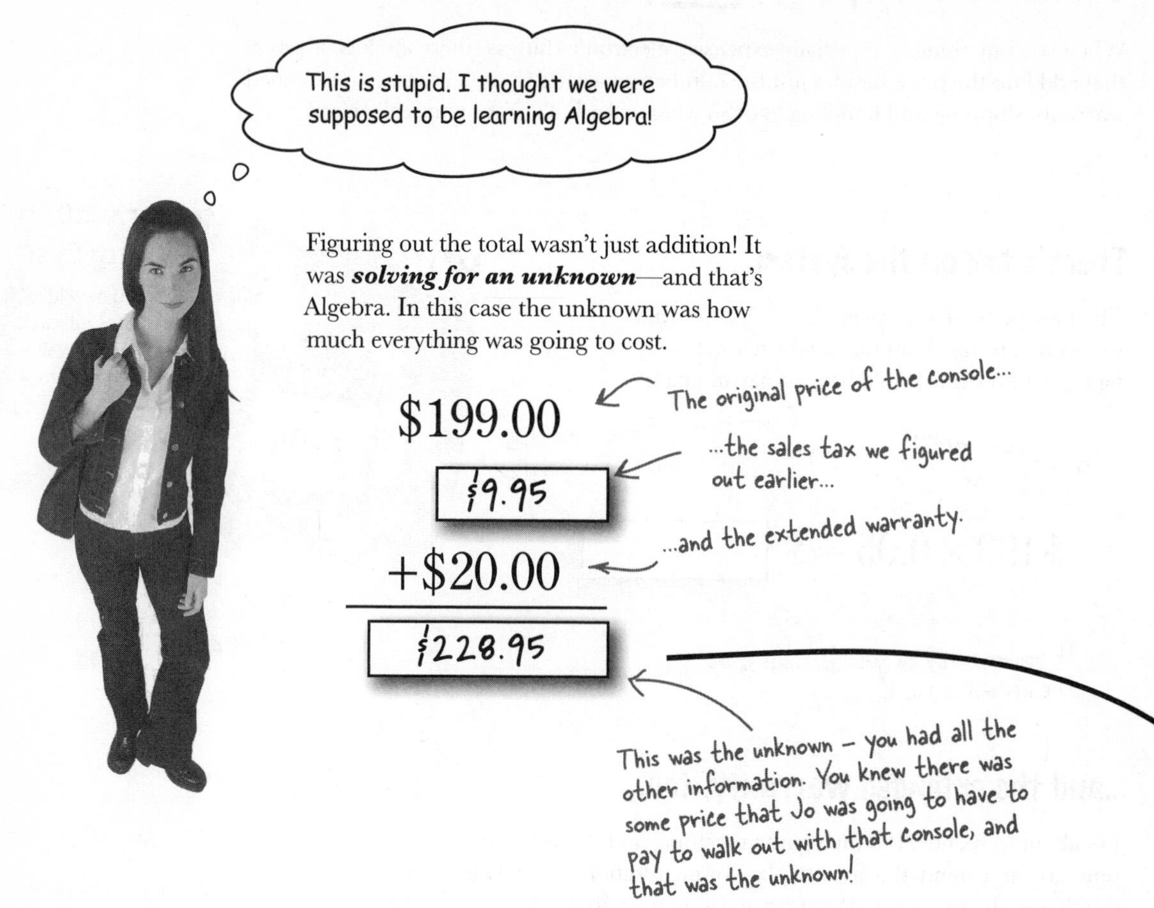

This is stupid. I thought we were supposed to be learning Algebra!
Figuring out the total wasn't just addition! It was ***solving for an unknown***—and that's Algebra. In this case the unknown was how much everything was going to cost.
$199.00
The original price of the console...
$9.95
...the sales tax we figured out earlier...
+$20.00
...and the extended warranty.
$228.95
This was the unknown - you had all the other information. You knew there was some price that Jo was going to have to pay to walk out with that console, and that was the unknown!

Algebra is about solving for unknowns

Algebra is about finding the **missing information** that you're looking for by using the information you already have. The unknown could be the cost of a car loan, the quantity of soda you need, or how high you can throw a water balloon. If you don't know it, it's an **unknown**.

All the other things that you'll learn in Algebra are just ways to jiggle things around to help you find a piece of missing information. There are rules about when you can multiply things or when you can bump something from one side of an equals sign to another, but at the end of the day, they're all just tricks to help you find that missing piece of information you're looking for.

Jo's got more unknowns

So Jo knows how much it will take to buy an awesome gaming system, including an extended warranty. But she still doesn't have any games... or another controller... or a headset.

Jo started with $315.27 in her bank account. Now that she's paid for the console, how much can Jo spend on accessories? Let's start by writing this out in words:

Account balance – Cost of console = Money for accessories

Writing a problem out verbally is a great way to get started. You don't need to worry about numbers at this stage.

We know how much the console costs ($228.95), and we know how much Jo has in her account ($315.27). Now just fill in the blanks, and we can figure out Jo's accessory budget:

$$\$315.27 - \$228.95 = \boxed{}$$

Here's what's in Jo's account.

Here's the cost of the console you figured out earlier.

Fill in the unknown!

With money to spend on accessories, Jo is a happy gamer...

Console Pricing Up Close

Let's take a closer look at what we just did. First, we're going to swap out the unknown box for the standard Algebra version of an unknown—an ***x***.

There's no magic to the letter x – that's just the most common letter used in Algebra.

x = money for accessories

$$\$315.27 - \$228.95 = \$86.32 = x$$

Here's what's in her account.

Here's the cost of the console you figured out earlier.

...and here's our unknown.

$$\$86.32 = x$$

We can also turn this around to be:

$$x = \$86.32$$

We'll talk more about why you can swap these around and other tricks you can do later in the chapter. The important thing for now is that these are the same, regardless of whether the x is on the right or left.

Understanding a problem and finding the unknown, ***x***, is working with Algebra. Using tricks like writing the problem out with words and flipping things around are just ways to make finding that unknown possible.

Cool! $86.32 is still plenty to spend on games and a headset.

Solving for any unknown is Algebra.

X marks the ~~spot~~ unknown

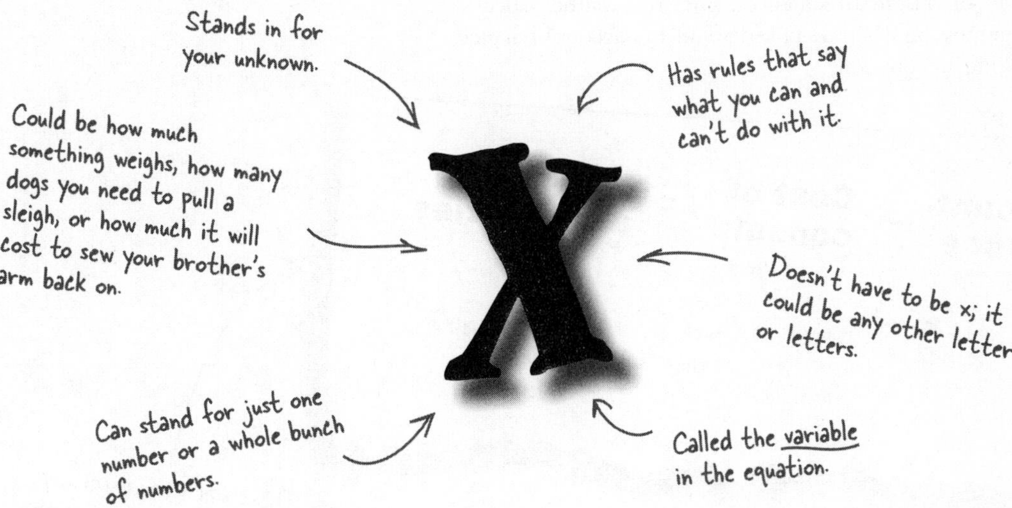

Stands in for your unknown.
Has rules that say what you can and can't do with it.
Could be how much something weighs, how many dogs you need to pull a sleigh, or how much it will cost to sew your brother's arm back on.
X
Doesn't have to be x; it could be any other letter or letters.
Can stand for just one number or a whole bunch of numbers.
Called the variable in the equation.

x is just a user-friendly stand-in for the unknown box we used earlier. ***x*** is easier to write and it's what you're looking for when you solve an equation. The unknown in any given situation is called a ***variable***. In the real world, problems present themselves every day; translating them into mathematical equations allows you to solve them.

there are no Dumb Questions

Q: Will the unknown always be x?

A: Nope. As you progress with math, you'll see x, y, and z pretty often. You can use any letter that you want, though.

Q: Back on page 6, how come you could just flip that equation?

A: All we really did was switch the same equation around, called ***manipulating the equation***. There are rules about exactly how you can work with equations without changing any values, and we'll learn lots more about them in the rest of this book.

Equations are math sentences

Equations, like the one you used earlier to figure out how much Jo could spend on accessories, are just math sentences. They're a mathematical way of saying something. So when we talked about Jo's account balance, we were actually using an equation:

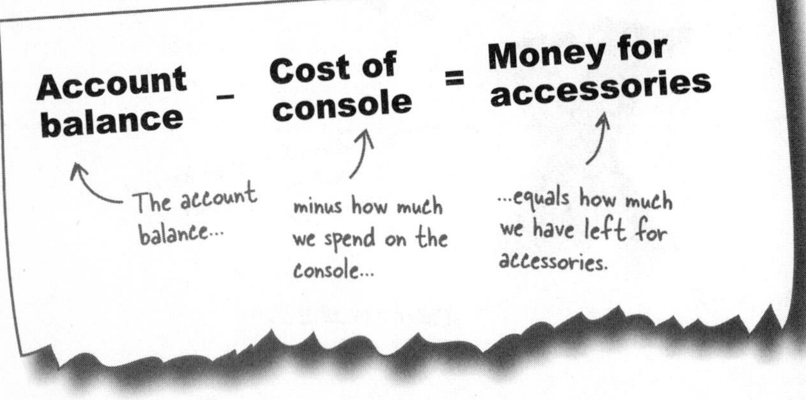

These math sentences say the same thing. They're two different forms of the same problem.

Our equation means *"The account balance minus how much we spend on the console equals how much we have left for accessories."* So, that means that *the account balance must equal the cost of the console plus the money for accessories.* If we write that sentence as an equation it looks like this:

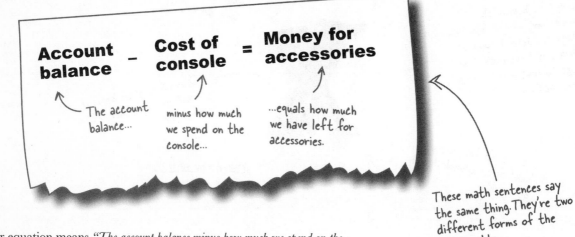

Account balance = Cost of console + Money for accessories

The account balance...

equals how much we spend on the console...

...plus how much we have left for accessories.

Both sentences mean the same thing; they're just phrased differently. Over the next few pages, you'll learn how to rearrange math sentences and make sure that you don't change any values.

Equations can be rearranged like sentences.

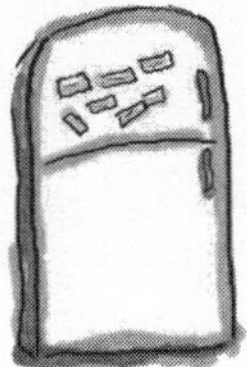

Math Magnets

Below are some word problems and magnets. Your job is to make equations from the magnets that say the same thing as the word problems. Once you have the equation put together, circle the unknown—the value you need to figure out. Then, write out your equation in a complete phrase.

1. Jo and her 3 brothers are thinking about upgrading their LIVE subscription to the Platinum membership, which is $12 per person. How much will it cost them in total?

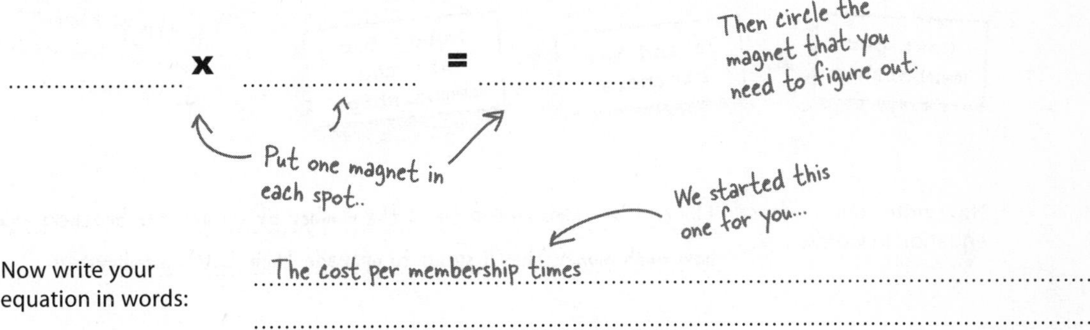

x
=
Then circle the magnet that you need to figure out.
Put one magnet in each spot..
We started this one for you...

Now write your equation in words: *The cost per membership times* ....

....

2. Jo started playing a hot new game, but she only has two hours before she has to go out. She spent 20 minutes on level 1, 37 minutes on level 2, and 41 minutes on level 3. How much time does she have left to play level 4?

This time you build the whole equation...

Now write your equation in words: ....

....

`Time for level 1` `Cost per membership` `number of games` `Time for level 3` `Level 5`

`LIVE` `money for all the memberships` `Jo` `Jo and her brothers` `Time for level 2` `Time for level 4` `Time until Jo leaves`

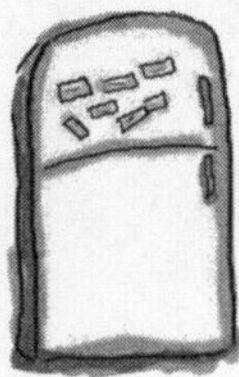

Math Magnets Solutions

Below are some word problems and magnets. Your job is to make equations from the magnets that say the same thing as the word problems. Once you have the equation put together, circle the unknown—the value you need to figure out. Then, write out your equation in a complete phrase.

1. Jo and her 3 brothers are thinking about upgrading their LIVE subscription to the Platinum membership, which is $12 per person. How much will it cost them in total?

Now write your equation in words: The cost per membership times the number of Jo and her brothers equals how much money they'll spend to upgrade their LIVE memberships.

2. Jo started playing a hot new game, but she only has two hours before she has to go out. She spent 20 minutes on level 1, 37 minutes on level 2, and 41 minutes on level 3. How much time does she have left for level 4?

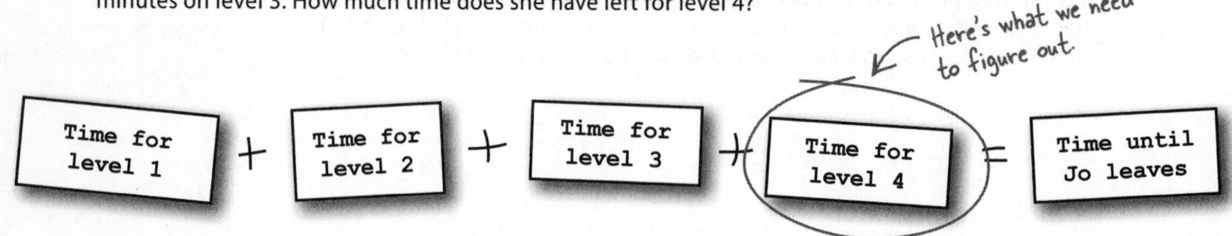

Now write your equation in words: The time for level 1 plus the time for level 2 plus the time for level 3 plus the time for level 4 equals the time until Jo leaves.

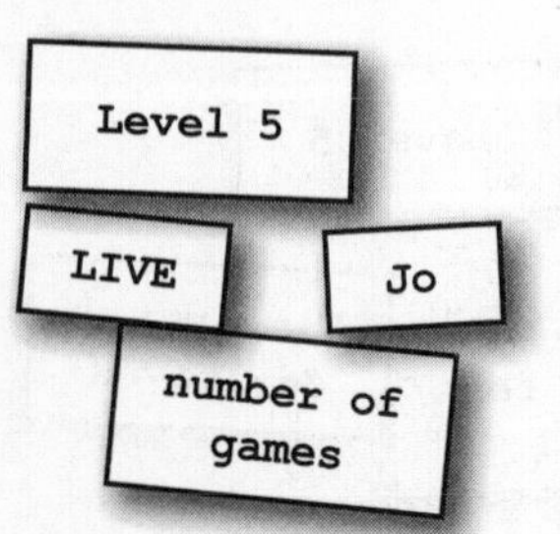

You could also have made an equation that says, "Time until Jo leaves minus the time for level 1 minus the time for level 2 minus the time for level 3 equals the time for level 4." Would that be any better? Why?

EQUATION CONSTRUCTION

Now you build the equations! Take the "math sentence" you wrote and translate it into an equation. Use the numbers that were given and **x** for the unknown.

1. Jo and her 3 brothers are thinking about upgrading their LIVE subscription to the Platinum membership, which is $12 per person. How much will it cost them in total?

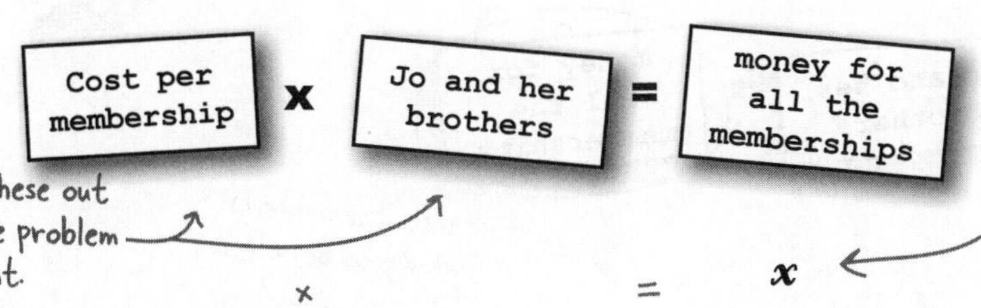

Figure these out from the problem statement.

We need a variable to stand in for what we're trying to find, so we'll use x.

.... × .... = x

This means multiplication.

The same thing for this problem – you fill in the x for the unknown.

2. Jo started playing a game that just came out, but she only has two hours before she has to go out. She spent 20 minutes on level 1, 37 minutes on level 2, and 41 minutes on level 3. How much time does she have left for level 4?

Watch this one! Hours or minutes?

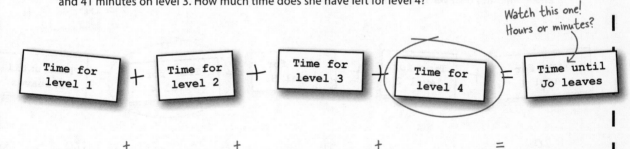

Time for level 1 + Time for level 2 + Time for level 3 + Time for level 4 = Time until Jo leaves

.... + ... + ... + ... = ...

.... + ... = ...

Use this line to combine the numbers and write a shorter equation.

EQUATION CONSTRUCTION SOLUTION

Now you can build the equations! Take the "math sentence" you wrote and translate it into an Algebraic equation. Use the numbers that were given and **x** for the unknown.

1. Jo and her 3 brothers are thinking about upgrading their LIVE subscription to the Platinum membership, which is $12 per person. How much will it cost them in total?

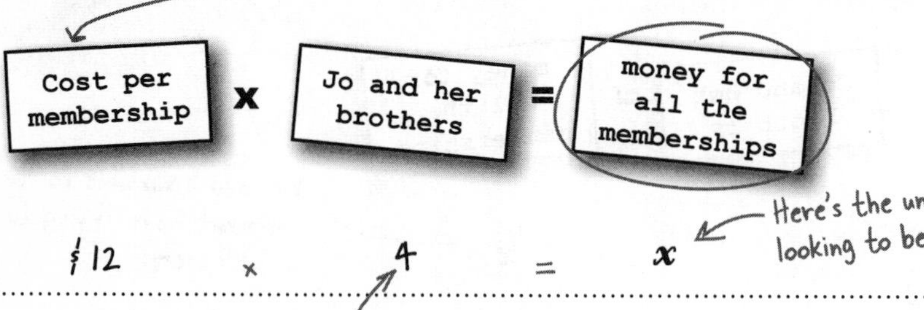

$12 × 4 = x

Here's the unknown looking to be solved.

3 brothers + Jo = 4

2. Jo started playing a game that just came out - but she only has two hours before she has to go out. She spent 20 minutes on level 1, 37 minutes on level 2, and 41 minutes on level 3. How much time does she have left for level 4?

We need minutes here since the time for the levels is in minutes.

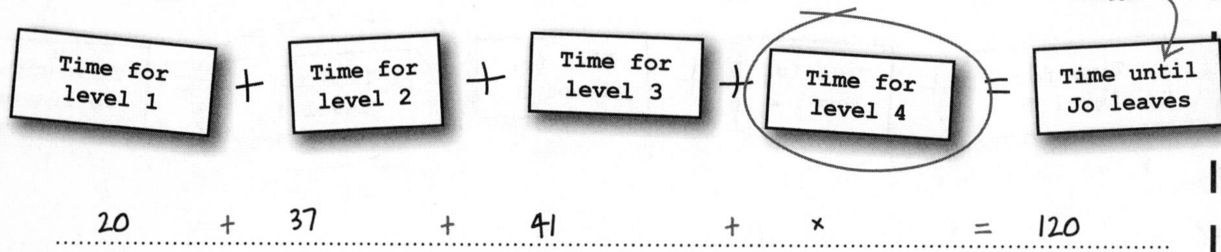

20 + 37 + 41 + x = 120

Good work – you've simplified the equation here – that will make it easier to solve.

98 + x = 120

Now SOLVE for the unknown

Jo is trying to decide if it's worth it for her to buy a LIVE subscription. She has 10 games, and 7 of them don't have any online play. How many does she have that can be played online? Does it make sense for her to buy the subscription?

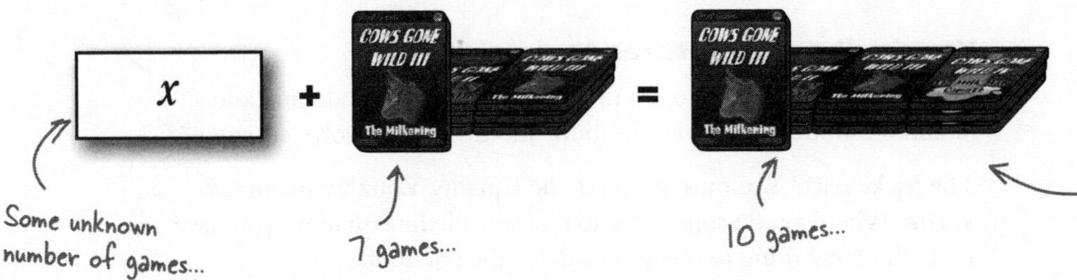

We're going to jump right into the equation without the sentence. If it helps, you can write it out, but just thinking about the equation in words first should be enough to help.

What we really care about here is what **x** is—the unknown number of games. We don't really care about the seven games on the left side of the equation. In fact, we can get rid of that seven as long as we make sure we do the ***same thing*** to ***both sides*** of the equation.

An ***equals*** sign means that both sides are the ***same***. So if we take 7 away from one side, we ***have to do the same thing*** to the other side of the equation:.

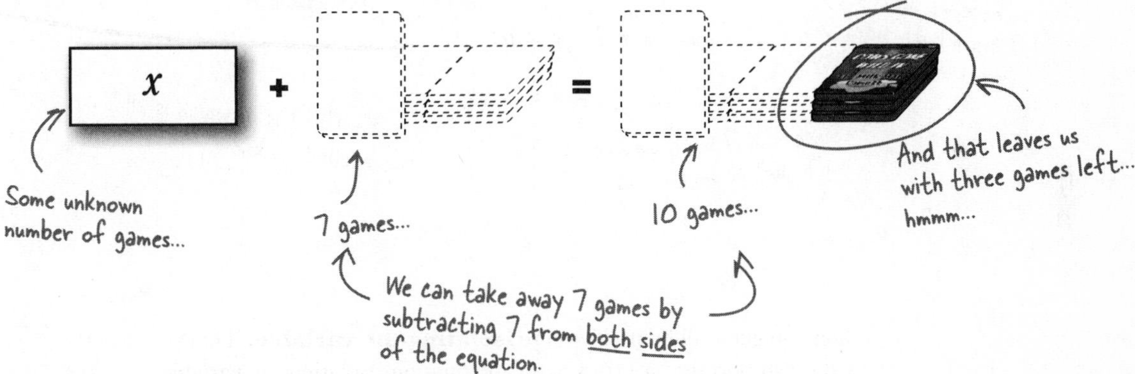

So here's what we have left:

How does that help! I'm not going to spend the rest of my life drawing pictures of all my problems.

You don't need pictures to do algebra.

What you need is a way to use the operations that you already know (addition, subtraction, multiplication, and division) to solve equations.

The tricky part? You must preserve the equality. Equality means ***the same***. When you do something to one side of the equation, you have to do the same thing to the other side of the equation.

Here's another way to look at Jo's online problem without pictures:

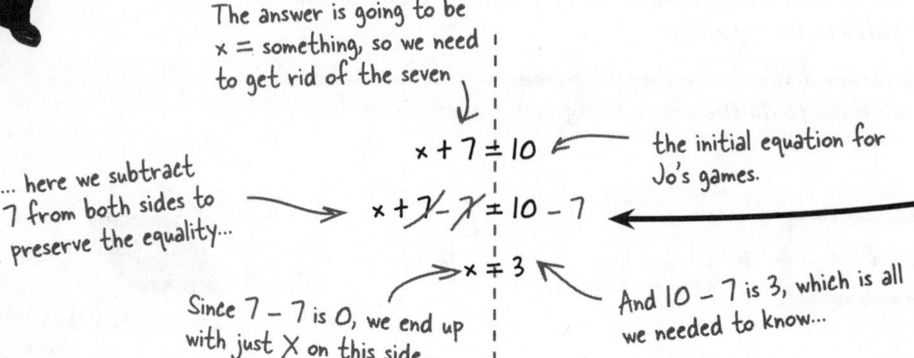

When you get x all by itself, you're **isolating the variable.** That's the most important part of solving an equation. Isolating the variable means you've gotten the variable by itself on the left side of the equation and everything else stacked up over on the right. If you can isolate a variable, then you've solved the equation—the answer just pops out, like **x = 3**.

Knowing that your goal is to isolate the variable means that you know which numbers to move away from the left side. Since you're trying to get **x** alone, that means that you move the seven, not the 10!

So which operation do you use when?

The opposite of addition is subtraction. So, if some number is being added on one side of the equation, and you want to move that number to the other side, you can subtract that number from both sides. The math term that describes opposite operations is ***inverse operations***.

The basic math operations are addition, subtraction, multiplication, and division. An inverse operation is the operation that undoes an operation (like addition undoes subtraction). Inverse operations let you shift a number or variable from one side of the equation to the other by "undoing" that number on one side of an equation.

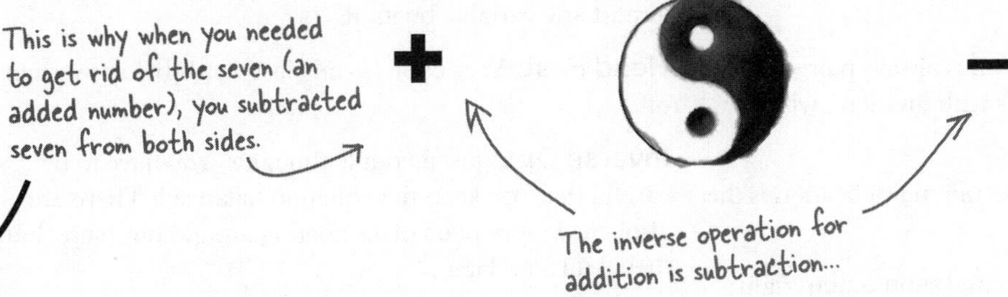

This is why when you needed to get rid of the seven (an added number), you subtracted seven from both sides.
+
−
The inverse operation for addition is subtraction...

When you want to solve an equation:

1. **Look at the equation and figure out what numbers to move.**
   Using Jo's equation, we had to get rid of the 7. That's because we're trying to isolate the variable, the x.

2. **Figure out which operation to use.**
   You need to use the inverse operation for the number to remove it. For a subtracted number, add. For a divided number, multiply, and so on.

3. **Preserve equality.**
   Whatever you decide to do to one side of the equation, you must do to the other. That keeps the equation the same.

There are other inverse operations out there. Can you think of other operation pairs that work?

Inverse Operations Exposed

This week's interview:
Just who are the inverse operations?

Head First: And welcome back to Algebra at Night. Tonight's guest... or guests... are inverse operations. So do you guys always travel in pairs?

Inverse Ops: Well, yes. We're not inverse operations unless both of us are here. We're about maintaining balance.

Head First: Ah, right. So addition is always paired with subtraction, multiplication always with division.. why is that?

Inverse Ops: Opposites attract, and multiplication is the opposite of division.

Head First: Same with addition and subtraction, right?

Inverse Ops: Yeah, and we're all opposites because we all undo each other.

Head First: When you say undo each other, do you mean if there's a multiplication, than division can make it go away?

Inverse Ops: Well, not really go away—remember, our job is to keep everything in balance. We just move things around. If you have a multiplication you need to move, you can undo that multiplication with a division—on both sides of the equation.

Head First: Ok, I think I get it—you can move numbers from one side of the equation to the other. So you're pretty useful for getting a variable by itself?

Inverse Ops: Absolutely! That's what we do best. A little addition here or multiplication there, and you can get almost any variable by itself.

Head First: Very cool! So any last words before we sign off?

Inverse Ops: Just a couple thoughts. You have to be careful that you keep the equation balanced. There are also a few more pairs of us floating around out there, but they'll turn up later.

Head First: Well, it's been great talking to you—all of you—and I appreciate you coming by. Until next time, may your multiplications always have a division, and your additions subtract.

Inverse operations help you isolate the variable.

BULLET POINTS

- Algebra is about solving for **unknowns**.
- You use other information from your problem to setup an equation with the unknown.
- The unknown is called a **variable**.
- In order to solve for a variable (like ***x***), you need to **isolate the variable**.
- You can isolate the variable by using **inverse operations** to **manipulate the equation**.
- **Addition** is the inverse of **subtraction**, and **multiplication** is the inverse of **division**.

Sharpen your pencil

Below are equations that have unknowns and numbers on both sides of the equations. Use inverse operations to isolate the variable and solve the equation.

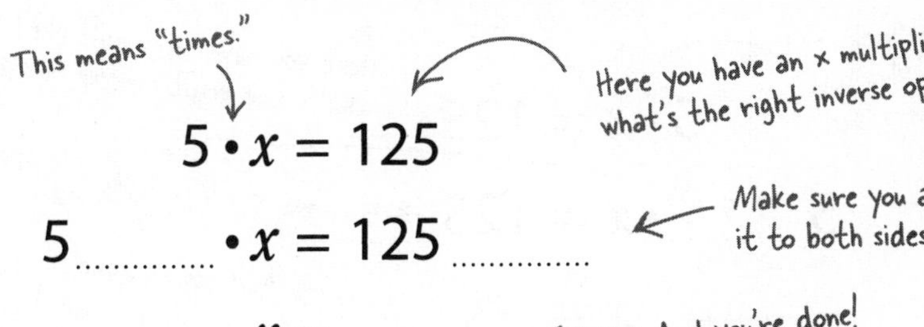
This means "times."
Here you have an x multiplied by 5... what's the right inverse operation?
Make sure you apply it to both sides...
And you're done!

$$5 \cdot x = 125$$

$5 \ldots\ldots \cdot x = 125 \ldots\ldots$

$x = \ldots\ldots$

$$x - 13 = 29$$

$x - 13 \ldots\ldots = 29 \ldots\ldots$

$x = \ldots\ldots$

$$x \cdot 6 = 47 - 11$$

$x \cdot 6 \ldots\ldots = \ldots\ldots$

$x = \ldots\ldots$

$$x + 22 = 25$$

$x + 22 \ldots\ldots = 25 \ldots\ldots$

$x = \ldots\ldots$

This also means times.

$$3(x) = 5$$

$3(x) \ldots\ldots = 5 \ldots\ldots$

$x = \ldots\ldots$

Why are there dots and parentheses for multiplication? What's wrong with x?

Whoever thought it was a good idea to use ***x*** to mean the typical unknown apparently didn't mind the confusion it might cause with the multiplication sign, x. However, lots of other people did.

They gave up on using ***x*** for multiplication and came up on with a few easier to read options:

Each of these mean multiplication.

A simple dot: $5 \cdot x = 125$

Parentheses: $5(x) = 125$

Nothing at all: $5x = 125$

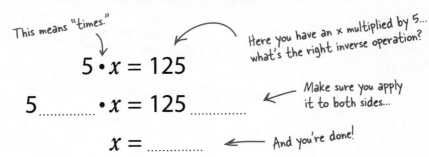

Sharpen your pencil Solution

Below are equations that have unknowns and numbers on both sides of the equations. Your job was to use inverse operations to isolate the variable and solve the equation.

Don't forget to check your work – just plug your answer back in for x and make sure it works.

$5 \times 25 = 125$

Here you have an x multiplied by 5... what's the right inverse operation?

$$5 \bullet x = 125$$

Make sure you apply it to both sides...

$$5 \div 5 \bullet x = 125 \div 5$$

$$x = 25$$

And you're done!

Check it....

$42 - 13 = 29$

Here you need to get that 13 out of the way, so use the inverse operation and add 13 to both sides.

$$x - 13 = 29$$

$$x - 13 + 13 = 29 + 13$$

$$x = 42$$

This one was a little tricky. There's a "– 11" over here, but you can figure out "47 – 11," there's no need to move it anywhere.

$$x \bullet 6 = 47 - 11$$

$$x \bullet 6 \div 6 = 36 \div 6$$

$$x = 6$$

The real problem is this 6, which you can get rid of by dividing both sides by 6.

Check this one too...

$6 \times 6 = 47 - 11$

$$x + 22 = 25$$

$$x + 22 - 22 = 25 - 22$$

$$x = 3$$

Almost done...

$3 + 22 = 25$

$$3(x) = 5$$

The new "divided by" symbol

$$3(x) / 3 = 5 / 3$$

$$x = 5/3 \text{ or } 1.6667$$

And check this one, just to make sure that fraction is right.

$5/3 \times 3 = 5$

And a change for division too...

The division sign you're used to seeing was tossed away, too. Instead you'll see things like this:

A slash: $125/x = 5$

Stacked: $\dfrac{125}{x} = 5$

These are just two other ways to show division.

Match each example of something you've learned in this chapter to it's name. Be careful, some of the names are used **twice**!

Operation example	Operation name
Inverse operation for addition	Division
Opposite of multiplication	Equation
$7/3$	Manipulating the equation
x	Variable
$2x = 10$	Subtraction
$x \cdot 3$ .... $= 5$	

WHO DOES WHAT?
SOLUTION

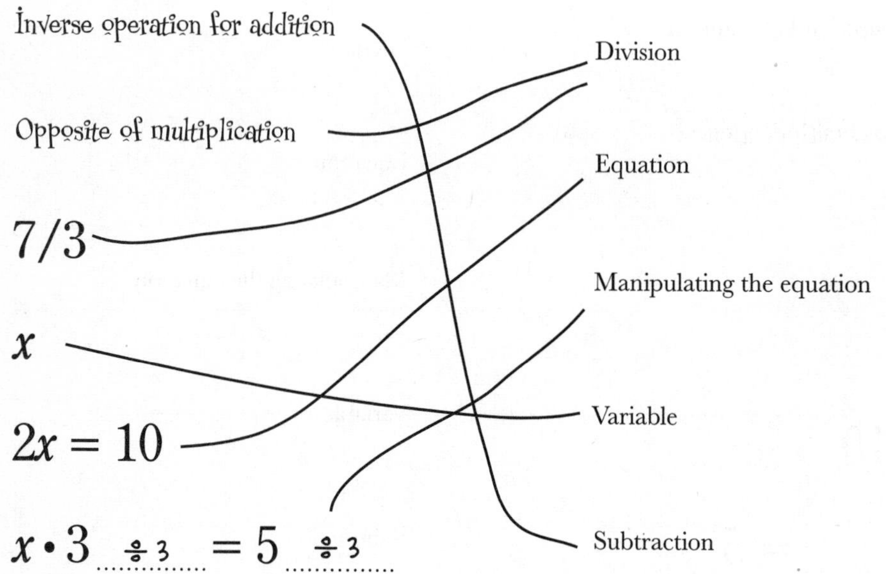
Operation example
Operation name
Inverse operation for addition
Opposite of multiplication
7/3
x
2x = 10
x • 3 ÷ 3 = 5 ÷ 3
Division
Equation
Manipulating the equation
Variable
Subtraction

Jo is ready to accessorize!

Jo figured out that she had \$86.32 left in her account for accessories. She decided that she wants to get more games and not worry about the headset just yet.

Jo did some quick Algebra to figure out how many game she can buy:

Gaming Headset
Headset with microphone, ideal for online gaming
(HS-AL1-867)
special value
$39

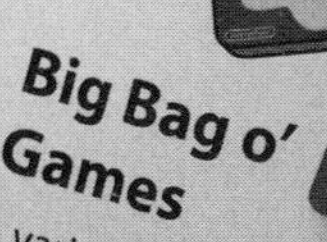

Jo's going to buy a couple of games on special.
COWS GONE WILD IV
The Milk Man Cometh
COWS GONE WILD III
special value
$49
Big Bag o' Games
A variety of games for the oPOD system
(HD-ISH-5309)

After I figured out I could buy 2 games, I went to buy them and it came up to more than $86.32! I must have done something wrong - I thought I had enough money!

Jo's work

$49x = 86.32$

$49x/49 = 86.32/49$

$x = 2.007$ ← Jo *thinks* she has just enough for 2 games.

Sharpen your pencil

Fix Jo's problem! Figure out where she went wrong.

....

....

....

Sharpen your pencil Solution

Fix Jo's problem! Figure out where Jo went wrong.

$49x = 86.32x$ — Jo messed up working with the sales prices. Could she have prevented her mistake?

$\not{4}\not{9}x/\not{4}\not{9} = 86.32/49$

$x = 1.76$ — The division was wrong!

~~Cecking~~ Checking your work...

You'll find as you go forward with Algebra that the problems become more complicated, and it's pretty easy to make a mistake. Jo didn't divide correctly, and that got her! Checking your work doesn't mean just looking over what you did. It also means using a specific technique called **substitution**.

Substitution uses your solution in the original equation

Substitution means putting something in for something else. A substitute teacher is in the place of a regular teacher, right? To check your work, you substitute in the answer you found for the variable in the original equation.

Substitution is a process that can be used not just for checking your work, but for other things too. When we get to more complex equations, and equations with more than one variable, you'll want to use substitution as part of the solving process.

Jo's work

$49x = 86.32$

$\not{4}\not{9}x/\not{4}\not{9} = 86.32/49$

$2.007 = x$

Checking Jo's work:

$49x = 86.32$

$49(2.007) = 86.32?$

$89.343 \neq 86.32$

Take Jo's answer and substitute it for x in the original equation.

The answers don't match, so Jo's wrong!

there are no Dumb Questions

Q: So about this "checking your work" thing...

A: Do it. It's easy to do, and ***it will tell you if you have the right answer!!*** Seriously. If you have an equation like $5 + x = 11$, and you say $x = 2$, when you plug 2 back in you end up with $5 + 2 = 11$, which is seriously wrong. ***x*** does not equal 2. It's easy to check your work after you've solved for the unknown.

Q: When else will we use substitution?

A: You'll see it come up again and again—always with checking your work, but also as a starting point to solve equations with two variables, for graphing lines, figuring out inequalities ...keep reading, we're getting there!

Q: Why are there different notations for multiplication and division?

A: It's really more convenient and a lot less confusing than the traditional multiplication and division signs. As you get to later chapters, you'll see more complicated equations where being able to show division as a single line really makes a huge difference. Multiplication (especially with parentheses) is the same way—sometimes what's inside that parentheses can get pretty complicated. Finally, multiplying a number by a variable is so common that just writing them next to each other is a lot less confusing than having a multiplication symbol in between.

Q: When should I use parentheses versus dots versus just bumping the number and variable together?

A: There's no difference between the different notations; it's just whatever is easier and looks cleanest. If you have a number times a bunch of things, you can use parentheses. We'll talk a lot more about that in chapter 2. If you have a number times a variable, just push them together. As for the dot... well, it's good for variety if you're bored with the other notations.

With division, you almost always see the stacked form, unless you're typing an equation in a word processor or an email, where the stacked form is a pain. In those cases, you can use the slash.

Q: Do addition and subtraction have other notations, too?

A: Nope, they stay the same. Plus means addition, and the minus means subtraction, but...

Q: What's the difference between a negative number and subtracting a positive number?

A: As far as working with them, none. That means that when you have a -4, it's the same thing as + (- 4).

Q: There seem to be a lot of elements that go into solving an equation, how do I keep track?

A: There are a lot, but they'll soon become second nature. Once you get used to working with equations, you'll automatically use inverse operations to move numbers around, you'll simplify the equation you end up with, and then you'll just keep going until you get that variable by itself.

Later we'll follow the exact steps, but really they are just "what you do" when you get an equation. Probably the easiest one to forget is checking your work...make sure you do it!

Substitution means putting a new value back into the original equation.

Jo's perfecting her set up with that new game she bought with what she had left of her savings. Help her figure out the details!

During Jo's embarrassing trip the first time she tried to buy the games, she put back the headset and just bought the new game, so she has $33.55 left. The new game is networked, so it's time to invest in that LIVE subscription ($12) and the headset ($39). How much does she need to save up to buy all of these accessories?

Make sure to check your work...

LIVE subscription + headset = money Jo needs + savings balance

Manipulate your equation and solve for x.

Check you work!

Jo wants to buy an extra level for her game on LIVE, and it's 720 points. It costs $1 for 60 points. How much will it cost for the new level?

60 points (number of dollars) = total point cost of the level

Substitute back in for x in the original equation.

Now how much does Jo need to come up with? She wants all the accessories and the new level for her game...

This one is pretty straightforward...

Substitute back in...

Jo has figured out that she can sell some used games that she's already beaten to pay for the headset, subscription, and extra level. She can get $8 a game. How many games does she need to sell to cover the new stuff?

Jo's perfecting her set up with that new game she bought with what she had left of her savings. Help her figure out the details!

During Jo's embarrassing trip the first time she tried to buy the games, she put back the headset and just bought the new game, so she has $33.55 left. The new game is networked, so it's time to invest in that LIVE subscription ($12) and the headset ($39). How much does she need to save up to buy all of these accessories?

Make sure to check your work...

LIVE subscription + headset = money Jo needs + savings balance

$$12 + 39 = x + 33.55$$

$$51 - 33.55 = x + 33.55 - 33.55$$

$$\$17.45 = x$$

$$12 + 39 = 17.45 + 33.55$$

Substitute 17.45 in for x in the original equation to check it.

Check the work :-)

$$51 = 51 \checkmark$$

Jo wants to buy an extra level for her game on LIVE, and it's 720 points. It costs $1 for 60 points. How much will it cost for the new level?

60 points (number of dollars) = total point cost of the level

$$60x = 720$$

Divide both sides by 60.

$$\frac{60x}{60} = \frac{720}{60}$$

This is the new division notation.

$$x = 12$$

Substitute back in for x in the original equation

$$60(12) = 720$$

$$720 = 720 \checkmark$$

Now how much does Jo need to come up with? She wants all the accessories and the new level for the game...

This one is pretty straightforward...

New accessories + new level = total money Jo needs

$17.45 + 12 = x$

$\$29.45 = x$

Substitute back in...

$17.45 + 12 = 29.45$

$29.45 = 29.45$ ✓

Now we know the total that Jo needs.

Jo has figured out that she can sell some used games that she's already beaten to pay for the headset, subscription, and extra level. She can get $8 a game. How many does she need to sell to cover the new stuff?

$$\frac{\text{total money Jo needs}}{\text{amount she can get per game}} = \text{number of games to sell}$$

$$\frac{29.45}{8} = x$$

$3.68125 = x$

$$\frac{25.45}{8} = 3.68125$$

$3.68125 = 3.68125$

That 3 point whatever number works – but this is whole games we're selling – so Jo actually needs to sell 4 games to get enough money to get all of this stuff!

Equation training

Let's put all your mad equation solving skills together to solve a real world problem using Algebra:

2 **Write the equation to solve it**
Once you understand what you need to find, write out the equation in algebraic form. Use x (or some other letter) for your unknown.

1 **Understand the problem statement**
Each problem will come with hints and an unknown. Figure out what you're looking for and what other numbers in the problem will help you solve it. Think of it verbally first.

7 **Check your work!!**
Check that you have the right answer by plugging your number back into the original problem in place of the unknown.

6 **Write the equation as unknown = number**
Once you've gotten the variable by itself, you've found the solution!

x = something

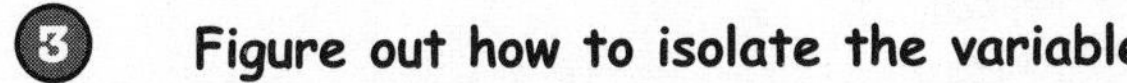

3 Figure out how to isolate the variable

Use inverse operations and work the arithmetic when you have actual numbers to get the variable by itself on one side of the equation.

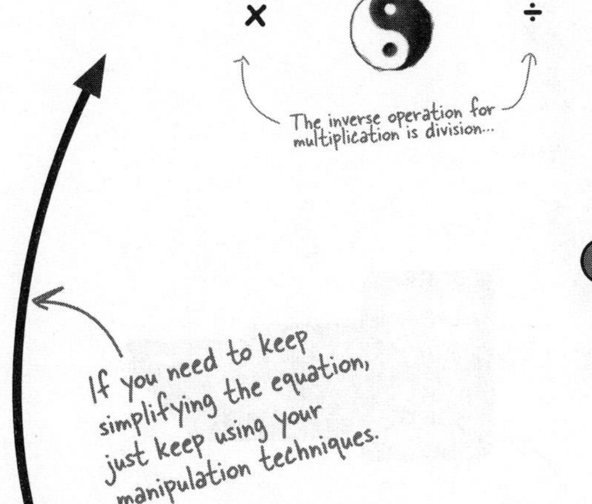

×
÷
The inverse operation for multiplication is division...
If you need to keep simplifying the equation, just keep using your manipulation techniques.

4 Manipulate the equation

Apply your techniques to actually move the numbers around in the equation to get that variable by itself. Make sure you always keep the equation equal by doing the same thing to both sides.

5 Rewrite the equation

Clean up the equation by performing any arithmetic you've setup after you moved things around and see if you've gotten the variable by itself. If not, apply another technique to get the variable by itself.

Jo has an awesome setup!

After a trip to sell off 4 games and buy a headset, Jo signed into LIVE and bought that new level, and she is ready to play!

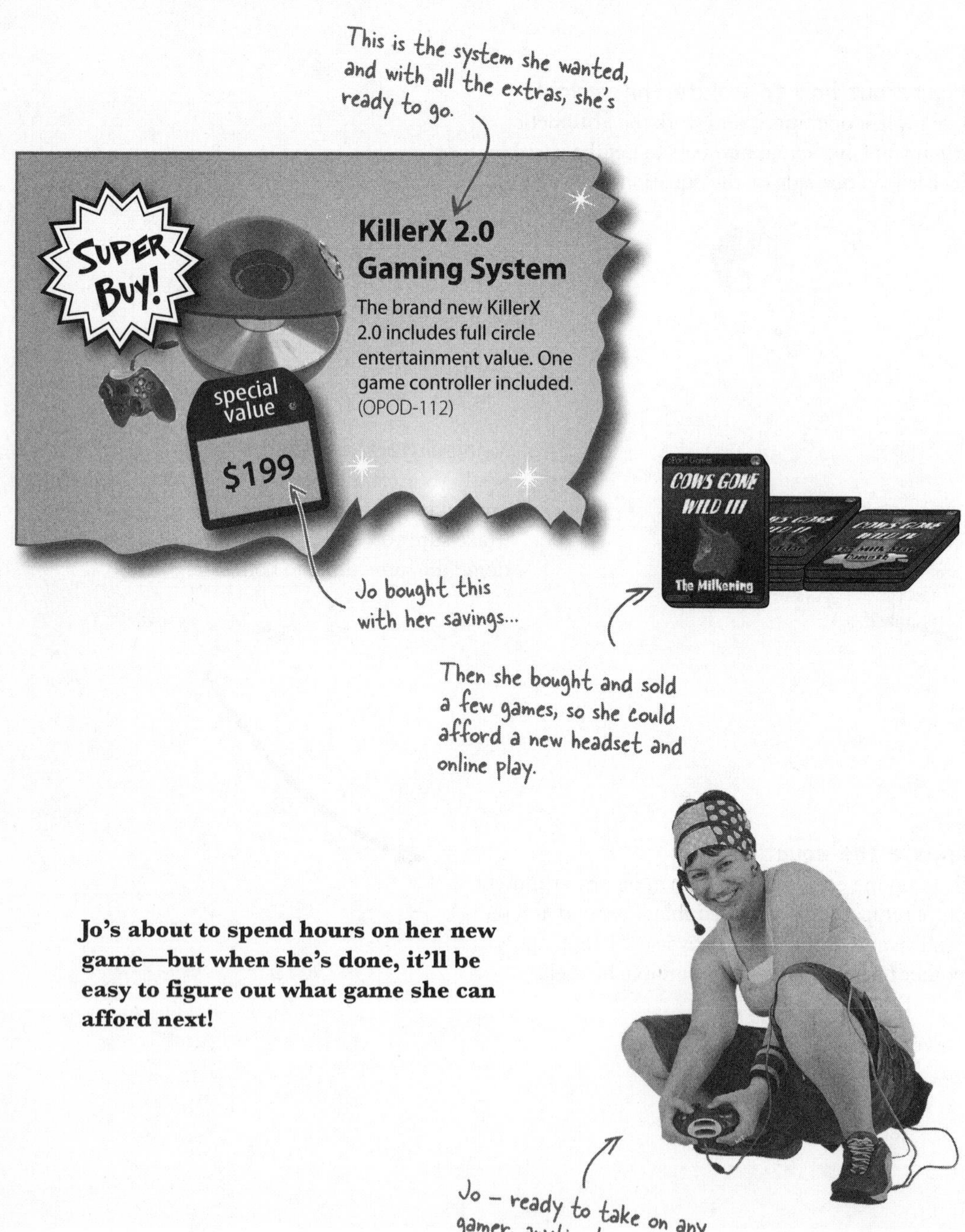

This is the system she wanted, and with all the extras, she's ready to go.
SUPER Buy!
KillerX 2.0 Gaming System
The brand new KillerX 2.0 includes full circle entertainment value. One game controller included. (OPOD-112)
special value
$199
COWS GONE WILD III
The Milkening
Jo bought this with her savings...
Then she bought and sold a few games, so she could afford a new headset and online play.
Jo – ready to take on any gamer, anytime!.

Jo's about to spend hours on her new game—but when she's done, it'll be easy to figure out what game she can afford next!

Equationcross

Take some time to sit back and give your right brain something to do. It's your standard crossword; all of the solution words are from this chapter.

Across

2. The operation that undoes some other operation.
5. Use this to do away with addition.
6. Checking your work is really just ....
8. Plug your answer back in to .... it.
10. A variable, by itself.
11. Jiggling the equation around to solve it.

Down

1. The anti-multiplication.
3. Stands for an unknown.
4. Math sentences.
7. Algebra is solving for ....
9. The key thing an equation promises.

CHAPTER 1

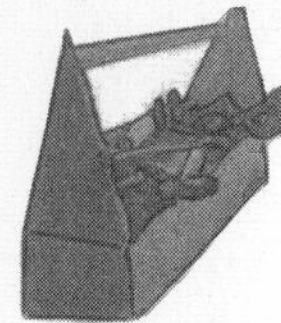

Math Toolbox

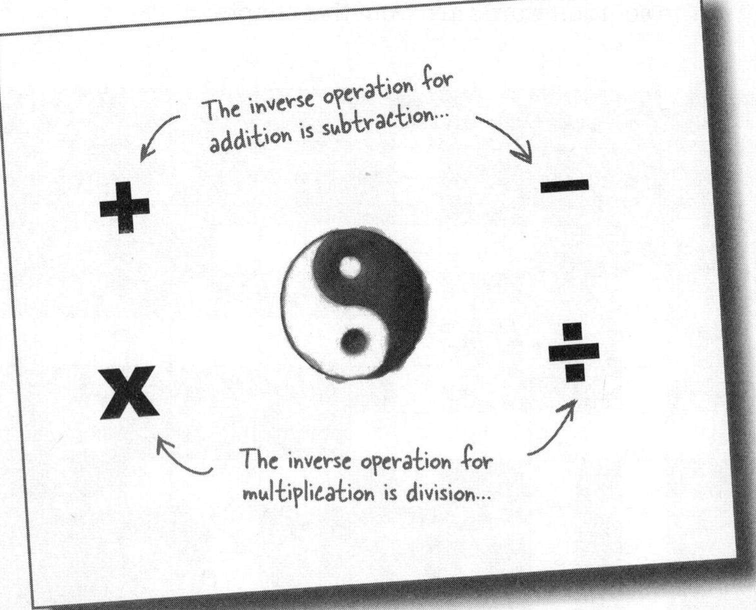

The inverse operation for addition is subtraction...
+
−
x
÷
The inverse operation for multiplication is division...

BULLET POINTS

- Algebra is about solving for **unknowns**.
- You use other information from your problem to setup an equation with the unknown.
- The unknown is called a **variable**.
- In order to find the solution to the equation, you need to **isolate the variable**.
- You can isolate the variable by using **inverse operations** to **manipulate the equation**.
- **Addition** is the inverse of **subtraction,** and **multiplication** is the inverse of **division**.

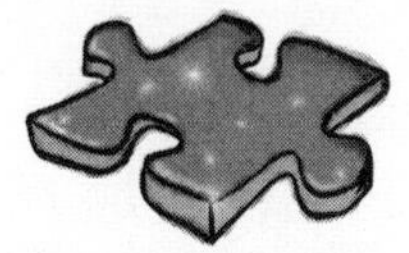

Equationcross solution

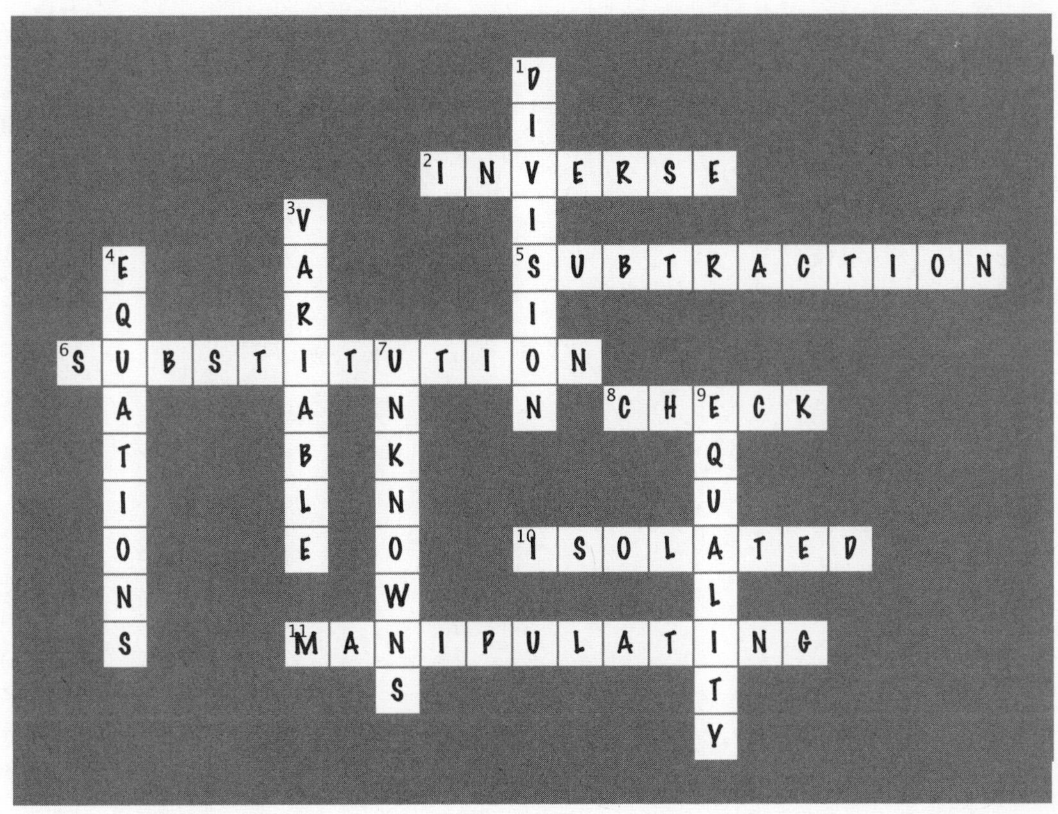

2 (more) complicated equations

Taking Algebra on the road

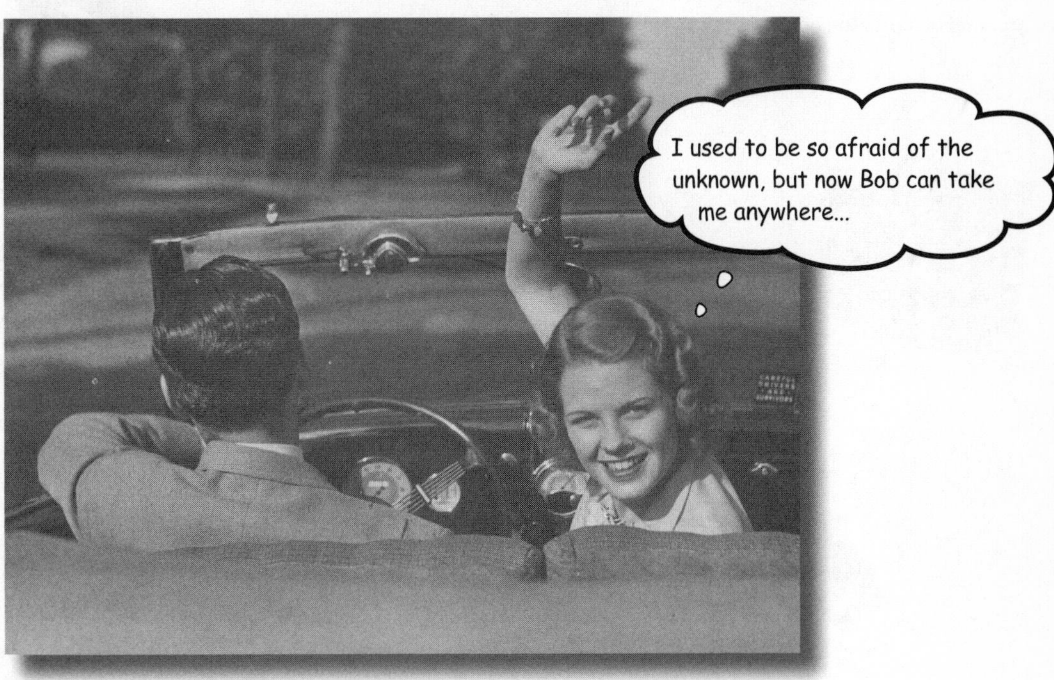

I used to be so afraid of the unknown, but now Bob can take me anywhere...

Imagine a world where there is more than ONE thing you don't know. Yes, it's hard to imagine... but there are problems out there with **more than one unknown**. Not only that, but sometimes you've got **one unknown** that appears ***multiple times in the same equation***! No worries, though... you already know how to manipulate your equations. Add that knowledge to the tools you'll learn in this chapter, and you'll be solving more complicated expressions in no time at all.

Paul loves "Pajama Death"

Paul is a huge fan of the punk band Pajama Death. The band's kicking off their US tour this week in Florida, and Paul's determined to be there. Paul's got his savings ready to blow, but he's got no idea how much cash he needs to pull out of his account.

Can you help Paul?

Paul has $1,330 in his bank account and is willing to spend it all. In fact, Paul wants to bring his buddies and really blow it out this weekend. But how many friends can he bring? Not only that, but there are a *lot* of costs to keep up with:

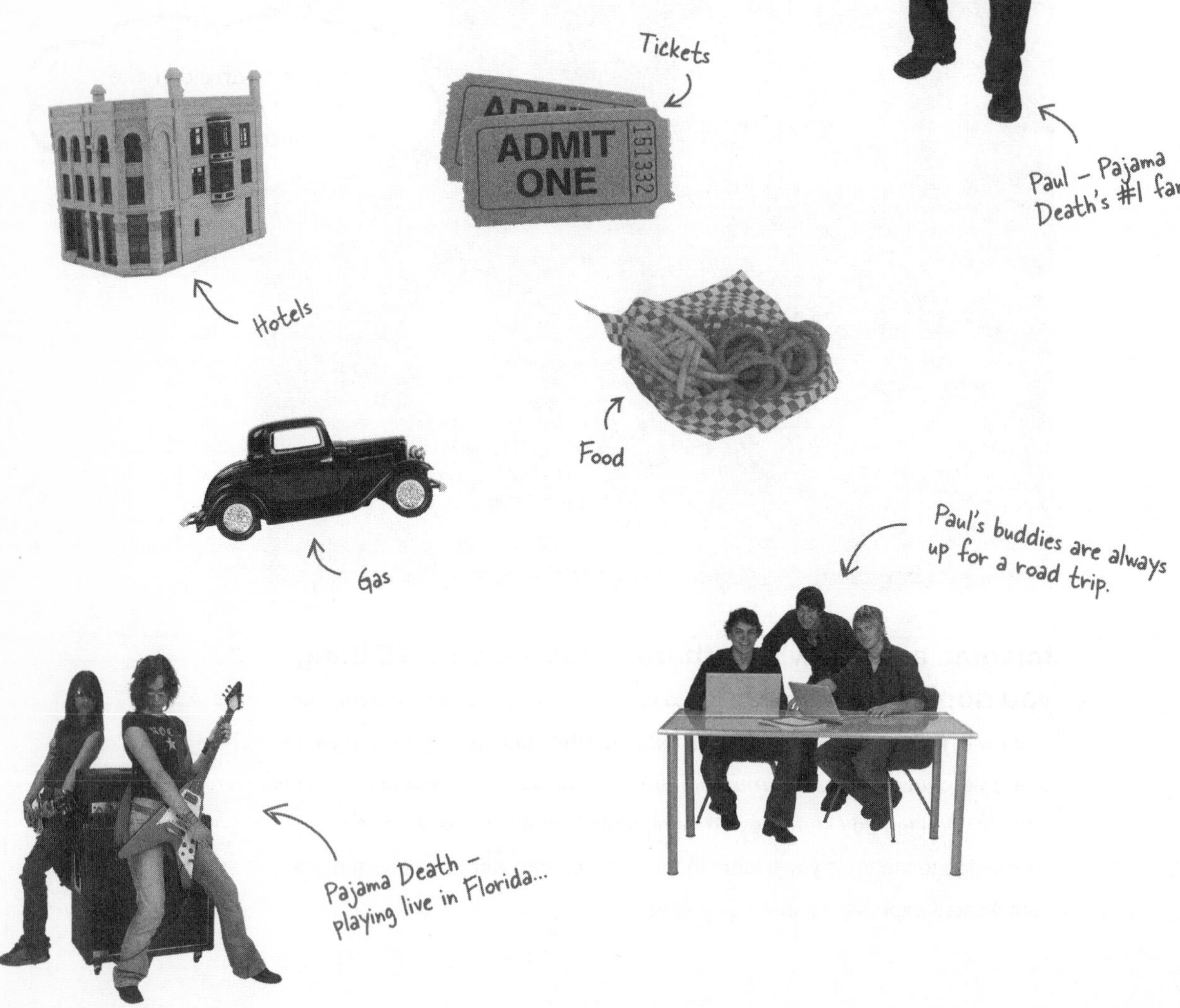
Paul - Pajama Death's #1 fan!
Tickets
ADMIT ONE
151332
Hotels
Food
Gas
Paul's buddies are always up for a road trip.
Pajama Death - playing live in Florida...

Always start with what you know

The best way to work any problem is to figure out what things you know and what things you don't know. The big unknown in this problem is how many friends Paul can bring. Let's call this ***g*** for guys. We also know Paul's got $1,330 in his bank account. Paul can spend up to that amount on the trip.

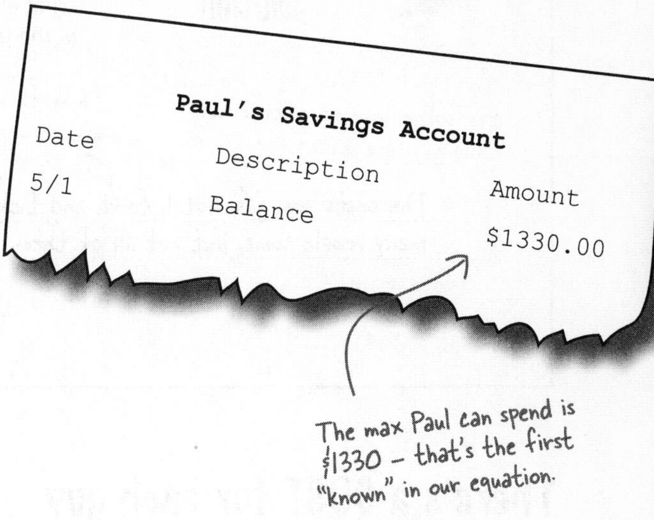

Paul's Savings Account
Date
Description
Amount
5/1
Balance
$1330.00
The max Paul can spend is $1330 - that's the first "known" in our equation.

Here's what we know:

But there's still a lot of things missing. We can't simply set these things equal to each other... that doesn't make any sense.

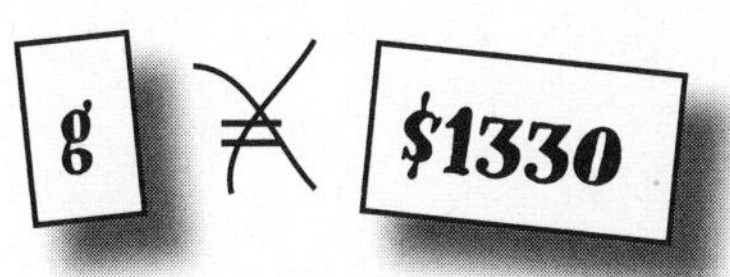

Sharpen your pencil

What else has to go into the equation to add up to the total cost of the trip? Does it matter how many people come?

....

....

....

Sharpen your pencil Solution

What else has to go into the equation to add up to the total cost of the trip? Does it matter how many people come?

Gas is the same no matter how many guys are in the car.

These three depend upon how many guys go.

What we were missing is the expense of the trip itself.

The costs are: gas, hotel, food, and tickets. Some depend on how many people come, but not all of them.

There's a COST for each guy

Paul is going to have to spend money on gas to get to Florida. But what about food? Tickets? Hotel rooms? Those all depend on how many guys come (and Paul is one of those guys).

So we've got to figure out the fixed costs, like gas, and then we've got to figure out how much each guy costs. That's got to be multiplied by each guy, and added to the fixed costs. And all of that has to be related to how much money Paul can actually spend.

So we've got something that looks like this:

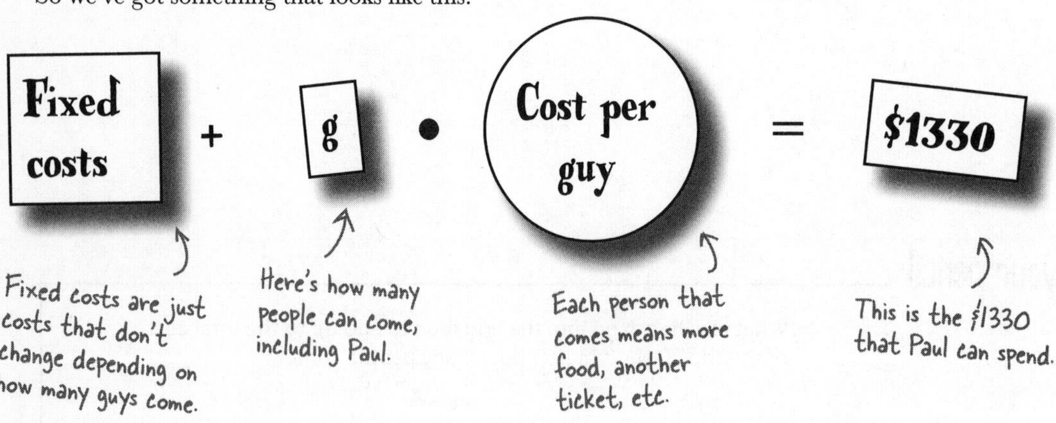

Fixed costs
+
g
•
Cost per guy
=
$1330
Fixed costs are just costs that don't change depending on how many guys come.
Here's how many people can come, including Paul.
Each person that comes means more food, another ticket, etc.
This is the $1330 that Paul can spend.

The fixed costs won't change, but the total cost goes up as we add more guys (***g***). The question is how do we figure out how much each guy costs?

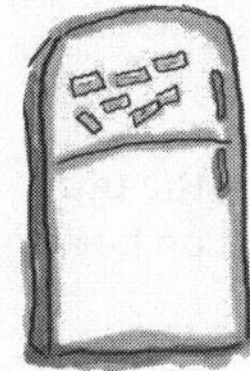

Costs Magnets

Now that you have the basic idea of the equation, use the magnets below to put together what the cost of the trip is based on the number of guys coming. Remember, some costs depend on how many people are there, and some don't.

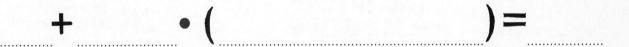

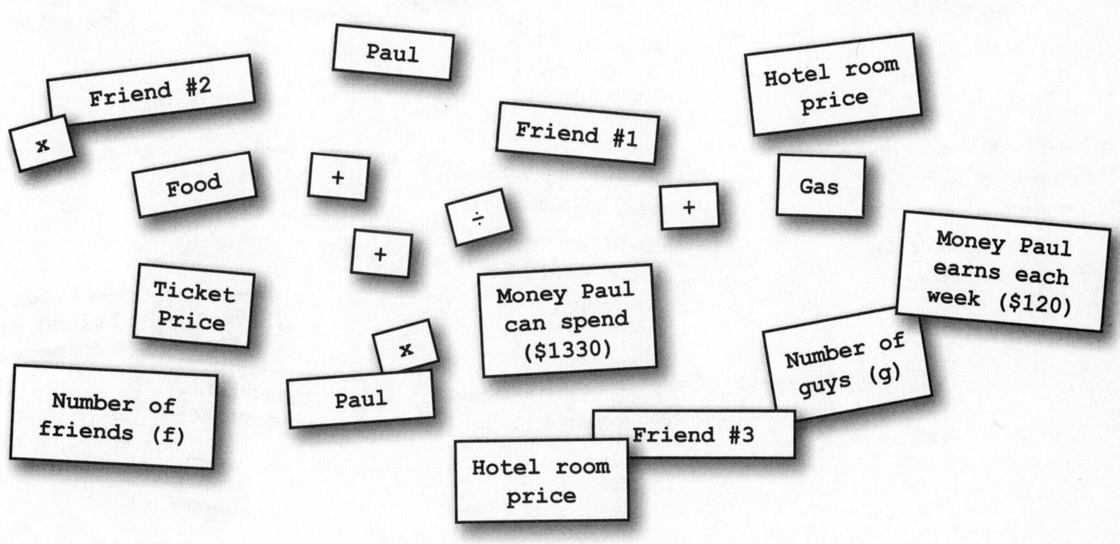

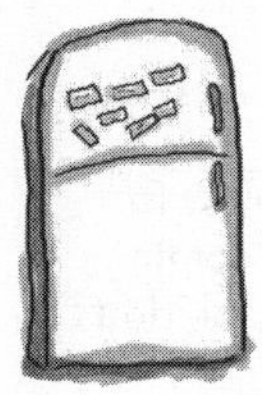

Costs Magnets Solutions

Your job was to use the magnets below to put together what the cost of the trip is based on the number of guys coming. Remember, some costs depend on how many people are there, and some don't.

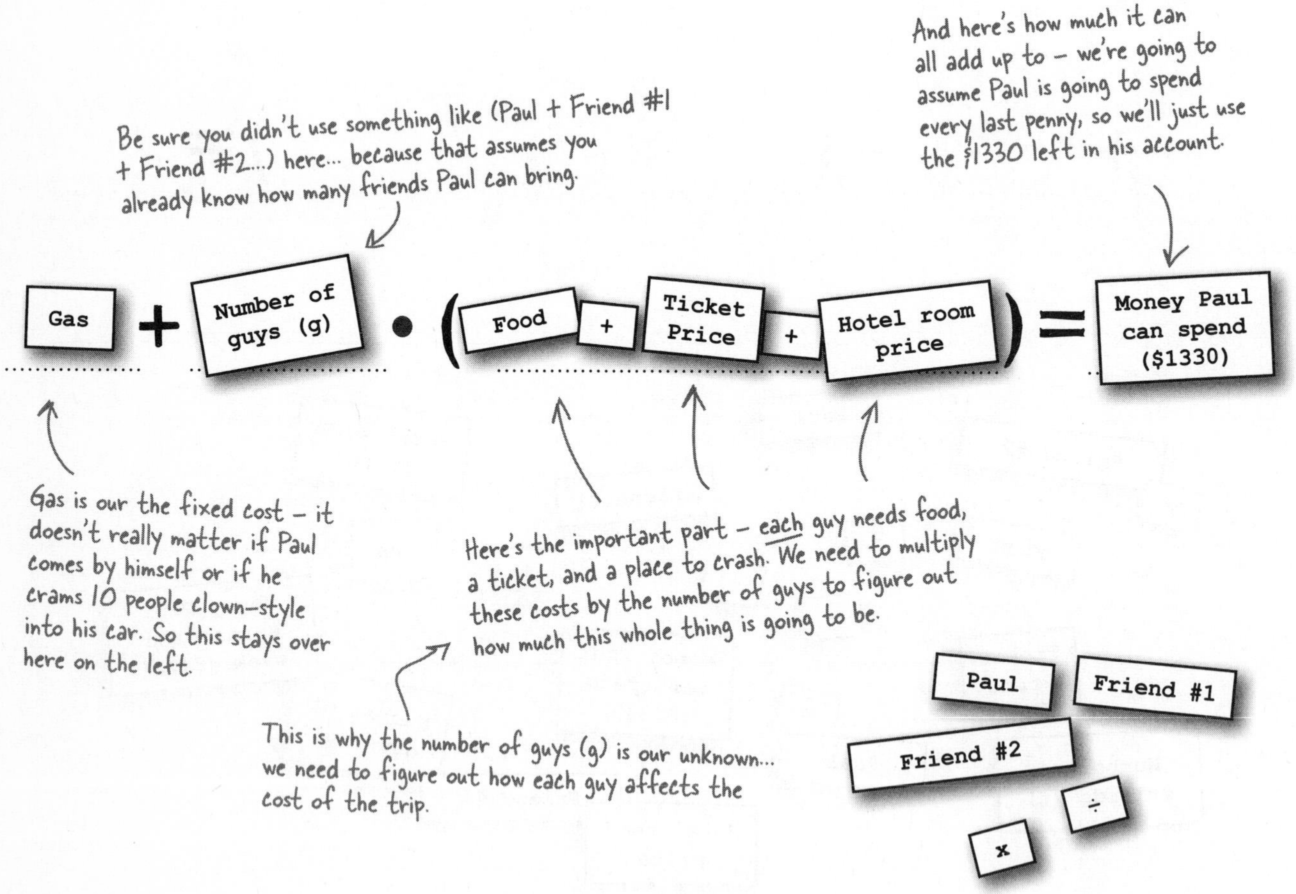

Be sure you didn't use something like (Paul + Friend #1 + Friend #2...) here... because that assumes you already know how many friends Paul can bring.

And here's how much it can all add up to – we're going to assume Paul is going to spend every last penny, so we'll just use the $1330 left in his account.

Gas + Number of guys (g) • (Food + Ticket Price + Hotel room price) = Money Paul can spend ($1330)

Gas is our the fixed cost – it doesn't really matter if Paul comes by himself or if he crams 10 people clown-style into his car. So this stays over here on the left.

Here's the important part – each guy needs food, a ticket, and a place to crash. We need to multiply these costs by the number of guys to figure out how much this whole thing is going to be.

This is why the number of guys (g) is our unknown... we need to figure out how each guy affects the cost of the trip.

Paul
Friend #1
Friend #2
÷
x

Replace your words with numbers

Now that you've got a general equation, you're ready to put some numbers in. You can replace the boxes below with actual numbers for each cost...

Fast food for most meals - say $60 for the trip, each?
Dude, gas is $4 a gallon now. Sucks. We're going to need about $160 for gas...
Here's a hotel for $100 a night, and it's a 3 night trip, so it's $300 for the trip.
Ok- tickets for the show are going to be $50 each.

Sharpen your pencil

Use the specific numbers for the costs to figure out the actual equation for the trip. (You don't need to solve the equation, just get it to a form you can work with.)

This is what you're trying to figure out – leave it as "g" in your equation.

Gas + Number of guys (g) • (Food + Ticket Price + Hotel room price) = Money Paul can spend ($1330)

First replace each box with the actual values from the conversation above...

Now clean up the numbers and simplify the equation a little.

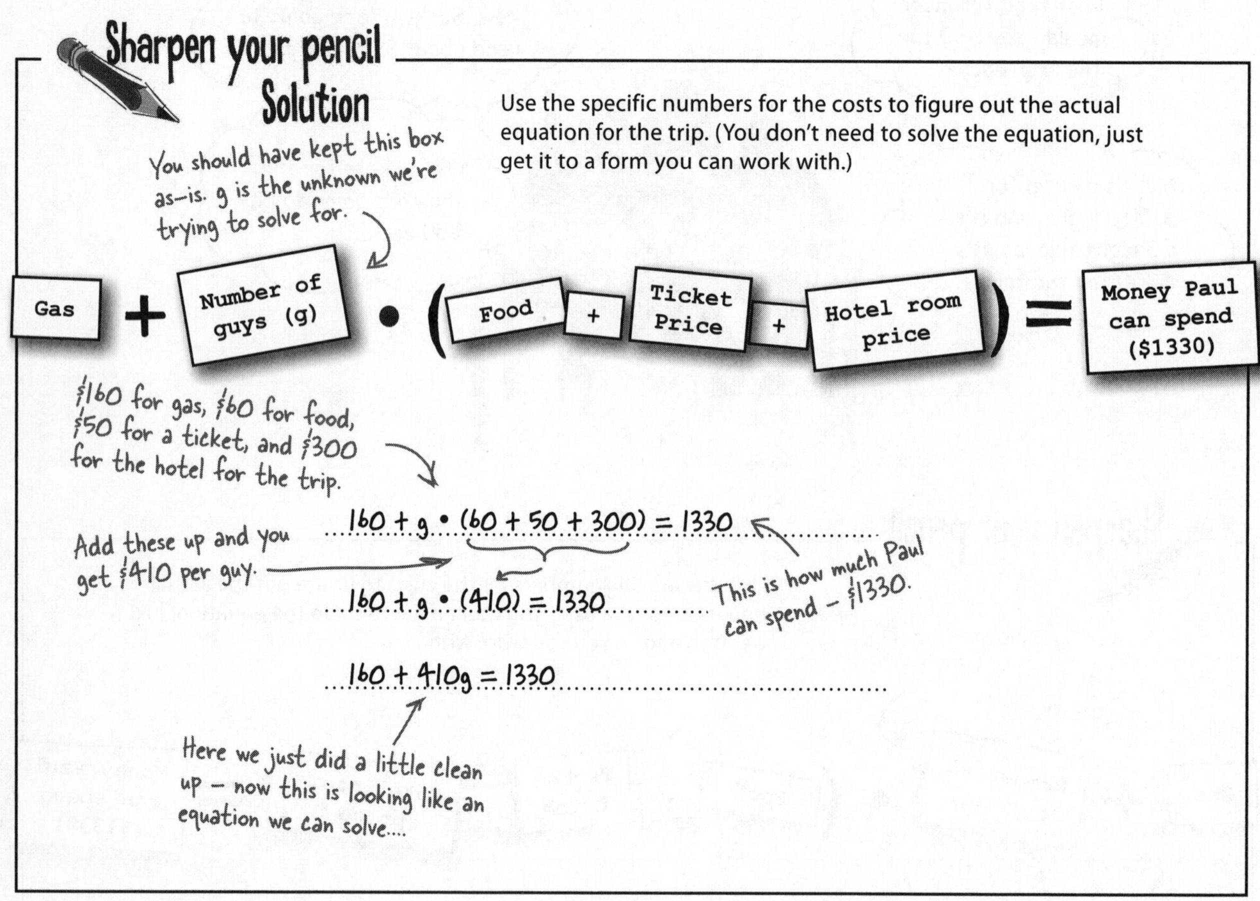
Sharpen your pencil Solution
Use the specific numbers for the costs to figure out the actual equation for the trip. (You don't need to solve the equation, just get it to a form you can work with.)
You should have kept this box as–is. g is the unknown we're trying to solve for.
Gas
+
Number of guys (g)
•
Food
+
Ticket Price
+
Hotel room price
=
Money Paul can spend ($1330)
$160 for gas, $60 for food, $50 for a ticket, and $300 for the hotel for the trip.
160 + g • (60 + 50 + 300) = 1330
Add these up and you get $410 per guy.
160 + g • (410) = 1330
This is how much Paul can spend – $1330.
160 + 410g = 1330
Here we just did a little clean up – now this is looking like an equation we can solve....

Now solve for g... one step at a time

You've got an equation (160 + 410***g***=1330) that can tell you how many guys Paul can take on his trip. Back in Chapter 1, you worked equations to **isolate the variable**. In this case, the variable is ***g***, and we can use inverse operations to get to just ***g***:

$$160 + 410g = 1330$$

We have to get 160 off of this side of the equation...

...and over to this side...

We'll need to get that 410 out of the way too...

...but you have to keep the equation equal!

We need to isolate the variable (***g***) in this equation to figure out how many guys can go on the trip. But where should you start? There are *two* things you need to do to get that ***g*** buy itself: you've got to get rid of the 160 on the left side that's added to 410g, and you've also got to get rid of the 410 that's multiplied by ***g***. And no matter what you do, you've got to keep the sides of the equation equal.

$$160 + 410g = 1330$$

Do we move the 160 first?

Or the 410?

The cost equation is a **multistep equation**. You have to deal with both the 160 and the 410 in order to solve for g. Which should you do first? Will you get different answers if you do one before the other? Which is right?

If you follow the rules, you'll ALWAYS get the right answer

The most important part of equations is that they stay equal on both sides. So you could first move the 160 by using subtraction, if you wanted to, or you could deal with the 410***g*** part first by using division. So the question is really, which is ***easiest*** to do first?

Seriously... there's nothing wrong with picking the easy way out when it comes to Algebra.

If you try to get rid of the 410 first, you'll need to divide both sides of the equation by 410. Since 410***g*** isn't the only thing on the left side of the equation, we have to divide ***everything*** by 410, like this:

We have to divide everything by 410...

$$\frac{160}{410} + \frac{410g}{410} = \frac{1330}{410}$$

$$0.3902 + g = 3.2439$$

These numbers aren't very fun to work with.

$$-0.3902 + 0.3902 + g = 3.2439 - 0.3902$$

Now we subtract .03902 from both sides.

$$g = 2.8537$$

Working with 410 first was fine, but we were left with all those nasty decimal numbers to work with. That's not bad, but without a calculator, it might be harder than it really needs to be.

Even though you'll get the same answers doing things both ways, sometimes one particular way is easier. Let's try and work things out the other way, by subtracting the 160 from both sides first...

When you've got more than one way to solve an equation, look for the SIMPLEST way to work with your equation.

Whole numbers are usually easier to work with

Instead of dividing everything by 410 and getting that nasty 160/410 thing, let's try subtracting 160 from both sides of the equation, like this:

$$-160 + 160 + 410g = 1330 - 160$$

$$410g = 1170$$

$$\frac{410g}{410} = \frac{1170}{410}$$

$$g = 2.8537$$

We have to do the same thing to both sides of the equation.

Subtracting first left us easier numbers to work with...

...and only one trickier division, right at the end.

See! The same answer.

Doing the subtraction first was easier, and we got the same answer. It was easier for two reasons: you didn't have to deal with decimals until the end, and solving the equation took less steps!

It's not always easy to determine which steps you should do first, but the good news is that even if you don't pick the easiest strategy, you can still get to the right solution. Just follow the rules, and you'll get the same solution, no matter what order you work things out in.

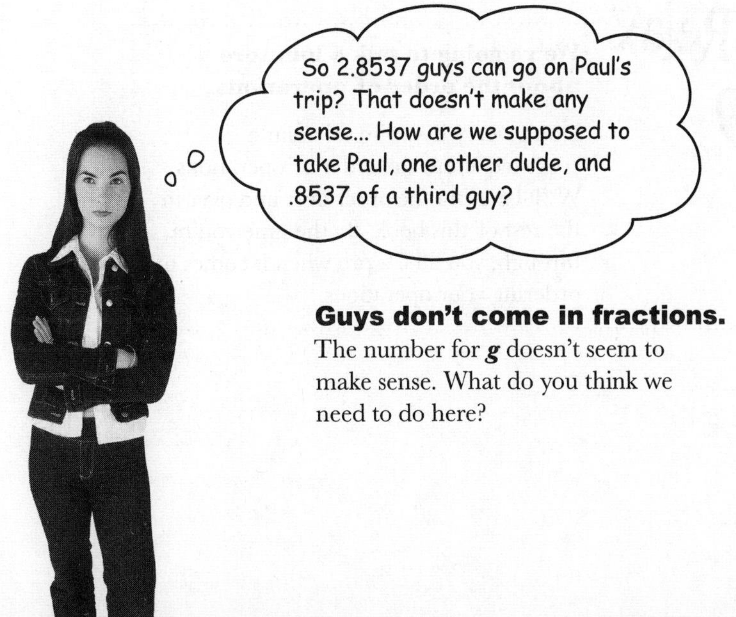

So 2.8537 guys can go on Paul's trip? That doesn't make any sense... How are we supposed to take Paul, one other dude, and .8537 of a third guy?

Guys don't come in fractions.

The number for ***g*** doesn't seem to make sense. What do you think we need to do here?

there are no Dumb Questions

Q: What about the order of operations? Don't I need to do multiplication and division before addition and subtraction?

A: If you have a bunch of additions, subtractions, multiplications, and divisions all together that you need to work out, then yes, you need to follow the order of operations. But what we're doing here is just *manipulating* the equation. We're keeping things equal by doing exactly the same thing to each side of the equation. When you manipulate an equation, it doesn't matter what you do first as long as you do that thing to both sides of an equation.

Q: Do I always have to solve my equations twice? How would I know which operation to do first on a problem?

A: You don't have to solve problems twice. We just did that here to show that either approach works. As for what to do first with different problems, well, it depends. When you have the option of doing addition or subtraction to get terms from the left to the right, they're usually the easiest to knock out first. Then you have fewer things to deal with for later manipulations, like multiplication or division.

Q: Can there be even more complicated equations? What if there are lots and lots of steps?

A: Equations can get really long, and we'll get to some of those later, but the general idea remains the same. You'll need to determine which variable you need to get by itself and then manipulate the equation to get that variable alone.

It doesn't matter how many times you apply inverse operations or multiply both sides of an equation or any other operation. As long as you follow the rules, you'll get the solution you're after.

Q: When you start with a big long problem, do you have to write the problem out with words first?

A: Not really. It's up to you, but using words is a helpful way to get the idea of the problem you're solving down on paper without getting bogged down with numbers.

Writing out a problem also means you have to step back for a second and think about the context of the problem.

We're going to talk a lot more about the order of operations.

Don't worry if you're still a little confused about the order of operations. We'll be using the order over and over in the rest of this book. By the time you're through, you'll be a pro when it comes to ordering your operations.

I can bring 2.8 guys, including myself. Hmm. O'Brien you're a bio major—can I bring 0.8 of one of you guys?

So, you technically got a correct solution for ***g***, but 2.8537 is probably not the right answer for this particular *problem*. Since Paul can't bring 0.8 of a person, the right answer to the problem is actually 2 guys (including Paul). Since you can't bring 3 guys (3 is more than 2.85, so more than Paul can afford), you've got to go down to 2 guys: that's Paul and one buddy.

When you're solving an Algebra problem, part of the problem is working the equation... but another part of the problem is keeping up with what the problem means. That's called the context of the problem. Math is not just about numbers or manipulating equations; it's about solving real-world problems.

Whoa - hang on. You can afford to bring more of us if you don't pay for a hotel room for each of us. We can put 2 people in a room.

Paul's buddies are right! Our equation charges each guy $300 for the hotel. But we can put two guys in each room. How do we fix this? What part of our equation has to change? Is the cost for each guy still based on the number of guys?

A variable can appear in an equation MORE THAN ONE TIME

The hotel rooms cost \$300. And right now, we're multiplying that \$300 by each guy on the trip.

$$160 + g \cdot (60 + 50 + 300) = 1330$$

This \$300 gets added in, then multiplied by the number of guys. So each guy costs \$300 in hotel fees.

But now, we need to take that \$300 and separate it out. Each guy doesn't have to pay \$300 because 2 guys can fit into a room. So let's pull out the \$300 hotel fee first:

$$160 + g \cdot (60 + 50) + 300 = 1330$$

Now the \$300 doesn't get multiplied by each guy... but this isn't right... the cost is still related to the number of guys going.

For every 2 guys, the room costs \$300. So for each guy, then, the room cost is \$150... like this:

$$160 + g \cdot (60 + 50) + 150g = 1330$$

Since each guy is paying half, we divide \$300 by 2 to get 150.

Then multiply by g to get each guy to pay \$150.

Wait - we have two g's in the equation now. What are we supposed to do with those? Are they the same?

YES! If the same variable shows up more than once in an equation, the variable has to have the same value. A variable just represents a number in an equation. So each time ***g*** shows up, it must represent the same value.

In fact, since ***g*** represents the same value each time it shows up, you can combine terms where ***g*** is the variable. For example, 2***g*** + 3***g*** = 5***g***. Let's see how we can use that to help us out...

Sharpen your pencil

Below is the new trip cost equation. Use the new costs to figure out how many guys can go if they share a room. Be careful, this equation will take multiple steps, and you'll probably need to combine some g's.

$160 + g \cdot (60 + 50) + 150g = 1330$

Here's the cost equation with the hotel cost broken out. Start by simplifying the equation.

First, work inside the parentheses.

Then combine like terms.

Next isolate the variable.

Then solve for the number of guys that can come..

So does this answer work? Really, how many guys can come?

BRAIN BARBELL

There's something tricky about this problem related to whether an even or odd number of guys come. Can you figure out what it is?

Sharpen your pencil Solution

Below is the new trip cost equation. Your job was to use the new costs to figure out how many guys can go if guys share a room.

First we'll get rid of the parentheses by adding up the values.

$$160 + g \cdot (60 + 50) + 150g = 1330$$

$$160 + 110g + 150g = 1330$$

Then combine the like terms (110 + 150 = 260)...

$$160 + 260g = 1330$$

$$-160 + 160 + 260g = 1330 - 160$$

Now we can get rid of the 160 by subtracting 160 from both sides of the equation.

$$260g = 1170$$

$$\frac{260g}{260} = \frac{1170}{260}$$

Now, to get the g by itself, we need to divide both sides by 260.

$$g = 4.5$$

And here's our new answer!

So does this answer work? Really, how many guys can come?

g works out to be 4.5 guys, which is the correct answer mathematically, but obviously Paul can't bring half a person. However, by putting two people in a room, now 4 people can come instead of just 2!

Wow - much better. So it looks like 4 of us can go if we double up in the rooms. Before I say anything to the guys, how can we be sure this is right?

Always check your work!

That's a great question, with an easy answer. You need to ***check your work.*** It's easy to do. Just plug your answer back into the equation wherever your variable appears, and make sure both sides of the equation come out to be the same.

Sharpen your pencil

Check that you got the right answer by plugging your answer back into your equation and making sure things are equal.

Make sure you use the real solution: 4.5.

$$160 + g \cdot (60 + 50) + 150g = 1330$$

Sharpen your pencil Solution

Check that you got the right answer by plugging it back into the equation and making sure things are equal.

$$160 + g \cdot (60 + 50) + 150g = 1330$$

$$160 + 4.5 \cdot (60 + 50) + 150(4.5) = 1330$$

Plug in 4.5 for g wherever it appears in the equation.

$$160 + 4.5\,(110) + 150(4.5) = 1330$$

$$160 + 495 + 675 = 1330$$

Use the order of operations to work out the rest of the math.

$$1330 = 1330$$

...and you end up with 1330 = 1330. Perfect!

Checking your work proves your answer

You solved for ***g***, substituted your answer back in to check it, and made sure you got the right answer. 4 people can go on the trip to see Pajama Death, twice as many as when each guy has his own room!

So now you've handled food, you've handled hotel rooms, and you've handled tickets. There's only one thing that could turn this whole trip on its head...

Dude, can my girlfriend come?

BULLET POINTS

- A **term** is a piece of an equation.
- No matter how many terms you have, you're still looking to **isolate the variable** in order to solve an equation.
- Whether you're manipulating an equation, combining like terms, or isolating a variable, you still have to follow the **order of operations** when you start to solve an equation.

there are no Dumb Questions

Q: Why did we plug 4.5 back into the equation when we said 4 was the answer to the problem?

A: The numerical answer to the equation is 4.5. That's the value that makes all of the numbers work out. But since Paul can't bring half a person, 4 is the answer he actually needs. But, when we go back to check our work, we need to use the mathematically correct answer, 4.5.

Q: Is it worth taking all that time to check your work?

A: Absolutely. It is crazy-frustrating to go through a whole problem and get an answer, but *get it wrong*. That is a totally preventable problem if you go back and check your work.

Q: My work looks a little different than the solution, but I got the right answer. Did I do something wrong?

A: Not at all. With Algebra, you just need to apply the rules consistently, and you'll get to the same answer. The solutions we've presented are the way we would go about getting the answer, but they're definitely not the only way.

Q: When we isolated the variable, how did you know to subtract before you divided?

A: We were trying to minimize the number of terms we'd have in our division. If you can reduce the number of terms in the equation by combining like terms, that usually makes things easier for later steps. There will be fewer terms to deal with. By subtracting both sides, we could get rid of the 160 altogether, and only have to divide one term by 260.

Q: What if I miss combining a term?

A: No problem. As long as you manipulate the equation using the rules properly, you will not get a wrong answer... but your problem may get more complicated.

If you do start to feel like you're going down the wrong road, don't feel like you have to keep going that way. Just go back to your original problem and start over.

Q: How can this be math if there is more than one way to do the problem? Isn't there supposed to be just one right answer?

A: There's a big difference between "one right answer" and "one way to get there." A big part of mathematics is that there are different ways to solve problems and still arrive at the same (correct) answer.

For example, you *could* do away with multiplication and just use lots of additions. But using multiplication is another way to get to the same answer. Different problems will need different techniques, but you should always end up at the same place when you're done.

Algebra is about solving for an unknown, but it's also about making smart choices about how to get to that solution.

This is your chance to look at some problems, write some equations, and do some manipulating.

Pajama Death's Profit

This tour for Pajama Death is pretty important. They make most of their money on the road (the music industry is a tough business). The deal Pajama Death has with the record company is that they get the same percentage of profit from all of the different revenue they make on tour. In order for the band to make $15,000 a show, what is the minimum percentage of the revenue they need to get from the record label?

Touring Revenue

1. Avg. Food Sales—$17,000/night
2. CDs (sold at the show)—$10/each, 100 per show.
3. T-shirts—$15/each, 800 sold per show
4. Tickets—$50/each, 4,000 seats per venue.

A couple of hints: a percentage will be a decimal or fraction in an equation, and each type of revenue will have the same profit percentage.

Pajama Death's Loss

One of Pajama Death's concerts got a little out of hand, and the venue has fined them $3600. They are charged the same amount for each problem (broken fan, bartender that resigned, etc.), but the band needs to know how much that per-incident amount really was.

Figure out how much they were charged per incident.

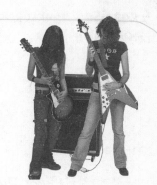

Concert Incidents

```
1. 4 ceiling fans were broken
2. 3 guitars lost their necks
3. 2 bartenders quit to become roadies
```

Exercise Solution

This is your chance to look at some problems, write some equations, and do some manipulating.

Touring Revenue

```
1. Avg. Food Sales—$17,000/night
2. CDs (sold at the show)—$10/each, 100 per show.
3. T-shirts—$15/each, 800 sold per show
4. Tickets—$50/each, 4,000 seats per venue.
```

These are each pieces of revenue that were listed above

$$\text{Food Profit} = 17{,}000 \cdot P$$
$$\text{CD Profit} = 10 \cdot 100 \cdot P$$
$$\text{T-shirt Profit} = 15 \cdot 800 \cdot P$$
$$\text{Ticket Profit} = 50 \cdot 4000 \cdot P$$

The total profit will just be the total revenue times the profit percentage.

We used "P" for percentage, and since it's the same for each source of revenue, it is in all the equations.

For CD's, T-shirts, and tickets, you have to figure out the revenue by taking the number of items sold times the price.

Total Profit= Food Profit + CD Profit + T-shirt Profit + Ticket Profit

In the problem statement it said that they want to make $15,000 or more.

$$15{,}000 = 17{,}000P + 1{,}000P + 12{,}000P + 200{,}000P$$
$$\frac{15{,}000}{1000} = \frac{17{,}000P}{1000} + \frac{1{,}000P}{1000} + \frac{12{,}000P}{1000} + \frac{200{,}000P}{1000}$$

Here we are dividing both sides by 1,000. Why? Because it'll make things easier, and it's totally legal – you just have to do the division to all of the terms on both sides of the equation.

$$15 = 17P + 1P + 12P + 200P$$

These are all in terms of "P," so they are like terms

$$15 = 230P$$
$$\frac{15}{230} = \frac{230P}{230}$$

Isolate the variable by dividing by 150.

$$\frac{15}{230} = P$$
$$0.0652 = P$$
$$6.52\% = P$$

Convert to a decimal, and then a percent.

Relax

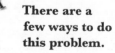

There are a few ways to do this problem.

You may not have divided by 1000 or added up all of the terms first, but they should all get you to the same answer.

Pajama Death's Loss

One of Pajama Death's concerts got a little out of hand, and the venue has fined them \$3600. They are charged the same amount for each problem (broken fan, bartender that resigned, etc.), but the band needs to know how much that per-incident amount really was.

Figure out how much they were charged per incident.

Concert Incidents

```
1. 4 ceiling fans were broken
2. 3 guitars lost their necks
3. 2 bartenders quit to become roadies
```

4 ceiling fans — 3 guitars — 2 bartenders — For a total of \$3600

$$4x + 3x + 2x = 3600$$

Since the fine is the same for each incident, we can use one variable: x.

$$9x = 3600$$

Since the variable is the same, we can combine like terms and get 9x...

$$\frac{9x}{9} = \frac{3600}{9}$$

...then we divide both sides by 9...

$$x = 400$$

...and end up with \$400 per incident. Wow, that's steep.

Now check your work by plugging your answer back in...

$$4(400) + 3(400) + 2(400) = 3600$$

$$1600 + 1200 + 800 = 3600$$

$$3600 = 3600$$

What's a road trip without some girls?

You've figured out how many guys can come, but what about girlfriends? Paul doesn't have any more money, so if girls are coming, Paul can't bring as many guys. Not only that, but girls might not cost as much as guys to bring...

So there are a few things we'll need to change with the equation to add girls to the mix:

- Girls don't want to share a room with a guy—they're going to need their own hotel rooms.
- Girls only need $30 for food.

The problem we have is that our current equation multiplies the number of guys coming on the trip by their costs:

Fixed costs + Number of guys to take (g) • Cost per guy = $1330

Here we multiply the number of guys times the cost per guy... but now we've got girls that cost a different amount. We need to make some changes...

I want to go too! And I'm not real excited about being the only girl, so I want to bring some friends..

This is Paul's girlfriend, Amanda.

We need another variable

Since girls have a different cost associated with them, we need to treat them separately in our equation. We're going to have to introduce another variable. Let's call this variable ***r*** for girls (since ***g*** is already taken...). Now our equation looks something like this:

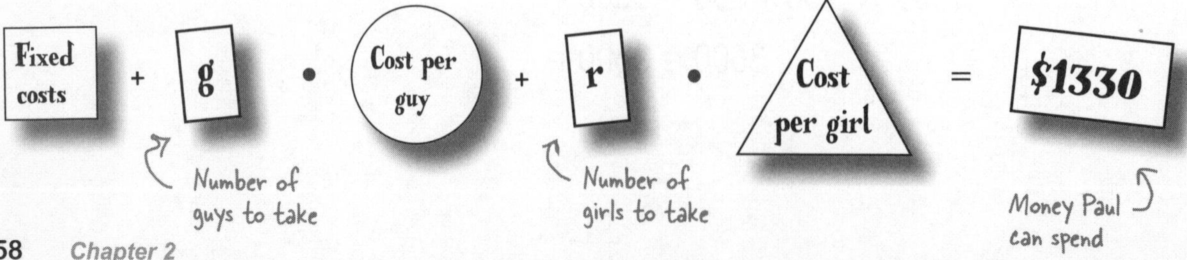

Sharpen your pencil

Use the information about cost per guy and the information about cost per girl to write up the new equation we need to solve to figure out who can come.

$60 for food, $50 for a ticket and $300 for the hotel for the trip.

$30 for food, $50 for a ticket and $300 for the hotel for the trip... but no rooms with guys.

Fixed costs + g • Cost per guy + r • Cost per girl = $1330

Number of guys to take

Number of girls to take

Money Paul can spend

Don't forget about two guys sharing a hotel room – the girls can do the same thing.

Be careful when you combine your variables – make sure you only mix g's with g's and r's with r's.

Sharpen your pencil Solution

Use the information about cost per guy and the information about cost per girl to write up the new equation we need to solve to figure out who can come.

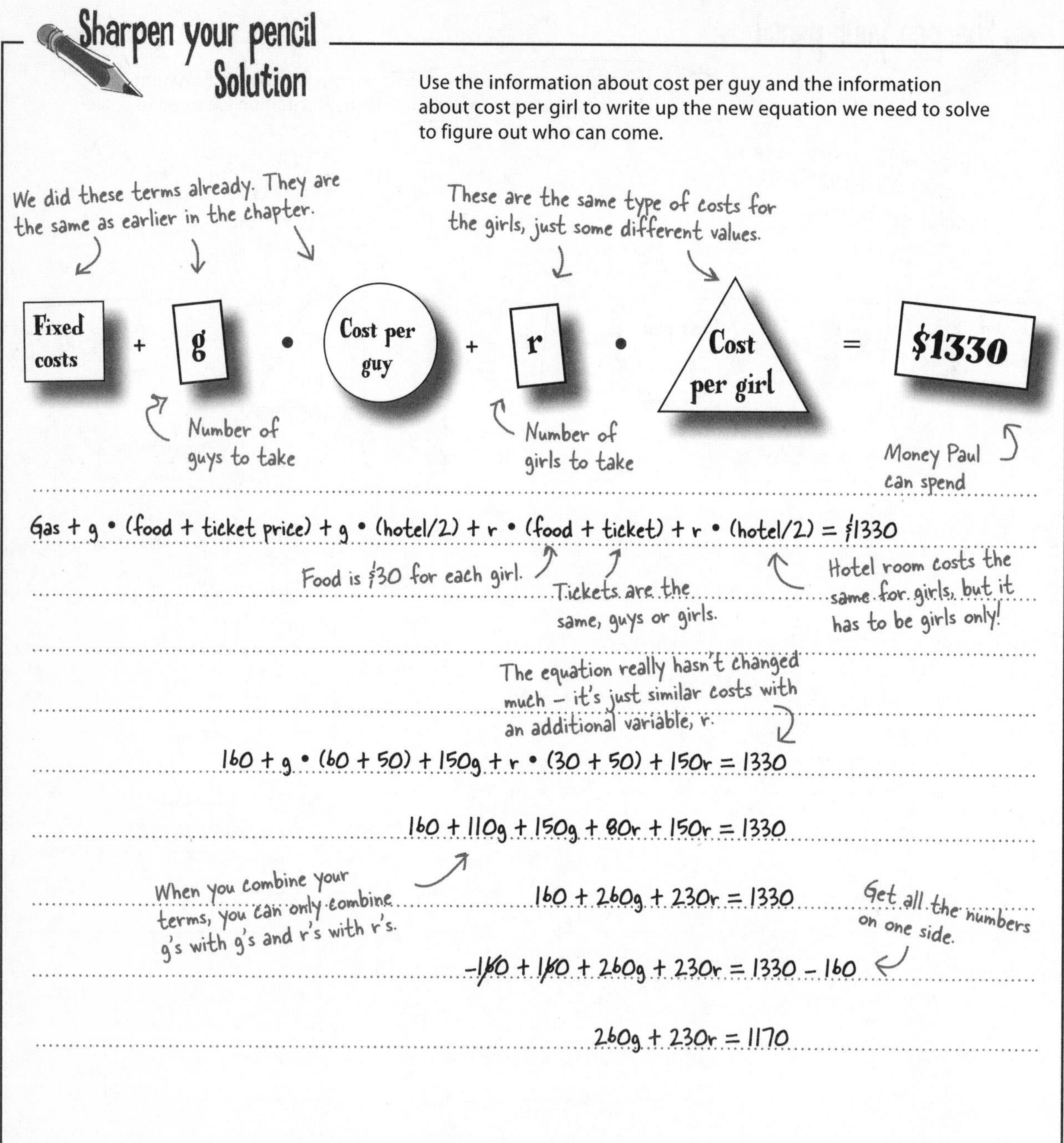

We did these terms already. They are the same as earlier in the chapter.
These are the same type of costs for the girls, just some different values.
Fixed costs
+
g
•
Cost per guy
+
r
•
Cost per girl
=
$1330
Number of guys to take
Number of girls to take
Money Paul can spend
Gas + g • (food + ticket price) + g • (hotel/2) + r • (food + ticket) + r • (hotel/2) = $1330
Food is $30 for each girl.
Tickets are the same, guys or girls.
Hotel room costs the same for girls, but it has to be girls only!
The equation really hasn't changed much – it's just similar costs with an additional variable, r.
160 + g • (60 + 50) + 150g + r • (30 + 50) + 150r = 1330
160 + 110g + 150g + 80r + 150r = 1330
When you combine your terms, you can only combine g's with g's and r's with r's.
160 + 260g + 230r = 1330
Get all the numbers on one side.
-160 + 160 + 260g + 230r = 1330 - 160
260g + 230r = 1170

Now what do we do? We have two variables in there, g & r!

Two variables seem to be a problem, but it's something we can handle. Just think about how you worked with one variable. You put together a equation to say how things relate to each other in terms of the variable.

So with just guys, we were putting the cost of the trip ***in terms of*** the number of guys going, the ***g*** variable. Even the individual parts of the trip were that way, like how we figured out the cost of the hotel:

Hotel Cost Up Close

How much the guys will be spending on hotels...

Hotel Cost = $150g$

...in terms of guys

The hotel costs are ***in terms of*** guys. The base hotel cost is $150, then for each guy that comes along the total hotel cost goes up by that amount.

Great question. For that, we need to talk a little bit more about what exactly a term is...

A term is a chunk of an algebraic equation

It's important to make a distinction between a **variable** and a **term**. A variable is some letter we use to stand in for something that's unknown: g for guys, r for girls, etc. A term, on the other hand, is just a piece of an equation. So, the equation $6g + 10g = 32$ only has one variable, g, but it has three terms: $6g$, $10g$, and 32. When we talk about **combining like terms**, it just means pick the terms that have the same variable, $6g$ and $10g$, and combine them: $6g + 10g$, so $16g$.

But how do you figure out how many terms there are in an equation? They're held together by multiplication (or division). So, $60b$ is a term, but $60 + b$ is two terms. How about $3(x+2)$? Well, that's just one term. Everything is glued together because everything is multiplied by 3.

"In terms of" is the secret to multiple variables in an equation

In Algebra, a lot of expressions have multiple variables. The most common equations you will see in multiple variables will be x and y, even though in our equation, we've got g and r.

When you get into multiple variable equations, this is where "in terms of" really starts to matter. If you have an equation in two variables:

- It's much easier to work with if you get one variable "in terms of" the other. That makes **substitution** possible.
- An equation in two variables establishes a **proportional relationship** between the two variables.
- A single equation in two variables can ***not*** be solved without an **additional** relationship.

This is the key to our guy-girl problem – we'll come back to this in just a second...

Sum it up

Term – A piece of an algebraic statement that is related by multiplication or division.

Match each equation to a description of which variable is ***in terms of*** which other variable. Some equations have two correct answers, and some descriptions will be used twice!

$T = 15d - 45 + 2^2$	Can be simplified to variable ***f*** in terms of variable ***r***
$h - e = a + 12 - \frac{5}{e}$	Variable ***Q*** in terms of variable ***x***
$Q = \frac{\sqrt{x-4}}{11}$	Variable ***h*** in terms of variable ***e***
$f - r = \frac{(r-6)^2}{8}$	One variable is not isolated (yet)
$h - 5 = 4 + 12 - \frac{5}{e}$	Variable ***T*** in terms of variable ***d***

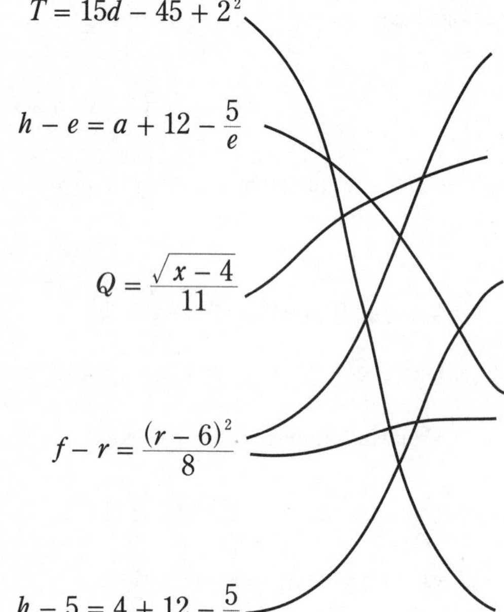

Can be simplified to variable ***f*** in terms of variable ***r***

Variable ***Q*** in terms of variable ***x***

Variable ***h*** in terms of variable ***e***

One variable is not isolated (yet)

Variable ***T*** in terms of variable ***d***

BULLET POINTS

- Keep in mind that "in terms of" may be all you need when solving a problem.
- Break a large problem into smaller pieces to make it easier.
- You need to fully understand a problem before you do anything else.
- If you have constants in your problem, you can use the actual number or use a letter to stand in for it, if that makes things easier.

I think we should have the same number of girls as guys.

What's the relationship between guys and girls?

From what we know about multiple variables in an equation, we have to find another relationship between guys and girls (***g*** & ***r***) in order to get any further with Paul's problem. To make things fair, let's say that we have to bring the same number of guys and girls. That's a new relationship!

If we say we have to have the same number of guys and girls coming on the trip, then we can say $g = r$. Once we say that, everywhere you see an ***r***, you can substitute in a ***g*** (since they are equal). Suddenly the equation looks a whole lot more solvable...

Sharpen your pencil

Finish up the road trip equation, and solve for ***g***. Then solve for ***r***.

$260g + 230r = 1170$

Since g = r, we can substitute g wherever we have an r...

Only do the division to one decimal place. (We can't bring less than a whole person, right?)

Sharpen your pencil Solution

Yor job was to finish up the road trip equation, and solve for g. Then solve for r.

$260g + 230r = 1170$

Combine like terms

$490g = 1170$

$\frac{490g}{490} = \frac{1170}{490}$

$g = 2.3$ and $r = 2.3$

We only show one decimal but the full answer is 2.387755102...

Going back to the new $g = r$ relationship, you also know that $r = 2.3$. That makes sense, because the total number of $g + r$ is still 4.6, just like when it was just guys.

there are no Dumb Questions

Q: Why did we only write part of the decimal, and not the whole thing?

A: Because for this problem, we can only bring whole people. You could round down to the nearest whole person and just show two decimal places, but to avoid confusion, we used one decimal place. Remember to always think about the context of the problem. The math might work out to 2.387755102 people, but who really wants to be that .387755102 part?

Q: So can we just say that g = r?

A: Sure, because that's the second relationship (equation) that we were given for this problem. It was part of the problem statement, and we just worked with it to solve our initial equation. Knowing that $g = r$ means that you can substitute g back in for r in the original equation and then solve for one variable.

Q: Can it sometimes be the end of the problem if you don' t have a number? What if you can only get an equation in terms of two variables?

A: Depending upon your problem, that may be all you need. It's another case of needing to think about the entire problem, not just the equation. The goal may just be the proportion of the variables and not a numerical answer.

Hey, we've got to get going. I have enough money to bring 4 of us. It's me, O'Brien, Amanda and her best friend!

Road Trip!

With a little Algebra, you've figured out how many can go and how much the whole thing is going to cost. Pretty impressive. Now there's just one more thing to sort out...

Unfortunately, this is a problem that even a great Algebra book can't solve...

We have to agree on music to listen to. There's this great new song by Fergi... it's really deep.

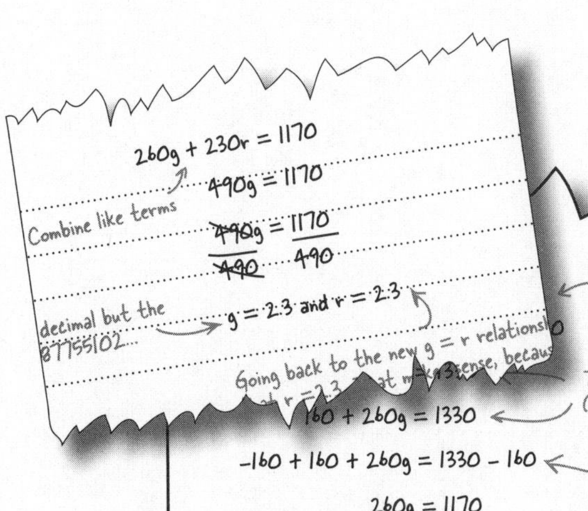

260g + 230r = 1170
490g = 1170
Combine like terms
490g = 1170
490 490
decimal but the
87755102...
g = 2.3 and r = 2.3
Going back to the new g = r relationship
160 + 260g = 1330
-160 + 160 + 260g = 1330 - 160
260g = 1170
260g = 1170
260 260
g = 4.5

First we'll get rid of the parentheses by adding up the values.

Then combine the like terms (110 + 150 = 260).

Now we can get rid of the 160 by subtracting 160 from both sides of the equation.

Now to get the g by itself we need to divide both sides by 260.

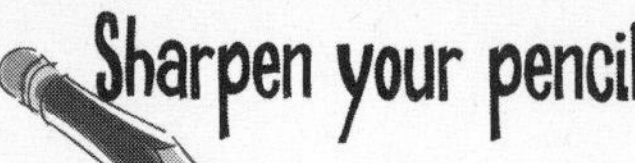

Simplify each of these mathematical expressions and combine like terms. So remember that a term needs to be held together only by multiplication or division.

Apply the order of operations first.

$$6 + 5x - 10y + 2y - 2x + 3y - 4$$

$$2x(y-1) + 4 - \frac{2x}{2}$$

$$-3xy + 4y(x-2) - \frac{12}{3}x$$

$$\frac{(9 \cdot 3)}{x}xy - 4^2 + \frac{1}{2} - \frac{1}{4}x - 0.75x - 2y$$

$$\frac{(9 \cdot 3)}{x}xy - 4^2 + \frac{1}{2} - \frac{1}{4}x - \frac{3}{4}x - 2y$$

This one is started for you.

$$3(3b - g) + 3^2 - \frac{16}{(10-2)} + g - 10$$

Multicross

Take some time to sit back and give your right brain something to do. It's your standard crossword; all of the solution words are from this chapter.

Across

2. Use ..... to check your work.
3. Plug your answer back in to ..... it.
4. Equations express a relationship in .... some variable.
5. Tells you what you're looking for and what you need to find it.
6. A piece of an algebraic statement lumped together through multiplication.
7. When solving a problem, it's always important to think about ..... .

Down

1. More than one variable.
3. You can do this to like terms.
4. In order to solve an equation, you have to do this to the variable.
6. An absolute value equation typically has this many solutions.

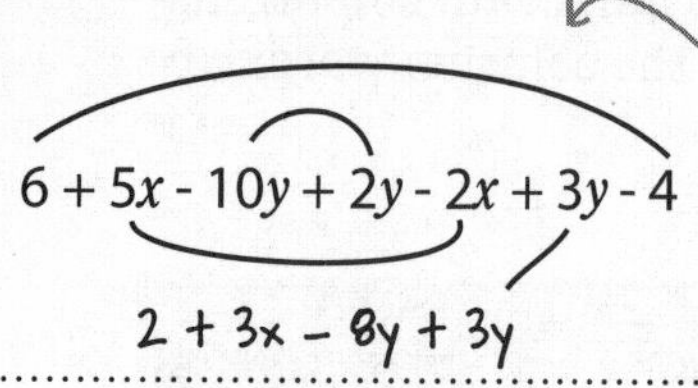

Simplify each of these mathematical expressions and combine like terms. So remember that a term needs to be held together only by multiplication or division.

$6 + 5x - 10y + 2y - 2x + 3y - 4$

Writing on the problem can help keep track of the like terms.

$2 + 3x - 8y + 3y$

$3 + 3x - 5y$

$-3xy + 4y(x - 2) - \frac{12}{3}x$

Cleaning up the fraction

$-3xy + 4y(x-2) - 4x$

$-3xy + 4yx - 8y - 4x$

the commutative property

$-3xy + 4xy - 8y - 4x$

$xy - 8y - 4x$

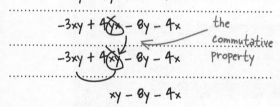

$3(3b - g) + 3^2 - \frac{16}{(10-2)} + g - 10$

parens

$3(3b - g) + 3^2 - \frac{16}{8} + g - 10$

exponent

$3(3b - g) + 9 - \frac{16}{8} + g - 10$

multiply

$9b - 3g) + 9 - \frac{16}{8}$ (2) $+ g - 10$

$9b - 2g - 3$

$2x(y - 1) + 4 - \frac{2x}{2}$

Multiplication using the distributive property

$2xy - 2x + 4 - \frac{2x}{2}$

and division

$2xy - 2x + 4 - x$

$2xy - 3x + 4$

This is not a like term with x because there is an additional variable (y) in the term.

$\frac{(9 \cdot 3)}{x}xy - 4^2 + \frac{1}{2} - \frac{1}{4}x - 0.75x - 2y$

Convert to a fraction

$\frac{(9 \cdot 3)}{x}xy - 4^2 + \frac{1}{2} - \frac{1}{4}x - \frac{3}{4}x - 2y$

parens

$\frac{(27)}{x}xy - 4^2 + \frac{1}{2} - \frac{1}{4}x - \frac{3}{4}x - 2y$

exponent

$\frac{27}{x}xy - 16 + \frac{1}{2} - \frac{1}{4}x - \frac{3}{4}x - 2y$

division

$27y - 16 + \frac{1}{2} - \frac{1}{4}x - \frac{3}{4}x - 2y$

$25y - (16 + \frac{1}{2}) - 1x$

Just a notation change

$25y - 16\frac{1}{2} - 1x$

Multicross Solution

Take some time to sit back and give your right brain something to do. It's your standard crossword; all of the solution words are from this chapter.

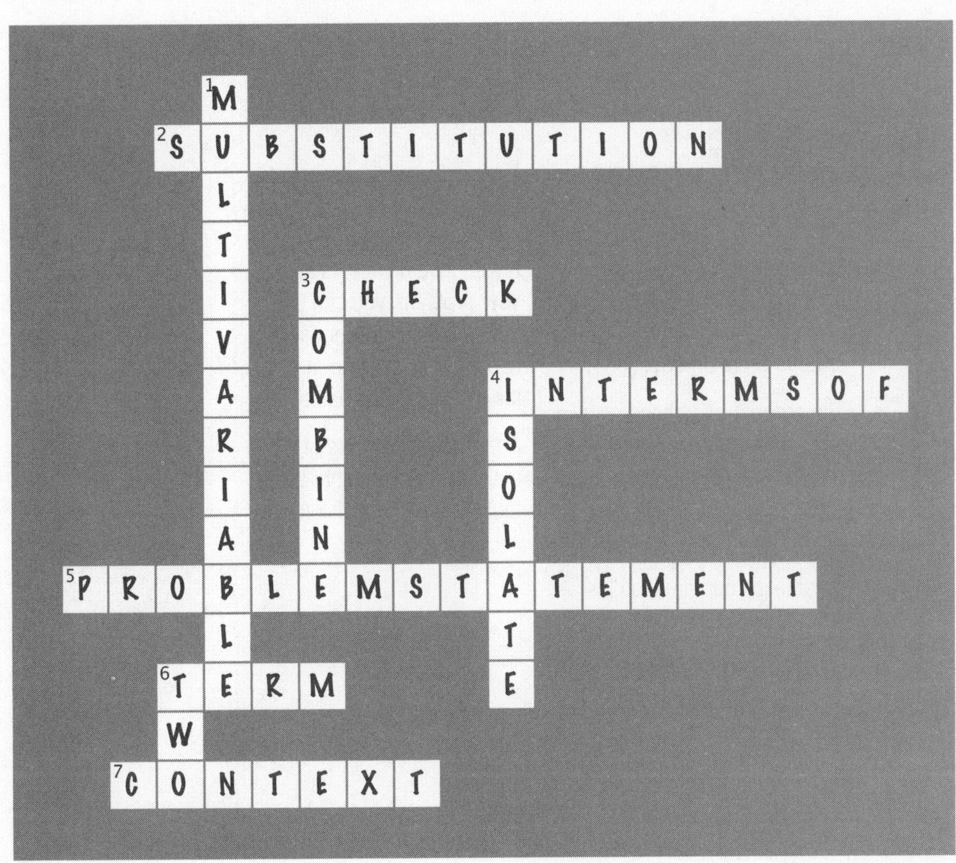

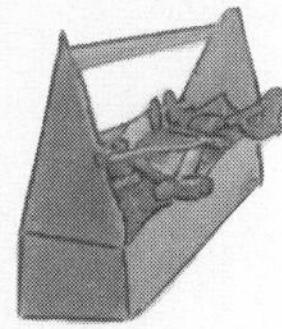

CHAPTER 2

Tools for your Algebra Toolbox

This chapter was about solving more complex algebraic equations.

BULLET POINTS

- No matter how many terms, you're still looking to isolate a variable in order to solve an equation.
- The order of operations says what order you should work on things.
- You'll need one relationship for each variable in your equation in order to find an answer.
- Make sure you're dealing with the same terms when you combine parts of an equation.
- If you have constants in your problem, you can use the actual number or use a letter to stand in for it if that makes things easier.
- A term is just a piece of the equation held together with multiplication or division.

3 rules for numeric operations

Follow the rules

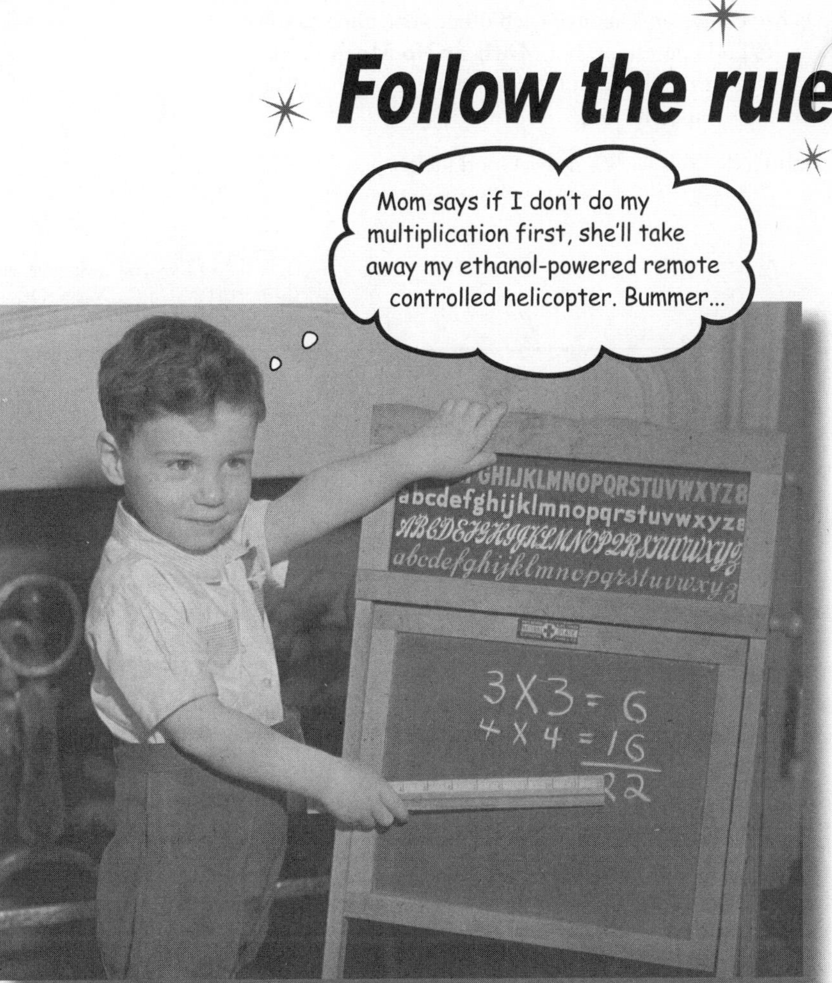

Mom says if I don't do my multiplication first, she'll take away my ethanol-powered remote controlled helicopter. Bummer...
3X3 = 6
4X4 = 16
22

Sometimes you just gotta follow the stinking rules.

Yes, there comes a time in everyone's life when mom and dad are out of the picture, and there's nobody making you clean your room or surrender your cell phone until your homework's done. But when it comes to Algebra, **rules are a good thing**. They'll keep you from getting the wrong answer. In fact, lots of times, rules will **help you solve for an unknown** without a lot of extra work. Leave your dunce cap behind for this chapter because we'll be following a few handy rules all the way to a perfect score.

Math or No Math

It's the quiz show that's sweeping the nation—**Math or No Math**. This new primetime hit pits two contestants against each other, struggling to solve math problems. It's easy to find contestant, but **Math or No Math** needs help. They don't have any judges to figure out if contestants are getting the problems right! That's where you come in....

You've been brought in to judge this week's show. Good luck...

Problem #1:

If you're not sure about exponents, turn to the appendix...

$$2 + (2 + 4) + (6 - 3)^2 - 5 - \left(\frac{2 + 3}{10}\right)$$

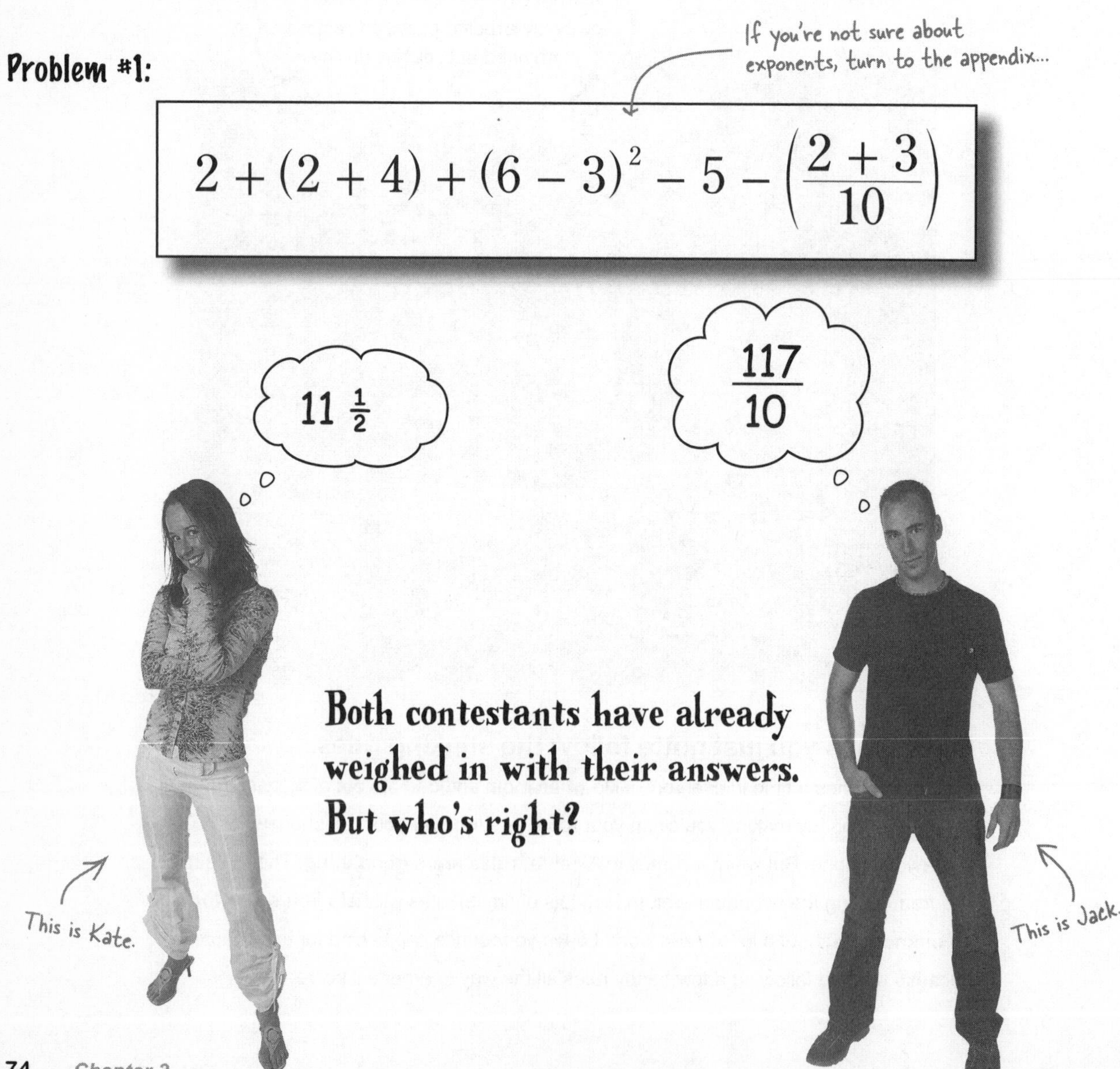

$11 \frac{1}{2}$
$\frac{117}{10}$
Both contestants have already weighed in with their answers. But who's right?
This is Kate.
This is Jack.

BE the judge

Your job is to judge what Kate and Jack did. Mark up their work step by step to show how they each got their answer.

The original expression

Kate:

$$2 + (2 + 4) + (6 - 3)^2 - 5 - \left(\frac{2 + 3}{10}\right)$$

Here's how Kate got her answer.

$$2 + (6) + (3)^2 - 5 - \left(\frac{5}{10}\right)$$

$$2 + (6) + (3)^2 - 5 - \left(\frac{1}{2}\right)$$

$$2 + (6) + 9 - 5 - \left(\frac{1}{2}\right)$$

$$8 + 9 - 5 - \left(\frac{1}{2}\right)$$

$$11\frac{1}{2}$$

Make notes about how each contestant worked the problem. See any mistakes?

Jack:

$$2 + (2 + 4) + (6 - 3)^2 - 5 - \left(\frac{2 + 3}{10}\right)$$

Here's what Jack did.

$$4 + 4 + (6 - 3)^2 - 5 - \left(\frac{2 + 3}{10}\right)$$

$$8 + (6 - 3)^2 - 5 - \left(\frac{2 + 3}{10}\right)$$

$$14 - 3^2 - 5 - \left(\frac{2 + 3}{10}\right)$$

$$11^2 - 5 - \left(\frac{2 + 3}{10}\right)$$

$$121 - 5 - \left(\frac{2 + 3}{10}\right)$$

$$116 - \left(\frac{2 + 3}{10}\right)$$

$$\frac{114 + 3}{10} = \frac{117}{10}$$

Who's right?

(Circle one) **Kate** **Jack**

BE the judge solution

Your job is to judge what Kate and Jack did. Mark up their work step by step to show how they each got their answer.

The original expression

Kate:

$$2 + (2 + 4) + (6 - 3)^2 - 5 - \left(\frac{2+3}{10}\right)$$

Kate combined the parentheses first...

$$2 + (6) + (3)^2 - 5 - \left(\frac{5}{10}\right)$$

then, cleaned up the fraction

$$2 + (6) + (3)^2 - 5 - \left(\frac{1}{2}\right)$$

simplified the squared term

$$2 + (6) + 9 - 5 - \left(\frac{1}{2}\right)$$

added 2+6

$$8 + 9 - 5 - \left(\frac{1}{2}\right)$$

added & subtracted what was left

$$11\frac{1}{2}$$

Kate decided to work with groups first, then exponents, then addition and subtraction....

Jack:

$$2 + (2 + 4) + (6 - 3)^2 - 5 - \left(\frac{2+3}{10}\right)$$

2+2 = 4

Jack went left to right

$$4 + 4 + (6 - 3)^2 - 5 - \left(\frac{2+3}{10}\right)$$

$$8 + (6 - 3)^2 - 5 - \left(\frac{2+3}{10}\right)$$

$$14 - 3^2 - 5 - \left(\frac{2+3}{10}\right)$$

Then, 14–3 = 11, so square that

$$11^2 - 5 - \left(\frac{2+3}{10}\right)$$

$$121 - 5 - \left(\frac{2+3}{10}\right)$$

...Jack worked his entire problem from left to right.

$$116 - \left(\frac{2+3}{10}\right)$$

116–2

$$\frac{114+3}{10}$$

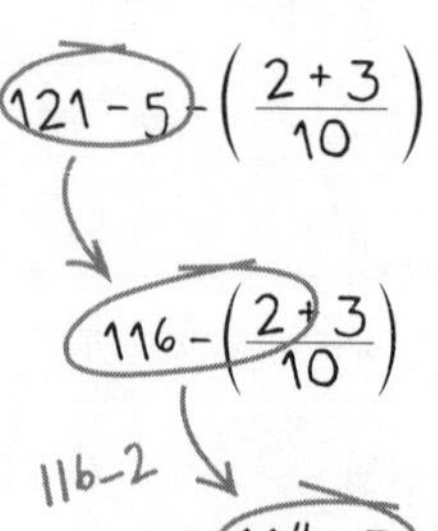

add what's left.

$$\frac{117}{10}$$

Who's right?

(Circle one) **Kate** (circled) Jack

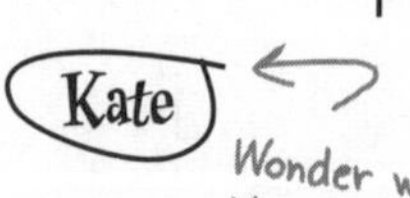

Wonder why? Keep reading to find out more...

How come Kate's right? Who says? Jack's answer makes sense too!

Kate is right because Kate's answer follows the rules for working with numbers.

Kate solved the expression properly because she used the ***order of operations***. She got the correct answer because she followed that order—which is really just a rule for working with numbers—precisely.

If you don't follow the rules nothing's going to work out the way it's supposed to.
Strangely menacing authority figure.

Jack DIDN'T follow the rules, so Jack got the wrong answer.

Jack worked his problem from left to right. That seems pretty logical, but since that's not what everyone else doing Algebra has agreed on, he's not going to get the right answer to problems.

Equations and expressions are written to communicate an order. That order needs to be the same for ***everyone who works with the expression***, or there wouldn't be any right answers. Hello math chaos!

You need to learn and use the ***order of operations*** to solve problems, and that's what we're going to do next. That way, you can be sure you—and the **Math or No Math** contestants—are following the rules.

The order of operations is one of the ways everyone can get the same answer to the same problem.

There's an order for working expressions

The order you're supposed to work with numbers in a math expression is called the **order of operations**. If you always follow the order of operations, you'll get the same answers to problems that everyone else does. Here's the order:

The Order Of Operations

First, do...

Parentheses

$(\;)$ ← Anything in here

You need to work through and get all of those pieces simplified first and write those down.

Parentheses include everything that is grouped in the expression.

This is Kate and Jack's original equation. →

$$2 + (2 + 4) + (6 - 3)^2 - 5 - \left(\frac{2 + 3}{10}\right)$$

Since these are all expressions inside parentheses, they get simplified in the first step.

Then, do...

Exponents

Everything in the expression that is raised to a power, any power (that includes roots, too).

This is an exponent.

$$2 + (6) + (3)^2 - 5 - \left(\frac{1}{2}\right)$$

After simplifying inside the parentheses, the exponent needs to go next.

Next, do...

Multiplication & Division

These operations are equal in order, so work left to right, simplifying both the multiplication and division parts of the expression:

Multiplication and division are the exact opposite of each other, so they are the same "strength" operation.

We don't have any multiplication or division for this problem, so we leave the expression as-is.

$$2 + (6) + 9 - 5 - \left(\frac{1}{2}\right)$$

Last, do...

Addition & Subtraction

Addition & subtraction are also opposites.

Addition and subtraction are equal in terms of the order, so work from left to right doing all the addition and subtraction. Once you've finished that, the expression should be simplified.

$$2 + (6) + 9 - 5 - \left(\frac{1}{2}\right) = 11\frac{1}{2}$$

Work what's still there left to right.

Addition & subtraction are the "weakest" operations, so they go last.

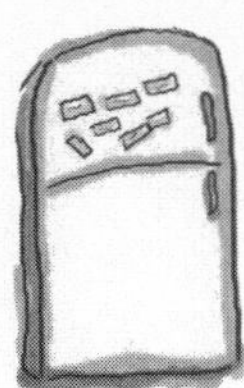

Math Magnets

Time for a little extra judging. Below are several problems that are partially worked out. Your job is to figure out what the next thing to do in the problem is. Use the order of operations, and place the correct magnet for the operation you'd do next.

For this one, what is the first thing you do inside the parentheses?

$$\frac{(6 - 3 \cdot 2 + 4^2)}{2}$$

.... What goes next?

$$\frac{8}{12} - 1 - \frac{1}{3} - \left(\frac{2}{3}\right)$$

.... What goes next?

$$-1\left(\frac{5+3}{12}\right) - \frac{1}{3} - 2^3$$

$$-1\left(\frac{8}{12}\right) - \frac{1}{3} - 2^3$$

$$-1\left(\frac{8}{12}\right) - \frac{1}{3} - 8$$

....

Remember, a root is just an exponent.

$$2 \cdot (-1) + \sqrt{4} - 3$$

....

$$-0.4 + 0.1\left(6 + \sqrt{9}\right)^3$$

.... inside ....

$$\frac{(12 + 13)^{1/2} + 7}{6}$$

This one is tricky - take your time, and you'll figure it out.

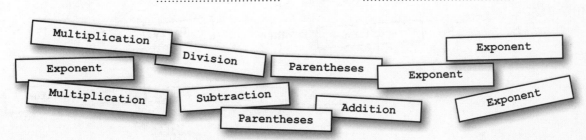

$$\frac{(25)^{1/2} + 7}{6}$$

$$\frac{5 + 7}{6}$$

.... then ....

Multiplication

Division

Parentheses

Exponent

Exponent

Exponent

Multiplication

Subtraction

Addition

Exponent

Parentheses

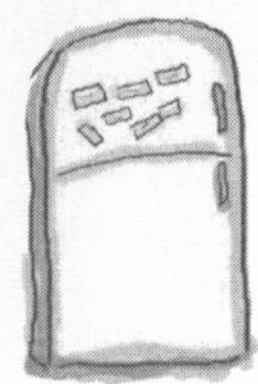

Math Magnets Solution

Your job is to figure out what the next thing to do in the problem is. Use the order of operations, and place the correct magnet for the operation you'd do next.

Since everything is grouped over the 2, you need to simplify inside the parentheses —and in there the exponent goes first.

$$\frac{(6-3\cdot 2+4^2)}{2}$$ **Exponent**

$$\frac{8}{12}-1-\frac{1}{3}-\left(\frac{2}{3}\right)$$ **Subtraction**

You should drop the parentheses, and then do the exponent (the square root).

$$2\cdot(-1)+\sqrt{4}-3$$ **Parentheses**

$$-1\left(\frac{5+3}{12}\right)-\frac{1}{3}-2^3$$

$$-1\left(\frac{8}{12}\right)-\frac{1}{3}-2^3$$

$$-1\left(\frac{8}{12}\right)-\frac{1}{3}-8$$ **Multiplication**

Inside the parentheses is simplified, so the next thing is multiplication.

You still need to work inside the parentheses – but you need to handle the exponent inside first – and don't forget that roots are just another way of writing exponents.

$$-0.4+0.1(6+\sqrt{9})^3$$ **Exponent** inside **Parentheses**

$$\frac{(12+13)^{1/2}+7}{6}$$

$$\frac{(25)^{1/2}+7}{6}$$

$$\frac{5+7}{6}$$

The addition goes first because it's grouped over the division sign.

Addition then **Division**

These magnets are left over. **Exponent** **Multiplication** **Exponent**

there are no Dumb Questions

Q: Where did the order of operations come from?

A: It was established by early mathematicians (geeky people who do Algebra for fun) who were trying to compare their work. The order let those folks talk to each other and get the same answers for their problems, which is a pretty big deal.

Q: Why did the order of operations get set up this way?

A: The strongest operations go first. Parentheses are a way to say, "Do this first!" Then, exponents, and then multiplication and division. Finally, addition and subtraction. And we work on what's what's left by moving from left to right because that's the way we read.

Q: Are roots exponents?

A: Yes, which means that they go second in the order of operations. If you need a refresher on the details, just turn to the appendix, where exponents and roots are discussed. A root is just an expression raised to a fractional power, like 1/2 or 1/3 (for square root and cube root respectively).

Q: Do inverse operations always go together in the order of operations?

A: They do. Addition, subtraction, multiplication and division are pretty straightforward. Exponents and roots are also inverse operations.

Q: Do I have to memorize this?

A: Yes, but if you just think of the operations going in order of strength that should help.

Q: Do I need to reduce fractions right away?

A: The fractions are up to you. If you want to work with large numerators and denominators, you can (but it's not recommended).

Q: So are fractions really division, or can you leave fractions alone?

A: Both. In the case of a fraction, you're not changing the value of the number, just how it's expressed (1/2 vs. 0.5), so you can work with it either way. If you want to divide your fractions to get a number like 0.5, you can... or you can leave fractions as they are.

Q: When can you drop the parentheses? Do they need to stay after you did what was inside?

A: That's up to you. Just like the fractions, when you've combined whatever was in a set of parentheses and you're down to the most simplified form, you're done. Some people like to keep the parentheses to indicate multiplication or to clarify exponents, but it's not required.

Q: This seems like a lot of steps. Is it hard to keep track of all this?

A: It can be, that's why you should write down your work as you're solving a problem. In fact, since Jack wrote down his steps, we were able to figure out where he made his mistakes, and why he got a wrong answer.

It's a good idea to write down what your expression looks like at each step. You can keep track of what you did and check your work.

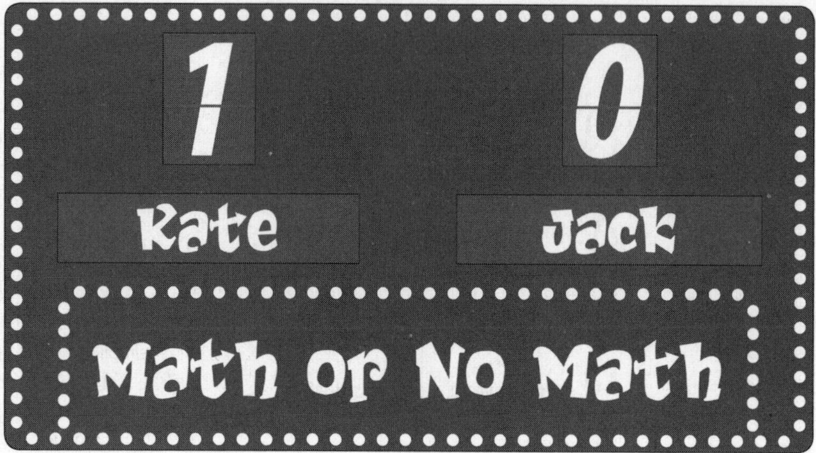

Back to Math or No Math...

Here we go: it's time for round 2. The rules have changed a bit, too. Now contestants get one point for getting a problem's answer right, and another point for getting the right answer ***first***. So speed is definitely a factor.

Let's see how Kate and Jack do, especially now that everyone knows the order of operations...

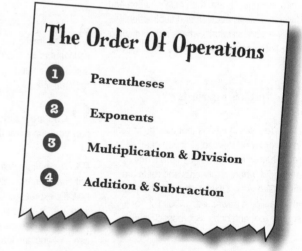

The Order Of Operations

1. **Parentheses**
2. **Exponents**
3. **Multiplication & Division**
4. **Addition & Subtraction**

Problem #2:

$$\left(\frac{1}{2} + .6\right) + \frac{1}{5} + \left(\frac{1}{2} \cdot 8\right) + \frac{1}{2}$$

BE the judge

Your job is to judge what Kate and Jack did. Mark up their work step by step to show how they each got their answer.

Kate's Work

original expression

$$\left(\frac{1}{2} + .6\right) + \frac{1}{5} + \left(\frac{1}{2} \cdot 8\right) + \frac{1}{2}$$

$$\left(\frac{1}{2} + \frac{3}{5}\right) + \frac{1}{5} + \left(\frac{1}{2} \cdot 8\right) + \frac{1}{2}$$

$$\frac{1}{2} + \left(\frac{3}{5} + \frac{1}{5}\right) + 4 + \frac{1}{2}$$

$$\frac{1}{2} + \frac{1}{2} + \frac{4}{5} + 4$$

$$1 + 4 + \frac{4}{5}$$

$$5\frac{4}{5}$$

Jack's Work

$$\left(\frac{1}{2} + .6\right) + \frac{1}{5} + \left(\frac{1}{2} \cdot 8\right) + \frac{1}{2}$$

$$\left(\frac{1}{2} + \frac{3}{5}\right) + \frac{1}{5} + \left(\frac{1}{2} \cdot 8\right) + \frac{1}{2}$$

$$\left(\frac{5}{10} + \frac{6}{10}\right) + \frac{1}{5} + \left(\frac{1}{2} \cdot 8\right) + \frac{1}{2}$$

$$\frac{11}{10} + \frac{2}{10} + 4 + \frac{1}{2}$$

$$\frac{11}{10} + \frac{2}{10} + \frac{40}{10} + \frac{5}{10}$$

$$\frac{13}{10} + \frac{45}{10}$$

$$\frac{58}{10}$$

$$5\frac{4}{5}$$

Who's right? (Circle one or both)

Kate   Jack   Neither

Whose solution was fastest? (Circle one)

Kate   Jack

Relax

Fractions are in the appendix.

Rusty on fractions? Just flip to the appendix in the back for a little help.

BE the judge solution

Your job is to judge what Kate and Jack did. Mark up their work step by step to show how they each got their answer.

original expression

Kate's Work

$$\left(\frac{1}{2} + .6\right) + \frac{1}{5} + \left(\frac{1}{2} \cdot 8\right) + \frac{1}{2}$$

Kate started converting everything to fractions right away...

$$\left(\frac{1}{2} + \frac{3}{5}\right) + \frac{1}{5} + \left(\frac{1}{2} \cdot 8\right) + \frac{1}{2}$$

...then grouped the fifths.

$$\frac{1}{2} + \left(\frac{3}{5} + \frac{1}{5}\right) + 4 + \frac{1}{2}$$

Next she reordered what was left to get the common denominators together.....

$$\frac{1}{2} + \frac{1}{2} + \frac{4}{5} + 4$$

One more reordering and she can add the whole numbers...

$$1 + 4 + \frac{4}{5}$$

$$5\frac{4}{5}$$

Jack's Work

$$\left(\frac{1}{2} + .6\right) + \frac{1}{5} + \left(\frac{1}{2} \cdot 8\right) + \frac{1}{2}$$

Jack's got it now – parentheses first so he needs common denominators...

$$\left(\frac{1}{2} + \frac{3}{5}\right) + \frac{1}{5} + \left(\frac{1}{2} \cdot 8\right) + \frac{1}{2}$$

$$\left(\frac{5}{10} + \frac{6}{10}\right) + \frac{1}{5} + \left(\frac{1}{2} \cdot 8\right) + \frac{1}{2}$$

Now the parentheses are handled, it's more common denominators...

$$\frac{11}{10} + \frac{2}{10} + 4 + \frac{1}{2}$$

$$\frac{11}{10} + \frac{2}{10} + \frac{40}{10} + \frac{5}{10}$$

$$\frac{13}{10} + \frac{45}{10}$$

$$\frac{58}{10}$$

Reduce the fraction and make it proper

$$5\frac{4}{5}$$

Who's right? (Circle one or both)

They both got it right

Kate Jack Neither

Whose solution was fastest? (Circle one)

Kate Jack

Kate was way faster – she saved a ton of steps over Jack.

What? Kate totally blew off the order of operations, but got the right answer. What's that about?

There are properties as well as rules.

Kate didn't ignore the order of operations; she just used some other properties of numbers first. Kate used the ***associative and commutative properties*** to work with her equation, and then applied the order of operations.

Properties like the associative and commutative properties are really just another type of rule... and you can apply these properties before, during, or after applying the order of operations.

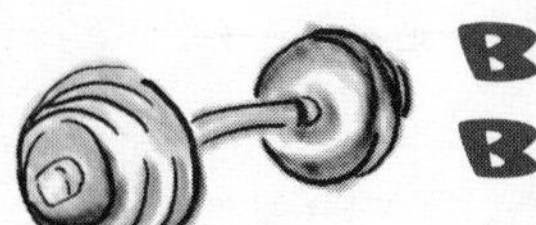

Go back and look at Kate's work. Circle where you think she used a special property. Don't turn the page until you think you know where Kate did something sneaky.

You can re-group your equations

The **associative property** lets you change the *grouping* of numbers in addition or multiplication operations. Suppose you've got a bunch of numbers you need to add. You can change the groupings of those numbers around all you want. In fact, that's what Kate did:

Kate's Work

$$\left(\frac{1}{2}+\frac{3}{5}\right)+\frac{1}{5}+\left(\frac{1}{2}\bullet 8\right)+\frac{1}{2}$$

$$\frac{1}{2}+\left(\frac{3}{5}+\frac{1}{5}\right)+4+\frac{1}{2}$$

Kate used the associative property to re-group parts of her problem. So she worked with common denominators next to each other first, and then did the rest.

What's going on here? Because all of the operations are addition, the parentheses *don't affect the outcome*. The **associative property** says that when you're performing addition or multiplication, grouping does not affect the outcome, so you can regroup those types of problems all you need.

You can take a problem like 10 x (4.2 x 0.225) and reorder it to something easier, like (10 x 4.2) x 0.225. It's much easier to multiple things by 10 than 0.225, so it's better to rearrange the grouping of this problem some.

Sum it up

The Associative Property – Changing the groupings of a set of numbers being added or multiplied does not change the outcome of the operation.

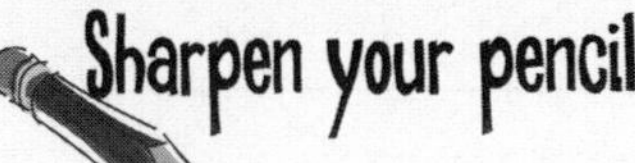

Here's your chance to put the associative property to work. There are two expressions next to each other, regrouped for you. Do the answers come out to be the same?

We've already changed the grouping for you - see how it works both ways!

$\left(\frac{1}{2}+\frac{3}{5}\right)+\frac{1}{5}$ vs. $\frac{1}{2}+\left(\frac{3}{5}+\frac{1}{5}\right)$

Are the answers the same?

Yes No

$\frac{1}{3}(9 \cdot 2)$ vs. $\left(\frac{1}{3} \cdot 9\right)2$

Are the answers the same?

Yes No

$12 - (5 - 3)$ vs. $(12 - 5) - 3$

Are the answers the same?

Yes No

$12 \div (4 \div 2)$ vs. $(12 \div 4) \div 2$

Are the answers the same?

Yes No

Why did the answers come out the same for some of the problems and not for the others?

Sharpen your pencil Solution

Your job was to solve both sets of problems, see if the answers came out the same, and figure out why you got the results you did.

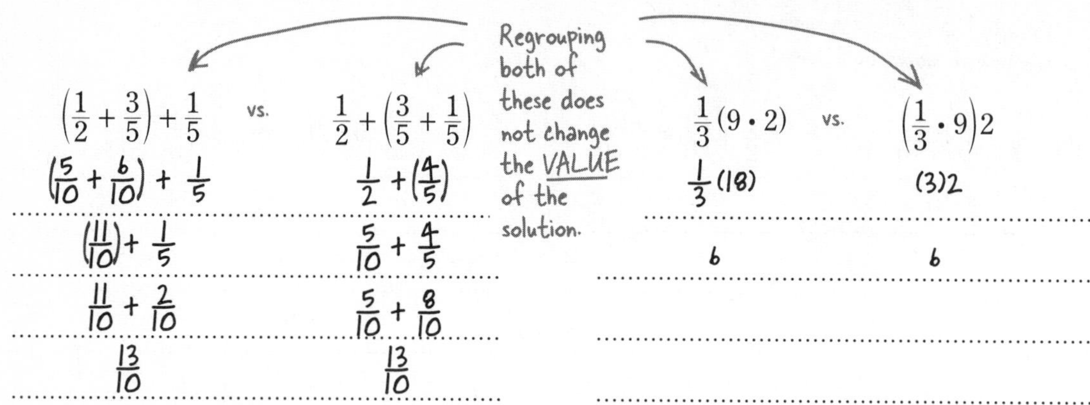

Are the answers the same?

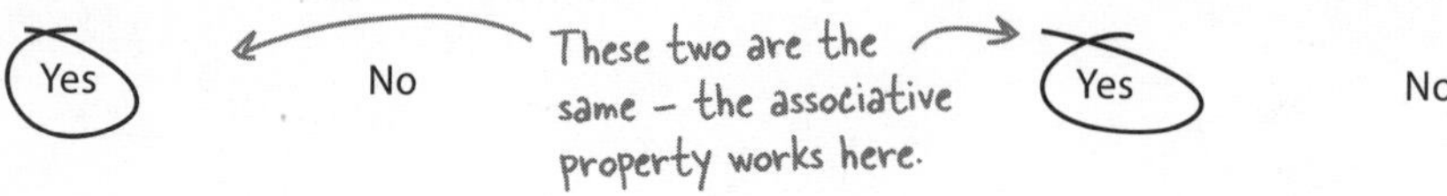

Yes
No
These two are the same – the associative property works here.
Yes

Are the answers the same?

No

$12 - (5 - 3)$ vs. $(12 - 5) - 3$

12 – (2) (7) – 3

10 4

Whoa – those don't match. Regrouping here didn't work.

Regrouping these totally changed the value of the solution – the property doesn't work for these!

$12 \div (4 \div 2)$ vs. $(12 \div 4) \div 2$

12 ÷ (2) (3) ÷ 2

6 $\frac{3}{2}$

This didn't work either – the associative property has some limits!

Are the answers the same?

Yes No

Are the answers the same?

Yes

Why did the answers come out the same for some of the problems and not for the others?

You can change the grouping for addition or multiplication – but it doesn't work for division and subtraction.

The associative property only works for addition or multiplication - NOT subtraction and division.

This means you cannot regroup subtraction or division problems without changing the value of the solution. You have to solve expressions with subtraction and division as written.

there are no Dumb Questions

Q: So do we need the order of operations or not?

A: Yes - the order of operations (parentheses, exponents, multiplication & division, addition & subtraction) is the order in which you need to simplify a problem. With the associative property, you are **not** changing the order of operations—you'll still do parentheses first—you're just moving certain parts of a problem around.

Q: What's the point of the associative property anyway? So what if I can move groupings around?

A: The associative property means you can work through an expression in the easiest, fastest way. Grouping together fractions that are easy to work with saves tons of common denominator time, and you can do the same thing with decimals, too.

Grouping things in terms of how you want to work on them can sometimes help you get started on a tough problem, too!

Q: Are there more properties?

A: Yes - we're going to talk about two more in the next few pages, the commutative property and the distributive property. The commutative property lets you reorder items in an equation. The distributive property helps spread multiplication and division out across the equation (or pull it together into a single term, but we'll get to that later).

Q: So the associative property lets me change the order of numbers, right?

A: No - the associative property just says you can change the **grouping** of numbers that are added or multiplied. But you can't change any orderings, or move numbers from one part of your problem to another.

However, all is not lost, There is a property that will help you out with ordering: the commutative property. Once you get that figured out, you'll be able to reorder *and* regroup. Keep reading...

The associative property says you can change groupings in addition or multiplication, but NOT with subtraction or division!

It looks like Kate did more than just re-group things... she re-ordered things too. So is there a property that lets you move numbers around, too?

$\frac{1}{2} + \left(\frac{3}{5} + \frac{1}{5}\right) + 4 + \frac{1}{2}$
$\frac{1}{2} + \frac{1}{2} + \frac{4}{5} + 4$
$1 + 4 + \frac{4}{5}$

Kate did something tricky with the order of these numbers.

You can re-order numbers as well as re-group them... using the commutative property.

The **commutative property** deals with the *order* of the terms in addition or multiplication operations. The commutative property says that you can add the numbers involved in addition operations or multiply the numbers in multiplication operations ***in any order*** and not affect the value of your answer.

You still have to follow the order of operations and do multiplications before additions though!

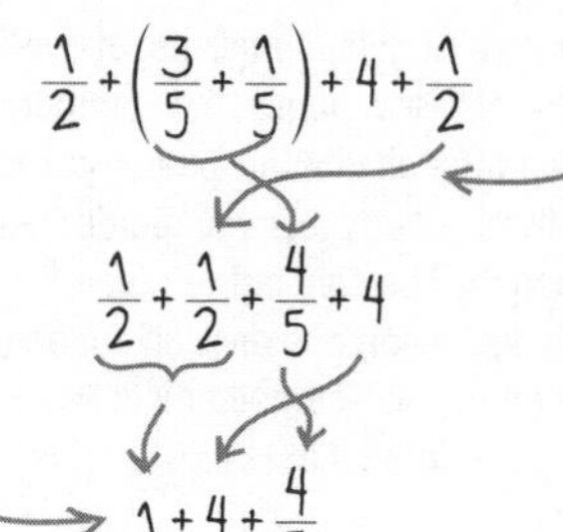

$\frac{1}{2} + \left(\frac{3}{5} + \frac{1}{5}\right) + 4 + \frac{1}{2}$
$\frac{1}{2} + \frac{1}{2} + \frac{4}{5} + 4$
$1 + 4 + \frac{4}{5}$

Kate moved the halves together to make the addition easier.

She moved the whole numbers together, and she didn't have to mess around with improper fractions either.

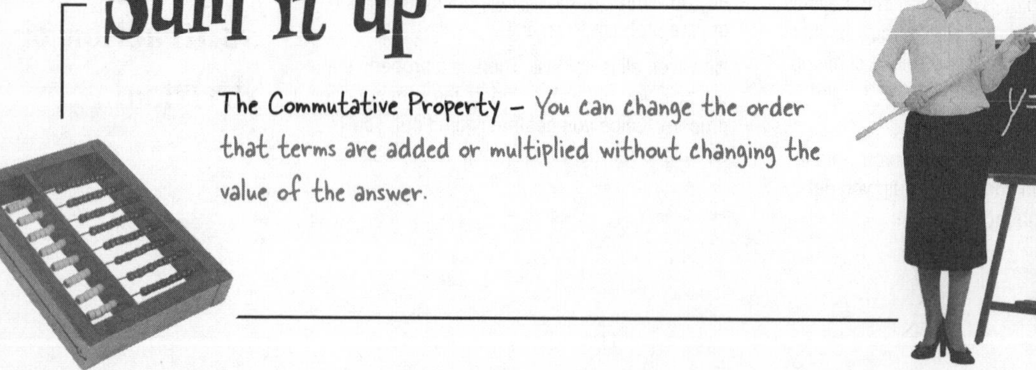

Sum it up

The Commutative Property – You can change the order that terms are added or multiplied without changing the value of the answer.

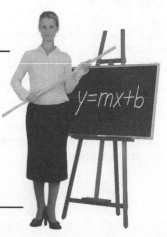

A bunch of equivalent expressions, in full costume, are playing a party game, “Who am I?” They’ll give you a clue, and, based on what they say, you try to guess what property they use. Assume they always tell the truth about themselves. Fill in the blanks to the right to identify the attendees.

Tonight’s attendees:

Any of the charming properties you’ve seen so far just might show up... and they may even work together!

What property was used?

$(15 + 14) + 2$ is equivalent to $15 + (14 + 2)$ ....

$2 \cdot 8 \cdot 16$ is equivalent to $8 \cdot 2 \cdot 16$ ....

$(3 \cdot 4) + \left(\frac{1}{2} + \frac{1}{3}\right)$ is equivalent to $\left(\frac{1}{2} + \frac{1}{3}\right) + (3 \cdot 4)$ ....

$5(0.5 \cdot 0.12)$ is equivalent to $(5 \cdot 0.5)0.12$ ....

$127(16 \cdot 0.177) + 16 + (4 + 0.23)$

is equivalent to ....

$0.177(16 \cdot 127) + (16 + 4) + 0.23$

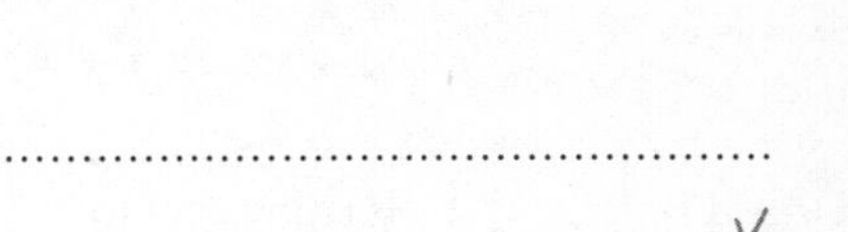

You may need more than one for certain problems.

BULLET POINTS

- The **associative property** says you can move parentheses around in addition or multiplication expressions without changing the answer.
- The **commutative property** says you can change the order of the terms around in addition or multiplication without changing the answer.
- You can’t use the associative or commutative properties with division or subtraction.
- The order of operations always tells you what order you need to work through an expression.

Who am I? Solution

A bunch of equivalent expressions, in full costume, are playing a party game, "Who am I?" They'll give you a clue, and, based on what they say, you try to guess what property they use. Assume they always tell the truth about themselves. Fill in the blanks to the right to identify the attendees.

Tonight's attendees:

Any of the charming properties you've seen so far just might show up... and they may even work together!

What property was used?

Only the parentheses changed... this is associative.

$(15 + 14) + 2$ is equivalent to $15 + (14 + 2)$ .... The associative property

The order changed here... this is commutative.

$2 \cdot 8 \cdot 16$ is equivalent to $8 \cdot 2 \cdot 16$ .... The commutative property

This one's a little tricky. The order of the two groups changed... but the actual groups stayed the same.

$(3 \cdot 4) + \left(\frac{1}{2} + \frac{1}{3}\right)$ is equivalent to $\left(\frac{1}{2} + \frac{1}{3}\right) + (3 \cdot 4)$ .... The commutative property

The parentheses moved, but nothing else did: associative.

$5(0.5 \cdot 0.12)$ is equivalent to $(5 \cdot 0.5)0.12$ .... The associative property

$127(16 \cdot 0.177) + 16 + (4 + 0.23)$

is equivalent to

$0.177(16 \cdot 127) + (16 + 4) + 0.23$

.... Both the associative and commutative properties

Here, the grouping changed...

...but so did the order of things.

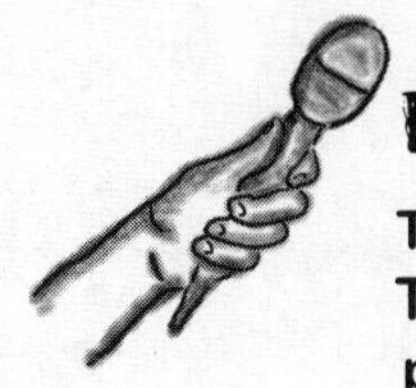

Properties Exposed

This week's interview:
The associative and commutative properties - who does what?

Associative Property: Hi, commutative. Is everything all right? You look a little mixed up.

Commutative Property: Ha, I get it. Mixing it up is my specialty. If you have some additions or multiplications, I can move the numbers around without causing any problems.

Associative: Nice. I work with addition or multiplication too, but I'm not allowed to jiggle numbers around. I just work with parentheses.

Commutative: Wait, aren't parentheses the top of the food chain in the order of operations?

Associative: Yes, and that's who I work with. There are strict rules, though. I can't mess with the order of operations or change the answer, so I can only move parentheses around if they are all around additions or multiplications.

Commutative: Yeah, I have the same rules. I can't change the answer, so I can only reorder numbers in addition or multiplication, too. I guess you're not allowed to do anything with division or subtraction either, right?

Associative: Right. Subtracting or dividing changes the answer, so I can't change those types of groupings.

Commutative: Same problem over here. Order is really important for subtraction and division, I guess, so I have to keep my hands off.

Associative: You know, I think we need to clear something up, as long as we're chatting here.

Commutative: What?

Associative: That both of us can be used outside of the order of operations without changing the answer.

Commutative: Sure. I'm used to it, but I guess that can seem confusing. We work any time! Addition or multiplication, they can be regrouped or reordered at any point when you're simplifying.

Associative: See, we're always helpful. Oh... before I go, did you hear the one about the addition expression that was wrongly accused?

Commutative: It had its sentence commuted.

Associative: No, what about it?

Jack got one point last round for getting the problem right, but Kate got 2 for being right and fast.

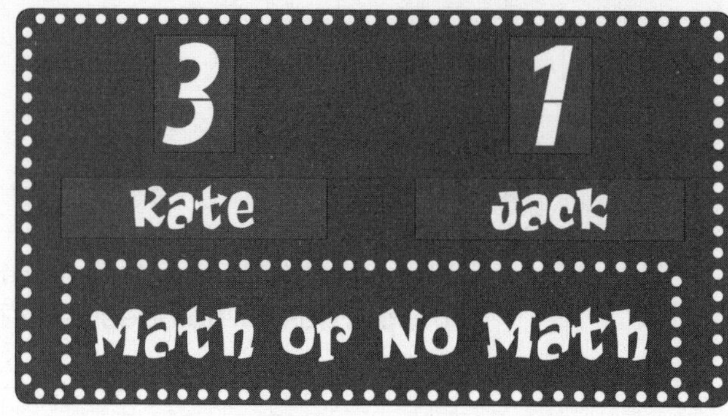

It's an important round...

The next round's worth two points, which means the pressure's on: you've got to judge the next problem correctly, or there's going to be a real brawl over who wins tonight's **Math or No Math**.

The problem's worth a single point again, and there's a bonus point for solving the problem as quickly as possible. Not only that, but both Jack and Kate know about the order of operations *and* the commutative and associative properties.

The Order Of Operations

1. **Parentheses**
2. **Exponents**
3. **Multiplication & Division**
4. **Addition & Subtraction**

The Commutative Property

You can change the order of the terms with addition or multiplication without changing the results.

The Associative Property

You can change the grouping of terms with addition or multiplication without changing the outcome.

Problem #3 - The final round

Simplify this:

$$12\left(\frac{1}{3} + \frac{5}{6} + \frac{11}{12}\right) + 3^2 - 15$$

BE the judge

Your job is to judge what Kate and Jack did. Mark up their work step by step to show how they each got their answer.

Kate's Work

$$12\left(\frac{1}{3}+\frac{5}{6}+\frac{11}{12}\right)+3^2-15$$

$$\left(\overset{4}{\cancel{12}}\frac{1}{\cancel{3}}+\overset{2}{\cancel{12}}\frac{5}{\cancel{6}}+\cancel{12}\frac{11}{\cancel{12}}\right)+3^2-15 \quad (4,\ 2(5),\ \text{circled } 11)$$

$$(4+10+11)+3^2-15$$

$$(25)+3^2-15$$

$$25+9-15$$

$$19$$

Time: 45s

Jack's Work

$$12\left(\frac{1}{3}+\frac{5}{6}+\frac{11}{12}\right)+3^2-15$$

$$12\left(\frac{4}{12}+\frac{10}{12}+\frac{11}{12}\right)+3^2-15$$

$$12\left(\frac{25}{12}\right)+3^2-15$$

$$\cancel{12}\left(\frac{25}{\cancel{12}}\right)+9-15 \quad (\text{circled } 25)$$

$$25+9-15=19$$

Time: 1m 20s

Who's right? (Circle one or both)

Kate Jack Neither

Who's fastest? (Circle one)

Kate Jack

BE the judge solution

Your job is to judge what Kate and Jack did. Mark up their work step by step to show how they each got their answer.

Kate's Work

$$12\left(\frac{1}{3} + \frac{5}{6} + \frac{11}{12}\right) + 3^2 - 15$$

These numbers are what's left after the denominators are canceled out

Instead of simplifying inside the parentheses, Kate multiplied each term inside by 12.

$$\left(\cancel{12}\frac{\cancel{1}}{\cancel{3}}\,4 + \cancel{12}\frac{5}{\cancel{6}}\,2(5) + \cancel{12}\frac{11}{\cancel{12}}\right) + 3^2 - 15$$

$$(4 + 10 + 11) + 3^2 - 15$$

All that was left was to do the exponent and then add and subtract.

$$(25) + 3^2 - 15$$

$$25 + 9 - 15$$

$$19$$

Time: 45s

Jack's Work

$$12\left(\frac{1}{3} + \frac{5}{6} + \frac{11}{12}\right) + 3^2 - 15$$

Jack had to get all the fractions with the same denominators.

Here's where Jack lost: dealing with all these fractions.

$$12\left(\frac{4}{12} + \frac{10}{12} + \frac{11}{12}\right) + 3^2 - 15$$

$$12\left(\frac{25}{12}\right) + 3^2 - 15$$

The exponent is next.

$$\cancel{12}\left(\frac{25}{\cancel{12}}\right) + 9 - 15$$

Jack canceled things out here.

$$25 + 9 - 15 = 19$$

Time: 1m 20s

Who's right? (Circle one or both)

Kate

Jack

Neither

Who's fastest? (Circle one)

Kate

Jack

Are you kidding me? Every time we figure out how to judge these things, Kate pulls out another trick.

The distributive property lets you multiply over several numbers. **(And it's not a trick, really.)**

Kate got rid of all the fractions in one step by multiplying ***all*** her fractions by 12. That canceled her denominators. When you multiply everything in a group by the same numbers, you're using the ***distributive property.*** Let's take a closer look at what exactly Kate did... and how you can do the same thing.

Kate's Work Way Up Close

Here's Kate's work, broken down even further. So what's really going on?

$$12\left(\frac{1}{3}+\frac{5}{6}+\frac{11}{12}\right)+3^2-15$$

Kate multiplied each term inside the parentheses by 12.

$$\left(12\cdot\frac{1}{3}+12\cdot\frac{5}{6}+12\cdot\frac{11}{12}\right)+3^2-15$$

Kate didn't change anything else at this step.

Here, Kate simplified each fraction by dividing out the denominators.

$$\left(\frac{12\cdot 1}{3}+\frac{12\cdot 5}{6}+\frac{12\cdot 11}{12}\right)+3^2-15$$

$$\left(\frac{4\cdot \cancel{3}\cdot 1}{\cancel{3}}+\frac{\cancel{6}\cdot 2\cdot 5}{\cancel{6}}+\frac{\cancel{12}\cdot 11}{\cancel{12}}\right)+3^2-15$$

These factors are all that was left after the denominators were canceled out.

$$(4+2\cdot 5+11)+3^2-15$$

$$(4+10+11)+3^2-15$$

She went on with the order of operations.

$$(25)+3^2-15$$

$$25+9-15$$

$$19$$

Distributing a value over a grouping doesn't change a problem's value

When you take a value and multiply over a grouping, you're distributing that value. The **distributive property** says that if you have two groups multiplied together, you can simplify the groups, then multiply; or multiply first and then simplify.

Here's the piece of Kate's work where she used the distributive property:

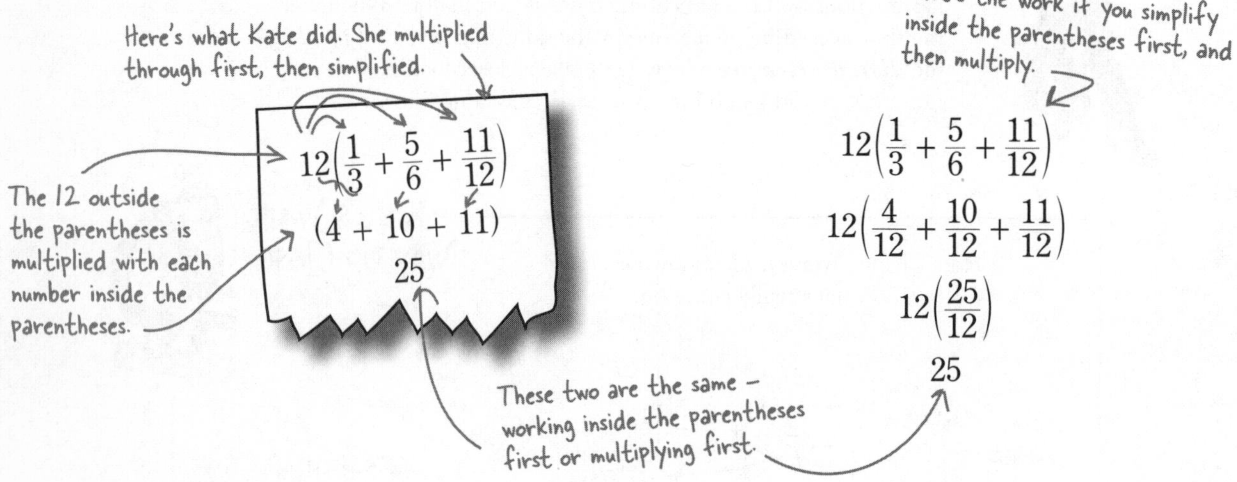

Here's what Kate did. She multiplied through first, then simplified.
$12\left(\frac{1}{3} + \frac{5}{6} + \frac{11}{12}\right)$
$(4 + 10 + 11)$
25
The 12 outside the parentheses is multiplied with each number inside the parentheses.
Here's the work if you simplify inside the parentheses first, and then multiply.
$12\left(\frac{1}{3} + \frac{5}{6} + \frac{11}{12}\right)$
$12\left(\frac{4}{12} + \frac{10}{12} + \frac{11}{12}\right)$
$12\left(\frac{25}{12}\right)$
25
These two are the same - working inside the parentheses first or multiplying first.

there are no Dumb Questions

Q: Can we multiply before we do the parentheses?

A: Yes. If you have a situation where you're multiplying two groups together, you can multiply and then simplify; or you can simplify and then multiply.

Q: Isn't the distributive property ignoring the order of operations?

A: No, it's just knowing when you can work around the order of operations. Just like with the associative and commutative properties, the distributive property is about working with problems in a simpler, more efficient way. And these properties work with the order of operations, not against them.

Q: What if there is subtraction or division inside the parentheses?

A: It doesn't matter. You just have to keep the same operators after you distribute a value that's outside the parentheses. If a number is subtracted inside the parentheses, it's still subtracted after the distribution.

Q: So parentheses don't have to go first?

A: Not if you're multiplying what's in the parentheses by a number. In that case, you can distribute the number over the grouping.

It's time for you to practice your distributions. Take the expressions below and simplify them in two ways: by distributing, and by following the order of operations. Which way do you think is faster for each problem?

$2(4 - 2 + 11)$

Distributive Property | Order of Operations

Which way was faster? Why?

$4\left(\frac{1}{20} + \frac{9}{20} + \frac{7}{20} + \frac{3}{20}\right)$

Distributive Property | Order of Operations

Which way was faster? Why?

$24\left(\frac{3}{8} - \frac{1}{12} + \frac{5}{6} - \frac{3}{4}\right)$

Distributive Property | Order of Operations

Which way was faster? Why?

$18(110 - 80 + 3 - 22 - 10)$

Distributive Property | Order of Operations

Which way was faster? Why?

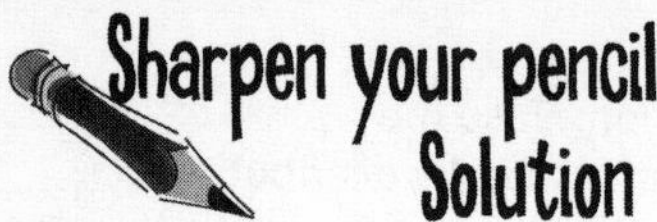

Your job was to take the expressions below and simplify them in two ways: by distributing, and by following the order of operations. Which way did you think was faster for each problem?

$2(4-2+11)$

Distributive Property

$2 \cdot 4 - 2 \cdot 2 + 2 \cdot 11$

$8 - 4 + 22$

26

Order of Operations

$2(13)$

26

Which way was faster? Why?

I think that using the order of operations was faster since there were fewer steps.

There's no right answer... it's up to you.

$4\left(\frac{1}{20}+\frac{9}{20}+\frac{7}{20}+\frac{3}{20}\right)$

Distributive Property

$4\left(\frac{1}{20}\right)+4\left(\frac{9}{20}\right)+4\left(\frac{7}{20}\right)+4\left(\frac{3}{20}\right)$

$\frac{1}{5}+\frac{9}{5}+\frac{7}{5}+\frac{3}{5}$

$\frac{20}{5}=4$

Order of Operations

$4\left(\frac{20}{20}\right)$

$4(1)$

4

Which way was faster? Why?

The order of operations was faster because the fractions went away.

Sometimes the order of operations is quicker.

$24\left(\frac{3}{8}-\frac{1}{12}+\frac{5}{6}-\frac{3}{4}\right)$

Distributive Property

$24\frac{3}{8} - 24\frac{1}{12} + 24\frac{5}{6} - 24\frac{3}{4}$

$9-2+20-18$

9

Order of Operations

$24\left(\frac{9}{24}-\frac{2}{24}+\frac{20}{24}-\frac{18}{24}\right)$

$\cancel{24}\left(\frac{9}{\cancel{24}}\right)$

9

Which way was faster? Why?

I think that the distributive way was faster since all the fractions went away.

$18(110-80+3-22-10)$

Distributive Property

$(1980-1440+54-396-180)$

18

Order of Operations

$18(1)$

18

Which way was faster? Why?

The order of operations was the faster way to go since the numbers worked out in the parentheses.

But, if the number inside had been huge, the distributive property would've been the way to go.

The Commutative Property

You can change the order of the terms with addition or multiplication without changing the results.

The Distributive Property

A number being multiplied by a group produces the same result by either simplifying the group and then multiplying, or multiplying each term within the group and then simplifying.

The Associative Property

You can change the grouping of terms with addition or multiplication without changing the outcome.

How am I supposed to remember all this? First there was the order of operations, and now there are all these rules... it's like a paragraph for each one. I'm supposed to memorize these?

You can memorize some general equations, not a lot of text.

Remembering all those sentences for each property is a pain... and this is math, not composition! Fortunately, we can turn these rules into some simple equations.

But to do that, we need a way to represent numbers... we need **constants**. But what exactly is a constant?

A constant stands in for a number

A **constant** is a term used to describe an unknown in an equation that represents a number that doesn't change. In other words, a letter "a" on one side of an equation is the same number as an "a" on the other side of an equation. The constant just represents a number.

Constants are great for turning specific problems into more general ones because we can use letters instead of specific numbers. For example...

not easy to remember

The Commutative Property

You can change the order of the terms with addition or multiplication without changing the results.

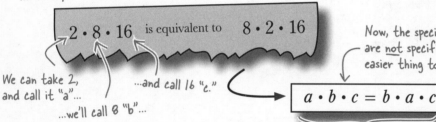

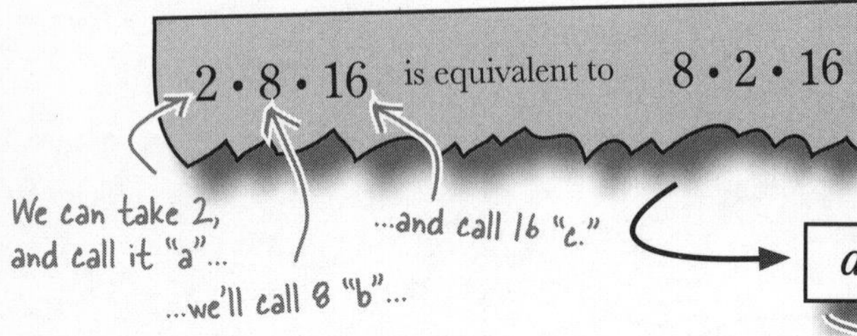

This is a very specific problem, and isn't easy to remember.
2 • 8 • 16 is equivalent to 8 • 2 • 16
We can take 2, and call it "a"...
...we'll call 8 "b"...
...and call 16 "c."

Now, the specifics of the problem are not specific. This is a much easier thing to remember.

$$a \cdot b \cdot c = b \cdot a \cdot c$$

So now we can remember that we can rearrange numbers if they're multiplied. This is the commutative property without any words!

At first, it might seem like these letters are just as hard to remember as a bunch of sentences. But let's make things even simpler:

$$a \cdot b = b \cdot a$$

This is the commutative property for multiplication.

$$a + b = b + a$$

This is the commutative property for addition.

A general equation is just a way to remember rules that apply to all numbers in a certain situation.

You know more than you think... try and match each property statement to it's name. Be careful, though... some of the property names are used **twice**!

$$a(b + c) = ab + ac$$

$$a + (b + c) = (a + b) + c$$

$$a \cdot b = b \cdot a$$

$$a + b = b + a$$

$$a(b \cdot c) = (a \cdot b)c$$

The commutative property

The distributive property

The associative property

Match each property statement using constants to it's name. Some of the property names are used **twice**!

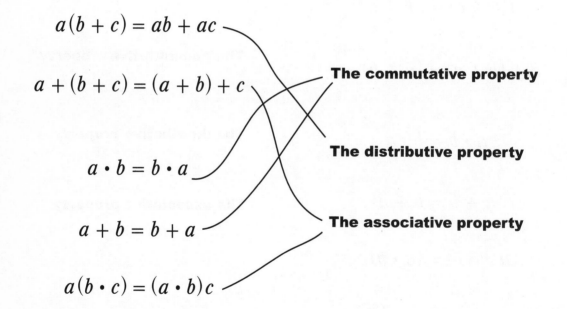

Property

Property name

$a(b + c) = ab + ac$ → The distributive property

$a + (b + c) = (a + b) + c$ → The associative property

$a \cdot b = b \cdot a$ → The commutative property

$a + b = b + a$ → The commutative property

$a(b \cdot c) = (a \cdot b)c$ → The associative property

The Commutative Property

You can change the order of the terms with addition or multiplication without changing the results.

The Commutative Property

multiplication

$$a \cdot b = b \cdot a$$

addition

$$a + b = b + a$$

The Associative Property

You can change the grouping of terms with addition or multiplication without changing the outcome.

The Associative Property

multiplication

$$a(b \cdot c) = (a \cdot b)c$$

addition

$$a + (b + c) = (a + b) + c$$

The Distributive Property

A number being multiplied by a group produces the same result by either simplifying the group and then multiplying, or multiplying each term within the group and then simplifying.

The Distributive Property

$$a(b + c) = ab + ac$$

Roll the credits...

After going through Kate's work, we figured out that she got all the problems right and was the fastest.

1. **Kate solved the first question using the order of operations.**
   After seeing what happened with Jack, everybody can't just go their own way, so the order of operations is important.

2. **Kate solved the second question using the associative and commutative properties before applying the order of operations.**
   Kate got done faster and smarter and could make the fractions much easier.

3. **Kate got the third question right, faster, by using the distributive property first, and then the order of operations.**
   Kate made fractions much easier by distributing.

Jack was the runner up. He learned the order of operations from his first problem, but he needed some more properties up his sleeve to compete with Kate.

1. **Jack got the first question wrong because he didn't follow the order of operations.**
   Jack tried to solve the easiest parts of the question first, regardless of the order of operations... and completely missed the question.

2. **Jack got question two right by using the order of operations.**
   Unfortunately, Jack didn't use the associative or commutative properties, so it took him longer than Kate to solve question two.

3. **Jack got question three right, too, but he lost the speed competition again.**
   The order of operations never let him down, but it slowed him down!

Keep those judging skills polished... you never know when Math or No Math might need you again!

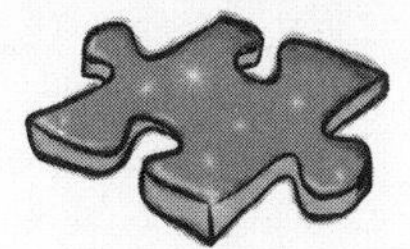

Propertycross

Take some time to sit back and burn these properties into your brain. It's your standard crossword; all of the solution words are from this chapter.

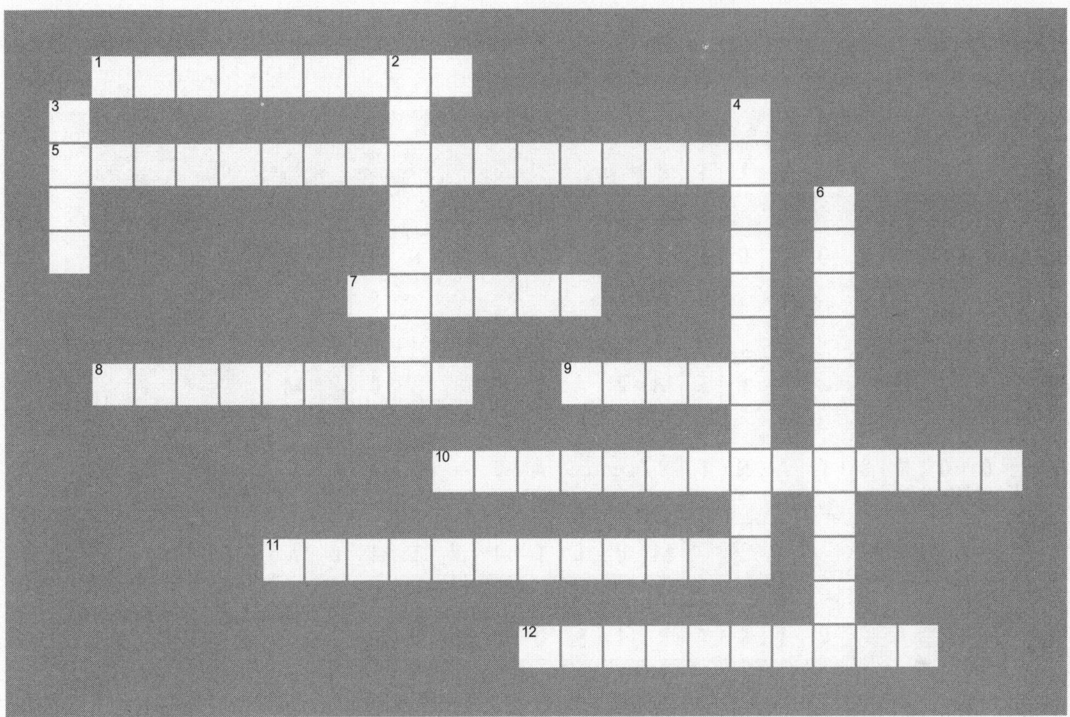

Across

1. x,y,z but subject to change
5. Tells you which computation you do first
7. Ord. of Opers.
8. The a,b,c's of Algebra
9. Constantly commutative (includes =)
10. Associative property only works with addition and _______________
11. Property that lets you multiply each term in a group by the value being multplied with the whole group
12. Different representation, same result

Down

2. Means how many times you multiply a number by itself
3. A fractional exponent
4. Property that lets you change the grouping without changing the results
6. Property, addition, terms, multiplication, operations, order

Propertycross Solution

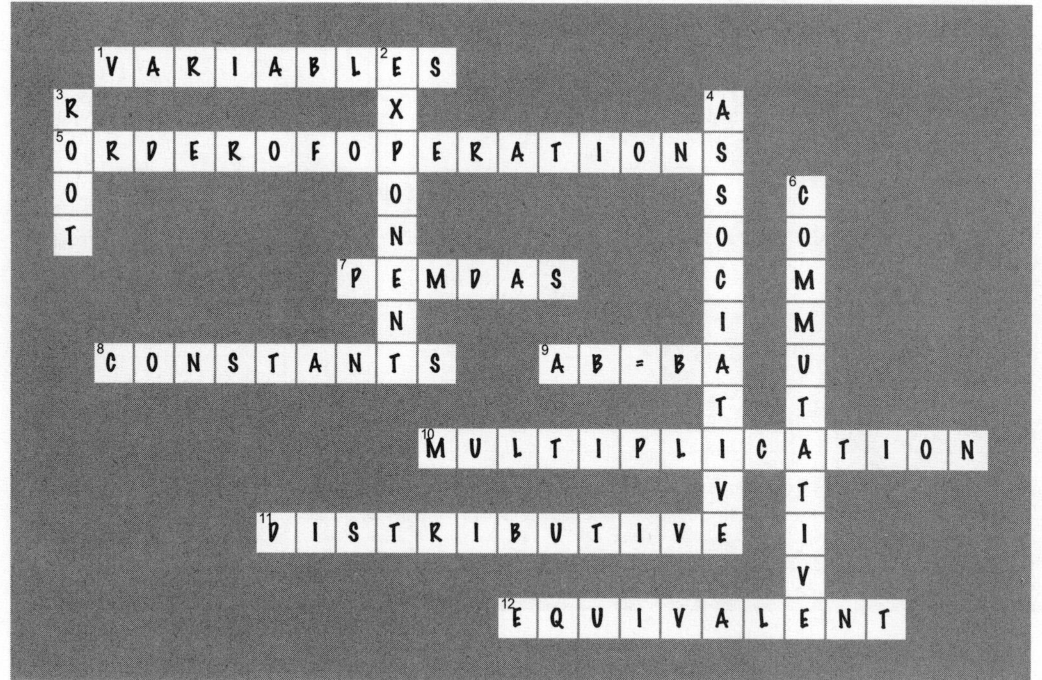

Tools for your Algebra Toolbox

This chapter was about numeric properties that are important to understand in order to work with equations.

BULLET POINTS

- The **associative property** deals only with grouping.
- The **commutative property** deals with order.
- The **distributive property** handles multiplication of groups.
- The **order of operations** will always correctly simplify an expression.
- A **variable** is an unknown that can change with the problem situation.
- A **constant** is a known or unknown that **does not** change.

CHAPTER 3

The order of operations

1. Parentheses
2. Exponents
3. Multiplication and Division
4. Addition and Subtraction

The associative property

$$a + (b + c) = (a + b) + c$$

$$a(b \cdot c) = (a \cdot b)c$$

The commutative property

$$a \cdot b = b \cdot a$$

$$a + b = b + a$$

The distributive property

$$a(b + c) = ab + ac$$

All of these properties are good for numbers and unknowns.

4 exponent operations

Podcasts that spread like the plague

(that's a good thing...)

iTunes totally sucked me into podcast subscriptions. Now I only turn the TV on to catch the latest Lost episode.
Craft:

Could you multiply that again? Could you multiply that again?

There's another way to express multiplication that's repeated over and over and over again, without just repeating yourself. **Exponents** are a way of **repeating multiplication**. But there's more to exponents, including some smaller-than-usual numbers (and we don't just mean fractions). In this chapter, you'll brush up on **bases**, **roots**, and **radicals**, all without getting arrested for any sit-in protests. And, as usual, ***zero*** and ***one*** come with their own problems... so jump into a podcasting exponentiation extravaganza.

Addie's got a podcast

I've been producing my own Podcast, but now I need better quality gear... and new equipment is expensive!

Addie, podcast producer extraordinaire.

Ahem, eccentric

Addie podcasts about ~~crazy~~ celebrities.

Addie's been getting a lot more listeners lately. To take it to the next level, she needs new equipment to deliver an even better podcast... but that takes a lot of cash.

Addie's got a website to host her podcasts, and she wants to advertise on her site to raise funds for new gear. She's lined up some possible sponsors, but they won't help out until Addie proves she can get some real traffic on her site. Addie needs to:

- ...monitor the daily hits on her website over the next 2 weeks.
- ...prove her site can generate at least **5,000,000 hits** in the next 2 weeks!

Wow, that's a lot.

Addie's computer. Can you say 1987?

Let's mobilize Addie's listeners

Addie knows she's got big fans. Here's a letter she's worked up to send out to her **3 top listeners**:

```
Dear TOP 3 Listeners,

I'm trying to get some sponsors for the podcast
so we can buy some new equipment. There will be
some advertisers tracking the hits to the site
over the next two weeks, and we need to have
5,000,000 hits in a single day.

Please make sure that you visit the site today
and send this email out to 3 more people.  If
everyone keeps visiting the site and emailing 3
new people for 14 days, I think we'll make it!

Thanks,

Addie from the StarTalk Podcast
```

Sharpen your pencil

Write down the equation to figure out how many hits Addie will get to by the end of 14 days, assuming each of her 3 top listeners tells 3 more people each day to visit her site. Don't worry about solving the equation for now.

....

....

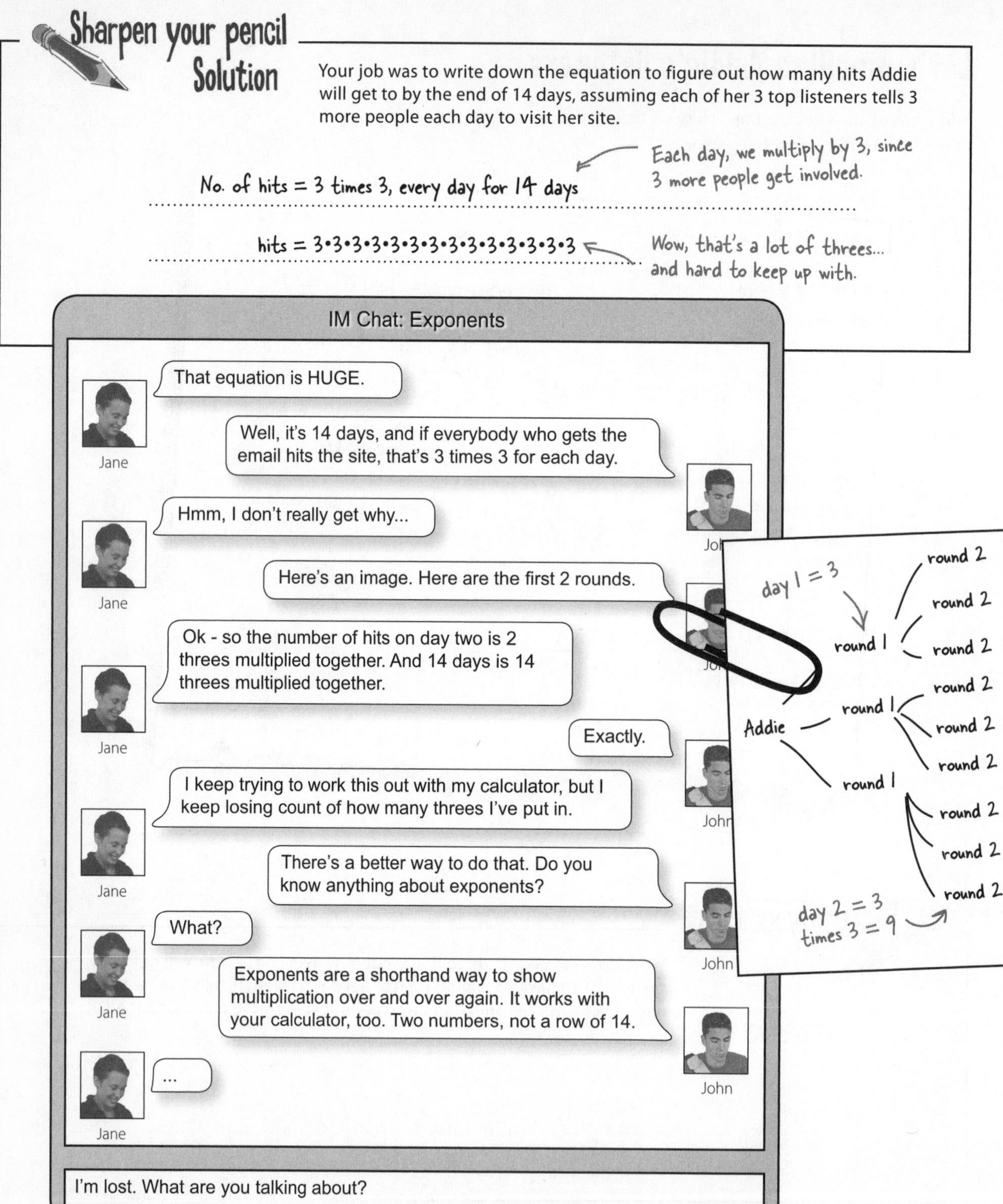

Sharpen your pencil Solution
Your job was to write down the equation to figure out how many hits Addie will get to by the end of 14 days, assuming each of her 3 top listeners tells 3 more people each day to visit her site.
No. of hits = 3 times 3, every day for 14 days
Each day, we multiply by 3, since 3 more people get involved.
hits = 3•3•3•3•3•3•3•3•3•3•3•3•3•3
Wow, that's a lot of threes... and hard to keep up with.
IM Chat: Exponents
Jane: That equation is HUGE.
John: Well, it's 14 days, and if everybody who gets the email hits the site, that's 3 times 3 for each day.
Jane: Hmm, I don't really get why...
John: Here's an image. Here are the first 2 rounds.
Jane: Ok - so the number of hits on day two is 2 threes multiplied together. And 14 days is 14 threes multiplied together.
John: Exactly.
Jane: I keep trying to work this out with my calculator, but I keep losing count of how many threes I've put in.
John: There's a better way to do that. Do you know anything about exponents?
Jane: What?
John: Exponents are a shorthand way to show multiplication over and over again. It works with your calculator, too. Two numbers, not a row of 14.
Jane: ...
I'm lost. What are you talking about?
day 1 = 3
Addie
round 1
round 1
round 1
round 2
round 2
round 2
round 2
round 2
round 2
round 2
round 2
round 2
day 2 = 3 times 3 = 9

Exponents Up Close

Exponents are a special notation used to express repeated multiplication. That's just what we need to figure out Addie's number of hits without counting a lot of threes: a way to show multiplying by 3 over and over.

When you see a number with an exponent, it looks like this:

$$x^{a} = x \cdot x \cdot x \ldots \cdot x$$

exponent

base

This means multiply x by itself "a" times.

The base is the number being multiplied (in Addie's case, 3), and the exponent is the number of times you repeat it (in Addie's case, 14). Those two numbers are all you need to put in your calculator, and you'll get the answer.

EQUATION CONSTRUCTION

Rewrite Addie's equation using exponent notation, and solve (using a calculator to get the number is a good idea).

This is the equation for Addie's website hits.

hits = 3•3•3•3•3•3•3•3•3•3•3•3•3•3

EQUATION CONSTRUCTION SOLUTION

Your job was to rewrite Addie's equation using exponent notation, and solve.

That's 14 threes!

hits = $3 \cdot 3 \cdot 3 \cdot 3 \cdot 3 \cdot 3 \cdot 3 \cdot 3 \cdot 3 \cdot 3 \cdot 3 \cdot 3 \cdot 3 \cdot 3$

3 is the base because it's the number being multiplied.

hits = 3^{14}

14 is the exponent because that's how many times you multiply 3 by itself.

$h = 3^{14} = 4{,}782{,}969$

Over 4 million hits on the 14th day. That's awesome... but not enough.

That's not going to cut it. I need 5,000,000 hits to get sponsorship. I need some help!

Addie's brother, Alex

Whaddup, girl?
I can help you out... I've got tons of friends, you know. Have you seen my Facebook page?

Can Addie and Alex get enough hits?

Alex has offered to send off another round of emails for Addie. He'll start with 3 friends, just like she did, and try to help get 5,000,000 hits in 14 days.

To figure out the total number of hits, you'll need to figure out how to add up both groups that Addie's working with. In chapter 2, you combined like terms to help Paul on his road trip, and this is the same idea. You may remember from chapter 2 that a **term** is any part of an equation held together with multiplication or division. Since an exponent is just a shorthand version of multiplication, that means ***exponential terms with the same base and the same exponent are like terms.***

With exponents, you can combine terms that have the same base. Let's try that out and see how it works:

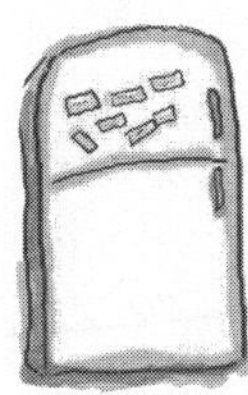

Math Magnets

Write the new equation for the number of hits that Addie and Alex will get. Will she reach 5,000,000 now?

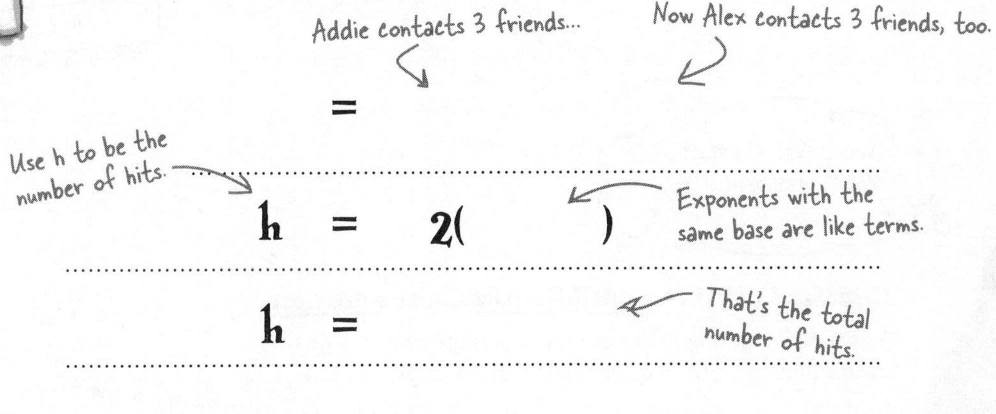
Addie contacts 3 friends...
Now Alex contacts 3 friends, too.
=
Use h to be the number of hits.
h = 2( )
Exponents with the same base are like terms.
h =
That's the total number of hits.

Did they get 5,000,000 hits?

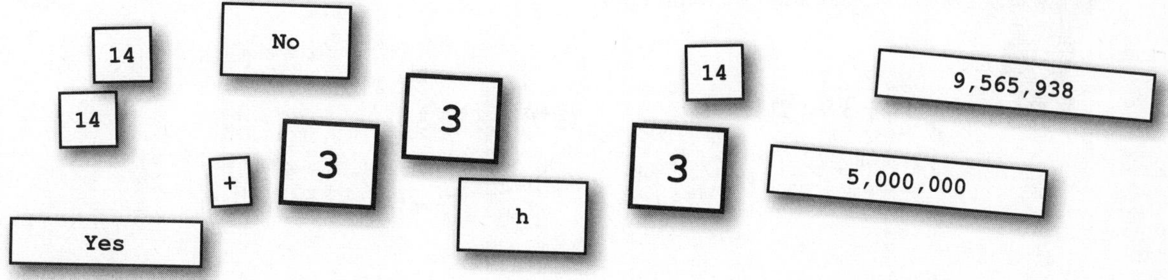
14
No
14
3
14
9,565,938
+
3
3
5,000,000
h
Yes

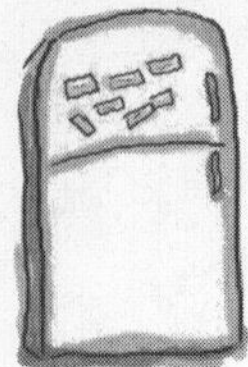

Math Magnets Solution

Write the new equation for the number of hits that Addie and Alex will get. Will she reach 5,000,000 now?

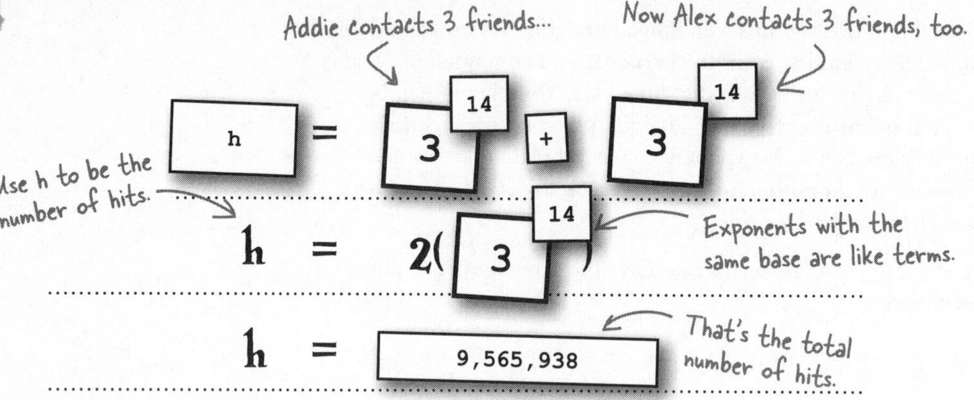
Addie contacts 3 friends...
Now Alex contacts 3 friends, too.
h = 3 14 + 3 14
Use h to be the number of hits.
h = 2(3 14)
Exponents with the same base are like terms.
h = 9,565,938
That's the total number of hits.

Did they get 5,000,000 hits? Yes

5,000,000

No

Wait. Why is that $2(3^{14})$ and not $(3^{14})^2$?

Because $(3^{14})^2$ is multiplication, not addition.

A term is something held together by multiplication, which means that the entire exponential term is treated as a group.

3^{14} is one term, not two.

When you group two like terms together, you're adding those terms together. But if you take those two terms and use exponents (that ^{2} at the end of $(3^{14})^2$), then you're multiplying, and that's not what we want. Look:

$3^{14} + 3^{14} = 2(3^{14})$

This term plus this term means two of the same term.

Remember – the exponent means you multiply.

$(3^{14})^2 = 3^{14} \cdot 3^{14}$

This is just another way to write multiply 3^{14} by itself 2 times.

Ok, so $(3^{14})^2$ isn't the right answer, but how would you work that sort of problem out, anyway? I guess we could write out a bunch of 3's in a big line?

Well, that would work, but that's a lot of threes...

Writing out the multiplication by hand will work; it's just not very convenient. Look how long this thing turns out to be:

So many 3's...

$$(3^{14})^2 = 3^{14} \cdot 3^{14}$$
$$= (3 \cdot 3 \cdot 3 \cdot 3 \cdot 3 \cdot 3 \cdot 3 \cdot 3 \cdot 3 \cdot 3 \cdot 3 \cdot 3 \cdot 3 \cdot 3)(3 \cdot 3 \cdot 3 \cdot 3 \cdot 3 \cdot 3 \cdot 3 \cdot 3 \cdot 3 \cdot 3 \cdot 3 \cdot 3 \cdot 3 \cdot 3)$$
$$= 3^{\square}$$

What's the final exponent? Count them up and fill this box in.

But look, there's a pattern! Here's what this means:

$$(x^a)^b = x^{a \cdot b}$$

Multiply 2•14, it's the same thing you got when we did it the long way.

We combined those exponents with like bases before. Is there anything else we can do with like bases?

Exponents with the same base are LIKE TERMS.

That means they can be added, subtracted, multiplied, and divided.

Try division:

$$\frac{3^{14}}{3^{12}} = \frac{3 \cdot 3 \cdot 3 \cdot 3 \cdot 3 \cdot 3 \cdot 3 \cdot 3 \cdot 3 \cdot 3 \cdot 3 \cdot 3 \cdot 3 \cdot 3}{3 \cdot 3 \cdot 3 \cdot 3 \cdot 3 \cdot 3 \cdot 3 \cdot 3 \cdot 3 \cdot 3 \cdot 3 \cdot 3}$$
$$= 3 \cdot 3$$

If you write it out, you can see how many factors you can divide out.

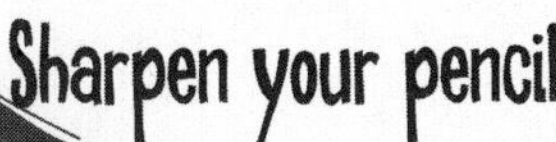

Write the general form to combine like terms. We've gone ahead and done the first one for you.

$3^{14} + 3^{14} = x^a + x^a = 2x^a$

$\frac{3^{14}}{3^{12}} =$ ______________ $=$

$(3^{14})^2 =$ $=$

$3^{14} \cdot 3^2 = x^a \cdot x^b =$

Go back to writing the whole thing out if you need to, but you can figure it out!

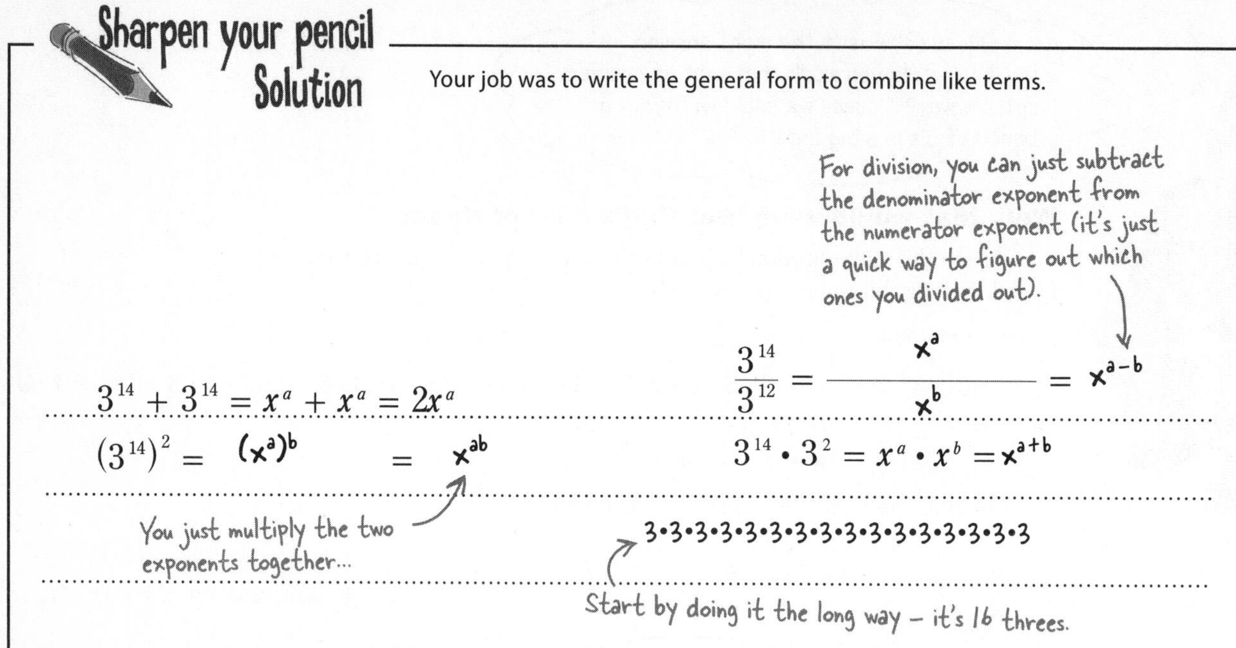

Sharpen your pencil Solution

Your job was to write the general form to combine like terms.

$3^{14} + 3^{14} = x^a + x^a = 2x^a$

$\frac{3^{14}}{3^{12}} = \frac{x^a}{x^b} = x^{a-b}$

For division, you can just subtract the denominator exponent from the numerator exponent (it's just a quick way to figure out which ones you divided out).

$(3^{14})^2 = (x^a)^b = x^{ab}$

You just multiply the two exponents together...

$3^{14} \cdot 3^2 = x^a \cdot x^b = x^{a+b}$

$3 \cdot 3 \cdot 3 \cdot 3 \cdot 3 \cdot 3 \cdot 3 \cdot 3 \cdot 3 \cdot 3 \cdot 3 \cdot 3 \cdot 3 \cdot 3 \cdot 3 \cdot 3$

Start by doing it the long way – it's 16 threes.

Alex is flaking out on his sister

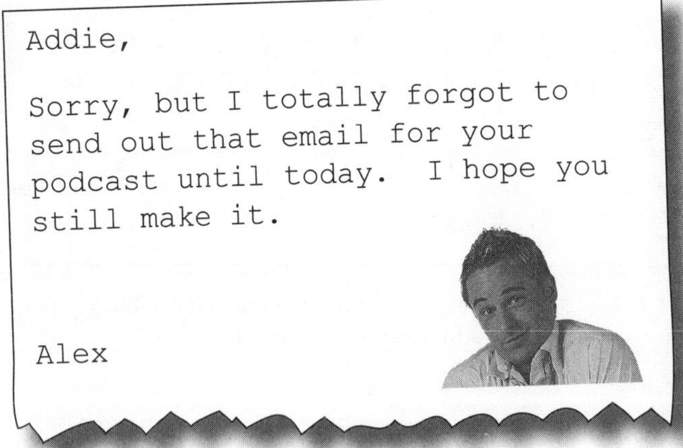

Addie,

Sorry, but I totally forgot to send out that email for your podcast until today. I hope you still make it.

Alex

Alex didn't send out any emails to his friends until the third day. That means he only has 12 days to get the word out. Will Addie still make it?

Sharpen your pencil

Now will Addie make it? Figure out how many hits she'll get since Alex's email only has 12 days to work, not 14.

Write the new equation and solve it: ....

....

....

....

What is the general form of this equation? ....

Use x and y as bases and a and b as exponents.

....

....

....

Do the exponential terms have the same base? **Yes** **No**

Circle one

Will Addie still have enough hits to make 5,000,000? **Yes** **No**

Sharpen your pencil Solution

Now will Addie make it? Figure out how many hits she'll get since Alex's email only has 12 days to work, not 14.

Write the new equation and solve it: $h = 3^{14} + 3^{12}$

Alex has two fewer days, so that's 12.

Addie's email is the same.

$h = 4{,}782{,}969 + 531{,}441$

$h = 5{,}314{,}410$

What is the general form of this equation? $h = x^a + x^b$

Use x and y as bases and a and b as exponents.

These can't really be combined easily. Since they don't have the same exponent, they're NOT like terms.

Do the exponential terms have the same base? **Yes** (circled) **No**

But just by 314,410 hits. It's pretty close.

Will Addie still have enough hits to make 5,000,000? **Yes** (circled) **No**

Phew - Alex didn't blow it. So now, I really can just wait to hit 5,000,000. As soon as the ad company sees that in a couple of weeks I'm finally going to snag some gear. Hello, Apple store.

there are no Dumb Questions

Q: : Why use exponents and not just multiplication?

A: Because it can save you a bunch of work. Writing out a value to multiply over and over again is tedious and leads to error. And when the numbers start to get really big (like an exponent of 14), they are just impossible to deal with otherwise.

Q: Why do you need the same base and the same exponent to do addition and subtraction?

A: Because they need to be **like terms**. Remember that exponents are a shorthand for multiplication. Because of the order of operations, you can't add two multiplication expressions together without doing the multiplications first... unless you've got like terms. If the expressions are like terms, then you can collect them together into a single term. That's exactly what adding exponential terms with the same base and exponent does!

Q: Where are exponents in the order of operations?

A: They're second. Since exponents are just a more powerful form of multiplication, they go ***before*** multiplication. So it's parentheses, exponents, ***then*** multiplication and division.

Q: How do you work with exponents with different bases?

A: We're going to be looking at those next. But fair warning, there's not much you can do to make those problems simpler. If you have two bases, then there are two things that need to be multiplied, divided, or whatever. There's not a good way to combine terms like that since you've got to keep track of both bases separately.

Q: What happens if I'm dividing exponential terms and the exponent becomes negative?

A: Great question! When you divide exponential terms, you subtract the exponents. This means that you could end up with a negative exponent. The good news is that this is easy to deal with. A negative exponent just means 1 over the positive exponential term. So:

2^{-1} is $1/2$,

x^{-25} is $1/x^{25}$

...and so on.

There's always a villain...

The Movie Podcast heard about Addie's plan to increase subsribers, and they don't like it. The sponsorship Addie's trying to get... well, it's money out of the Movie Podcast's pocket. So now they're fighting back.

Dear **Top 4** Movie Podcast subscribers,

The StarTalk Podcast is trying to steal advertisers from us! If they hit 5,000,000 hits in the next 10 days, our sponsor is going to leave our show.

We need to fight back! Don't go to the Startalk page, and email 4 Startalk users, telling them not to either. If everybody emails 4 people, we'll pull enough hits so that she won't make it!

Thanks,

Movie Podcast

Addie's had a head start before this mail went out.

Every person that hits Movie Podcast's page instead of Startalk Podcast's page is taking away potential hits. What does this mean for Addie's chances to score a new sponsorship deal?

EQUATION CONSTRUCTION

Since Movie Podcast is going to take away hits, how many will be left? Is Addie going to make it or is she in trouble?

Write the new equation and solve it: ....

Don't forget about what Addie and Alex already did.

....

....

Will Addie still have enough hits to make 5,000,000? **Yes** **No**

Circle one

If No, how many more hits does Addie need to get to 5,000,000? ....

Write the equation in general form:

....

....

....

How many different bases are involved? **1** **2** **3**

How many different exponents? **1** **2** **3**

EQUATION CONSTRUCTION SOLUTION

Since Movie Podcast is going to take away hits, how many will be left? Is Addie going to make it, or is she in trouble?

Addie's orignial hits — Alex's email (two days late)

Write the new equation and solve it: $h = 3^{14} + 3^{12} - 4^{10}$ ← Movie Podcast's email: 10 days left and 4 emails each.

$h = 4{,}782{,}969 + 531{,}441 - 1{,}048{,}576$

Uh oh. Movie Podcast made enough of a dent to push Addie below the numbers she needs. → $h = 4{,}265{,}834$

Will Addie still have enough hits to make 5,000,000? **Yes** **No** (circled: No)

If No, how many more hits does Addie need to get to 5,000,000?

The number of hits she needs — Less what she has now (thanks to the folks at Movie Podcast)

$5{,}000{,}000 - 4{,}265{,}834 = 734{,}166$

Addie needs to come up with over 700,000 new hits!

Write the equation in general form: $h = 3^{14} + 3^{12} - 4^{10}$

$h = x^a + x^b - y^c$

The new term has a different base AND a different exponent.

These are the same from earlier – same base but different exponent

How many different bases are involved? **1** **2** **3** (circled: 2)

Since Addie and Alex sent theirs out to the same number of folks, they have the same base. Movie Podcast sent it out to more people with less time.

How many different exponents? **1** **2** **3** (circled: 3)

That's why we have three different variables listed for the exponents. It also means that they can't be easily combined.

Since those terms have different bases, they can't be combined as variables, right?

Different bases = NOT like terms.

Terms with different bases are not like terms (regardless of the exponent). They just don't have anything in common. As exponential terms, they're not multiplying the same number, regardless of how many times.

As we saw earlier, they're only like terms if the base AND the exponent are the same.

You can't add exponents with different bases

If we just talk about the bit of Addie's equation that has two terms, it looks like this:

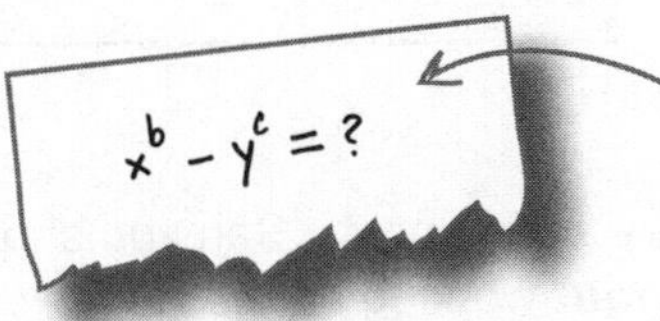

This is just a piece of Addie's equation.

You know that you can't add or subtract these two because they're not like terms. Multiplication and division don't work either. Exponential terms being multiplied together just get written together, like this:

$$x^b(y^c) = x^b \cdot y^c = x^b y^c$$

These are all same thing, just written differently, but you can't combine them at all.

The order of operations says exponents FIRST

You can't split up bases and combine different exponents because each base has to stay with its own exponent. The order of operations says that exponents go *before* multiplication. That means the exponents have to be **simplified** before they can be combined with something else.

OK

$$x^b\left(y^c\right) = x^b \cdot y^c = x^b y^c$$

These are all ok because the exponent stays with the base.

This is NOT ok.

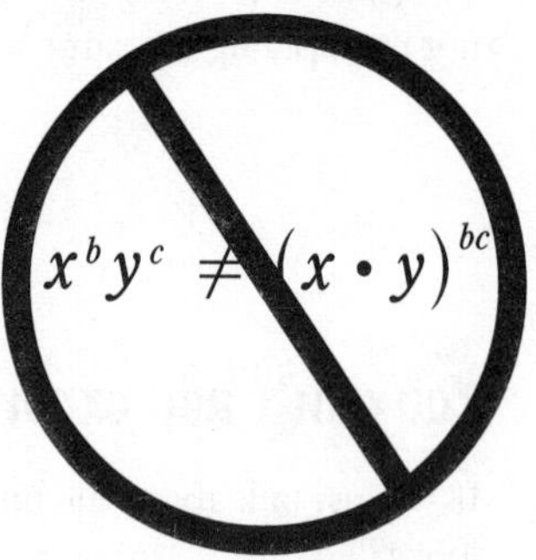

Test it out with real numbers - try 3^2 and 4^3. Can you show that $(3^2)(4^3)$ is not the same as $((3)(4))^{(2)(3)}$ ***without*** working things all the way out to the answer?

there are no Dumb Questions

Q: Do I really have to memorize all of these rules for working with exponents?

A: No, because you can always work through these equations by working out each term separately. But if you can remember these rules, you'll be able to combine like terms and solve equations more quickly. It's much easier to combine terms and do one calculation. That's a lot better than working out a ton of terms separately, especially if the terms can be combined because they're like terms.

Q: What if the bases are different and the exponents are the same?

A: Well, there's a *little* bit you can do there. If the exponents are the same, then each term is being multiplied the same number of times, so you CAN mush them together, like this: $x^a \bullet y^a = (xy)^a$
It only works because of the commutative and associative properties. This is all just multiplication, so you can mix up the order, and it will still work.

Q: So what about that Brain Power? How could you show it without solving the math?

A: You can do it with variables, so instead of 3 and 4, let's use x and y. Then we have $x^2y^3 = x \bullet x \bullet y \bullet y \bullet y$ and
$(xy)^{(2)(3)} = (xy)^6 = xy \bullet xy \bullet xy \bullet xy \bullet xy \bullet xy = x \bullet x \bullet x \bullet x \bullet x \bullet x \bullet y \bullet y \bullet y \bullet y \bullet y \bullet y$
You can just look at those two and see they're not equal.

But what about all those hits I need? I need 734,166 more hits, and I only have 9 days left. I can't lose that sponsorship, or who knows how many subscribers I'll lose, too.

Addie needs another round of emails.

But how many does she need to send out? Addie has only 9 days left and she needs to figure out how many emails she needs to start with today to make up for the campaign that Movie Podcast's running.

We've got to work our exponent "backward"

Let's go back to Addie's equation. We've got different information this time: the number of hits we need and the number of days left:

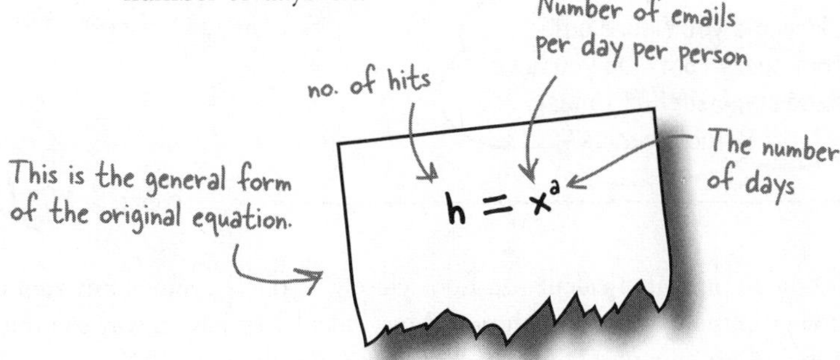

Now we fill in the things we know:

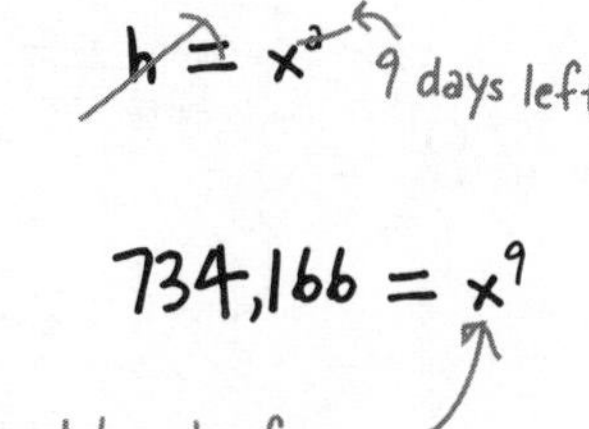

734,166
h = x^a
9 days left
734,166 = x^9
Now we need to solve for x.

What do I do with that? I can't solve for a base.

A root is the INVERSE of an exponent

We need an operation that can unravel an exponent. So when we have the exponent, what's the base to get a certain answer? Well, that's the **root**. When you find a root of something, you're finding the **number that can be multiplied** over and over to reach the final number.

Addie's equation.

$734{,}166 = x^9$

We need the 9th root here.

For Addie, we need take the ninth root of both sides of her equation. That will isolate x and give us a numeric value on the other side.

This is the equation we started with

$$734,166 = x^9$$

We take the ninth root of both sides. This undoes the exponent of nine on the right.

$$\sqrt[9]{734,166} = \sqrt[9]{x^9}$$

$$\sqrt[9]{734,166} = x$$

We know this side needs to be x because the root and the exponent are the same, and they cancel each other out.

How did you figure out that crazy root? Do you have to be some sort of times-table genius?

Roots Up Close

It's time to pull out a calculator. Look closely—you can punch any root of any number in there and get a solution. Most calculators have a way to punch in both a root key and the root you want (like 9 or 3).

Ask a teacher, or look this up in your calculator's instructions.

Let's look at a little closer at roots:

The little (index) number shows how many times the root needs to be multiplied (in this case, 2).

This symbol is called the radical. It means find the root

$$\sqrt[2]{9} = 3$$

$3^2 = 9$ (See the connection?)

This is the root.

If you wanted to read this, you'd say, "The second root of nine is three." The actual root is three, and it's the number that you multiply twice by itself to find the number under the radical. So, to find the ninth root for Addie, all you need is a calculator.

Sharpen your pencil

It's the moment of truth. How many more emails does Addie need to send out? Is there any way she can pull this off?

Solve Addie's root: ....

....

....

How many emails does Addie need to send out? ....

....

Is there any way she can pull it off? **Yes** **No**

Why? ....

It's the moment of truth. How many more emails does Addie need to send out? Is there any way she can do it?

Punch this into your calculator and you'll get a number you can use.

Solve Addie's root: $\sqrt[9]{734{,}166} = x$

$4.4849 = x$

This number goes on and on, but this is enough to get the idea.

How many emails does Addie need to send out? She needs to send out 5 emails in the first round.

She needs more than 4.4, so that's 5.

This is another one of those situations where you need to think about the context of the problem. The answer here isn't 4.4849 emails.

Is there any way she can pull it off? **Yes** (circled) **No**

Why? Sure – she's got to know 5 more people!

Addie can also recruit more of Alex's friends if she needs people to send mail to.

5 friends? No problem... I'll get those mails right out.

9 days later...

You've helped Addie land a big check!

Addie's site cleared 5,000,000 hits, no problem. Her sponsorship deals on, the subscribers are pouring in, and Addie's off to get some great new gear from her local Apple store. Next up... a video campaign on YouTube!

there are no Dumb Questions

Q: Just put the problem in a calculator? Is that for real?

A: There are actually several ways to find roots of numbers. There are tables where you can look them up, there's even a way to find them by hand that looks like long division. But honestly, they're all really old school. For most folks, a calculator is perfect.

Another way to get near the root of a number is to remember the perfect squares (2x2=4, 3x3=9,etc.). Then you can get an idea of what numbres might at least be close to what you're looking for.

Q: What's the inverse operation of exponents? The radical?

A: Not quite. It's finding the root. The radical is the symbol for the operation. It's just like the dot symbolizing multiplication.

Q: What if I see a radical without an index number?

A: Assume an index of 2. That's the square root. It's convention that if there isn't an index, then the equation is talking about the square root.

Q: Can you have a fractional exponent?

A: Yes. That simply means you should take the root of the base. For example, if you see 1/2 as an exponent, it means square root. 1/3 would be the third root, and so on.

Q: My calculator doesn't have a 9th root button, what do I do?

A: You can write a root as fractional exponent. So, a ninth root can be written as $\sqrt[9]{734{,}166}$ or $734{,}166^{(1/9)}$

Most calculators have an exponent button. So you could just put in a root of (1/9) and get the same answer.

Q: Will I ever need to solve for an exponent and not the base?

A: Not anytime soon. There are more operations out there that you can use to do this sort of problem, but they're well beyond this book. Don't worry about it for now. (Isn't that good to hear!)

Q: What about an exponent of 0?

A: Any number raised to the 0 power is one. Why? If you go back to the division of exponents, you subtract the bottom exponent from the top exponent. If you end up with the same term on the top and the bottom, then it's the base to the 0 power. That's always the number "1."

Q: What about an exponent of 1?

A: Any number raised to an exponent of 1 is itself. That means an exponent of one is implied over EVERY number and EVERY variable. It can come in handy sometimes to know that.

Q: Can an exponent be negative?

A: Yes - it means that it's the exponent in the denominator. So $X^{-2} = \frac{1}{X^2}$

That ties right in with subtracting exponents again. Since there's no exponent in the numerator, it's a negative exponent.

Q: Can you use negative exponents to get rid of fractions?

A: Yes. If you have an expression with fractions in it, you just rewrite the expression with the denominators as negative exponents. This really only helps if you find working with exponents easier than working with fractions. Of course, some people prefer that, and it's a perfectly okay way to work.

This also works the other way: if you think fractions are easier than exponents, just pull out all your negative exponents and rewrite them as fractions.

Q: I've heard of something called the principle root. What's that about?

A: When we talk about finding roots, we're actually talking about finding the **principle root**. That's the positive root of a value. There are other roots to numbers, too, though. The most common is the **negative root**. For example, the principle square root of 9 is 3, but -3 is a square root of 9, too, since (-3)(-3) = 9.

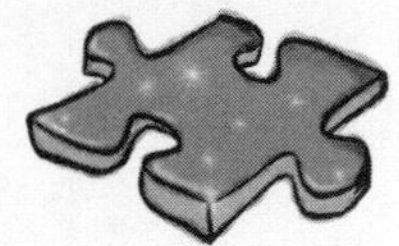

Exponentcross

Raise yourself! Can you get all the words?
They're all from this chapter.

Across

2. Any number raised to an exponent of zero is
4. Exponents are a faster form of
7. The number of times that the base gets multiplied is the
8. Any number raised to an exponent of one is
9. A fractional exponent is actually a
10. The inverse operation of an exponent is a

Down

1. A negative exponent means that the exponential term can be written as a
3. Another word used for exponent is
5. Exponents with the same base and exponent are
6. The number in an exponential term that gets multiplied

We've written the exponent operations that we've been talking about in terms of general variables. Match each expression to it's simplification.

Variable problem	The simplified version
$x^a \cdot x^b$	$x^{a \cdot b}$
$(x^a)^b$	$2x^a$
$x^a - x^b$	x^{a+b}
$\frac{x^a}{x^b}$	x^a
$x^a + x^b$	x^{a-b}
$2x^a - x^a$	The expression is already simplified.
$\frac{x^a}{y^b}$	

Who Does What? Solution

We've written the exponent operations that we've been talking about in terms of general variables. Match each expression to it's simplification.

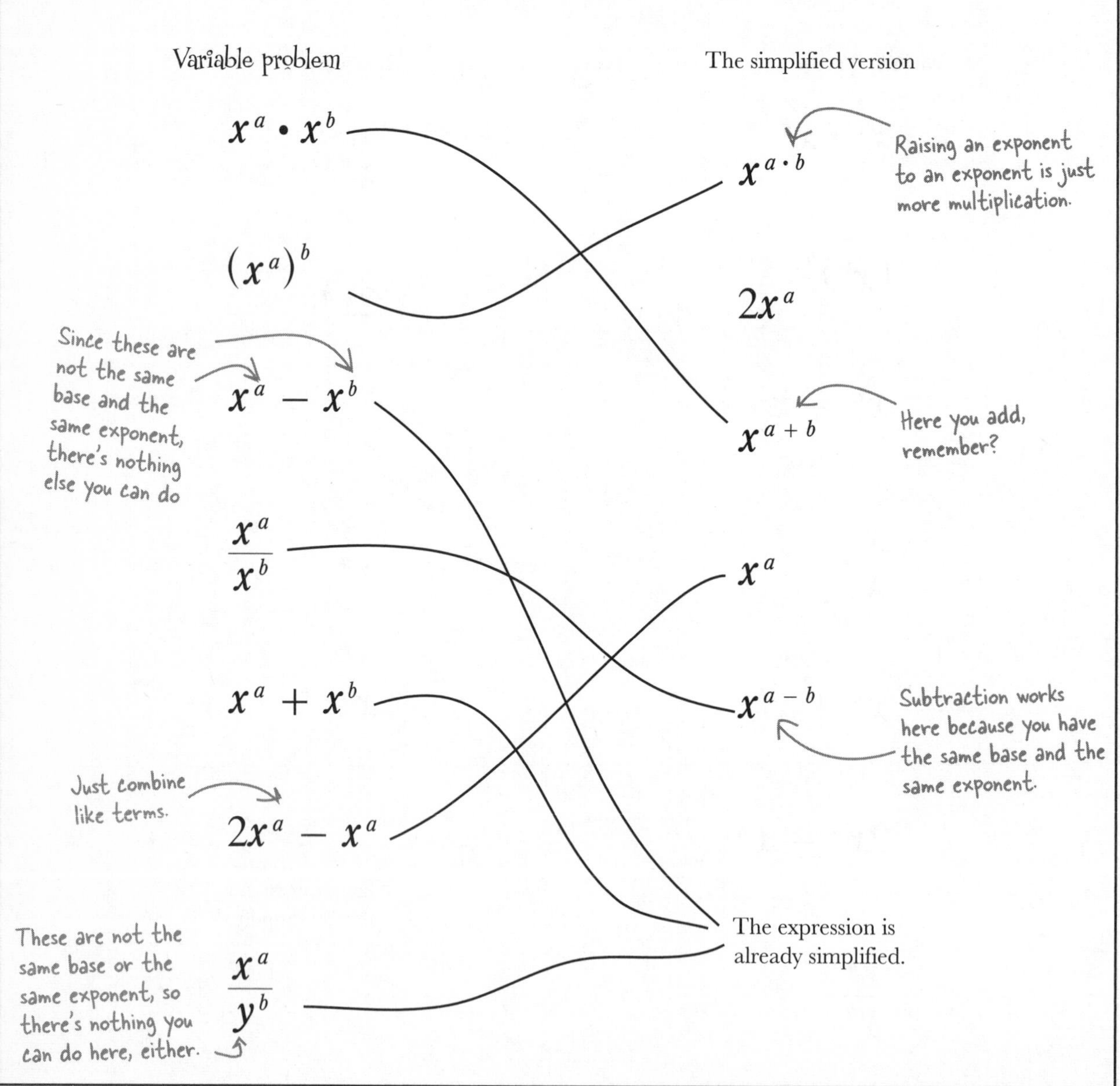

Variable problem
The simplified version
$x^a \cdot x^b$
$(x^a)^b$
$x^a - x^b$
$\frac{x^a}{x^b}$
$x^a + x^b$
$2x^a - x^a$
$\frac{x^a}{y^b}$
$x^{a \cdot b}$
$2x^a$
x^{a+b}
x^a
x^{a-b}
The expression is already simplified.
Raising an exponent to an exponent is just more multiplication.
Here you add, remember?
Subtraction works here because you have the same base and the same exponent.
Since these are not the same base and the same exponent, there's nothing else you can do
Just combine like terms.
These are not the same base or the same exponent, so there's nothing you can do here, either.

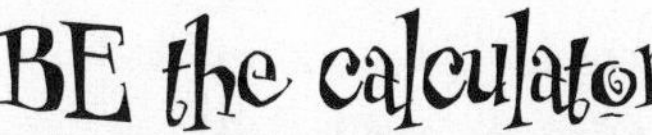

BE the calculator

Your job is to play calculator and crunch the numbers like your calculator would. You'll need to apply what you just learned about negative exponents and raising bases to zero. And since you're playing calculator, don't use one!

Just remember your exponent properties...

$1,467^0 + 1,856^1 =$ ....

....

....

Feel free to put in equals signs a few times – these could take a few steps.

$2^2 + 2^3 =$ ....

....

$2^2 \cdot 2^3 =$ ....

Try and come up with two ways to approach this one.

$\frac{5^7}{5^9} =$ ....

$=$ ....

There are a few different ways to do this one too – come up with two if you can.

$\frac{1}{3^3} + \frac{1}{3^3} =$ ....

$=$ ....

$=$ ....

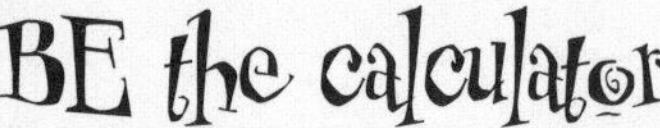

BE the calculator

Your job is to play calculator and crunch the numbers like your calculator would. You'll need to apply what you just learned about negative exponents and raising bases to zero. And since you're playing calculator, don't use one!

Any number to zero is one. Any number raised to one is itself.

$$1,467^0 + 1,856^1 = ?$$

$1 + 1,856 = 1,857$

These numbers can't be added without being the same exponent and the same base.

$$2^2 + 2^3 = 2 \cdot 2 + 2 \cdot 2 \cdot 2 = 4 + 8 = 12$$

All there is to do is simplify the exponents and then add them.

These terms are the same base and different exponents, but that's ok since we're multiplying.

$$2^2 \cdot 2^3 = 2^{2+3} = 2^5 = 32$$

When you multiply exponents, you can add them up first.

You didn't have to do all this, but there were a few right options.

The thing to remember to do here is subtract the exponents.

$$\frac{5^7}{5^9} = 5^{7-9} = \frac{1}{5^2} = \frac{1}{25}$$

$$= 5^{7-9} = 5^{-2} = 0.04$$

Then, just treat that negative exponent as a denominator (fractional notation or decimal both work).

The result is that you will get a negative exponent

There are a few different ways to do this one too.

Rewrite the fractions as negative exponents, and since they have the same base and the same exponent, they are like terms.

$$\frac{1}{3^3} + \frac{1}{3^3} = 3^{-3} + 3^{-3} = 2(3^{-3}) = 2\left(\frac{1}{3^3}\right) = \frac{2}{27}$$

$$= \frac{2}{3^3} = \frac{2}{27}$$

You can recognize them as like terms as fractions and add them first, then simplify.

$$= \frac{1}{27} + \frac{1}{27} = \frac{2}{27}$$

You could simplify them first, without taking advantage of exponent rules.

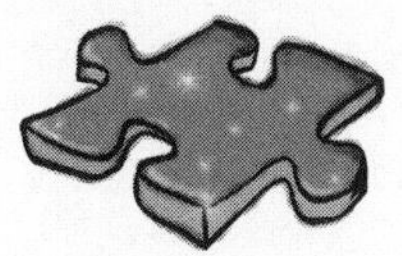

Exponentcross Solution

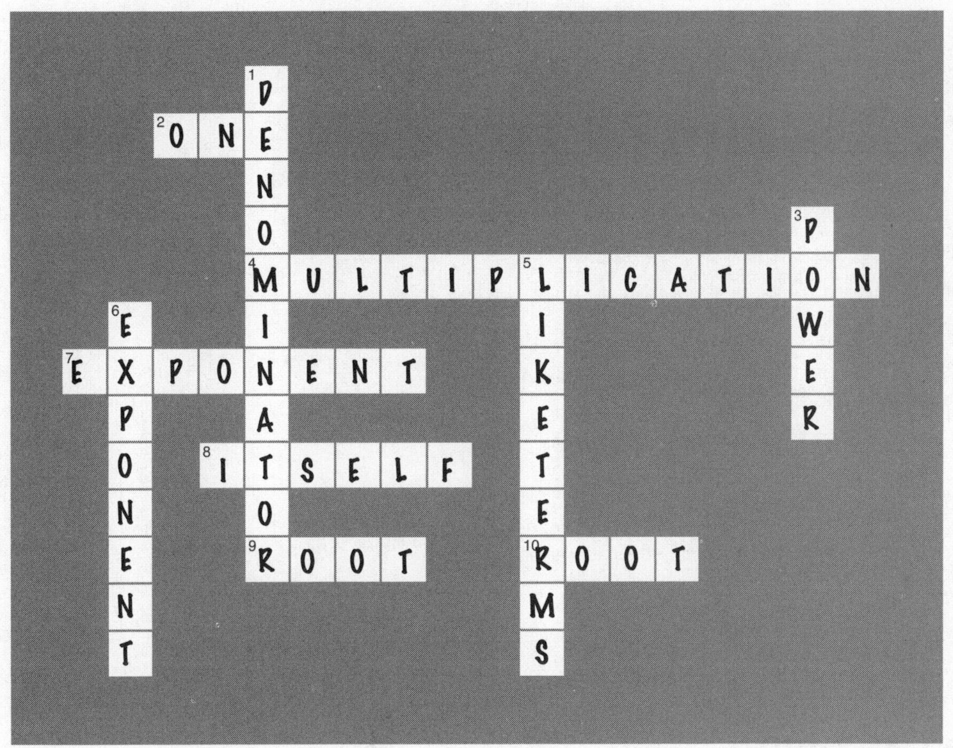

Tools for your Algebra Toolbox

This chapter was about numeric properties that are important to understand in order to work with equations.

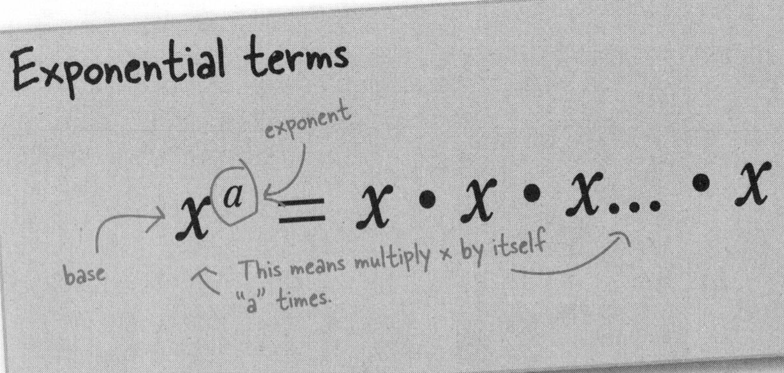

Exponential terms
exponent
$x^a = x \cdot x \cdot x \ldots \cdot x$
base
This means multiply x by itself "a" times.

$$x^a x^b = x^{a+b}$$

$$x^a y^a = (xy)^a$$

$$(x^a)^b = x^{ab}$$

$$\frac{x^a}{x^b} = x^{a-b} \text{ or}$$

$$\frac{x^a}{y^a} = \left(\frac{x}{y}\right)^a$$

$$x^0 = 1$$

$$x^1 = x$$

$$x^{-a} = \frac{1}{x^a}$$

These are the general forms for exponential operations for exponential terms of the same base and different bases.

BULLET POINTS

- Exponents are shorthand for **repetitive multiplication**.
- The **base** is the number that gets multiplied.
- The **exponent** is how many times the base is multiplied.
- To **add** or **subtract** terms with exponents, they must have the **same base** and the **same exponent**.
- Adding and subtracting those terms is just combining **like terms**.
- To **multiply** exponential terms with the same base, just **add the exponents**.
- To **divide** exponential terms with the same base, **subtract the exponents**.
- To **raise** an exponential term to an exponent, **multiply the exponents**.
- Rules for dealing with exponents apply to **numbers** and **variables**.

5 graphing

A picture's worth 1,000 words

I'll capture a lot more than a smile with this snapshot, just you watch.

Sometimes an equation might be hiding things.

Ever looked at an equation and thought, "But what the heck does that ***mean***?" In times like that, you just might need a **visual representation** of your equation. That's where **graphs** come in. They let you ***look*** at an equation, instead of just reading it. You can see where **important points** are on the graph, like when you'll run out of money, or how long it will take you to save up for that new car. In fact, with graphs, you can make **smart decisions** with your equations.

Edward's Lawn Mowing needs help...

Edward has been running his own lawn mowing and bush trimming business for a couple of years now.

Here's what his business looks like now:

* **Edward charges $12 per lawn.**
* **Edward has 7 current weekly customers.**
* **Edward mows each yard once a week.**
* **Edward gets paid weekly.**

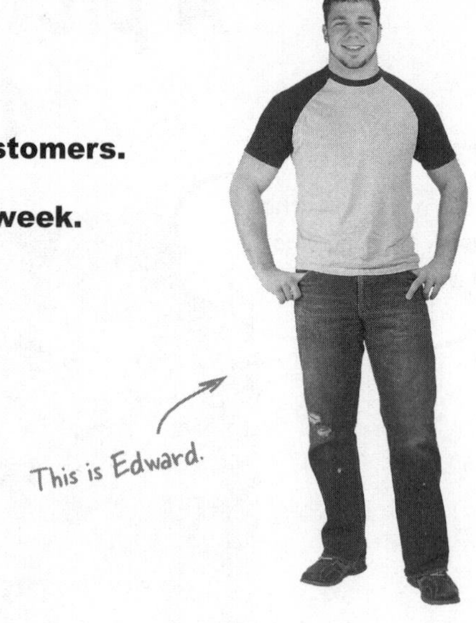

This is Edward.

Edward's Lawn Service

Edward has a list of new things that he wants to buy so he can expand his business—he's thinking long term. He'd like to find out when he'll be able to purchase each of these items:

* **Blade sharpener: $336**
* **Hedge trimmer: $168**
* **Bagger attachment: $504**

Edward's dying for some enhancements to his mower.

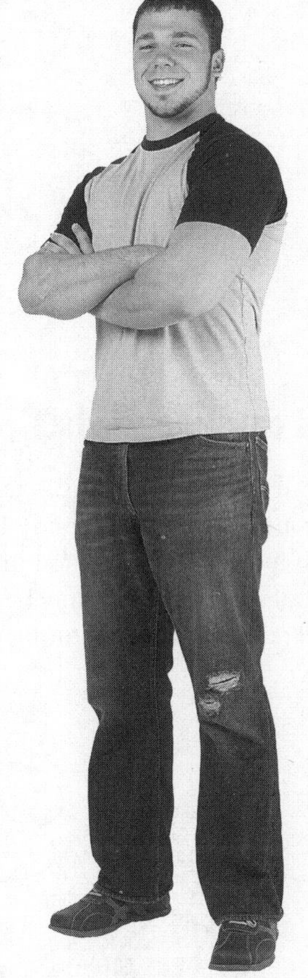

I need some help. I've got to do some planning, but I've got no idea how to do that...

Edward needs help to SEE what his financial future looks like.

Ed wants you to help him plan out when he can make future purchases, help him decide how fast he needs to add clients, and get his business financials organized.

You have all of Ed's information, his income and his clients, and a list of things he wants to buy. Sounds like an equation just waiting to happen...

BRAIN BARBELL

Using the information that Ed has given you, figure out the general equation for his income over the next weeks and months.

Use the letter "C" for Ed's cash and "t" for the number of weeks.

....

....

....

....

Just do your best, and turn the page when you're through.

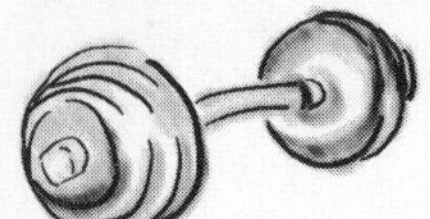

Brain Barbell Solution

Using the information that Ed has given you, figure out the general equation for his income over the next weeks and months.

This is how much Ed has made over some length of time, call it "C."

This is what changes, we'll call it "t" (for time).

Ed's cash = all lawns in a week times cost per lawn times weeks

This is known - he makes $12 per lawn, and he cuts 7 lawns.

$C = (7 \cdot 12)t$

$C = 84t$

Now Ed can know what his cash is at ANY time

The general equation that you've written can work in two different ways because you have two variables. If Ed has a time when he wants to know how much money he'll have, you just substitute that time for t, and solve for C. Or, if Ed knows how much money he wants, you can tell him when he'll get there by substituting the amount in for C and solving for t.

So when can I get that hedge trimmer?

Substitute for C, solve for t.
Just substitute the value of the hedge trimmer in for C and solve for t.

Ed's general cash flow equation

The hedge trimmer costs $168.

$$C = 84t$$

$$\frac{168}{84} = \frac{84t}{84}$$

fill this in

$\square = t$

This is how many weeks Ed needs to cut lawns to get his hedge trimmer.

Answer: 2

Ok, great. But what about the
blade sharpener? Or the bagger?
I need to know when I can start
thinking about those, too.

Sharpen your pencil

Figure out how many weeks it will take for Ed to earn enough for the blade sharpener or the bagger attachment. Then Ed will have an idea of what he can do.

Time for a blade sharpener: ....

Time for a bagger attachment: ....

Sharpen your pencil Solution

Figure out how many weeks it will take for Ed to earn enough for the blade sharpener or the bagger attachment. Then Ed will have an idea of what he can do.

Time for a blade sharpener: The blade sharpener will cost $336, so that's C.

$$C = 84t$$

This is exactly the same process we did before. We plug in how much cash Ed needs, and then solve for number of weeks, t.

$$\frac{336}{84} = \frac{\cancel{84}t}{\cancel{84}}$$

Ok, so 4 weeks to save up for the blade sharpener.

$$4 = t$$

Time for a bagger attachment: The bagger is expensive, it goes for $504

Same thing again. Just put in 504 for C.

$$C = 84t$$

$$\frac{504}{84} = \frac{\cancel{84}t}{\cancel{84}}$$

Well, it makes sense that it would take the longest to save for this one, the bagger is $504.

$$6 = t$$

Great, we get it, okay? And if Ed wants a edger, we do it again. New blades, again. And the next thing, and the next thing... isn't there a way to NOT do this over and over again?

Why don't you just SHOW me the money?

What if we could come up with a way where we could look up a value, like the amount an accessory costs, and then *see* what t was for that amount? In fact, there *is* a way to show all the possible "what if's" that you can have for an equation. A **graph** allows you to draw all of the possible points for an equation and then look up different points as you need them.

Then you can see how much money Ed will have at any given time and tell him if he can afford something, without having to solve that same equation again and again. Let's start by taking the information we know and drawing it on a grid:

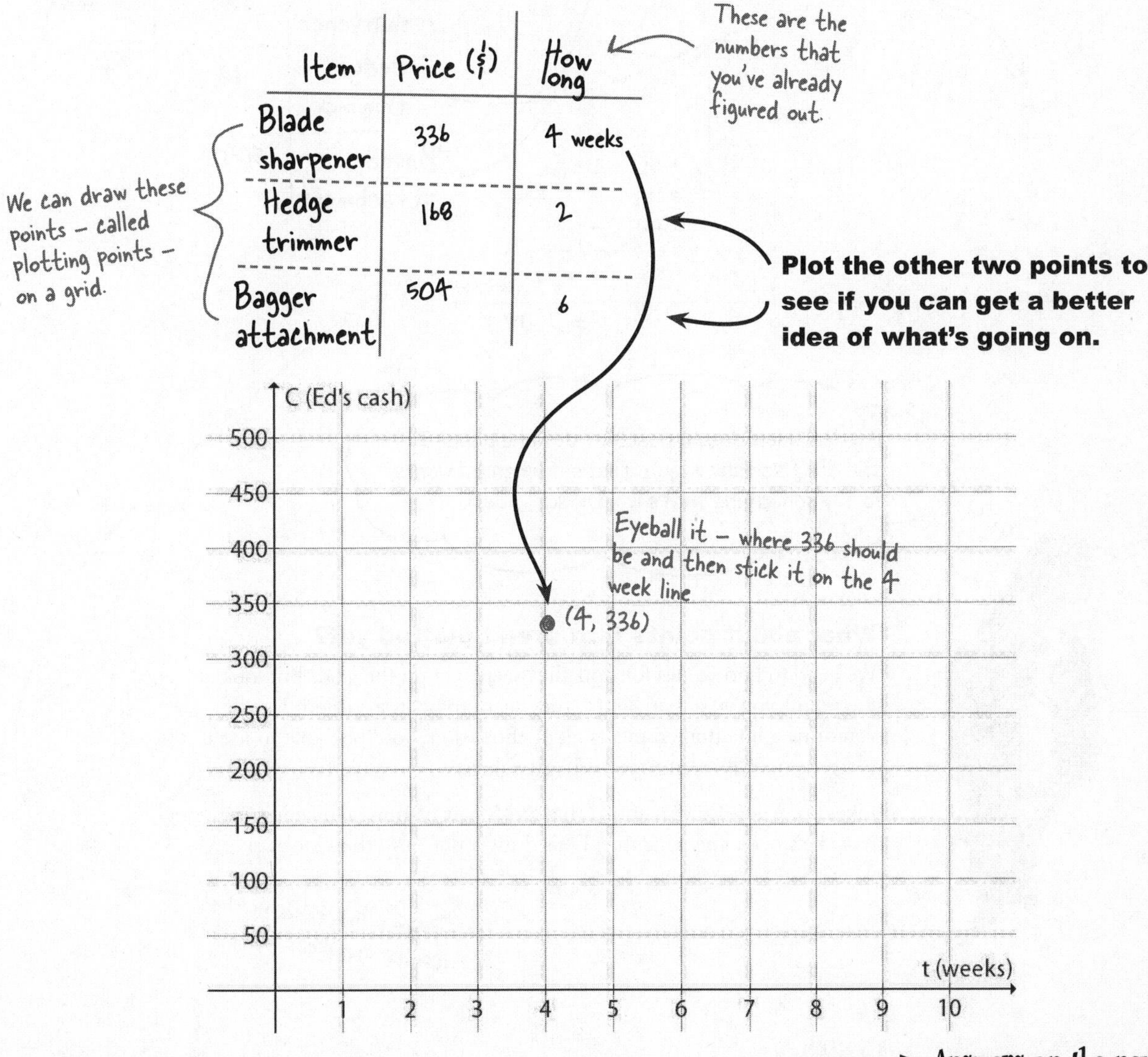

Answers on the next page.

Now we can LOOK at Ed's cash pattern

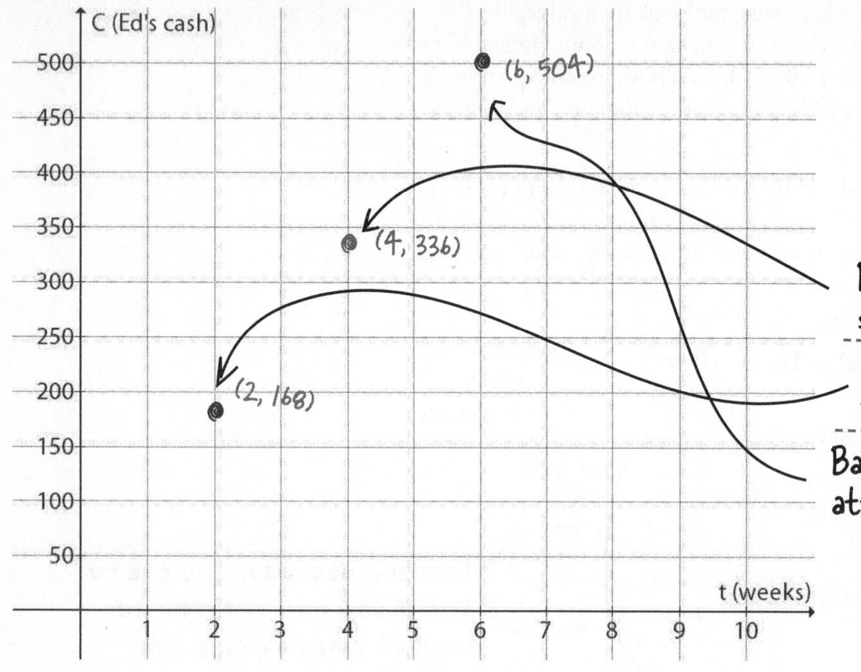

Big deal. How does a bunch of dots help me? I still don't see any way to find out when Ed wants to buy something that's worth, say, $245.

What about points that aren't plotted yet?

We need to find values for dots that aren't yet on the grid. But look at the points you've already got... they seem to be in a straight line. If you draw a line that connects all of those dots, you'll be able to use it to figure out different time and cost values.

Like, say, when Ed can afford that $245 pair of noise canceling ear buds he's been checking out. Let's draw a line, and then look up where $245 crosses our line.

Draw a line through the points. Go ahead...

Graphs show an ENTIRE relationship

Once we add the line to the picture of the points that we've figured out, it turns out we've drawn a graph of the relationship between ***C*** (Ed's cash) and ***t*** (the number of weeks he's been cutting grass):

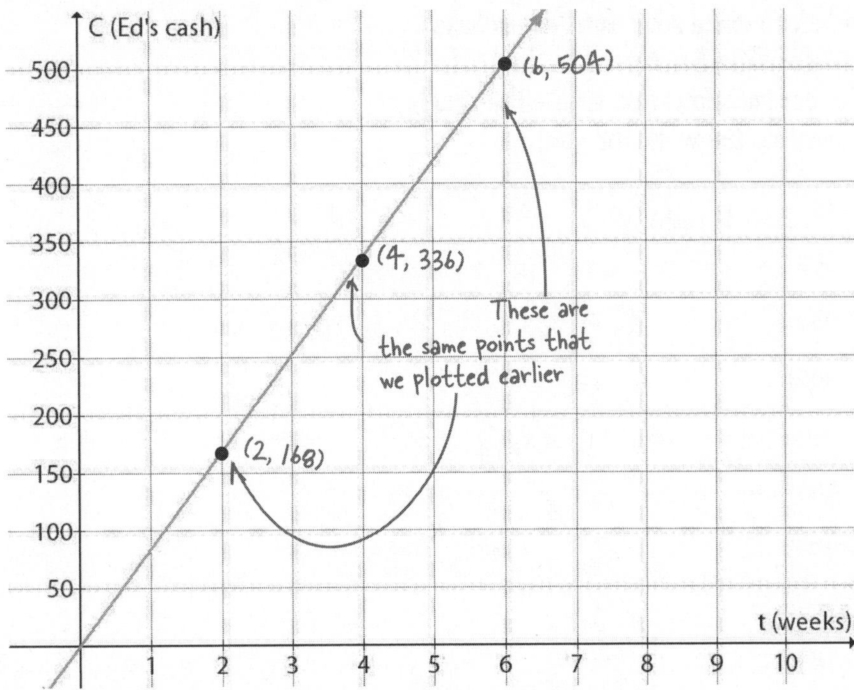

There's something else that covers the entire relationship

We also have an **equation** that covers the entire relationship between ***C*** and ***t:***

$$C = 84t$$

It turns out that what we have—the graph with the line and everything—is actually a ***graph of the equation***. It shows us the equation and lets us look at how ***C*** and ***t*** are related.

On top of that, this graph shows a **trend** for the equation: the general direction that the relationship is heading. Ed's graph is trending upward. That means he's going to have made *more* money as time goes on (meaning he'll keep moving lawns, save more cash, and so on).

Now, when Ed wants to buy something, he can just look up the value for ***C*** he wants. Let's see exactly how that works...

BE the planner

Your job is to play financial planner. Use your cash-flow graph for Ed to figure out when he can make some future purchases. You should not need to do ANY computations this time - the graph will do the work for you!

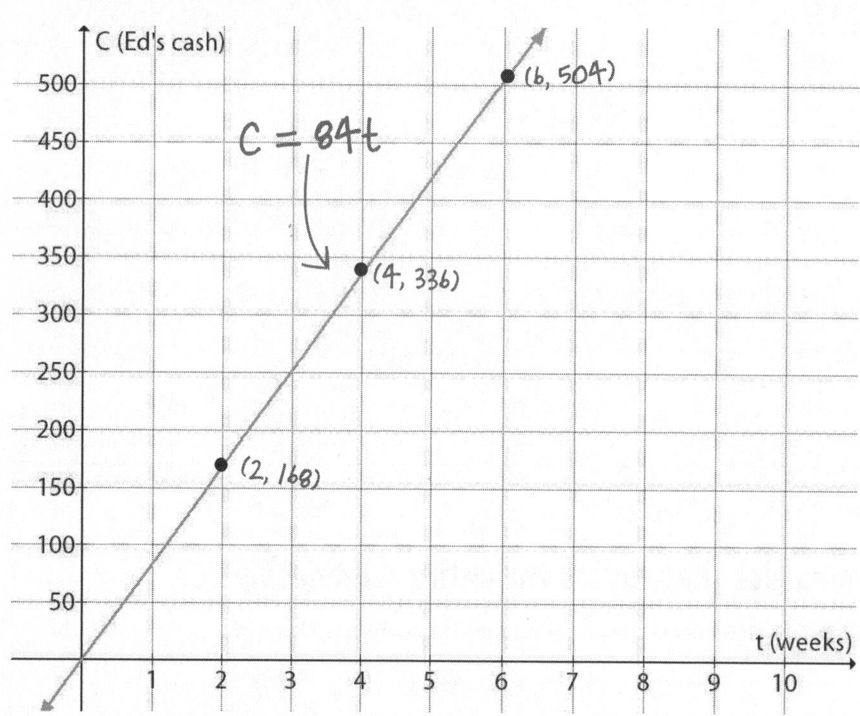

How long until Ed can get those $245 noise canceling ear buds?

Don't get freaked out about exact numbers here - just get to the closest week.

Ed is thinking he's going to need a new blade. It's $375. When can he get that?

there are no Dumb Questions

Q: Is the graph of the equation the line or the points?

A: Both. A line is made up of an infinite number of points. So the points that we computed for Ed were just a few from the relationship. Once we draw a line, that's the graph of the entire equation.

Equations and graphs both demonstrate a relationship between variables. In this case, the variables are ***C*** and ***t***. Graphs and equations are just different ways of showing the same thing.

Q: How do I know where to plot my points if they don't fall on an exact line on the grid?

A: Don't stress! Just look at the numbers on the **axis** (that's the line at the edge of the graph that tells you what numbers go on what grid line) and estimate where your point should go. As long as you're consistent about being close, the graph will be good enough to use.

Another thing to think about is that graphs won't always be as big as Ed's is. He's thinking pretty long term. If you've got a graph that you can work with just between 0 and 10, for example, it's going to be much easier to be exact with that smaller range of numbers.

Q: What's is exactly is a trend again?

A: A trend is just the general direction of a line. If a graph is heading upward, that means that as one variable increases, so does the other. And if the line is heading downhill, as one variable increases, the other variable decreases.

Q: How do I know which number to plot using the bottom axis, and which number to plot with the side axis?

A: Usually, each axis on your graph is labeled, like "time" or "number of weeks" or "Ed's cash." Once you see that, you'll be able to plot each value along the right axis line.

If not, you should label them! If your equation is in terms of ***x*** and ***y***, then the ***x*** is horizontal and the ***y*** is vertical. In the case where your variables are different, hang on—you'll be learning how to identify the structure of a linear equation, and then you'll be able to see which variable is acting like the ***x*** and is horizontal.

Q: Can a graph show *any* variables?

A: Just like with equations, you can use any variable you want. ***x*** and ***y*** are the most common, with ***x*** typically being on the horizontal axis, and ***y*** on the vertical axis, but you can use anything you like.

Q: Do we have to figure out the points first every time? Or can we just draw the line right away somehow?

A: You don't have to always plot points first. We're going to learn some methods that you can use that don't require ANY computations at all. Then, you'll be able to graph an equation by just looking at the equation. But you're going to need some more information first...

Q: Can a graph show any variables?

A: Just like with equations, you can use any variable you want. ***x*** and ***y*** are the most common, with ***x*** typically being on the horizontal axis, and ***y*** on the vertical axis, but it can be anything.

A graph and an equation are just different ways of looking at a relationship between two variables.

BE the planner solution

Your job is to play financial planner. Use your cash-flow graph for Ed to figure out when he can make some future purchases. You should not need to do ANY computations this time - the graph will do the work for you!

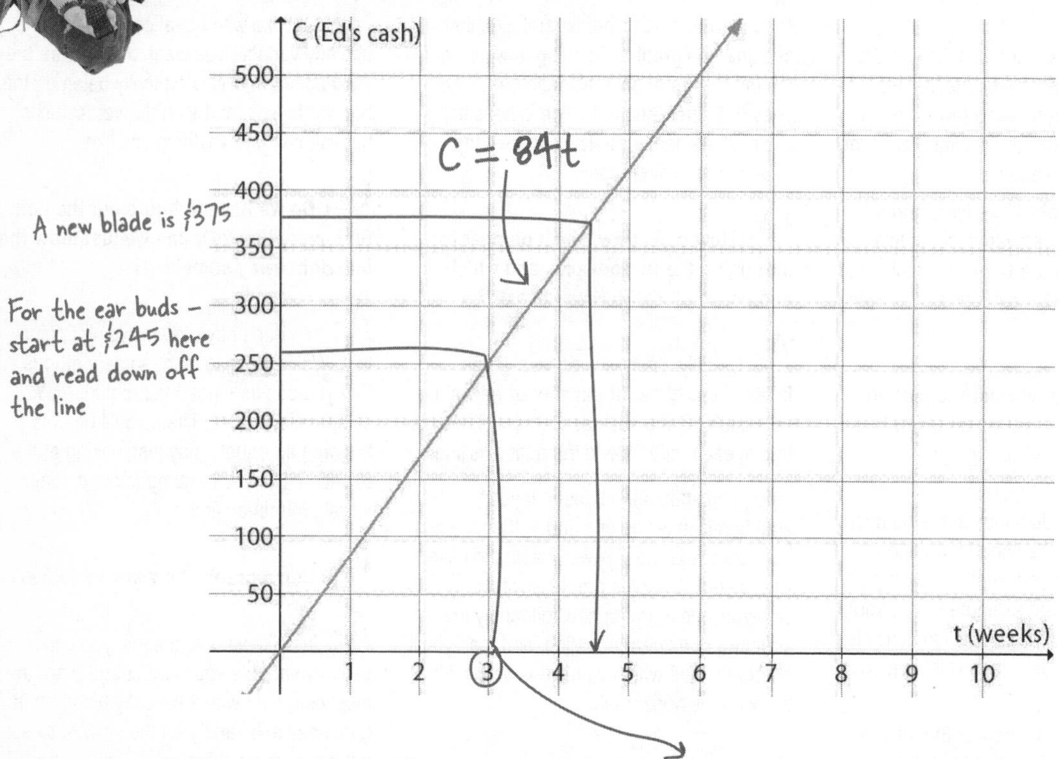

How long until Ed can get those $245 noise canceling ear buds?

After 3 weeks

Ed is thinking he's going to need a new blade. It's $375. When can he get that?

Between 4 and 5 weeks. So in the real world, it'll probably be after week 5

Ok great, now I can see what's coming up and plan some expansion.

The graph gives you all the answers.

Just by looking, you can let Ed know the details about when he can afford things. Now he's going to get started with his mowing season and save up to buy new lawn accessories.

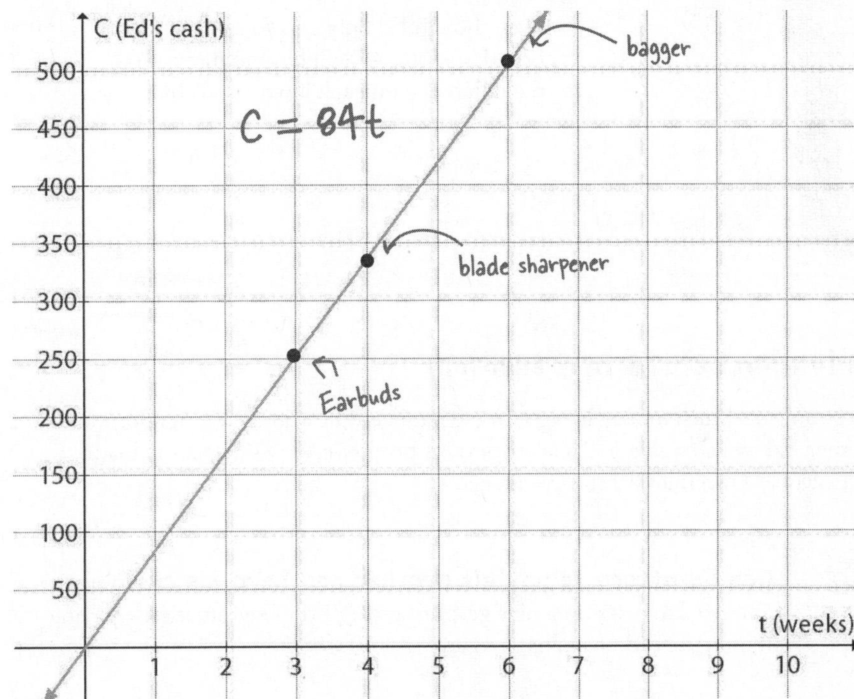

But sometimes things change... like hurting your leg in a freak weed whacker accident...

I broke my leg running over a pothole, and it's going to be 10 weeks 'till my cast comes off. But my customers need their grass cut!

Ed needs to subcontract.

Ed can't afford to lose all of his customers this early in the summer. He's only 3 weeks in and will be out of commission for 10 weeks. Luckily, his brother agrees to help him out... for $19 a lawn!

Ed's brother knows that Ed is stuck!

Even though Ed only gets $12 a yard, Ed's desperate, so he'll have to make up the difference on each lawn out of his own pocket.

A new situation needs a new equation

Ed has 3 weeks worth of lawn mowing money in the bank now...

Ed need to know how much money he's got in the bank and how long he can afford to pay his brother. Ed was charging $12 a lawn, and his brother costs $19, so it's going to cost Ed $7 extra per lawn until his cast comes off.

- **Can Ed afford to pay his brother and keep his customers?**
  You need to work up a new graph showing Ed's new situation—paying money instead of making money—for 10 weeks.

- **Will Ed run out of money before his cast comes off?**
  This is only a 10 week situation, but Ed doesn't have a ton of cash. At the end of the 10 weeks, what will Ed have left?

Work with Ed on a recovery plan. You need to figure out what Ed has, what it will cost him to hire out his brother, and when Ed will run out of money.

Write Ed's new cash equation: Ed's Cash = Ed's savings + Ed's income - Ed's brother's cost ....

Use C and t like we did before - but you also need to keep in mind how much money Ed started with and subtract what he'll spend.

....

....

....

When does Ed run out of money? ....

That will happen when C = 0

....

....

....

Will his cast be off? (circle one) **Yes** **No**

If Ed will still be in a cast, how can you figure out how far into debt he'll go? ....

Just jot down some ideas - no numbers or anything - of ways you could figure it out.

....

....

Work with Ed on a recovery plan. You need to figure out what Ed has, what it will cost him to hire out his brother, and when Ed will run out of money.

Ed's Cash = Ed's savings + Ed's income − Ed's brother's cost

This is "C."

Ed made 7•12 a week for 3 weeks, that's 3•7•12 = 252

84t

Ed's brother = 19 per lawn•7 = 133 lawns per week = 133t

We figured this out last time

$C = 252 + 84t - 133t$

$C = 252 - 49t$

When does Ed run out of money?

$0 = 252 - 49t$

If we solve the equation for C = 0, the t will be when that happens.

$49t + 0 = 252 - 49t + 49t$

$\frac{49t}{49} = \frac{252}{49}$

This decimal goes on for a while, but it doesn't matter. What you're after is the number of weeks, so 5 is the answer we need.

$t = 5.142...$

Yes **(No)** ← Ed's cast is staying on for 10 weeks, so he's only going to be halfway through that before he's out of money.

If he'll still be in a cast, how can you figure out how far into debt he'll go?

There's no wrong answer here – we just wanted to get you thinking...

How can I see how far into debt I'm going? Can you make me another graph?

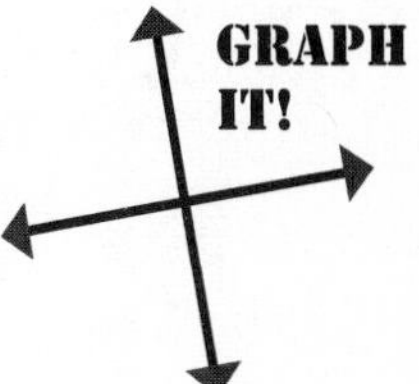

Draw a graph of Ed's new equation.

$C = 252 - 49t$
If $C = 0$, then $t = 5.142$

You need one more point, so you can draw a line. Try setting t = 0; it's easy to solve for. Once you plot those two points, you can just draw the line, and that's the graph of the equation.

....

← Here's some workspace if you need it.

....

C (Ed's cash)

500, 450, 400, 350, 300, 250, 200, 150, 100, 50

t (weeks)

1, 2, 3, 4, 5, 6, 7, 8, 9, 10

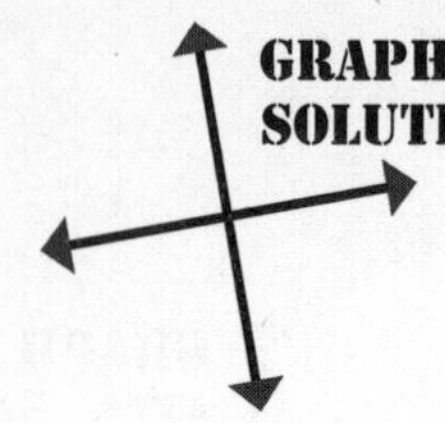

Draw a graph of Ed's new equation.

$C = 252 - 49t$

If $C = 0$, then $t = 5.142$

You need one more point so you can draw a line. Try setting $t = 0$; it's easy to solve for.

$C = 252 - 49t$

$C = 252 - 49(0)$ So, for $t = 0$, $C = 252$

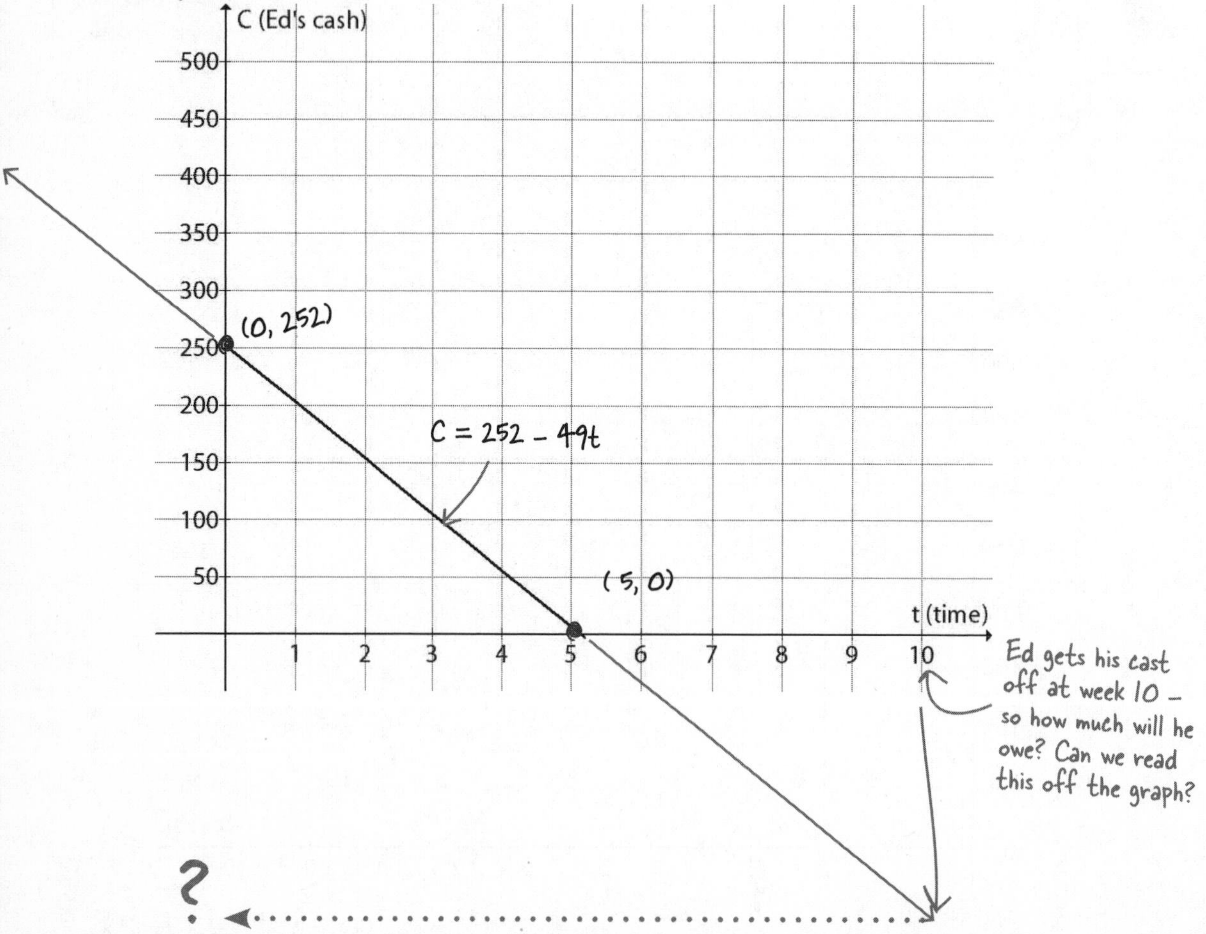

C (Ed's cash)

500
450
400
350
300
250
200
150
100
50

1 2 3 4 5 6 7 8 9 10

t (time)

(0, 252)

$C = 252 - 49t$

(5, 0)

Ed gets his cast off at week 10 – so how much will he owe? Can we read this off the graph?

?

We need to expand the graph to read that last value, right? If the C values went lower, that would work... but he's in debt, so C will be less than zero.

The Cartesian Plane allows values to go BELOW zero

Lots of graphs need to show negative numbers. The math standard for a graph is called the ***Cartesian Plane.*** Using the Cartesian Plane, both of the axis values can go negative, which means your values can also be negative, or less than zero.

Here's what the Cartesian Plane looks like:

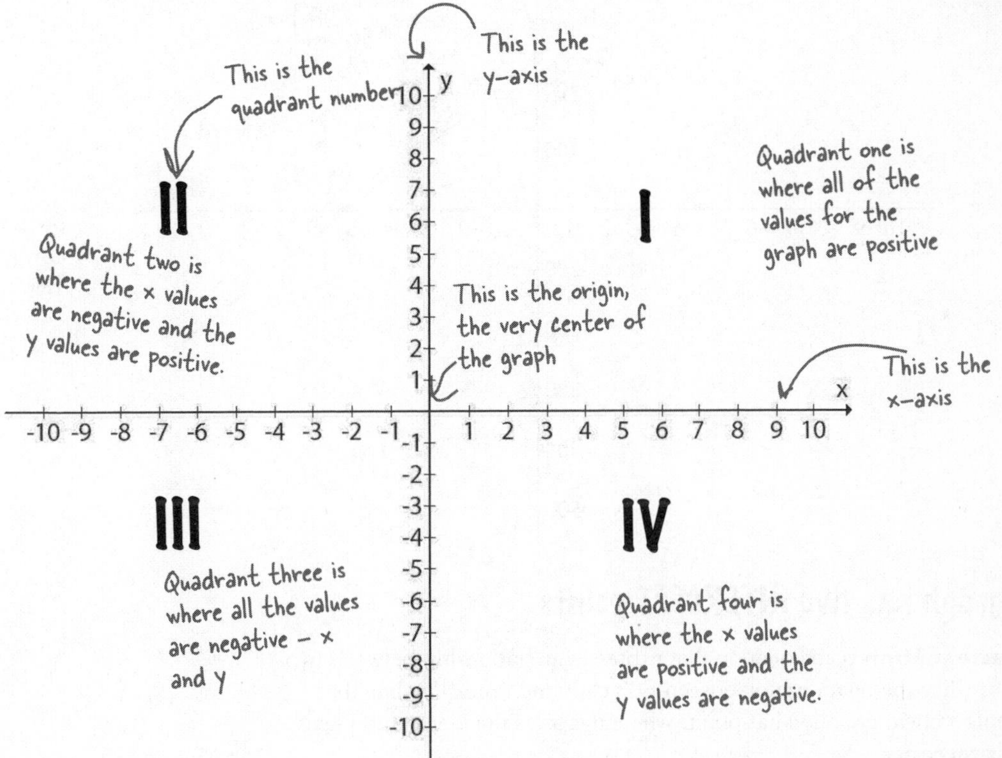

We need negative values for Ed's debt equation...

Let's graph Ed's equation on the Cartesian Plane

Just because we started with a smaller portion of the plane for Ed's graph doesn't mean we need to stay there. If we put Ed's graph on the Cartesian Plane, we'll be able to read off the value we need and figure out how far Ed is going to go into debt.

When we first started with Ed's graphs, we just made up a grid and plotted the points we knew. Each point was actually an **ordered pair**: one number followed by another number. We write those like this: (0, 252). The first number is the horizontal number, the second is for the vertical, and the parentheses say the numbers are connected. So each point for Ed's graph is (***t*, *C***), where ***t*** is time, and ***C*** is Ed's cash:

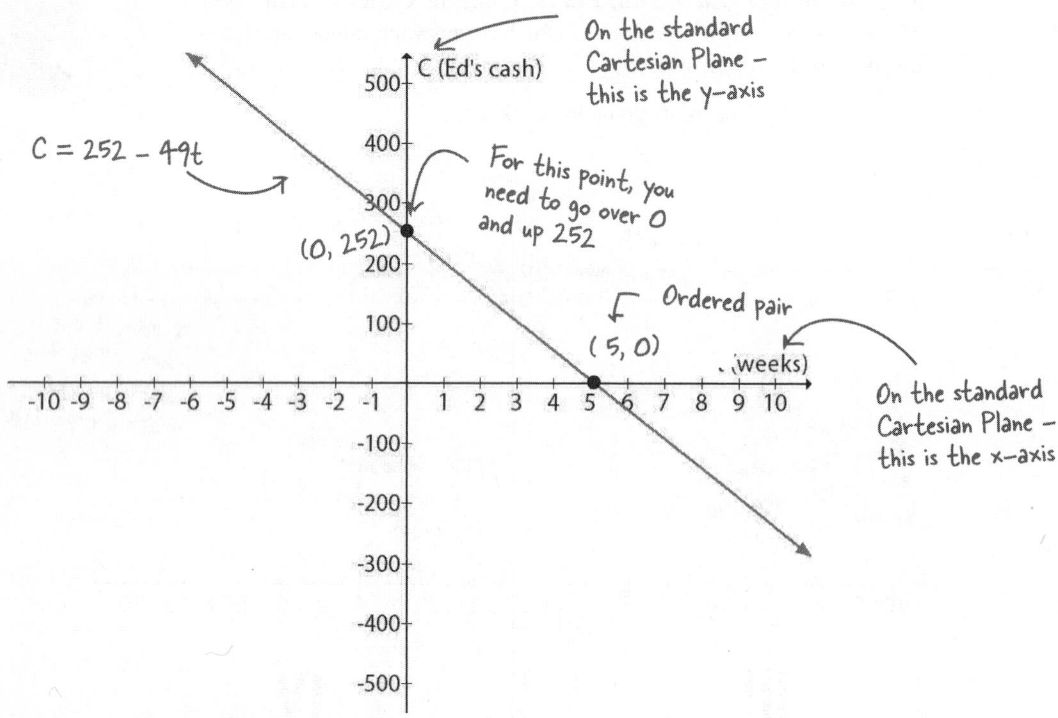

Ed's graph has two INTERCEPT points

A **linear equation** is an equation that expresses a relationship between two variables—like the relationship between Ed's cash and time. The line that represents a linear equation has points where it crosses the x-axis and y-axis, called **intercepts**.

Where ***C*** = 0 is the ***t***-intercept, which is where the line crosses the ***t***-axis. And where ***t*** = 0 is the value for the ***C***-intercept, where the line crosses the ***C*** axis. These intercepts are usually called ***x***- or ***y***-intercepts because ***x*** and ***y*** are the standard labels for the horizontal and vertical axis lines.

Hey, how can you just extend the lines into those other quadrants? We don't have any points to plot over there.

Lines go on forever.

After you've figured out any two points for an equation, you can draw a line through those points, and you've graphed your linear equation. But lines don't just stop. If nothing changes, they'll go on forever—to infinity.

It makes sense that two points make a line, but why? Because to be a specific line, you need a point and a direction. If you plot just one point, you can draw lines in all different directions that go through that point. Once you've picked a second point, you know which direction that line has to go to hit both of them.

To graph a straight line:

1. **Plot two points that satisfy your equation.**

2. **Draw a straight line through (and past) your two points.**
   Lines go on forever in both directions, so the line you draw has to go past both of your plotted points.

3. **Add arrows on both ends of the line.**
   The arrows indicate that the line goes past the part of the equation represented on your graph.

The truth about linear equations...

A ***linear equation*** is an equation that defines a line. That means any equation of this type, when you graph it, will produce a line. You can identify a linear equation by looking at it: if it has one or two variables, those variables have an exponent of 1, and all of the terms are constants or constants multiplied by variables, it's a linear equation.

Once you have looked at an equation and determined that it's linear, plot two points and draw your line. Start by setting one variable to 0, and solve for the other variable. Then flip the variables: assign the second one to zero, and solve for the first. This will give you your two intercepts. Then you can draw a line through the intercepts, and you've got your graph!

Ed's equation is linear:

$$C = 252 - 49t$$

The terms are all either one of the two variables multiplied by a constant or just a constant

there are no Dumb Questions

Q: Why are intercepts such a big deal?

A: Because they are one of those things that makes life easier. Ever noticed that when you throw a zero into the mix, equations seem to get easier? Since the ***x***- and ***y***- intercepts both allow you to set one coordinate in an equation to zero, they make finding a point to plot pretty easy.

Q: I've heard of a table of values, what is that all about?

A: A table of values is a more formal way of solving an equation to get points to use on a graph. Typically, you set up a table with columns for the ***x*** value, the ***y*** value, and the equation. You plug in values for ***x*** and solve for ***y***, and then vice versa. In fact, that's a lot like we did with Ed... just a little more formal.

The big difference between using a table of values and solving just for the intercepts is speed. You're only solving for two, easy-to-find points with intercepts, and that's usually pretty quick to do.

Q: What about equations with more than two variables?

A: That's a 3D graph, and we won't be doing any of that! You don't need to worry about those types of graphs in Algebra.

Q: Is there a way to check my graph?

A: Yes. The easiest way is to solve for another point and make sure that it's on your line. For our example, if you substitute x = -1 and solve for y, the y-value you come up with should be on your line. If it's not, something's wrong.

Q: Why is that grid called the Cartesian Plane?

A: This standard form of a grid was created by a guy named Rene Descartes in 1637 as part of his work to merge Algebra and Geometry. It works perfectly since we're going to be creating shapes (like lines) that can be described by Algebraic equations.

Q: Why are the quadrants written in roman numerals?

A: That's just the standard way all mathematicians talk about graphs—using roman numerals.

Q: Is there a standard variable for each axis?

A: Yes, typically ***x*** is the horizontal axis, and ***y*** is the vertical. That doesn't mean they have to be, though. Ed's equation used ***C*** and ***t***, and that was okay, too.

The Cartesian Plane just shows a relationship between two variables. You can either swap out the variables in your equation for ***x*** and ***y***, or you can re-label each axis in the graph.

BULLET POINTS

- The typical variables for a graph are ***x*** for the horizontal axis, and ***y*** for the vertical axis.
- ***x***-intercept is the point where the graph crosses the ***x***-axis (***x***, 0).
- To solve for the ***x***-intercept, set ***y*** = 0 and solve for ***x***.
- The ***y***-intercept is the point where the graph crosses the ***y***-axis, (0, ***y***).
- To solve for the ***y***-intercept, set ***x*** = 0 and solve for ***y***.
- Ordered pairs look like (***x***, ***y***). The horizontal axis goes first, and then the vertical.
- Lines are defined by **two points** and go on **forever**.

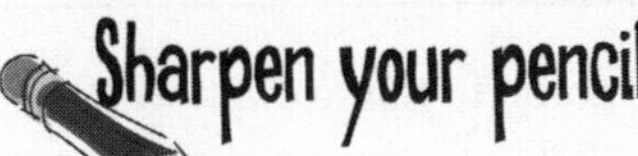

Using the full Cartesian plane, read Ed's graph to figure out how much he's going to owe 10 weeks from now.

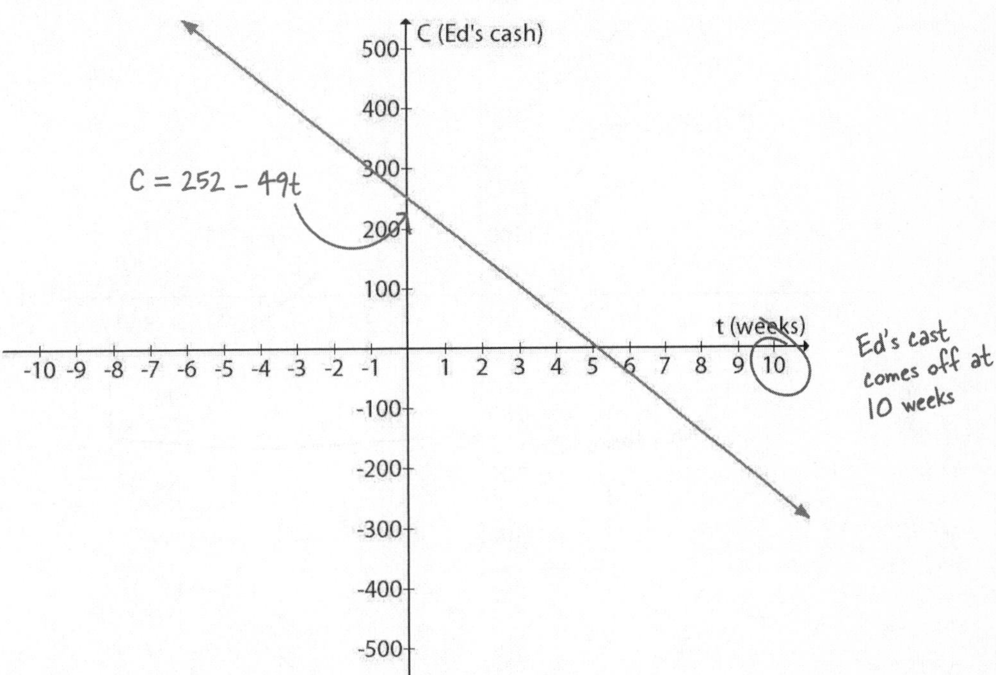

How much is Ed going to owe? ....

Just get this from the graph. You don't need a super-exact number.

....

....

After Ed's cast comes off, he's going to go back to earning $84 a week.
How long is it going to take him to get out of debt? ....

To solve this, you have to go back to Ed's original equation, $C = 84t$.

....

....

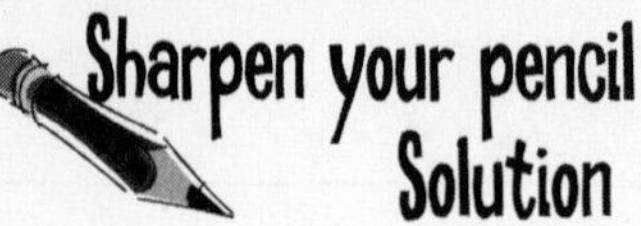

Sharpen your pencil Solution

Using the full Cartesian plane, read Ed's graph to figure out how much he's going to owe 10 weeks from now.

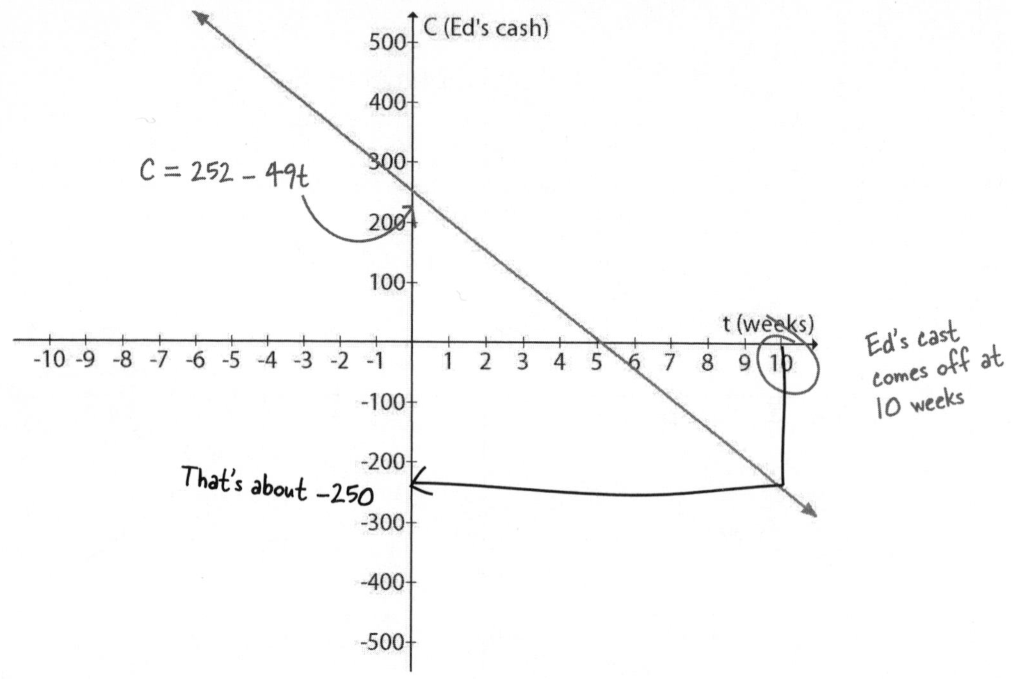

How much is Ed going to owe? Ed is going to be about $250 in debt.

$C = 84t$

After Ed's cast comes off, he's going to go back to earning $84 a week. How long is it going to take him to get out of debt?

$250 = 84t$

$250 is the new thing we're finding out time for, so we substitute it into the equation, just the way we did before.

$\frac{250}{84} = \frac{84t}{84}$

$2.97... = t$

That's just about 3 weeks, wow. That means Ed is going to be 16 weeks into summer (including cast time and everything) just to be back at 0.

Well, 3 weeks wasn't too long, and now that my cast is off, maybe I can get more customers and make up for my lame start.

Ed needs to take on new clients.

Ed needs to get his summer back on track. He's into June and healed up, but he owes money and wants to make up for lost time. Ed's created a new form where he details each lawn, so he can give potential customers new quotes.

Ed's focusing on how hilly lawns are, so he can charge more based on how steep a lawn is. Here's his new form:

Edward's Lawn Service

New Customer Estimate

Customer name
& address: ______________________

		Fee
Slope Evaluation:	**Slope = 0**	$12
	Slope > 0	$20
	Slope < 0	$15

Overall lawn cost: ______

How does he figure that out?

Ed's figuring out the SLOPE of lawns

To make sure he's consistent in figuring out how steep a lawn is, Ed's developed a system. He starts at the street and measures some key features of the lawn. Then he puts that information together and turns his information into a number for the ***slope*** of a lawn.

Here's how it all works:

Ed's slope = $\frac{\text{Rise}}{\text{Run}}$

Rise: How far up (or down) from the street the hill goes.

Run: How far over from the street the hill goes.

For the lawn below: Slope = $\frac{\text{Rise}}{\text{Run}} = \frac{\text{10 feet up}}{\text{5 feet over}} = 2$

That's greater than 0, so that means this lawn is going to cost $20, based on Ed's new pricing.

Edward's Lawn Service

New Customer Estimate

Customer name & address: **Mrs Ethel P. Humphries, 8 Infinite Loop**

Slope Evaluation:		Fee
	Slope = 0	**$12**
	Slope > 0	**$20**
	Slope < 0	**$15**
	Overall lawn cost:	**$20**

10 feet high

This is the street here.

0 feet high

5 feet

WHAT'S MY SLOPE?

Figure out the slope for each of Ed's new clients and how much he's going to charge. Remember: Rise over run!

Graph	Slope	Lawn Charge

Look back at Ed's chart.

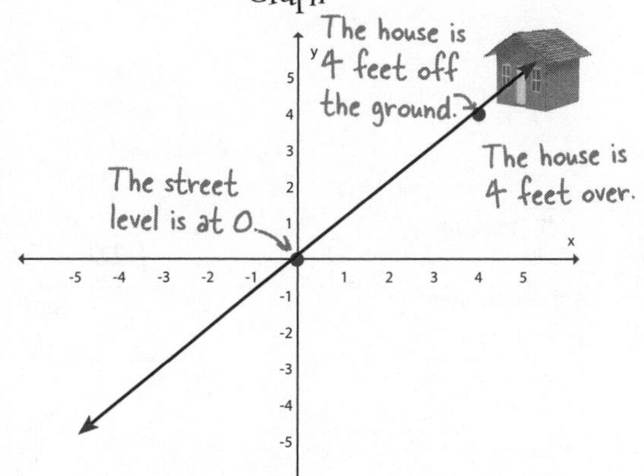
The house is 4 feet off the ground.
The house is 4 feet over.
The street level is at 0.

$\text{Slope} = \frac{\text{Rise}}{\text{Run}} = \frac{\ldots\ldots}{\ldots\ldots} = \ldots\ldots$ ....

The street level is at 4.

The house is 0 feet off the ground.

The house is 2 feet over.

Remember – up is positive and down is negative

$\text{Slope} = \frac{\text{Rise}}{\text{Run}} = \frac{\ldots\ldots}{\ldots\ldots} = \ldots\ldots$ ....

This point is 2 feet up.

The street here is 2 feet up.

4 feet over

$\text{Slope} = \frac{\text{Rise}}{\text{Run}} = \frac{\ldots\ldots}{\ldots\ldots} = \ldots\ldots$ ....

What's My Slope? Solution

Figure out the slope for each of Ed's new clients and how much he's going to charge. Remember: Rise over run!

Graph | Slope | Lawn Charge

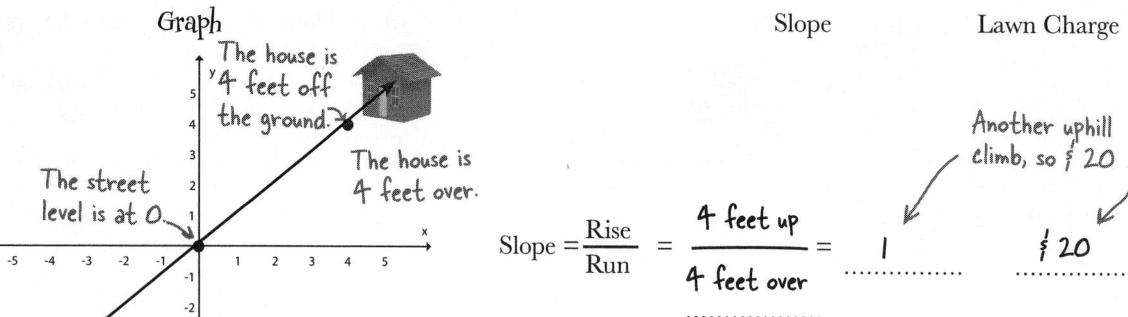
The house is 4 feet off the ground.
The house is 4 feet over.
The street level is at 0.
Another uphill climb, so $20
Slope = Rise / Run = 4 feet up / 4 feet over = 1
$20

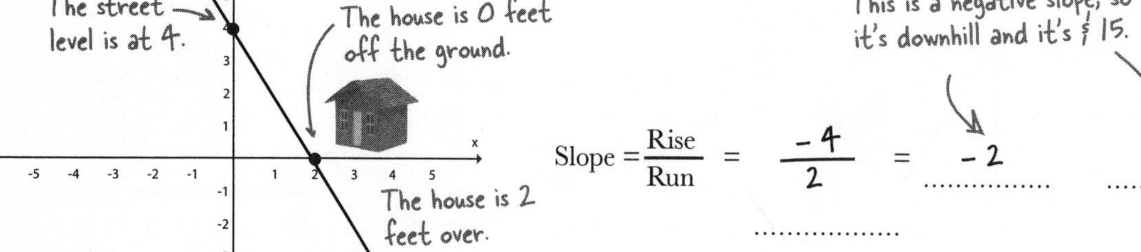
The street level is at 4.
The house is 0 feet off the ground.
The house is 2 feet over.
This is a negative slope, so it's downhill and it's $15.
Slope = Rise / Run = -4 / 2 = -2
$15

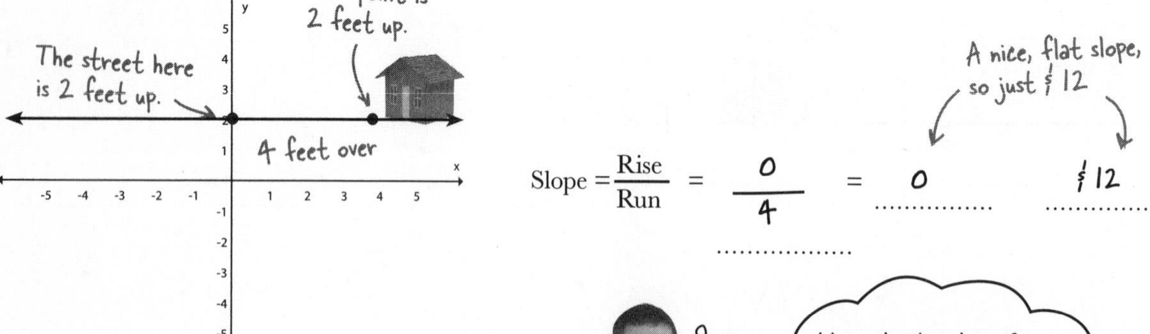
This point is 2 feet up.
The street here is 2 feet up.
4 feet over
A nice, flat slope, so just $12
Slope = Rise / Run = 0 / 4 = 0
$12
Hey, that's a lot of dough. I'm starting to make up some ground!

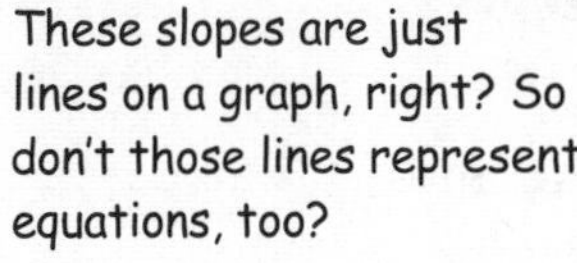

These slopes are just
lines on a graph, right? So
don't those lines represent
equations, too?

A line on a graph always represents an equation.

In fact, if you know one point on a line and the slope of a line, you can write out the equation for that line.

How can that be? That seems too easy.

An equation can take several different forms... and you can use the form that helps you out the most.

When linear equations are written a certain way, they have a particular **form**. The form is just the order of the variables, numbers, and operations. Sometimes a form has two points, and sometimes it has an intercept and a slope.

But no matter what form of a linear equation you're looking at, every line has a **slope**, **intercepts**, **no exponents greater than one**, and **two variables**. Understanding the forms just means knowing how to interpret the equations and write them in different ways.

Linear equations in point-slope form

The **point-slope form** of a linear equation represents an equation as a **point on the line** and the **slope of the line**. So as long as you know a line's slope and one point on the line, you can use the point-slope form:

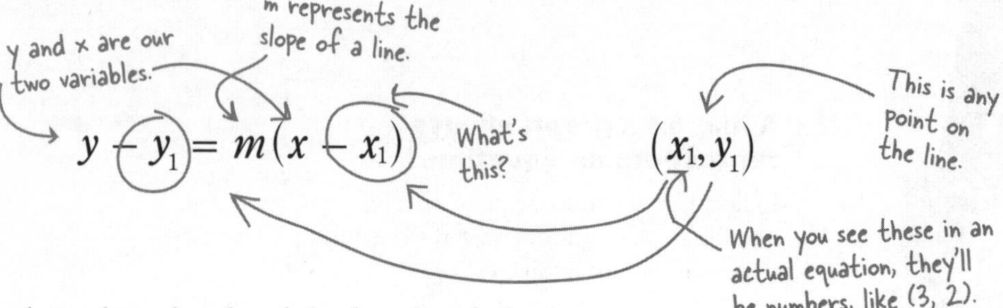
m represents the slope of a line.
y and x are our two variables.
y - y1 = m(x - x1)
What's this?
(x1, y1)
This is any point on the line.
When you see these in an actual equation, they'll be numbers, like (3, 2).

There's a lot going on here, though, so let's take a closer look.

Point-slope Way Up Close

The point-slope equation

$$y - y_1 = m(x - x_1)$$

These terms probably look a little weird and are a little confusing. Are they variables or constants? And what are those little numbers?

Subscripts

Here's the deal. The little 1's below and to the side are called **subscripts**. They indicate a specific value of x and y, and since the subscripts are the same, they indicate that those x and y values go together. So here's an ordered pair:

$$(x_1, y_1)$$

This pair represents a point somewhere along the line. It can be an intercept, but it doesn't have to be. With any single point you can still write the equation.

Since they're a point on a graph, they are also constants, not variables. So in the point-slope form, you're taking an ordered pair that represents a point on the line and splitting the two numbers of that pair up in the equation.

How does a point and a slope get you a line?

You've actually already seen how to figure out slope. We used the idea of rise over run to figure out slopes for Ed's yard estimates:

$$\text{Slope} = \frac{\text{Rise}}{\text{Run}}$$

The number of units up or down (negative rise means down).

The number of units right or left (negative run means go left).

And you've already got one point on the line, so here's what you do to draw the entire line:

1. **Plot the point.**
   If you start with the point-slope equation, that means your point is: (x_1, y_1)

2. **Interpret the slope and use it to plot the second point.**
   If you just go up the rise, and over the run you can plot the point you land on, and then draw your line.

3. **Connect your points to form a line.**

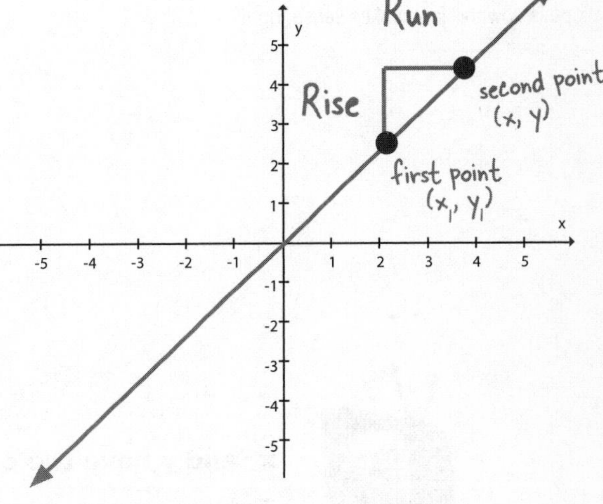

there are no Dumb Questions

Q: (x_1, y_1) can be any point?

A: Any point on the line. Ed knows two points for his lawns, so we could pick whichever one we want to use for the point-slope form of an equation.

Q: This slope thing seems a little fuzzy. "Downhill," "rise," "run"... are these really math words?

A: Sure. Just because it's Algebra doesn't mean it's complicated and weird. These concepts can be much easier to learn with more basic terms. Saying that a negative slope is downhill just makes sense, right?

Q: Why is the slope so important?

A: Every line is defined by some type of slope. If you have a negative slope or a positive slope, you'll know if ***y*** goes up or down with ***x***.

That translates into whatever two variables you're working with. For example, Edward saves more money (***C***) as time (***t***) goes on. So that equation has a positive slope.

Q: Do we need to memorize the form of the point-slope equation?

A: Yeah. It's pretty easy to do, though. The slope-intercept equation comes from the definition of slope.

$$slope = \frac{rise}{run}$$

The rise is the difference between two y-coordinates

m stands for slope

$$m = \frac{y - y_1}{x - x_1}$$

The run is the difference between the values of two x coordinates.

After some manipulation...

$$y - y_1 = m(x - x_1)$$

x_1 and y_1 have the opposite sign.

*The trickiest part of this equation is that positive **x_1** and **y_1** values are subtracted. That means if you see addition signs in the point-slope equation, you're dealing with a negative coordinate value, not a positive one.*

Sharpen your pencil

Try graphing a couple of point-slope equations using what you know about form to make it easy!

$$y + 1 = \frac{1}{2}(x + 8)$$

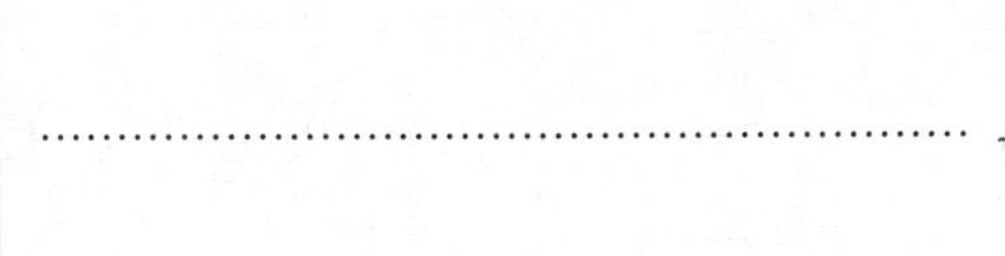

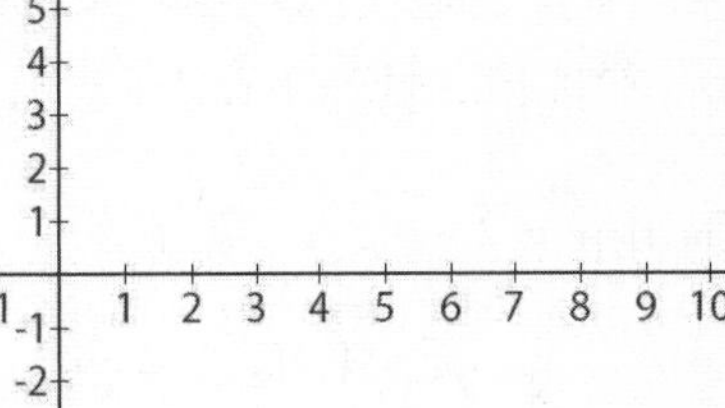

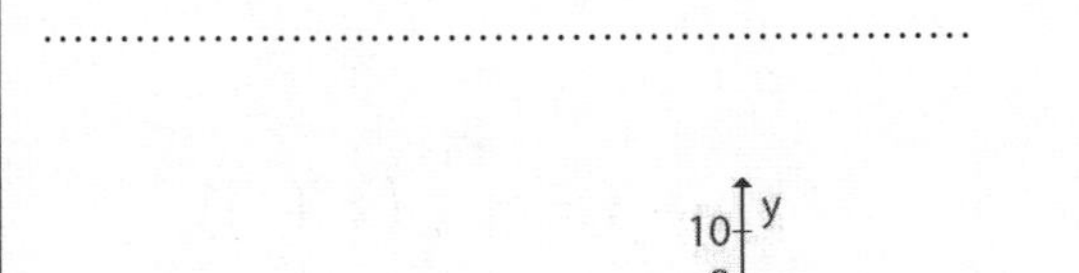

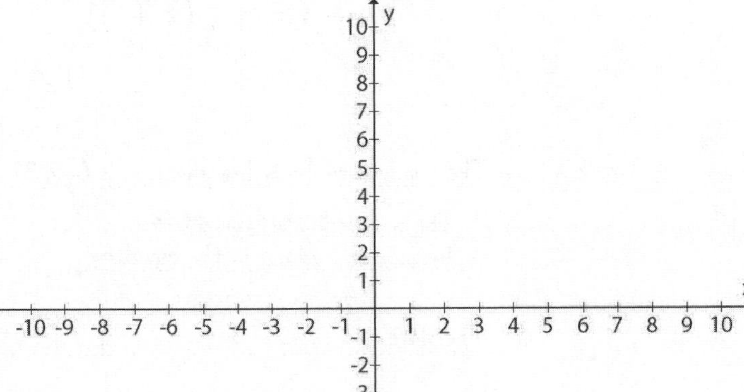

$$y - 3 = -\frac{2}{3}(x + 5)$$

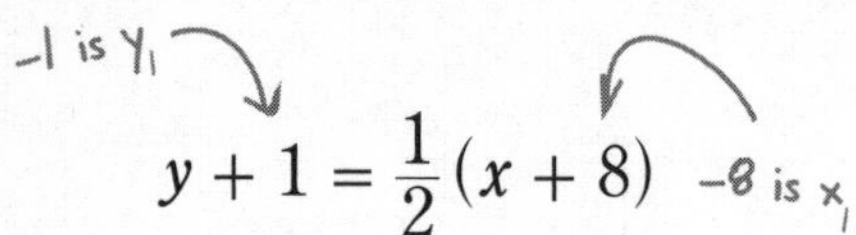

Try graphing a couple of point-slope equations using what you know about form to make it easy!

-1 is y_1

$$y + 1 = \frac{1}{2}(x + 8)$$

-8 is x_1

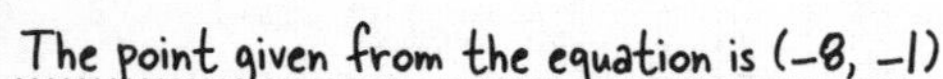

These values are negative because they are added in the equation

The slope is $\frac{1}{2}$

m is the number in front of the (x + 8) quantity

That means that the slope is up one, over 2

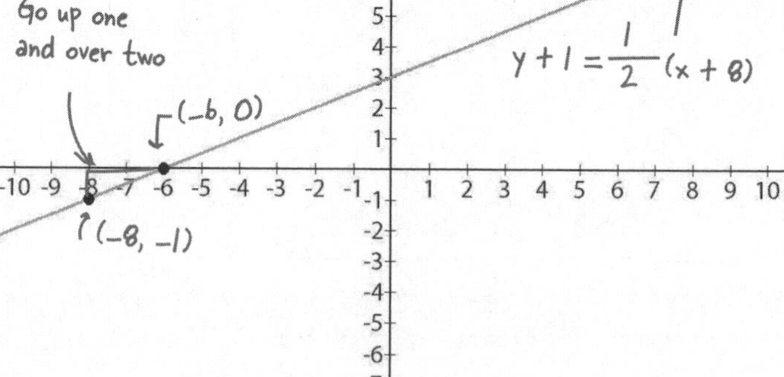

$x_1 = -5$

$$y - 3 = -\frac{2}{3}(x + 5)$$

$y_1 = 3$

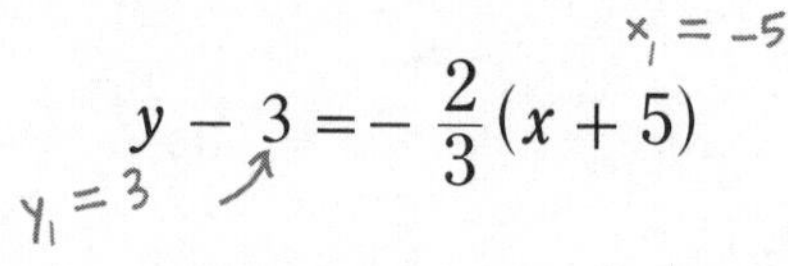

The point given from the equation is (-5, 3)

The x coordinate is negative because it's added in the equation

The slope is $-\frac{2}{3}$

That means that the slope is down 2, over 3

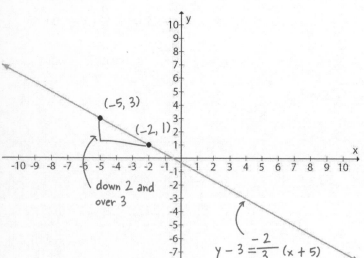

What about Ed's lawns? Do we know enough about to write an equation for them now?

We sure do!

With just one point and the slope that you've already figured out, we have enough to write equations for those lawns.

Let's do it...

Let's use the point-slope form

If you take Ed's first lawn and use the information you've already figured out, you can write the equation for that line with ***no*** additional calculations:

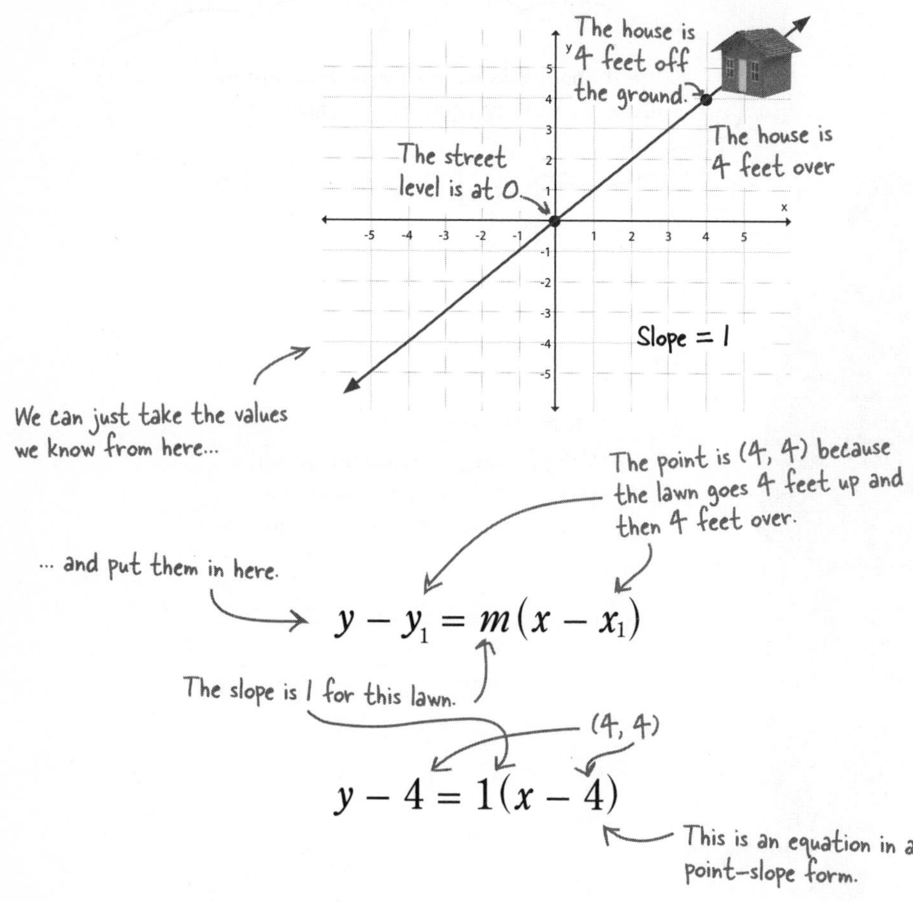
The house is 4 feet off the ground.
The house is 4 feet over
The street level is at 0.
Slope = 1
We can just take the values we know from here...
The point is (4, 4) because the lawn goes 4 feet up and then 4 feet over.
... and put them in here.
$y - y_1 = m(x - x_1)$
The slope is 1 for this lawn.
(4, 4)
$y - 4 = 1(x - 4)$
This is an equation in a point-slope form.

The same holds true for any line for which you know a point and the slope. Just start with the general point-slope equation, and substitute the values for m and x_1 and y_1 , and you have a valid equation for the line.

Watch the signs! If you have a negative value for either x_1 or y_1, put it in the equation with the negative and then simplify. You'll end up with an added value in the equation for the line.

Write the equations for the other two lawns that Ed evaluated in point-slope form.

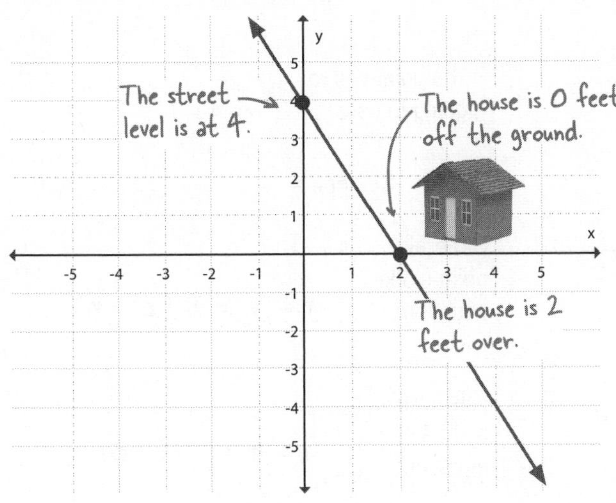

The point you're going to use: (..........)

Slope = –2

General point-slope equation: $y - y_1 = m(x - x_1)$

This lawn's point-slope equation: ....

The street here is 2 feet up.

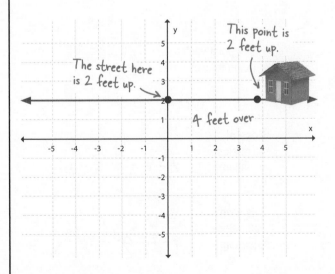

This point is 2 feet up.

4 feet over

The point you're going to use: (..........)

Slope = 0

General point-slope equation: $y - y_1 = m(x - x_1)$

This lawn's point-slope equation: ....

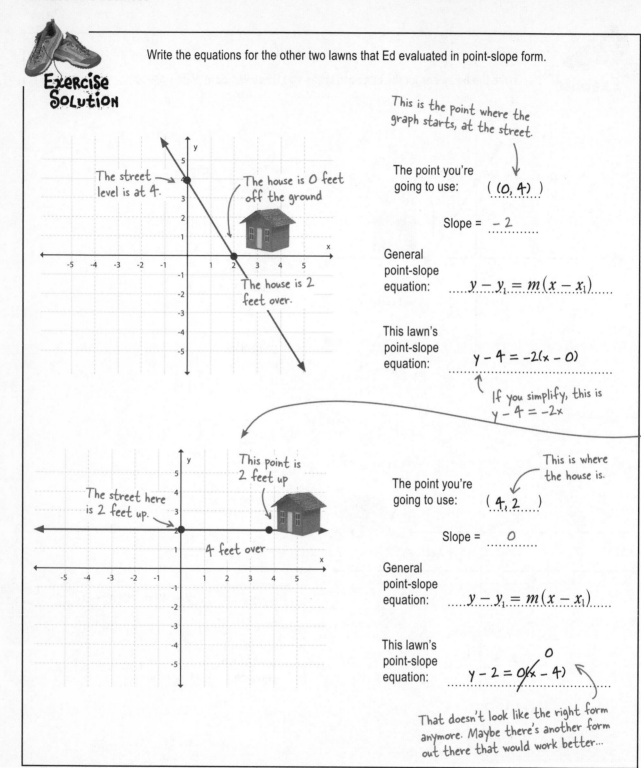

Exercise Solution

Write the equations for the other two lawns that Ed evaluated in point-slope form.

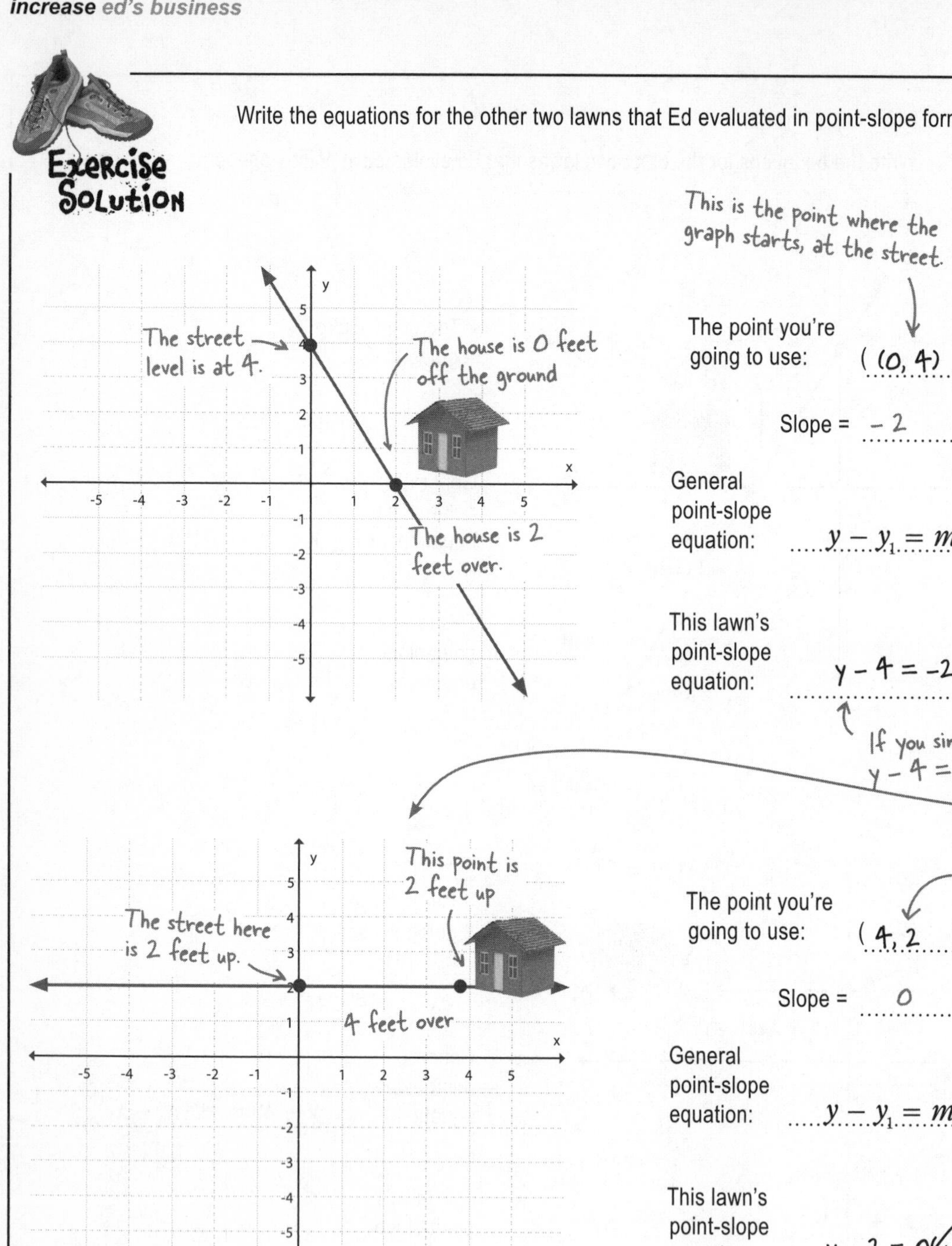

The street level is at 4.
The house is 0 feet off the ground
The house is 2 feet over.
This point is 2 feet up
The street here is 2 feet up.
4 feet over

This is the point where the graph starts, at the street.

The point you're going to use: ((0, 4))

Slope = – 2

General point-slope equation: $y - y_1 = m(x - x_1)$

This lawn's point-slope equation: y – 4 = –2(x – 0)

If you simplify, this is y – 4 = –2x

This is where the house is.

The point you're going to use: (4, 2)

Slope = 0

General point-slope equation: $y - y_1 = m(x - x_1)$

This lawn's point-slope equation: y – 2 = 0(x – 4) → 0

That doesn't look like the right form anymore. Maybe there's another form out there that would work better...

So now what do we do? With a 0 slope, everything's messed up.

With a slope of 0, things get weird.

There are cases where the point-slope form won't quite describe the line, like when a line has a slope of 0.

That last lawn is definitely a straight line. So how do we express an equation for that line? Well, we know that the equation will have to be in terms of ***x*** and/or ***y*** somehow, but what about slope? How does a zero-slope work?

Horizontal lines require a different form

Ed is looking at a nice, flat lawn. The rise of the line is 0, so it doesn't matter what the run is because the slope will still come out to be zero.

The slope of a horizontal line is always 0.

Writing an equation that describes a horizontal line like this is actually pretty simple. Since all of the ***y*** values are the same (2 in this case), **you can just write**:

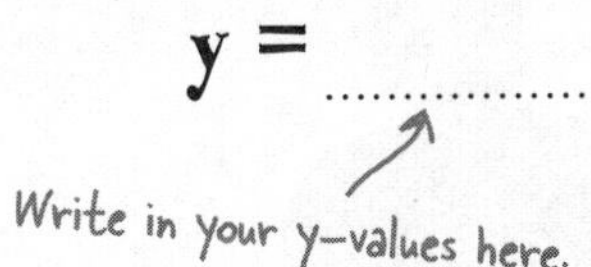
y = ...
Write in your y-values here.

So since the slope is zero, what happens to x? It can't be zero, too, right?

There's an x with a coefficient of 0.

If you have a coefficient of a variable that's zero, the variable will disappear. This equation can be written as:

$$y - 2 = 0x$$

It still doesn't really look quite right. Good thing there's another form that will work better...

Answer: 2

Equations also have a standard form

There's a form of a linear equation called the **standard form.** A horizontal line is a very specific case in which point-slope won't work, but there's a more general form. The standard form actually takes both y and x into account and works when the slope of a line is zero:

$$ax + by = c$$

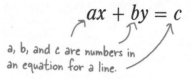

a, b, and c are numbers in an equation for a line.

Can you figure out what a and b would be for a line where the equation is y = 2.

In this form, there's no m. That means none of those variables mean slope. And a, b, and c do not stand for anything relating to a graph, either. So this form isn't quite so easy to turn into a graph. But that's okay because this equation handles ***every type of line***, no matter what.

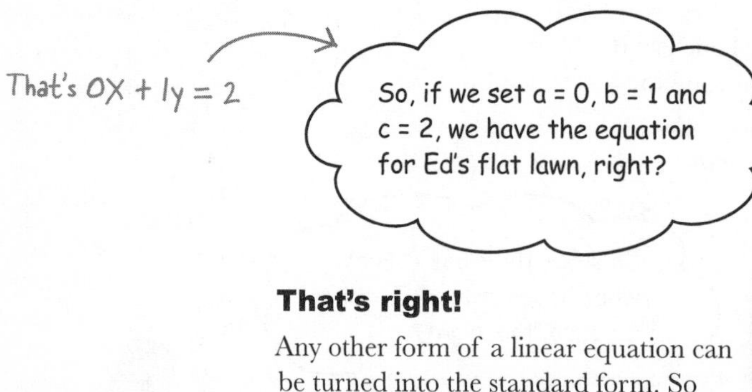

That's 0x + 1y = 2
So, if we set a = 0, b = 1 and c = 2, we have the equation for Ed's flat lawn, right?

That's right!

Any other form of a linear equation can be turned into the standard form. So we can turn the earlier equation, $y = 2$, into the $ax + by = c$ form.

Are there any other forms we need to know about?

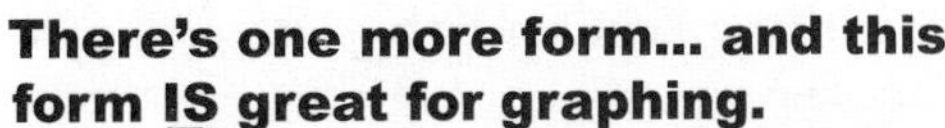

There's one more form... and this form IS great for graphing.

The last equation form that's left is similar to point-slope in what it contains. This last form comes with a point and a slope within the equation.

This form includes the *y*-intercept, so it's called the **slope-intercept form**.

The slope-intercept form is EASY to graph

Not only that, but slope-intercept comes with the same constant for slope, ***m***. So ***m*** is the slope, just like it was earlier. On top of that, ***x*** and ***y*** are still variables, like all the other forms of linear equations. The intercept is the part that's different. Here's what this format looks like:

$$y = mx + b$$

m = slope, like always.

This is the same slope we've been talking about rise over run.

(0, b) is the y-intercept. So b is where a line crosses the y-axis.

To draw this line, just plot the y-intercept at (0, ***b***) and then find your second point using the slope, ***m***, just like you did before.

Sum it up

Slope intercept equation – A linear equation of the form y = mx + b where m is the slope of the line and b is the coordinate for the y term of the y-intercept.

Linear Equations Up Close

Point-slope form

This form has a lot going for it. It gives you a quick point to plot, and it gives you a slope, so you can plot the rest of the line.

The downside is that if you need to manipulate this form, you'll have to use a lot of distribution and parentheses, and the constants aren't together, either.

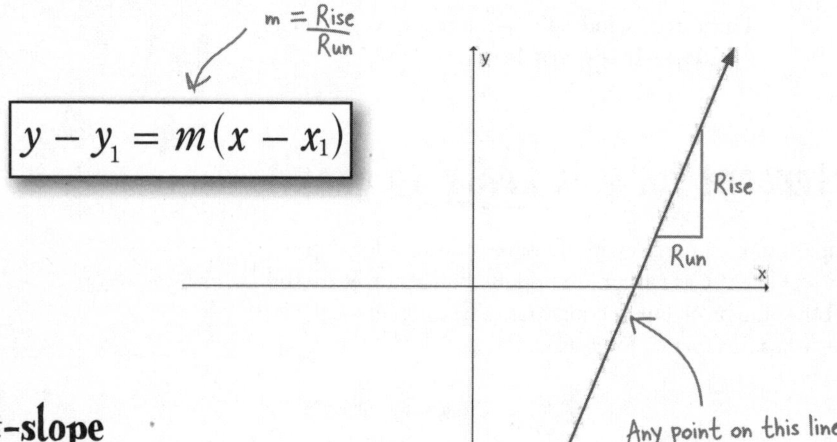

Point-slope is great for graphing, but not for manipulating the equation.

Standard form

Standard form is straightforward—***x*** and **y** both have a coefficient, and there's another extra constant. This form works for all lines, and it's an easy equation to manipulate.

Graphing is tough with standard form because you don't have a slope or a point to start with. The only way to graph this form is to solve for a couple of points and then plot those points and draw a line.

Slope-intercept form

This form gives you the ***y***-intercept, and it gives you a slope so you can plot the rest of the line.

Slope-intercept is a good middle ground. It's easy to manipulate, and it gives you a point right away. However, if the ***y***-intercept for the graph is really high or really low, it can make graphing the line really tough.

$$y = mx + b$$

$m = \frac{Rise}{Run}$

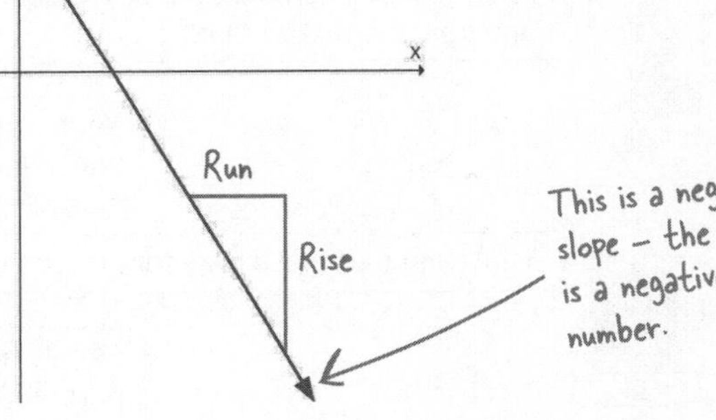

Slope-intercept is great for graphing lines that cross the y-axis near the origin and is pretty easy to manipulate.

$$ax + by = c$$

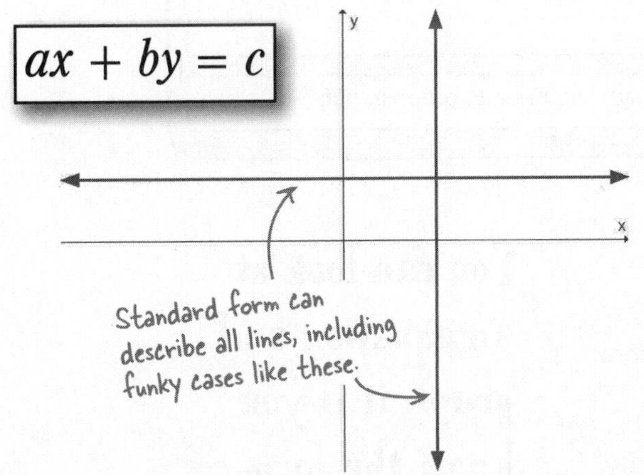

y
x
Standard form can describe all lines, including funky cases like these.

Standard form is easy to manipulate but hard to graph.

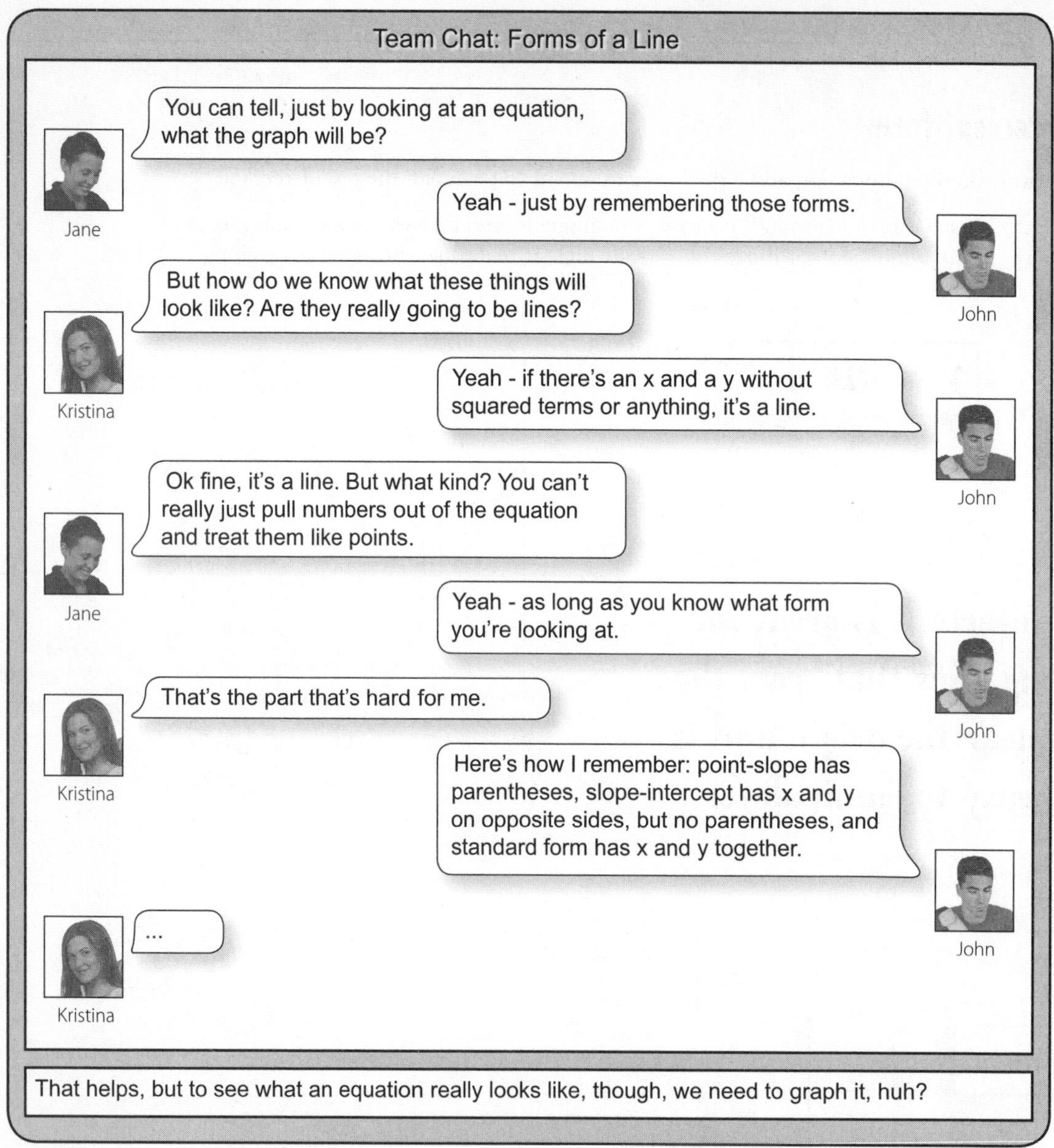

Team Chat: Forms of a Line
You can tell, just by looking at an equation, what the graph will be?
Jane
Yeah - just by remembering those forms.
John
But how do we know what these things will look like? Are they really going to be lines?
Kristina
Yeah - if there's an x and a y without squared terms or anything, it's a line.
John
Ok fine, it's a line. But what kind? You can't really just pull numbers out of the equation and treat them like points.
Jane
Yeah - as long as you know what form you're looking at.
John
That's the part that's hard for me.
Kristina
Here's how I remember: point-slope has parentheses, slope-intercept has x and y on opposite sides, but no parentheses, and standard form has x and y together.
John
...
Kristina
That helps, but to see what an equation really looks like, though, we need to graph it, huh?

Use the equation form to your advantage

You know enough now to look at an equation and graph it, if it's a point-slope or slope intercept equation. If it's in standard form, then you know to solve for some points ($y = 0$ and $x = 0$ are easy ones), and then plot your line.

That means most of the time no computation is required! Just draw.

You can look at an equation and graph it if you know the form.

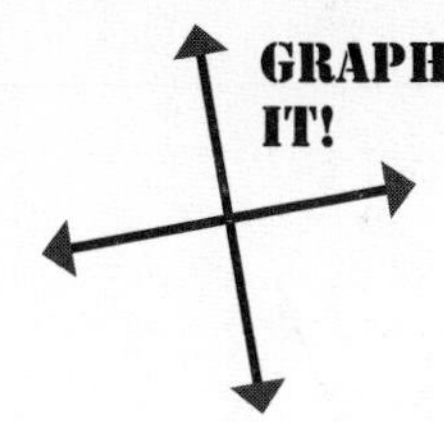

You're ready to graph some equations. Here are several equations in different forms... draw a visual representation of each one.

Equation: $y = 3x - 2$

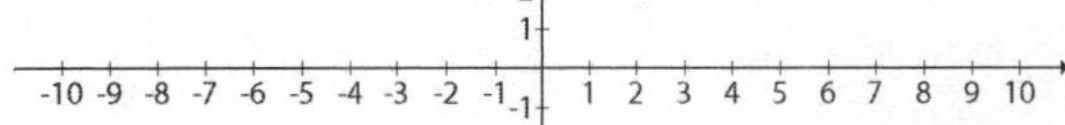

What form is the equation?

Equation: $1x + 0y = 7$

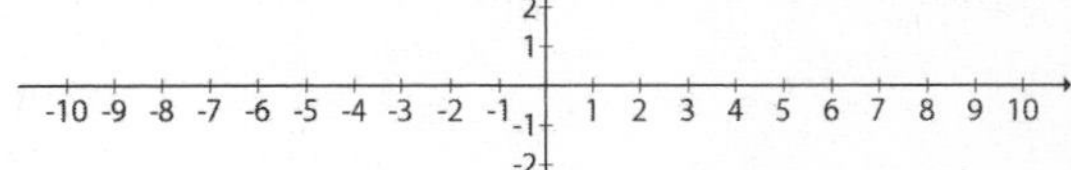

What form is the equation?

Equation: $y + 2 = -\frac{1}{2}(x - 1)$

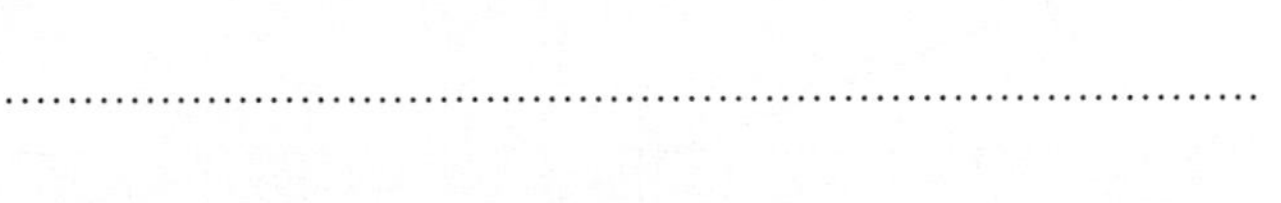

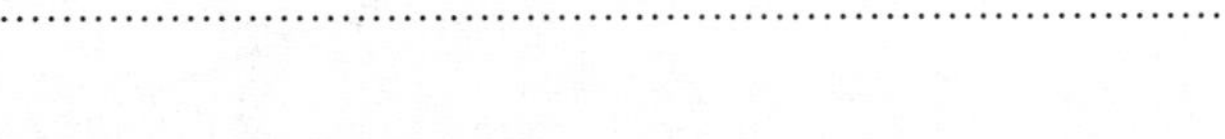

What form is the equation?

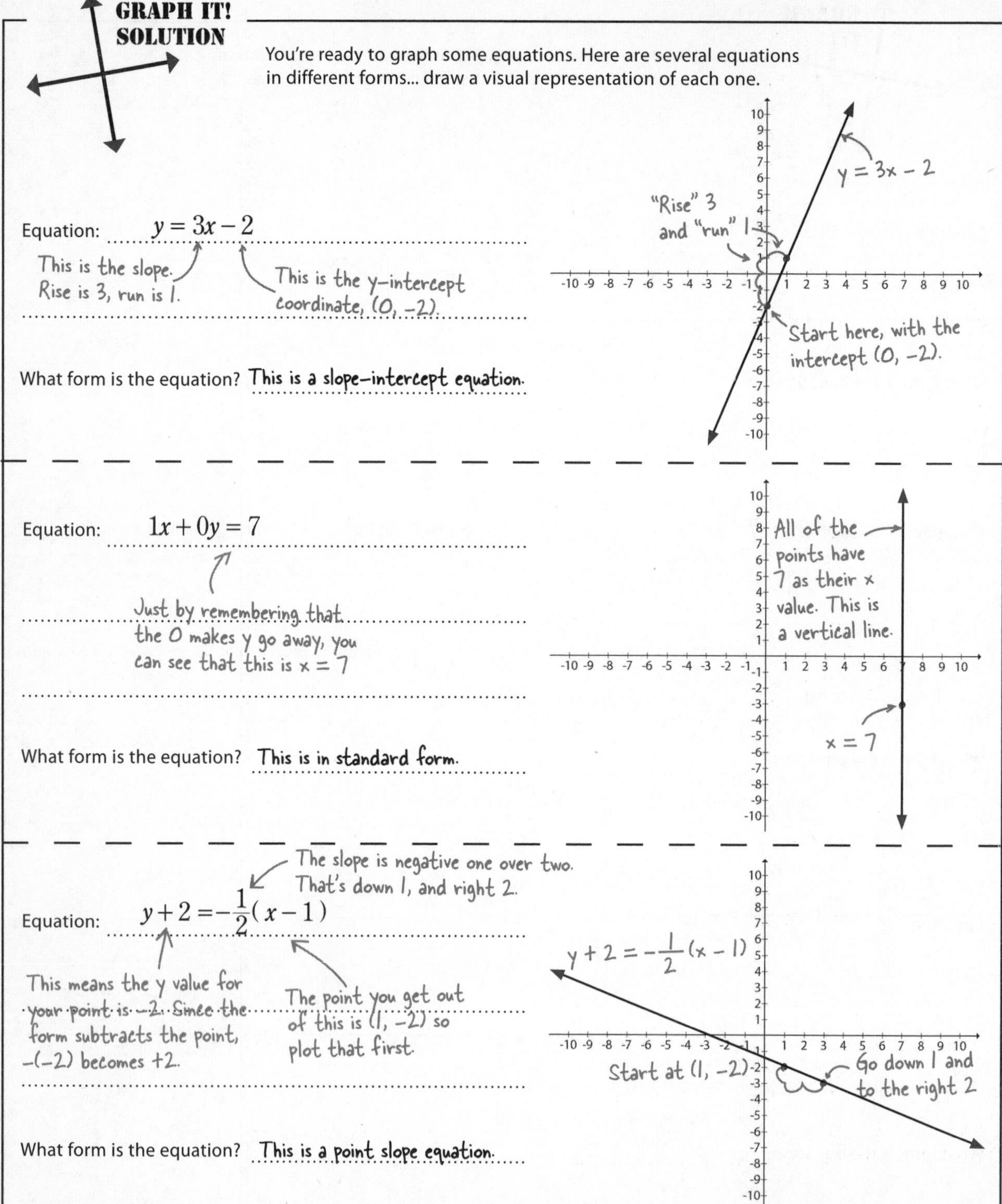
GRAPH IT! SOLUTION
You're ready to graph some equations. Here are several equations in different forms... draw a visual representation of each one.
Equation: $y = 3x - 2$
This is the slope. Rise is 3, run is 1.
This is the y-intercept coordinate, (0, -2).
What form is the equation? This is a slope-intercept equation.
"Rise" 3 and "run" 1
y = 3x - 2
Start here, with the intercept (0, -2).
Equation: $1x + 0y = 7$
Just by remembering that the 0 makes y go away, you can see that this is x = 7
What form is the equation? This is in standard form.
All of the points have 7 as their x value. This is a vertical line.
x = 7
The slope is negative one over two. That's down 1, and right 2.
Equation: $y + 2 = -\frac{1}{2}(x - 1)$
This means the y value for your point is -2. Since the form subtracts the point, -(-2) becomes +2.
The point you get out of this is (1, -2) so plot that first.
What form is the equation? This is a point slope equation.
$y + 2 = -\frac{1}{2}(x - 1)$
Start at (1, -2)
Go down 1 and to the right 2

there are no Dumb Questions

Q: Why does m stand for slope?

A: Nobody really knows. Descartes (Mr. Cartesian Plane himself) didn't use it. There are ideas about "modulus" or maybe it's just because it's a letter in the middle of the alphabet. At this point, it's just what everyone else uses.

Q: Why do we use "b" for the *y*-intercept?

A: Well, that's another mystery, like slope. In modern day math, several other countries use different letters for the *y*-intercept, like ***k***, ***n***, or ***h***.

Q: If x and y are on the same side of the equation, can the equation still be in slope-intercept form?

A: Sure, but you'll have to manipulate the equation to get it in the form ***y*** = m***x*** +b. The signs can be different, but that just means that the constant is a negative number.

Q: Why are m and b constants but x and y variables? How can you tell when a letter is a variable and not a constant?

A: The constants are what will appear as numbers in a typical situation. When you learn a standard form of an equation (like slope-intercept), you know now that m and b are constants because they'll always be the same in the equation.

And here's another clue when looking at a new equation: the coordinate plane is based on ***x*** and ***y***, making them variables. ***x*** can take on all sorts of values, just like ***y***, but m and b stay the same.

Q: If m is a whole number, then what's the "run" of "rise over run"?

A: Think back to your fractions! Any whole number is the same as a fraction with the whole number in the numerator and one in the denominator. So, if you have a slope of 5, then it's the same thing as 5 over 1, or, "rise 5, run 1."

Q: You keep saying that "form" is what's important. What does that really mean?

A: The form of the equation is just the arrangement of variables and constants. For example, in a point-slope equation, you have to have the ***y*** isolated on one side of the equation and an ***x*** term on the right for it to be in point-slope form.

Q: What if I have an equation like y = x? Is that in any special form?

A: Yes. ***y*** = ***x*** is exactly the same as ***y***=1***x*** + 0, which is slope-intercept form. So here, m = 1, and b = 0. And since you have an ***m*** and a ***b***, you're ready to plot your graph.

Q: When would you start with a graph and need to write the equation?

A: It happens a lot, actually. When you have a plot of actual data—financial data or experimental data—and you need to write a line to generalize what's happening, you'll have a graph before you have an equation.

Q: You can really just figure out the slope and the intercept and write an equation?

A: That's the beauty of having standard forms. Everybody knows that the coefficient for the ***x*** term is the slope, and the constant that's added onto the equation is the ***y***-coordinate of the ***y***-intercept.

If your slope is a whole number, the RUN of the slope is that number, and the RISE of the slope is 1.

I've been cutting those three new lawns for 2 weeks now. More people want quotes, and it's too much for me to keep up with, so my brother's going to help.
I can't work this way - I need his stuff in pictures so I can decide how much to charge.

Ed and his brother have gotten over their differences. To make up for taking all of Ed's money early in the summer, his brother is going to work for free to finish the season.

Ed has been evaluating his lawns based on drawings and deciding how steep the slope is, but his brother decided to get creative.

He's using a "new system" and when he turns the form in to Ed it has equations for the lawns instead of the pictures.

Your job is to convert each equation to a form that you can graph. Then, graph each equation so Ed can get these customers estimates.

Edward's Lawn Service

New Customer Estimate

Ed's brother wrote equations instead of drawing the lawn.

Ed likes to work with pictures. This is what his brother was supposed to fill in.

Lawn Evaluations:

$$y = -\frac{3}{2}x + 7$$

....

....

....

Smith's Lawn

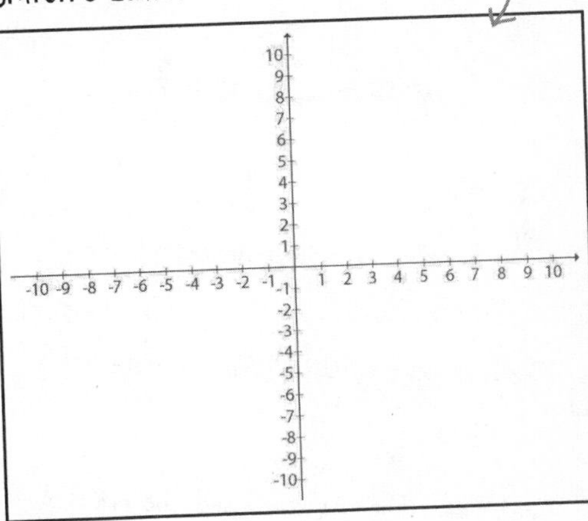

Chadwick's Lawn

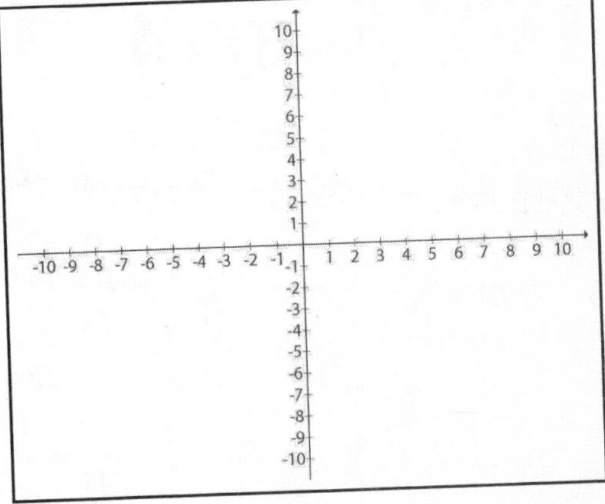

$$7x + 3y = 15$$

....

....

....

Your job is to convert each equation to a form that you can graph. Then, graph each equation so Ed can get these customers estimates.

Edward's Lawn Service

New Customer Estimate

Lawn Evaluations:

$$y = -\frac{3}{2}x + 7$$

This equation is in slope-intercept form.

The intercept is (0, 7) and the slope is $\frac{-3}{2}$

That means the rise is −3 and the run is 2.

Smith's Lawn

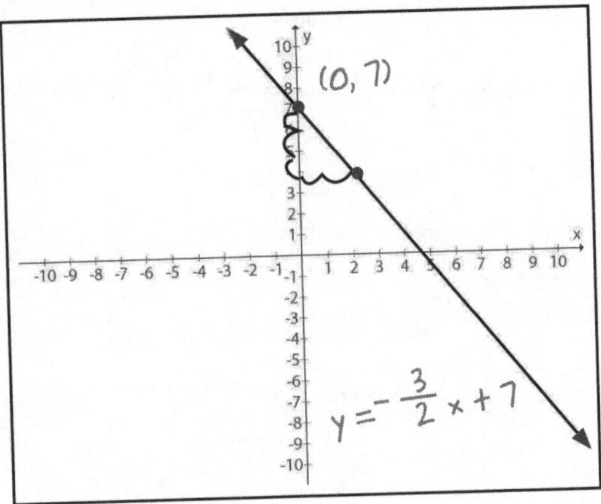

$$7x + 3y = 15$$

This equation is in standard form, no easy points.

$7(0) + \frac{3y}{3} = \frac{15}{3}$ $\quad \frac{7x}{7} + 3(0) = 15$

$y = 5$ $\quad x = \frac{15}{7}$

(0, 5) $\quad (\frac{15}{7}, 0)$

That's a little over 2, right?

Chadwick's Lawn

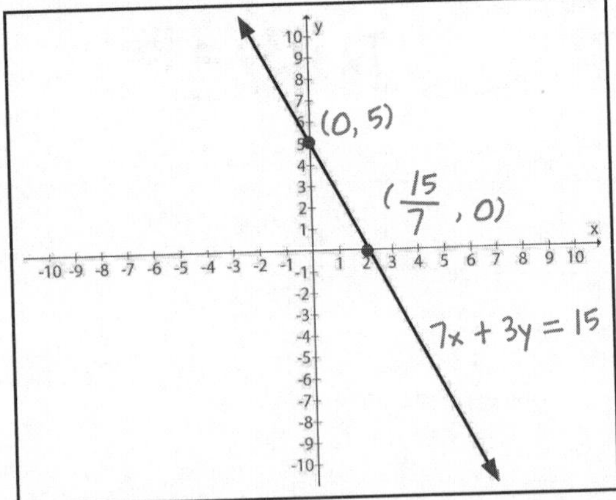

So what's the final amount I'll end up with after all these new yards?

Ed's had a busy summer—and not all of it good. He started out strong, broke his leg, had to pay his brother crazy amounts just to keep his customers, but then things started to turn around with a bunch of clients. Here's what happened:

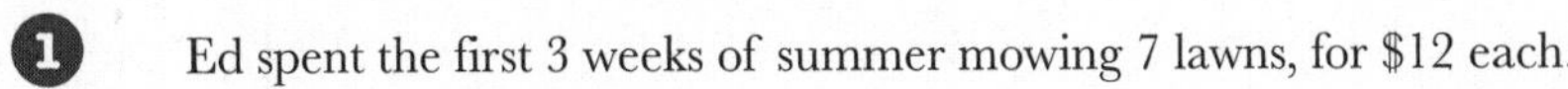

1. Ed spent the first 3 weeks of summer mowing 7 lawns, for $12 each.

2. Ed broke his leg and had to spend the following 10 weeks in a cast. His 7 clients were still paying $12 a cut, but Ed's brother was charging him $19 a lawn. He ended up $250 in debt.

3. For two weeks, he goes back to cutting the original 7 lawns for $12 each, and three new lawns, one for $20, one for $15 and one for $12.

4. Then his brother signed up these last 2 lawns at $15 a lawn. He's going to cut all of the lawns for the rest of the summer—6 weeks.

How much is Ed going to end up making? If you need some help, turn the page for a few hints...

Long Exercise

Figure out how much Ed is going to make by the end of the summer. Draw him a graph to show his cash projections for the next 22 weeks (that's how long his plan runs).

Fill out this chart to get started:

Week #	How many lawns?	How much per lawn?	How much did Ed make?	Ed's running total cash
1-3		$12		$252
4-14 Broken leg!				
15-16	10	Different rates: 8 at $12 each, 1 at $15, and 1 at $20	$131 per week at 2 weeks = $ 262	
17-22		Different rates – all of the lawns from 15 – 16 plus 2 more at $ 15	$131 + $30 per week = $ 161 per week	You need to write an equation and plot a graph for this part!

Write Ed's cash equation for the end of the summer:

....

....

....

Plot the new equation here:

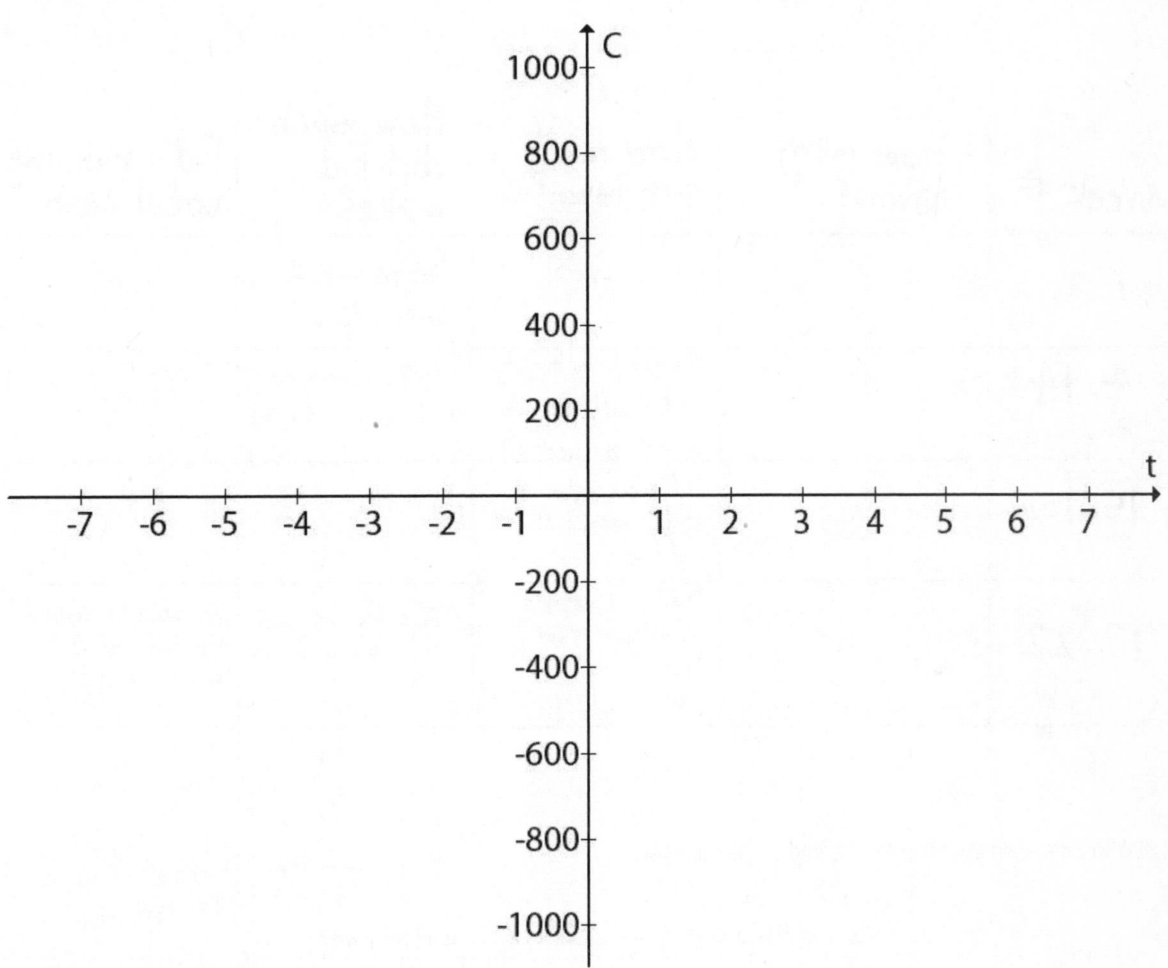

How much will Ed end up with at the end of the summer? ....

Read it from the graph – it's week 6..

....

Your job was to figure out how much Ed is going to make by the end of the summer, and draw him a graph to show his cash projections for the next 22 weeks (that's how long his plan runs).

Week #	How many lawns?	How much per lawn?	How much did Ed make?	Ed's running total cash
1–3	7	$12	$84 per week for 3 weeks = $252	$252
4–14 Broken leg!	7	$12 per lawn, but it cost him $19 per lawn = – $7 per week	–$49 per week for 10 weeks = – $490	– $238
15–16	10	Different rates: 8 at $12 each, 1 at $15, and 1 at $20	$131 per week at 2 weeks = $262	$24
17–22	12	Different rates – all of the lawns from 15 – 16 plus 2 more at $15	$131 + $30 per week = $161 per week	You need to write an equation and plot a graph for this part!

Write Ed's cash equation for the end of the summer:

Cash = $161 a week times weeks + $12 he starts with

Don't forget about the money Ed started with.

This is a lot like Ed's early equation, but he's now making more money a week.

$C = 161t + 12$

This is in the $y = mx + b$ form – easy to plot!

The y – intercept is (0, 12)

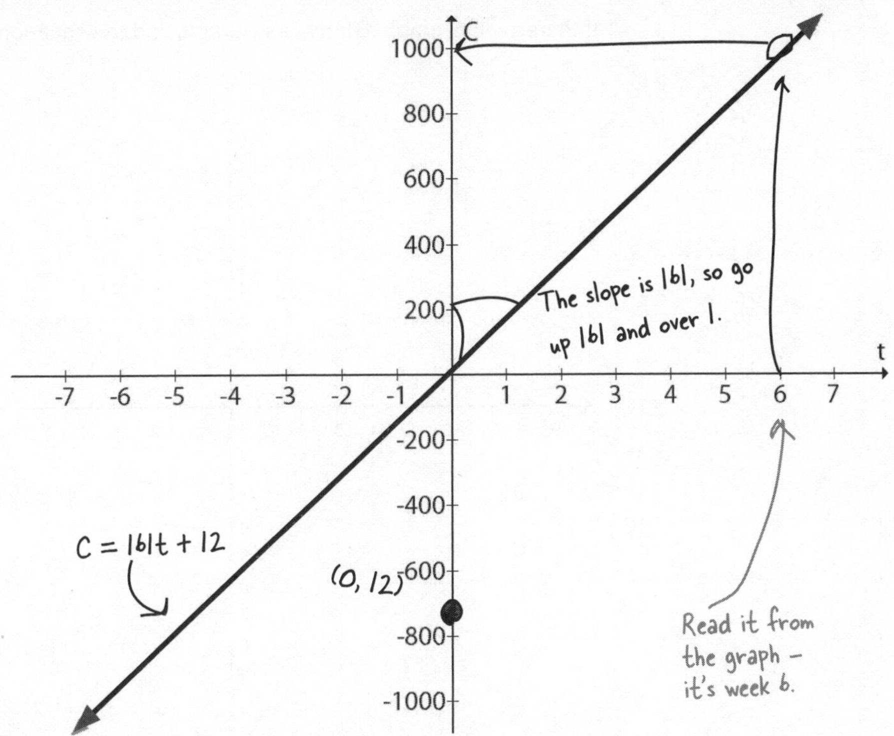

How much will Ed end up with at the end of the summer? $1000

$1000! Even after the broken leg - that's awesome. I'm going to take the graph you made and make some plans!

A picture might also be worth 1,000 bucks!

Sometimes an equation has a lot to say about the future. In fact, an equation can tell Ed how much to save, how many yards to mow, even how a particular yard will profit his business.

But sometimes you need more than a bunch of numbers and letters. In those cases, a graph can help you *see* your equation... and make informed decisions.

Sharpen your pencil

If there's a line, graph it! If there's a graph, find the equation for the line.

$$y + 3 = -\frac{3}{5}x$$

....

....

....

Point #1: (.......) ....

....

....

....

Point #2: (.......) ....

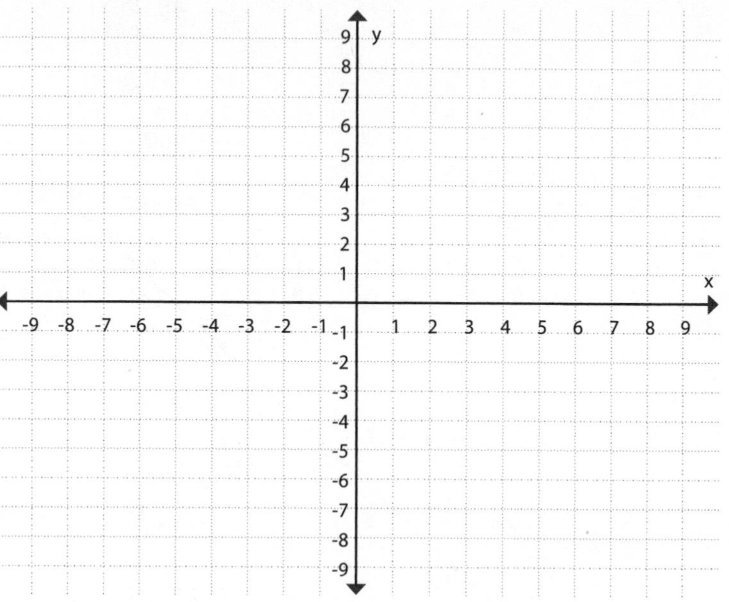

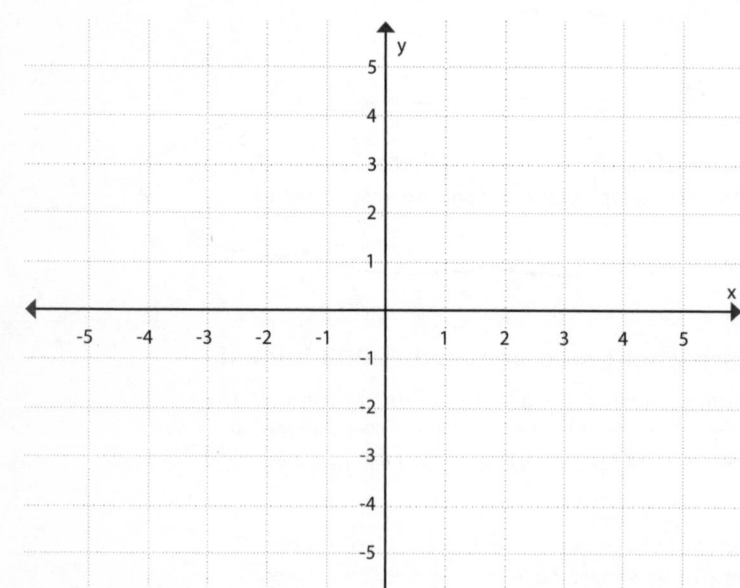

$$y = \frac{1}{2}x + 3$$

Slope = ...

Y- Intercept = ...

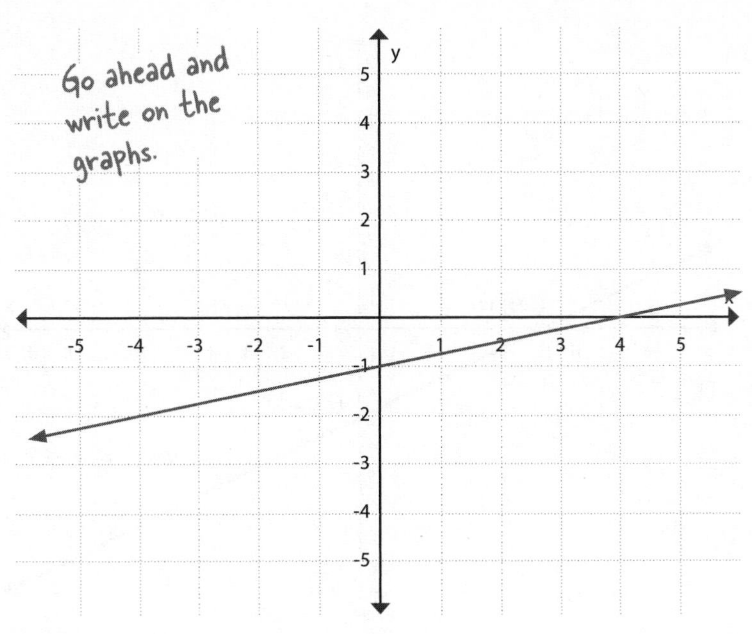

Slope = ...

Y- Intercept = ...

Y =...... x + ...

$m = \frac{1}{6}$ Point: $(-3, 4)$

....

....

....

....

....

....

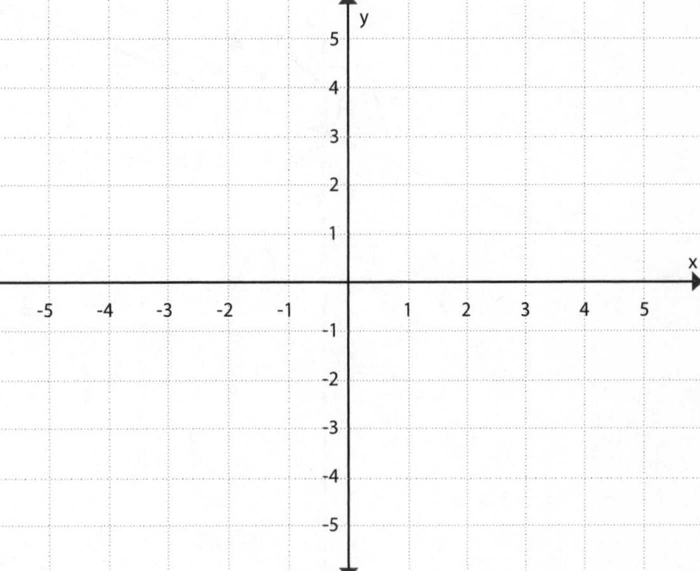

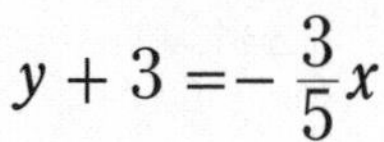

Sharpen your pencil Solution

If there's a line, graph it! If there's a graph, find the equation for the line.

$$y + 3 = -\frac{3}{5}x$$

y = 0 $\quad 5 \cdot (0 + 3) = -\frac{3}{5}x \cdot 5$

$\frac{15}{-3} = -\frac{3x}{-3}$

Point #1: (–5, 0) $\quad -5 = x$

x = 0 $\quad y + 3 = -\frac{3}{5}(0)$

$y + 3 - 3 = 0 - 3$

Point #2: (0, –3) $\quad y = -3$

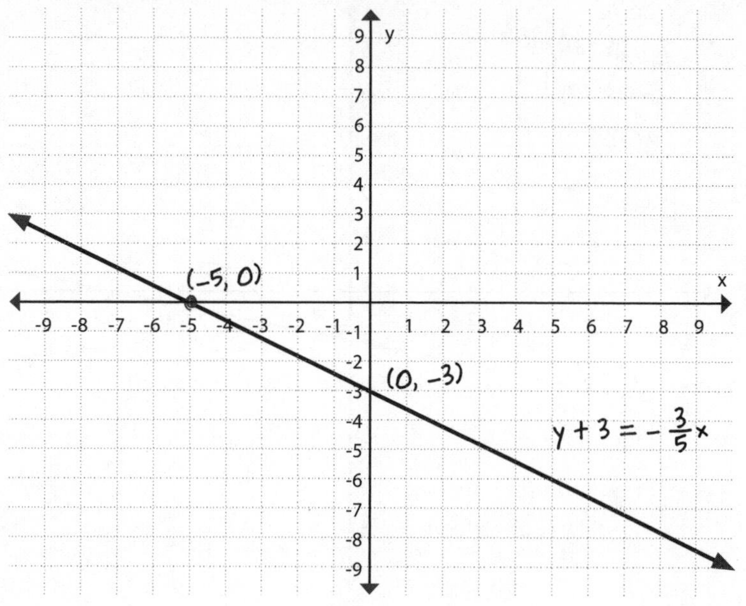

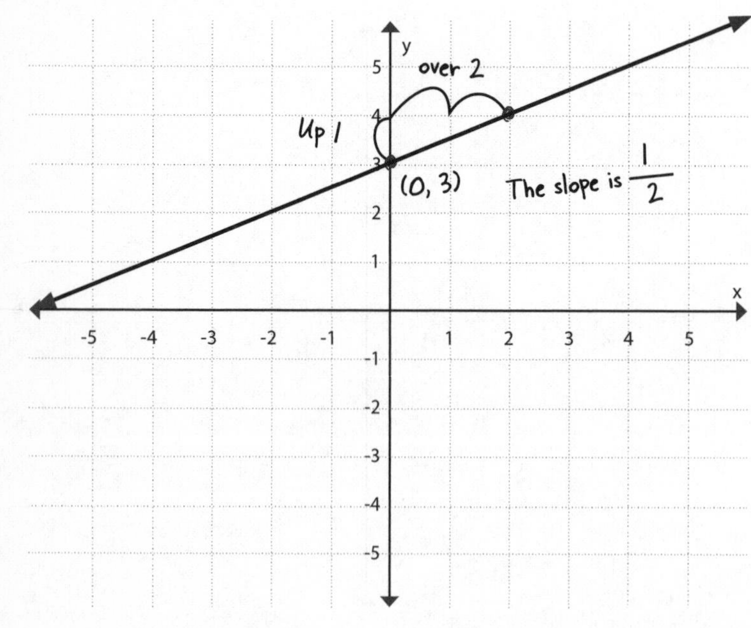

y = mx + b

slope → m, y - intercept → b

$$y = \frac{1}{2}x + 3$$

Slope = $\frac{1}{2}$

Y- Intercept = (0, 3)

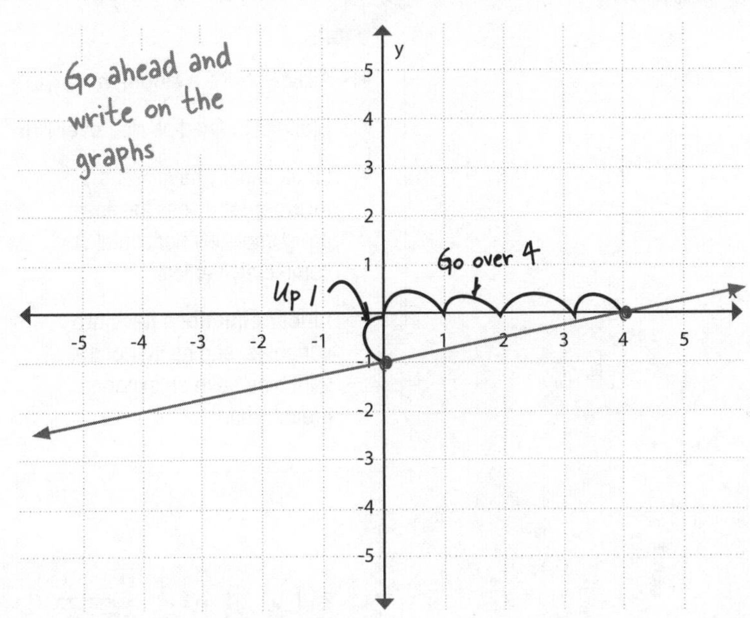
Go ahead and write on the graphs
Up 1
Go over 4

Slope = 1/4

Y- Intercept = (0, -1)

Y = 1/4 x + -1

Rise/Run = 1/4 = m

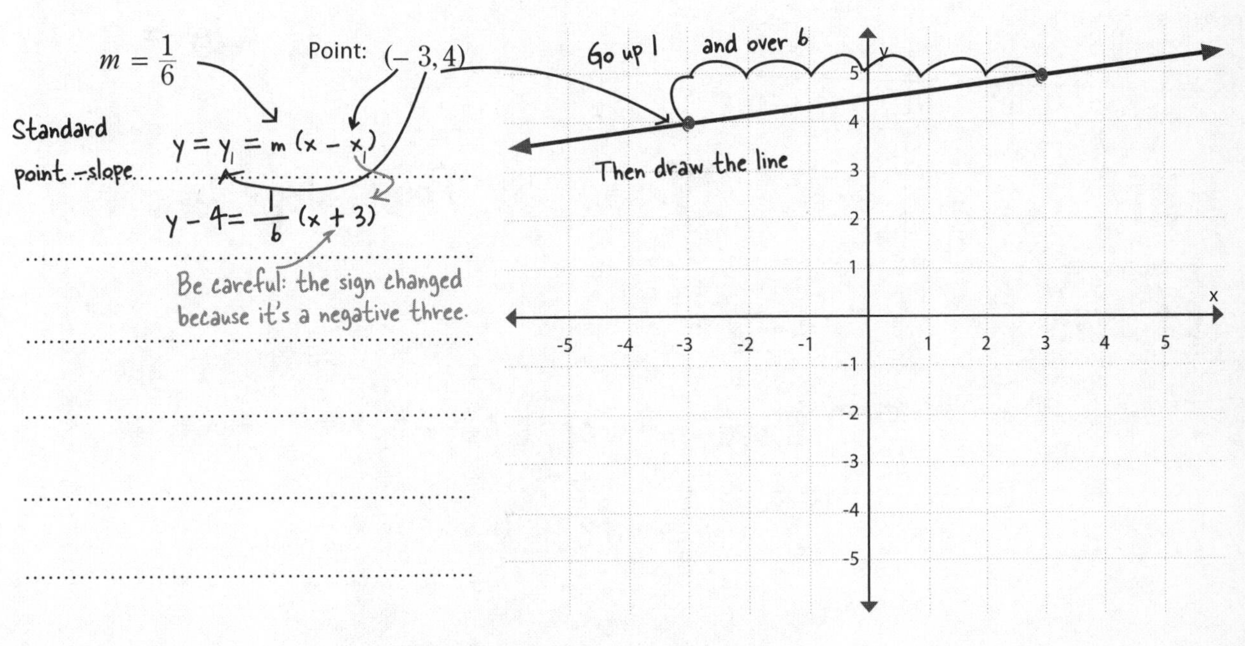
$m = \frac{1}{6}$
Point: $(-3, 4)$
Standard point-slope
$y - y_1 = m(x - x_1)$
$y - 4 = \frac{1}{6}(x + 3)$
Be careful: the sign changed because it's a negative three.
Go up 1 and over 6
Then draw the line

CHAPTER 5

Tools for your Algebra Toolbox

This chapter was about graphing equations.

BULLET POINTS

- All lines have a **slope**, ***m***.
- Slope is defined as **rise over run**.
- Lines usually have ***x*** and ***y* intercepts**, unless the lines are completely horizontal or completely vertical.
- **Linear equations** have two variables, and neither of the variables have an exponent greater than 1.

The Cartesian Plane

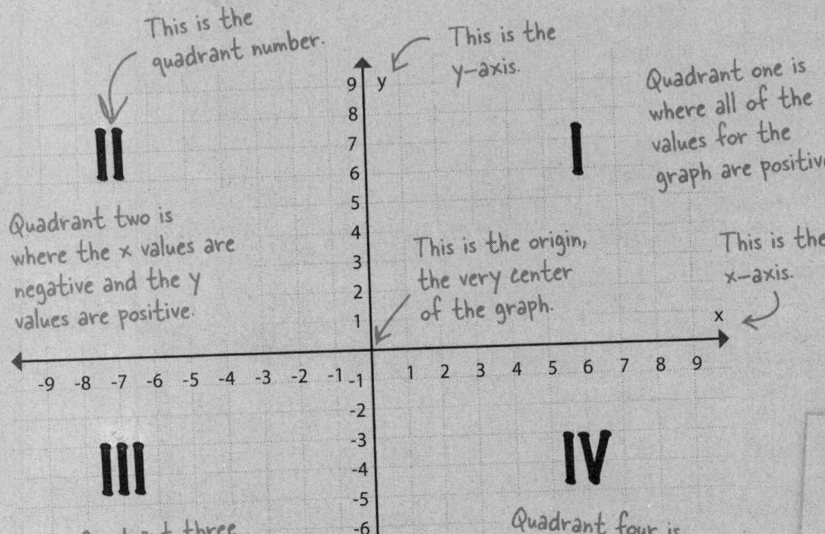

This is the quadrant number.
This is the y-axis.
y
x
II
I
III
IV
Quadrant one is where all of the values for the graph are positive.
Quadrant two is where the x values are negative and the y values are positive.
This is the origin, the very center of the graph.
This is the x-axis.
Quadrant three is where all the values are negative, both x and y.
Quadrant four is where the x values are positive and the y values are negative.

Standard form

$$ax + by = c$$

Point-slope form

$$y - y_1 = m(x - x_1)$$

$m = \frac{Rise}{Run}$

Slope-intercept form

$$y = mx + b$$

$m = \frac{Rise}{Run}$

6 inequalities

Can't quite get enough?

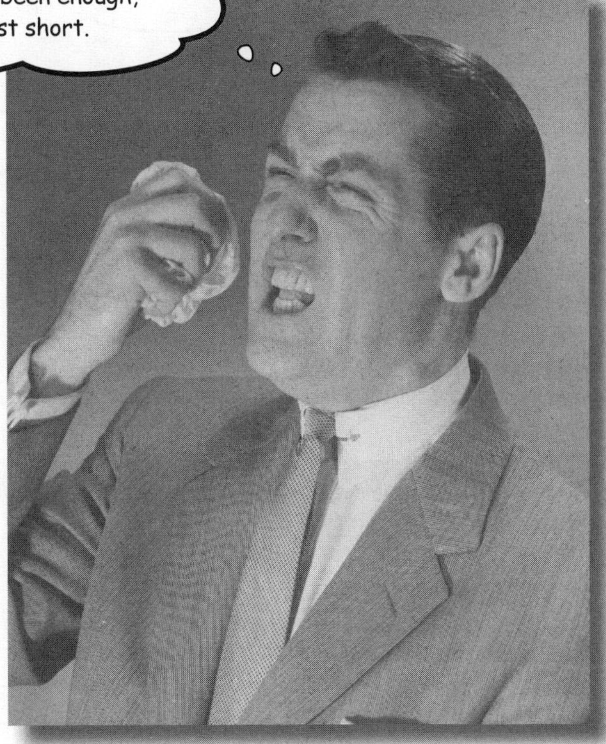

Oh, the inequality of it all... I should have been enough, but I came up just short.

Sometimes enough is enough... and sometimes it's not.

Have you ever thought, "I just need a little bit **more**"? But what if someone gave you ***more*** than just a bit more? Then you'd have **more than you need**...but life might still be pretty good. In this chapter, you'll see how Algebra lets you say, "Give me a little more... and then some!" With **inequalities**, you'll go ***beyond two values*** and allow yourself to get **more**, or **less**.

Kathleen really loves football

Kathleen wants to start her own fantasy football team but needs your help managing it. Each team is limited to spending no more than $1,000,000 on player salaries. Your job is to help Kathleen put together a balanced team.

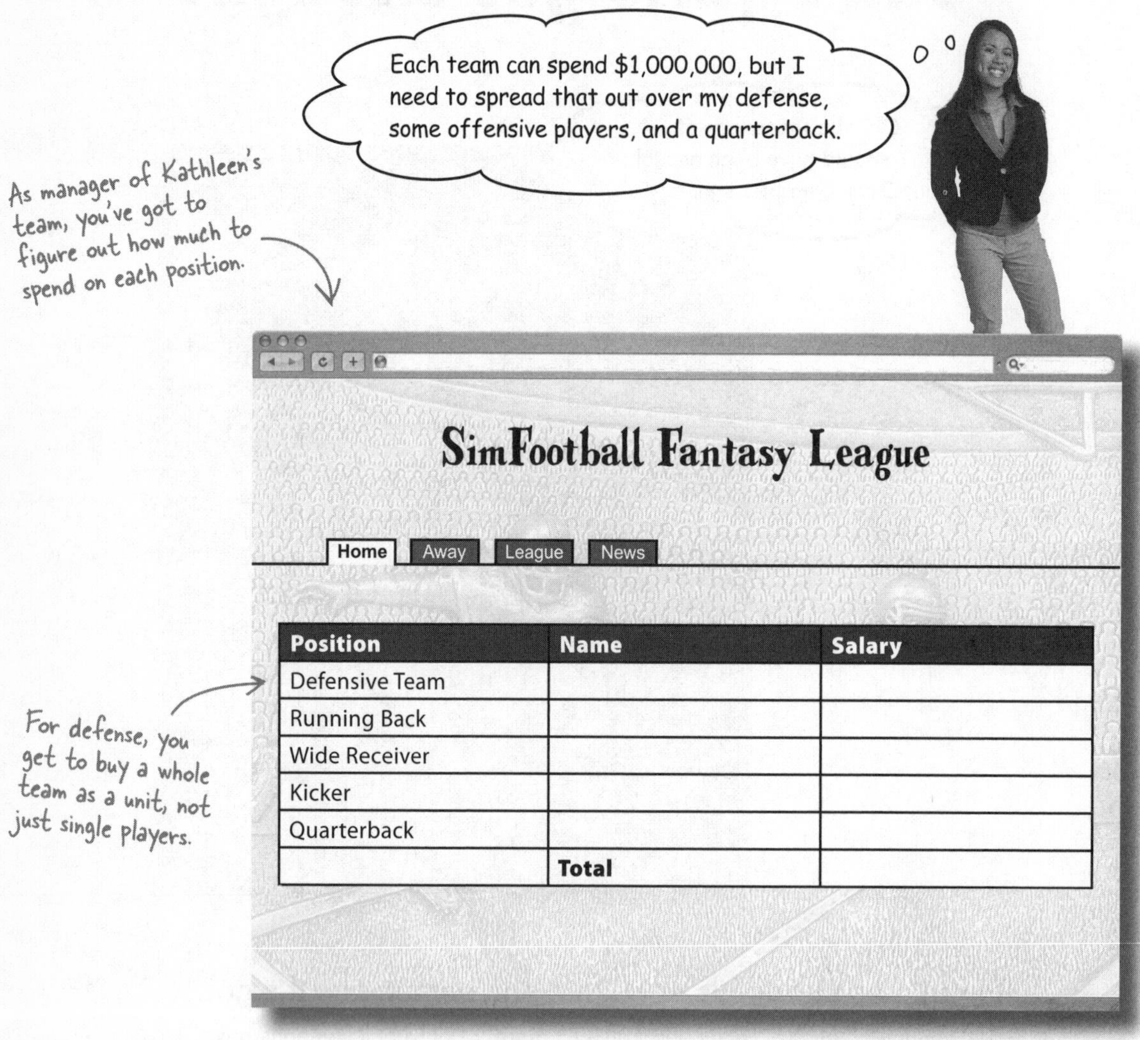
Each team can spend $1,000,000, but I need to spread that out over my defense, some offensive players, and a quarterback.
As manager of Kathleen's team, you've got to figure out how much to spend on each position.
SimFootball Fantasy League
Home
Away
League
News
For defense, you get to buy a whole team as a unit, not just single players.

Position	Name	Salary
Defensive Team		
Running Back		
Wide Receiver		
Kicker		
Quarterback		
	Total	

The cost of all players can't be more than $1,000,000

Kathleen needs to fill her team roster and keep her total costs under $1,000,000. Here are the choices Kathleen's got for her team... lots of decisions to make!

Defensive Teams

Team	Cost
Broncos	$300,000
Eagles	$200,000
Steelers	$333,000
Ravens	$250,000

Running Backs

Name	Cost
Mike Anta	$197,000
Bobby Hull	$202,187
Rick Timmer	$185,200
Ed Babens	$209,115

Wide Receivers

Name	Cost
Ben Toppy	$195,289
Eric Freidr	$212,000
Ron Jupper	$185,200
Mark Marten	$165,950

Here are the team and player lists you can pick from to put together Kathleen's team.

Kickers

Team	Cost
Joe Amten	$183,500
Rick Vuber	$155,000
Pete Hock	$203,200
Matt Eatens	$209,100

Quarterbacks

Name	Cost
Tony Jaglen	$208,200
Eric Hemal	$175,000
Pat Brums	$199,950
Dan Dreter	$202,400

Sharpen your pencil

Using the team price equation below, can you come up with a team that works? If not, why? Is there a problem with the equation?

Defensive Team Cost + Running Back Cost + Wide Receiver Cost + Kicker Cost + Quarterback Cost =1,000,000

Sharpen your pencil Solution

Using the team price equation below, can you come up with a team that works? If not, why? Is there a problem with our equation?

Defensive Team Cost + Running Back Cost + Wide Receiver Cost + Kicker Cost + Quarterback Cost =1,000,000

The problem with using an equals sign is that there isn't a combination of players that will equal exactly $1,000,000. We don't need exactly $1,000,000; the team just can't cost any more than that.

You're really working with a comparison...

We aren't looking for equality here. The cost of the entire team just needs to be ***less than or equal to*** $1,000,000. What we need is a way to show that the cost can be less than that amount, just not more than it.

When you have a comparison, you use something different than an equals sign. For that, use a comparison symbol like less than (<) or greater than (>).

Means "less than or equal to."

$$\text{Defensive Team} + \text{Running Back} + \text{Wide Receiver} + \text{Kicker} + \text{Quarterback} \leq \$1,000,000$$

All the players, added up.

The max that you can spend

So the < sign **compares** one side of the equation to the other, right?

The <, >, ≤ and ≥ symbols are comparisons. The first one, <, means the thing on the left is *less than* the thing on the right. The next one, >, means the thing on the left is *greater than* the thing on the right. The last two are *less than or equal to* and *greater than or equal to*, respectively. Statements using these comparison symbols are called ***inequalities*** since it's about two things ***not*** being equal.

More than and less than are COMPARISON WORDS. You'll see them all the time in word problems.

Inequalities Up Close

The = sign means "equal to," and you already know how it works. It means both sides of an equation are *the same*:

The equals sign says everything on this side...

...is equal to this side.

$$x + 7 = 10$$

With Kathleen's problem, we're dealing with **inequalities.** That means that the two sides aren't equal, but there is still a relationship. An inequality can be written with one of four symbols. They all express a different relationship: greater than, less than, greater than or equal to, and less than or equal to. They will appear in an equation exactly where you would expect to see an equals sign:

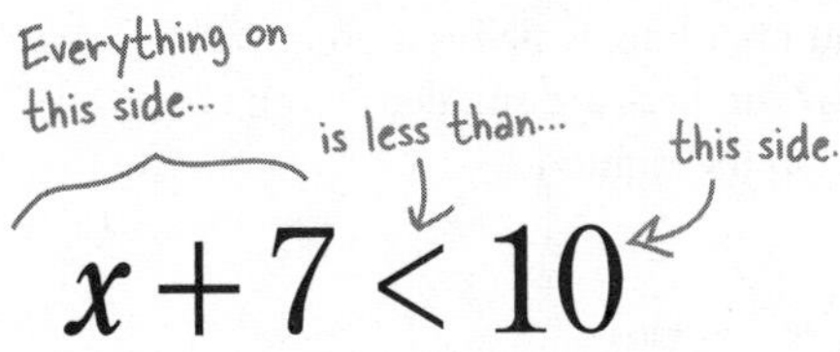

Sharpen your pencil

Below are some inequalities. Some are correct, some aren't. Write whether each inequality is true or false.

$5 < 10$	....	$1234 \geq 1233$	....
$8 > 4$	....	$101 \leq 101.5$	....
$4 < 8$	....	$-3 > 6$	....
$10 \leq 10$	....	$-10 < 10$	....
$1.23 > 3.2$	....	$-8 < -4$	....

Below are some inequalities. Some are correct, some aren't. Write whether each inequality is true or false.

$5 < 10$	true	$1234 \geq 1233$	true
$8 > 4$	true	$101 \leq 101.5$	true
$4 < 8$	true	$-3 > 6$	false
$10 \leq 10$	true	$-10 < 10$	true
$1.23 > 3.2$	false	$-8 < -4$	true

Remember, this is "less than or equal to," so since 10=10, this is true.

Negatives can be tricky! −4 is more than −8 even though 8 is bigger than 4.

Inequalities are COMPARISONS

The inequality symbols provide a way to **compare** two numbers, or two sets of numbers. Let's take a look a the number line, which is a great way to see the relationships between numbers. Numbers are smaller as you go to the left of the line, and they get bigger on the right.

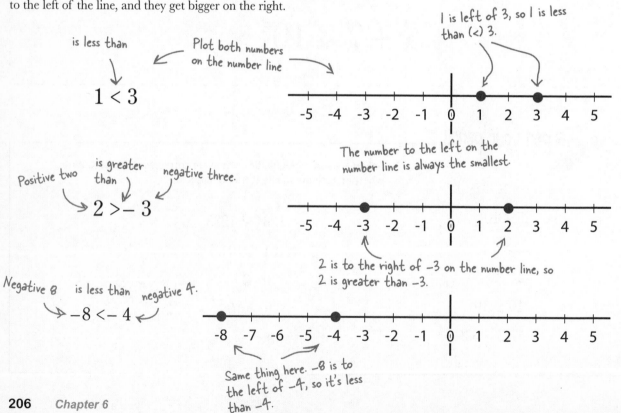

is less than

$$1 < 3$$

Plot both numbers on the number line

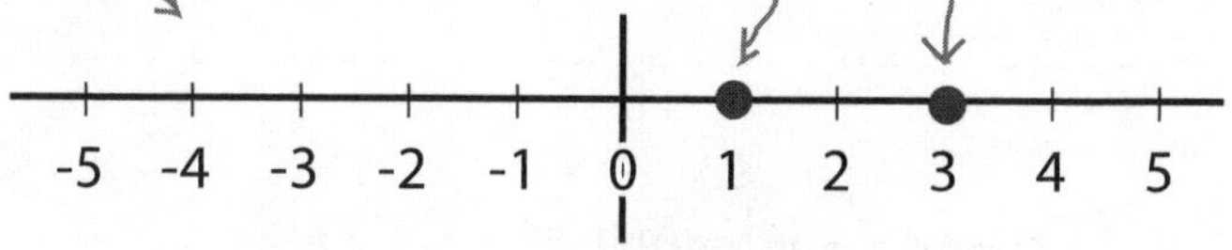

1 is left of 3, so 1 is less than (<) 3.

-5 -4 -3 -2 -1 0 1 2 3 4 5

The number to the left on the number line is always the smallest.

Positive two is greater than negative three.

$$2 > -3$$

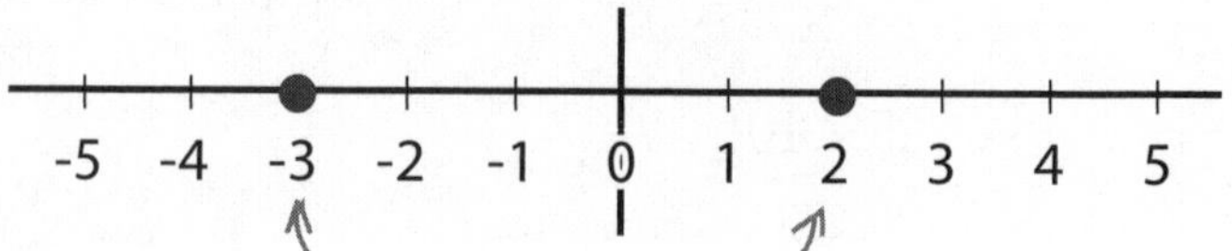

-5 -4 -3 -2 -1 0 1 2 3 4 5

2 is to the right of −3 on the number line, so 2 is greater than −3.

Negative 8 is less than negative 4.

$$-8 < -4$$

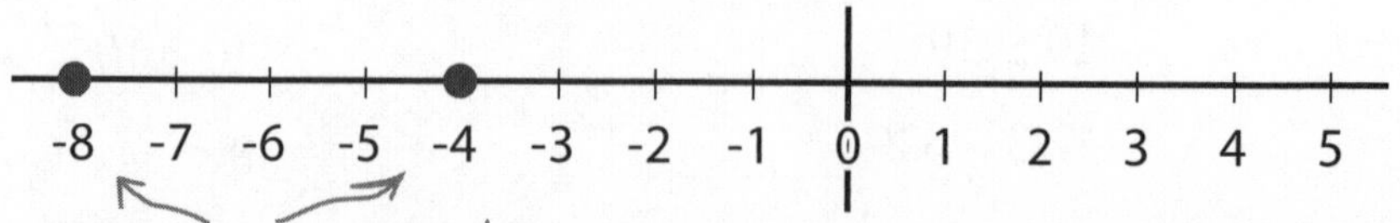

-8 -7 -6 -5 -4 -3 -2 -1 0 1 2 3 4 5

Same thing here. −8 is to the left of −4, so it's less than −4.

Hey - what about my team? I've made a bunch of picks, but I still need a quarterback.

SimFootball Fantasy League

Home | Away | League | News

Position	Name	Salary
Defensive Team	Steelers	$ 333,000
Running Back	Mike Anta	$ 197,000
Wide Receiver	Eric Freider	$ 212,000
Kicker	Rick Vuber	$ 155,000
Quarterback		
	Total	$ 897,000

Remember, Kathleen can only spend $1,000,000 total.

Sharpen your pencil

Write an inequality using the money Kathleen has already spent and figure out what's the most she could spend on her quarterback.

To isolate the variable, perform the inverse operation, just like you would with an equation.

Sharpen your pencil Solution

Write an inequality using the money Kathleen has already spent and figure out what's the most she could spend on her quarterback.

Defensive Team + Running Back + Wide Receiver + Kicker + Quarterback $\leq$ 1,000,000

Let q be how much Kathleen can spend on her quarterback.

$333{,}000 + 197{,}000 + 212{,}000 + 155{,}000 + q \leq 1{,}000{,}000$

$897{,}000 + q \leq 1{,}000{,}000$

$-897{,}000 + 897{,}000 + q \leq 1{,}000{,}000 - 897{,}000$

$q \leq 103{,}000$

I'm just not getting it. We did exactly the same math as if it had been an equals sign. Why are we going through this inequality stuff?

The big difference between an inequality and an equality is the MEANING of your ANSWER.

The solution to the equation means Kathleen can spend *at most* \$103,000 on a quarterback. But that doesn't mean she has to spend *exactly* that amount. If there was a quarterback for \$94,000, Kathleen could spend that, and her equation would be true (\$94,000 ≤ \$103,000).

Reading the solution of an inequality is the easiest way to make sense of it. For instance, "***q*** is less than or equal to 103,000." With inequalities, a range of answers work. The answers that work with an inequality are called the **solution set**.

A solution set is all of the values that satisfy an expression. In Kathleen's case, it's any number less than or equal to 103,000.

there are no Dumb Questions

Q: Why do we need inequalities?

A: An inequality is a lot more realistic in many situations. If you only need to find out if you have enough gas, *any* amount of gas over what you need is fine. So that's an inequality.

For Kathleen, there are a bunch of different ways she can spend \$1,000,000, and they all work...even if none of them add up to exactly 1,000,000.

Q: What's the difference between "less than" and "less than or equal to"?

A: "Less than" means that the solution set includes everything up to (but not including) the value on the other side of the inequality. That includes decimals and fractions, too.

So, if you have $x < 6$, x can be any number below 6. For example, 5.99999999999 is okay, but *not* 6 itself.

If you change the expression to $x \leq 6$, then 6 is an answer that works.

Q: Same deal with "'greater than" and "greater than or equal to"?

A: Yes, except the answers are higher. So if the inequality is $x > 6$, the solution set is above 6, for example, 6.000000001.

Q: We performed inverse operations around an inequality sign. So do inequality signs always work just like an equals sign?

A: Well, not always. We'll get into more detail later, but as long as you're only doing addition and subtraction, the inequality behaves just as an equality.

Q: What do you mean, not always?

A: With multiplication and division of negative integers, things can get interesting. In those cases, the inequality sign can change. We'll talk more about that in just a few pages.

Q: Does this inequality thing really help? I mean, having a whole mess of answers doesn't seem very exact.

A: It's not. What is important to understand is that when you're dealing with many real-world situations, you don't need a single number. You actually need to know *all* the numbers that would work.

Algebra is able to deal with many more complex situations than traditional arithmetic, and this is one of them. Part of Algebra is manipulating expressions to figure out the solution, but you also have to check that your solution actually makes sense.

Q: When do you use a number line?

A: Any time you feel like you're not quite sure about a comparison or how to analyze your answer. We'll revisit number lines later in the chapter as a way to show the entire solution set.

Q: An equation usually has a set number of solutions, right? How many solutions do inequalities have?

A: Inequalities have an ***infinite number of solutions***. You're trying to find the boundaries: the highest or lowest values that are allowed. Once you know that, you can solve your problem. And remember, just because there might be an infinite number of *mathematically* correct solutions, it doesn't always mean all those values make sense in the real world.

For example, there are infinite numbers between \$102,999.999 and \$103,000.00, but they probably don't make a whole lot of a difference to Kathleen. Since she's talking about money, \$102,999.999 is basically the same as \$103,000.00 for her problem.

Q: With equality, it means the same thing on both sides. What does inequality mean? Not the same?

A: Generally, yes. By reading the inequality symbol that is used in the expression, you can figure out exactly what the relationship is between two sides.

Q: What about that ≠ sign that I've seen in math books? Is that an inequality?

A: It means "not equal to." Like:

$$4 \neq 6$$

As symbols go, it's not very descriptive. It is an inequality symbol, though.

Inequalities involving some negative number operations need special treatment

Kathleen still needs a good quarterback, but there's another way she can stack the deck in her favor. The fantasy football league she's in allows you to handicap another team by buying "penalty points" before a game. For every penalty point she buys, the opposing team loses ten yards of offense when they play her team.

I'm playing my brother's team next, so I want a safety net for the upcoming game. I want at least 50 yards off my opponent.

What could Kathleen have said to make this an equality instead?

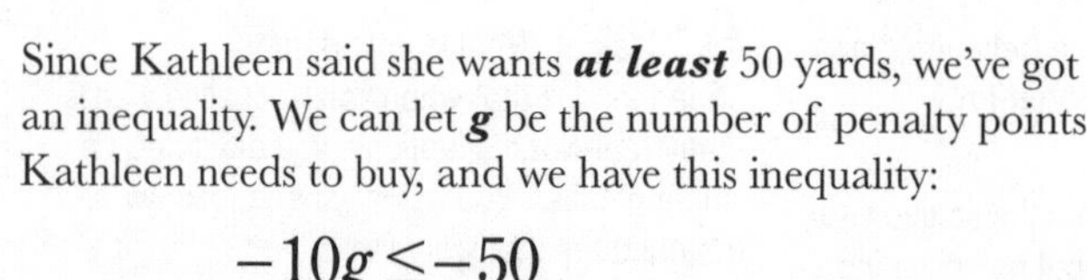

Since Kathleen said she wants ***at least*** 50 yards, we've got an inequality. We can let ***g*** be the number of penalty points Kathleen needs to buy, and we have this inequality:

$$-10g \leq -50$$

Take a close look at this. For every pentalty point (***g***), the opposing team loses 10 yards (-10). And Kathleen wants her brother to start out at -50 yards.

To solve this, we need to divide both sides by -10:

$$\frac{-10g}{-10} \leq \frac{-50}{-10}$$

$$g \leq 5...???$$

This can't be right... we must have something wrong here.

There is so much wrong here; I don't even know where to start. Why is -10g less than or equal to -50 when Kathleen said "at least"? Then, you get g ≤ 5? That works for g = 5, but anything less than that, and it's completely wrong! If g is 4, then we end up with -40. And that's **less than or equal to** -50?

Inequalities with negative numbers need some SPECIAL RULES.

Let's take a closer look at what's going on.

Negative inequalities work BACKWARD

First let's deal with the "at least" part of the problem. Kathleen wants at least 50 yards taken off of her opponent's offense. 51 yards, 52 yards, 60 yards... those all work. But 40 yards is not enough. Since these are actually negative yards for the opponent, that means -50 yards works, -60 yards works, but -40 yards doesn't:

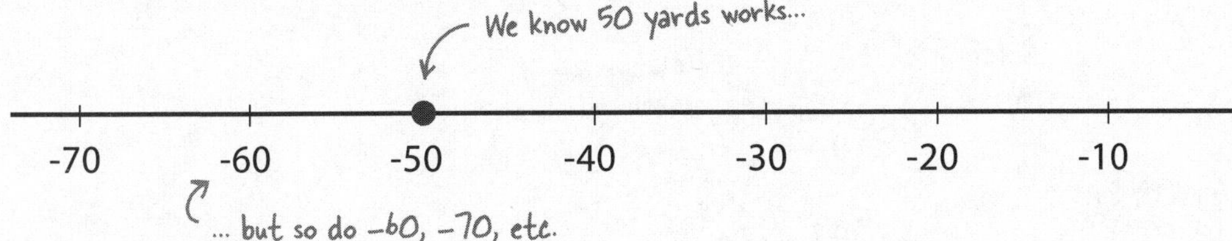

We know 50 yards works...
-70
-60
-50
-40
-30
-20
-10
... but so do -60, -70, etc.

Since numbers on the left of the number line are **less than** numbers on the right, the expression we're trying to solve has to be *less than* -50. So what we end up with is this:

$$-10g \leq -50$$

Whatever the left side of our inequality comes out to be, it has to be ***less than or equal to*** -50.

Multiplication and division of negative numbers causes problems for inequalities

So now we need to figure out what went wrong at the end when we solved for ***g***. The reason that the inequality doesn't work with negative numbers is simple: negative numbers are ***backward***. For example, -10 is actually *less than* -2, exactly the opposite of the positive numbers 10 and 2.

One million is a really BIG number

Ten is not a very big number...

$$-1{,}000{,}000 < -10$$

But, since it's negative, it's way less than zero, and it's value is small.

...but it's also pretty close to zero, so it's value is bigger.

The end result of all of this is that when you multiply or divide by a negative number, the relationship expressed with the inequality is ***reversed***. That's because you're changing the direction of the relationship of the numbers to zero.

So, how do you handle that?

If you multiply or divide by a negative number the inequality becomes backwards.

FLIP the inequality sign with negative multiplication and division

Since the **value** of negative numbers is the **opposite** of the size of the number, negative numbers reverse the relationship of the inequality. For an example, let's multiply both sides by -2.

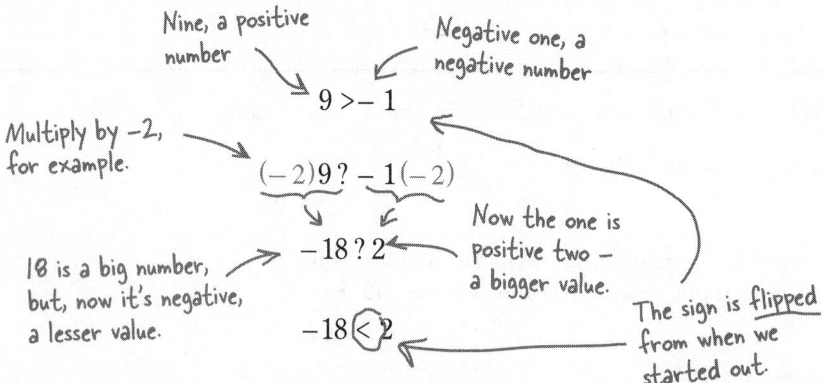
Nine, a positive number
Negative one, a negative number
9 >– 1
Multiply by –2, for example.
(– 2)9 ? – 1(– 2)
Now the one is positive two – a bigger value.
– 18 ? 2
18 is a big number, but, now it's negative, a lesser value.
– 18 < 2
The sign is flipped from when we started out.

The inequality started with a larger number on the left and ended with a larger number on the right—the opposite relationship. The solution to how to work with this is easy. When you multiple or divide by a negative number, flip your inequality sign..

Use the number line to visualize the relationship

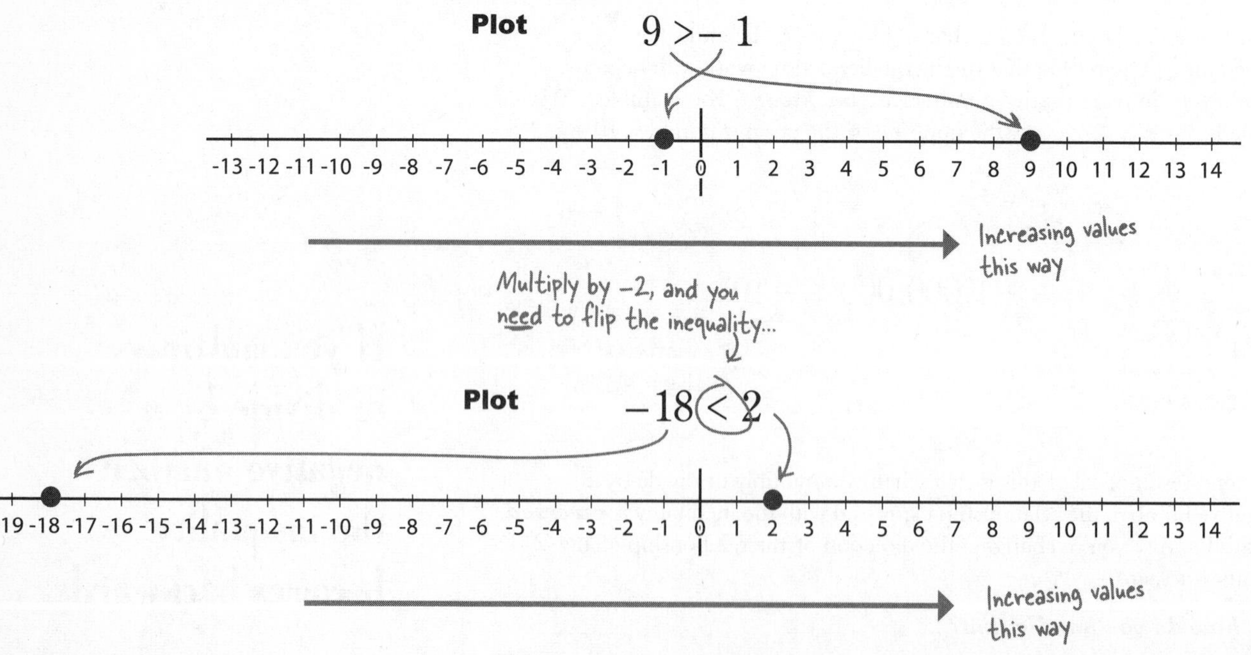
Plot
9 >– 1
-13 -12 -11 -10 -9 -8 -7 -6 -5 -4 -3 -2 -1 0 1 2 3 4 5 6 7 8 9 10 11 12 13 14
Increasing values this way
Multiply by –2, and you need to flip the inequality...
Plot
– 18 < 2
-19 -18 -17 -16 -15 -14 -13 -12 -11 -10 -9 -8 -7 -6 -5 -4 -3 -2 -1 0 1 2 3 4 5 6 7 8 9 10 11 12 13 14
Increasing values this way

When you're working with an inequality and negative multiplication or division...

1. **Start with a valid inequality.**
The inequality can contain numbers or unknowns, and it just needs to be true (and if you can check it with a number line, that's good too).

2. **Work with the equality just like an equation, as long as you don't need to multiply or divide by a NEGATIVE number.**

3. **Multiply or divide both sides by a NEGATIVE number.**
If you need to multiply or divide by a negative number, be sure and do that to both sides of your equation. But you're not done. Anytime you multiply or divide by a negative, you've got to immediately...

4. **Flip the inequality symbol in the equation.**
Do this right away! Greater-than becomes less-than, less-than-or-equal becomes greather-than-or-equal, and so forth. And you have to do this every time you multiply or divide by a negative number.

It's easy to visualize the location of your numbers on a number line and figure out how the inequality relationship changes. The inequality sign is just a way to keep track of the relationship in your equation. Your job is to work with your equation and the inquality sign in a way that preserves a correct relationship.

So, all you have to do is reverse the inequality when you multiply or divide by negative numbers. You're not changing the expression... you're actually ***preserving*** the expression.

Negative numbers get LARGER in value as they get CLOSER to zero.

there are no Dumb Questions

Q: How can we just flip the inequality? Isn't that changing everything?

A: Actually, flipping the inequality is how you ***keep things the same***. If you remember the number line,when you multiply or divide by a negative number, you change the relationship between the two sides of the equation.

The inequality symbol is just keeping track of which side is worth more. A high negative number is worth much less than a low positive number (and visa versa).

Q: How do you flip a ≤ ?

A: If your inequality has an "equal to" component, you replace it with the opposite "equal to" inequality. So, "less than or equal to" flipped is "greater than or equal to."

Q: Can you tell me again exactly what a solution set is?

A: The solution set for an inequality contains ALL of the numbers that make your inequality true.

Q: Will I ever have to flip an inequality sign more than once?

A: Possibly. If the inequality is written in such a way that you have to multiply or divide both with a negative number more than once, you'll flip your sign more than once. Just flip your sign every time you do the negative multiplication or division, and you'll be okay.

It's just like when you solve equations; if you apply all the rules correctly, you'll get the right answer every time.

Q: What if I need to multiply or divide by a fraction, or a decimal. Do I need to flip my inequality sign then?

A: Only if it's a *negative* decimal or fraction. It doesn't matter what form the negative number comes in; if it's negative, you'll need to flip the inequality sign.

Q: What if I add or subtract a negative number?

A: There's no flipping required. That's because you're preserving the direction things are going. If you add a negative number to both sides, both sides of the inequality will be moving to the left on the number line at the same rate. That means the sides of the equation have the same relationship to each other.

Remember, it's all about the relative relationship of the sides. If that doesn't change, then there's no reason to change anything.

Q: What if we're just simplifying an inequality by multiplying or dividing one side of an inequality by a negative number? Something like (-3)(2) > -10?

A: In that case, since you're not moving a negative number across the inequality, you don't change the sign. So, if you work the example, you get -6 > -10, which is correct.

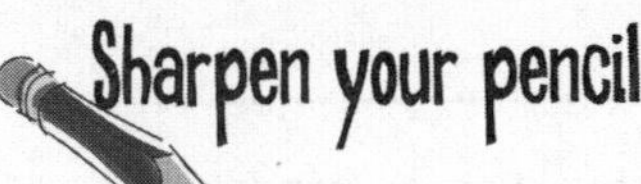

Now that you can handle negative inequalities, you can solve Kathleen's penalty points problem correctly. Here are a few other inequalities to try out, too. Use the number line at the bottom to check your work.

"I'm playing my brother's team next, so I want a safety net for the upcoming game. I want at least 50 yards off my opponent." Figure out how many penalty points Kathleen needs to buy.

The league gives 40 points for each game you win, and you have to have more than 260 points to get the Heisman Trophy at the end of the season. How many games does Kathleen need to win for her Quarterback to get the Heisman?

League rules say if you have more than 10 penalties in the season, your team is disqualified. Kathleen wants to make sure she spaces penalties out evenly over her 16 games. What's the maximum number of penalties she can get per game and not get disqualified?

-19 -18 -17 -16 -15 -14 -13 -12 -11 -10 -9 -8 -7 -6 -5 -4 -3 -2 -1 0 1 2 3 4 5 6 7 8 9 10 11 12 13 14 15 16 17 18 19

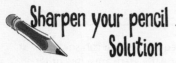

Sharpen your pencil Solution

Now that you can handle negative inequalities, you can solve Kathleen's penalty points problem correctly. Here are a few other inequalities to try out, too. Use the number line at the bottom to check your work.

"I'm playing my brother's team next so I want a safety net for the upcoming game. I want at least 50 yards off my opponent." Figure out how many penalty points Kathleen needs to buy.

$-10g \leq -50$

$\frac{-10g}{-10} \; ? \; \frac{-50}{-10}$

$g \geq 5$

Don't forget to flip the sign when you divide by −10.

Kathleen needs to buy 5 or more penalty points.

The league gives 40 points for each game you win, and you have to have more than 260 points to get the Heisman Trophy at the end of the season. How many games does Kathleen need to win for her Quarterback to get the Heisman?

$40w > 260$

$\frac{40w}{40} > \frac{260}{40}$

$w > 6.5$

No sign flipping here – you're dividing by a positive number.

Kathleen needs to win more than 6.5 games, but since you can't win half a game, she needs to win 7 or more games.

League rules say if you have more than 10 penalties in the season, your team is disqualified. Kathleen wants to make sure she spaces penalties out evenly over her 16 games. What's the maximum number of penalties she can get per game and not get disqualified?

$16p \leq 10$

$\frac{16p}{16} \leq \frac{10}{16}$

$p \leq 0.625$

The mathematical answer is less than or equal to 0.625 penalties per game, but since you can't get part of a penalty, she can't space her penalties evenly over all of the game. She'll have to decide which games she needs to play dirty in and which ones she'll need to be on good behavior for.

Math Magnets

Using the magnets below, fill in the blanks to solve the inequalities.

$12 > 2x - 4x + 16$

$-16 + 12 > 2x - 4x + 16$ ....

.... > ....

.... $> -2x$

$\frac{-4}{-2}$ .... $\frac{-2}{-2}x$

.... x

Check your work:

Choose any number that satisfies the inequality.

$12 > 2(5) - 4(5) + 16$

$12 >$ ....

$12 >$ ....

It's a good idea to pick a number that's easy to work with to test these out.

$15 - 3y > -9y + 18$

$-18 + 15 - 3y > -9y + 18$ ....

.... > ....

$3y$ .... $> -9y + 3y$

$-3 > -6y$

$\frac{-3}{-6}$ .... $\frac{-6}{-6}y$

.... y

Check your work:

$15 - 3(3) > -9(3) + 18$

.... > ....

.... > ....

Put your solution back into the original equation to make sure you got things right.

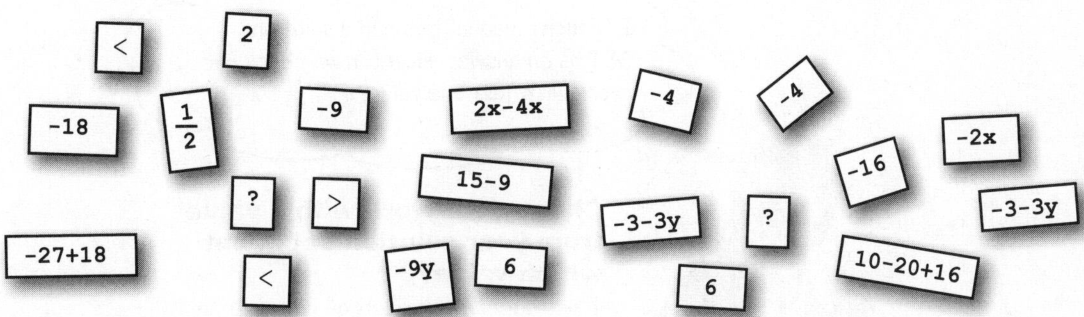

Math Magnets Solution

Your job was to fill in the blanks and solve the inequalities.

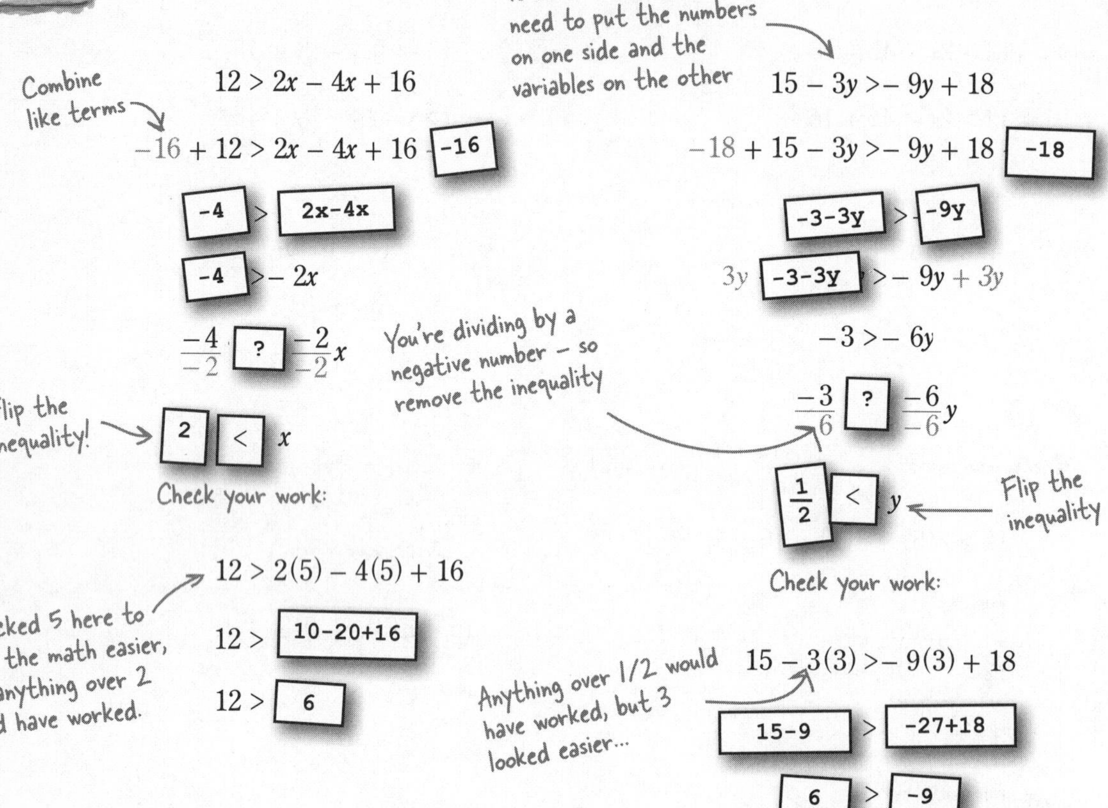

Combine like terms
12 > 2x – 4x + 16
–16 + 12 > 2x – 4x + 16 -16
-4 > 2x-4x
-4 >– 2x
–4/–2 ? –2/–2 x
Flip the inequality!
2 < x
Check your work:
12 > 2(5) – 4(5) + 16
We picked 5 here to make the math easier, but anything over 2 would have worked.
12 > 10-20+16
12 > 6
To combine like terms, you need to put the numbers on one side and the variables on the other
15 – 3y >– 9y + 18
–18 + 15 – 3y >– 9y + 18 -18
-3-3y > -9y
3y -3-3y >– 9y + 3y
–3 >– 6y
You're dividing by a negative number – so remove the inequality
–3/–6 ? –6/–6 y
1/2 < y
Flip the inequality
Check your work:
Anything over 1/2 would have worked, but 3 looked easier...
15 – 3(3) >– 9(3) + 18
15-9 > -27+18
6 > -9
-2x

I thought inequalities had a solution SET as an answer. How can we check our work with just one value?

Check your work with a value from your solution set.... but which value?

The solution set has lots of values. In fact, let's take a closer look at solution sets on our number lines...

You can visualize a solution set on a number line

The number line is a great way to visualize a solution set. You've been using number lines to show points, but how about a **range of points** or an **entire solution set**? That would help us pick numbers to check our work and also help to understand what numbers will satisfy an inequality.

Here's what we need to do:

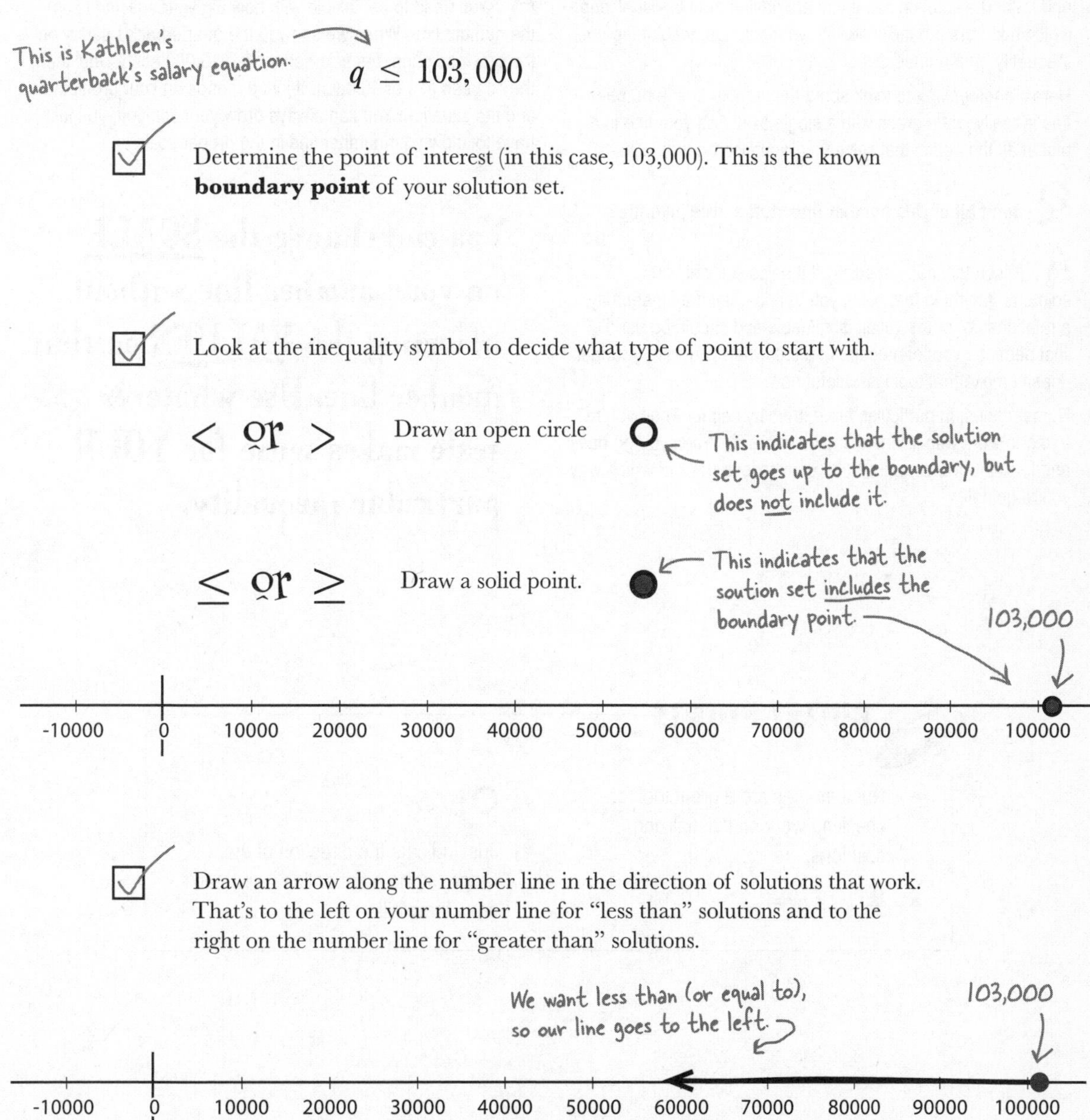

This is Kathleen's quarterback's salary equation.
$q \leq 103,000$
Determine the point of interest (in this case, 103,000). This is the known **boundary point** of your solution set.
Look at the inequality symbol to decide what type of point to start with.
$<$ or $>$
Draw an open circle
This indicates that the solution set goes up to the boundary, but does not include it.
$\leq$ or $\geq$
Draw a solid point.
This indicates that the soution set includes the boundary point.
103,000
-10000
0
10000
20000
30000
40000
50000
60000
70000
80000
90000
100000

Draw an arrow along the number line in the direction of solutions that work. That's to the left on your number line for "less than" solutions and to the right on the number line for "greater than" solutions.

there are no Dumb Questions

Q: Why bother with the number line?

A: The number line is useful as a tool to help you understand a solution set. If you are having trouble visualizing which numbers are included with your solution, try plotting your inequality on a number line.

Here's another way to think about the number line: a number line is really just a graph with a single axis. And your line is a plot of all the points that solve your equation on that axis.

Q: Isn't all of this number line stuff a little juvenile?

A: Absolutely not, especially if it helps you get your solutons. Anything that helps you to understand an inequality, a relationship, or a solution is valuable and should be used. Just because you learned about a tool when you were younger, doesn't make that tool less useful now.

Number lines, in particular, are extremely helpful when you're trying to work with integers and inequalities. When you go back and forth around zero, it can be easy to lose track of which way you're going!

Q: What do we do if there's a big number and I can't draw a line that goes up that high?

A: You need to be flexible with how big your spacing is on the number line. When we showed the quarterback's salary on the previous page, the tick marks were 10,000 each. Until then, they'd been at 1 or 10 each. It just depends on your problem and the situation. You can always draw your solution; you just have to show your number line in the proper scale.

You can change the SCALE on your number line without affecting the VALUES on that number line. Use whatever scale makes sense for YOUR particular inequality.

BULLET POINTS

- Number lines are a great tool for checking work and visualizing solutions.
- ● for ≤ or ≥
- ○ for < or >
- Just indicate the direction of the range on the number line—you can't draw them all!

Solve the following problems and graph the solution sets on the number lines.

The high school baseball league is trying to determine their MVP, and they're reviewing the stats for the season. To win MVP, a player has to have a batting average greater than 0.320. The batting average is the number of hits for the season divided by the number of at bats. There are 12 games in the season, and each player has 4 at bats a game.

How many hits does a player need to be a contender for the title of MVP?

-19 -18 -17 -16 -15 -14 -13 -12 -11 -10 -9 -8 -7 -6 -5 -4 -3 -2 -1 0 1 2 3 4 5 6 7 8 9 10 11 12 13 14 15 16 17 18 19

It's the last round of darts between Joe and Sam. The score is Joe, 18, and Sam, 12. Whoever has the high score after Joe's turn is over wins. There's also a rule that if you hit your opponent in the ear with a dart, it's an automatic deduction of 3 points. Sam said something about Joe's Mom, so now Joe wants to know how many times he can hit Sam in the ear and still tie or win the game.

-19 -18 -17 -16 -15 -14 -13 -12 -11 -10 -9 -8 -7 -6 -5 -4 -3 -2 -1 0 1 2 3 4 5 6 7 8 9 10 11 12 13 14 15 16 17 18 19

Exercise Solution

Solve the following problems and graph the solution sets on the number lines.

The high school baseball league is trying to determine their MVP, and they're reviewing the stats for the season. To win MVP, a player has to have a batting average greater than 0.320. The batting average is the number of hits for the season divided by the number of at bats. There are 12 games in the season, and each player has 4 at bats a game.

How many hits does a player need to be a contender for the title of MVP?

MVP batting average > 0.320

$\frac{\text{Number of hits}}{\text{No. of games} \cdot 4 \text{ bats a game}} > 0.320$

Here, we just wrote out the equation in words to get started.

$\frac{h}{12 \cdot 4} > 0.320$

$\frac{h}{48} > 0.320$

To isolate the variable, we need to multiply by 48

$\cancel{48} \cdot \frac{h}{\cancel{48}} > 0.320 \cdot 48$

$h = 15.36$

Think about the problem, we can't have 0.36 of a hit, so really it's h > 15.

This is the solution set

-19 -18 -17 -16 -15 -14 -13 -12 -11 -10 -9 -8 -7 -6 -5 -4 -3 -2 -1 0 1 2 3 4 5 6 7 8 9 10 11 12 13 14 15 16 17 18 19

It's the last round of darts between Joe and Sam. The score is Joe, 18, and Sam, 12. Whoever has the high score after Joe's turn is over wins. There's also a rule that if you hit your opponent in the ear with a dart, it's an automatic deduction of 3 points. Sam said something about Joe's Mom, so now Joe wants to know how many times he can hit Sam in the ear and still tie or win the game.

Joe's score after ≥ Sam's score now

Joe's score now – lost points ≥ Sam's score now

More getting started work

$18 - 3(E) \geq 12$

We used "E" for ear.

$\cancel{-18} + 18 - 3(E) \geq 12 \cancel{-18}$

$-3(E) \geq -6$

Divide by a negative, so lose the inequality symbol.

$\frac{\cancel{-3}(E)}{\cancel{-3}} \; ? \; \frac{-6}{-3}$

Flip the inequality.

$E \leq 2$

Joe can whack Sam twice in the ear and still tie!

-19 -18 -17 -16 -15 -14 -13 -12 -11 -10 -9 -8 -7 -6 -5 -4 -3 -2 -1 0 1 2 3 4 5 6 7 8 9 10 11 12 13 14 15 16 17 18 19

The whole solution set for the INEQUALITY goes on forever, but for our PROBLEM, it stops at 0. You can't whack a guy in the ear negative times!

Inequalities can have TWO variables

Kathleen wants to pick a different defensive team. That gives her more money to work with for the Quarterback, but now she has **two things** she doesn't know...

I need some help - $103,000 is not enough for a quarterback...

SimFootball Fantasy League

Home Away League News

Position	Name	Salary
Defensive Team		
Running Back	Mike Anta	$ 197,000
Wide Receiver	Eric Freider	$ 212,000
Kicker	Rick Vuber	$ 155,000
Quarterback		
	Total	$ 564,000

Kathleen removed her defensive team pick and freed up some more cash.

Kathleen can only spend $1,000,000.

Sharpen your pencil

Use the information above to work out the new inequality for the two missing salaries: Defensive Team and the Quarterback.

Sharpen your pencil Solution

Your job was to work out an inequality for the salaries of the Defensive Team and Quarterback.

Defensive Team + Running Back + Wide Receiver + Kicker + Quarterback ≤ 1,000,000

~~333,000~~ + 197,000 + 212,000 + 155,000 + q ≤ 1,000,000

Kathleen put these guys back.

d + 197,000 + 212,000 + 155,000 + q ≤ 1,000,000

Let's call this d.

d + 564,000 + q ≤ 1,000,000

~~−564,000~~ + d + 564,000 + q ≤ 1,000,000 − ~~564,000~~

So, here is our final inequality. d + q ≤ 436,000

So, what can I do with that? Subsitution?

Substitution is almost always a good idea.
For example, suppose you find a quarterback you like who makes $185,000. You can substitute back into the inequality, and then you'll always get another valid inequality that you can use for the defensive team, ***d***.

Put 185,000 in for q.

$$d + 185,000 \leq 436,000$$

$$-185,000 + d + 185,000 \leq 436,000 - 185,000$$

$$d \leq 251,000$$

You don't need to flip the inequality here, this is just subtraction.

Now we know how much is left for the defenseive team.

Trial and error every time? That seems really inefficient. Is there another way?

Remember graphing? With equations, a graph gives you an entire line worth of solutions (if the equation has two variables). So graphing is actually a way to see the solutions to an equation.

Let's try graphing a basic inequality. There are a couple of things to keep in mind:

1. **When you graph an inequality, start with a line.**
   Just graph the inequality the way you would any other equation.

2. **Figure out what range the inequality's solutions fall in.**
   Is your answer above or below the line? This is similar to graphing with the number line: think about how your answer relates to your line.

3. **Mark the side of the line your solutions are on.**
   You can used dashed lines, solid lines, shading... anything you want.

These are just some hints to get you started; we'll get into details soon!

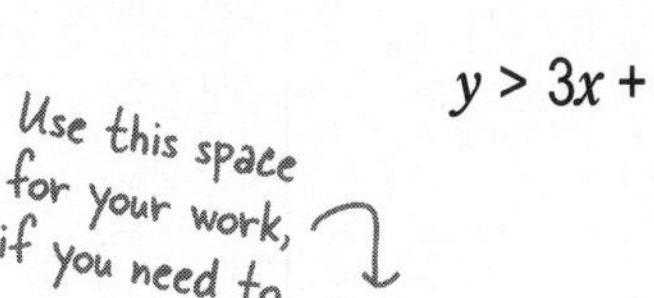

Sharpen your pencil

Try graphing the inequality on the graph below.

$y > 3x + 2$

Use this space for your work, if you need to.

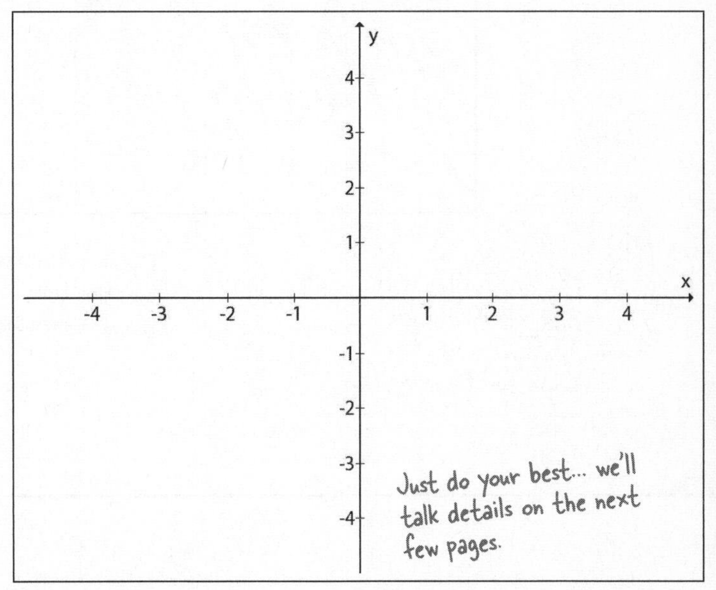

Sharpen your pencil Solution

Your job was to try and graph the inquality below. How did you do? Here's what we did:

This is in point-slope form!

$y > 3x + 2$

m (slope)

b (y-intercept)

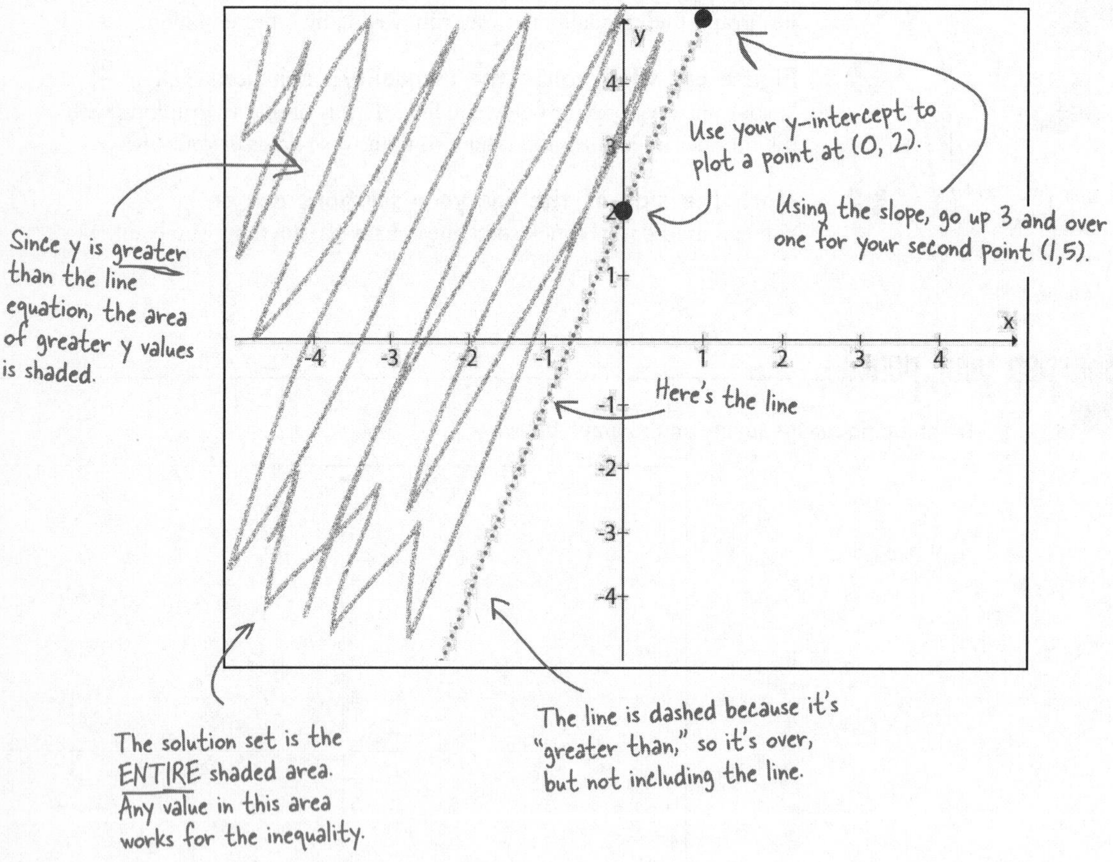

Use your y-intercept to plot a point at (0, 2).

Using the slope, go up 3 and over one for your second point (1,5).

Since y is greater than the line equation, the area of greater y values is shaded.

Here's the line

The solution set is the ENTIRE shaded area. Any value in this area works for the inequality.

The line is dashed because it's "greater than," so it's over, but not including the line.

Use a graph to visualize the solutions to an inequality

Graphing inequalities in two variables is like graphing equations, with some extra shading. (Just like solving inequalities is like solving equations.) If you have an inequality, to graph the whole solution, you need to do the following:

1. Start with a valid inequality in two variables. The variable on the vertical axis (usually y) needs to be isolated.

2. Decide the format for the line:

   $<$ or $>$ Draw a dashed line. – – – – –

   $\leq$ or $\geq$ Draw a solid line. ————

These boundaries are exactly the same idea as the ones that we used for the number line!

3. Draw the line the way you would an equality. Just make sure that it's dashed or solid, depending upon your inequality.

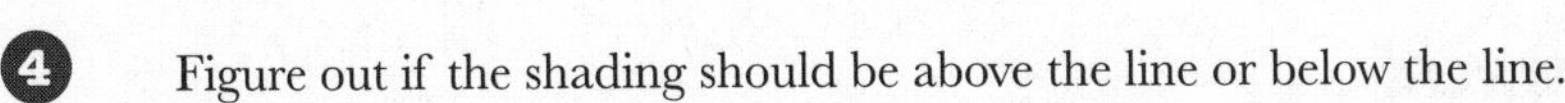

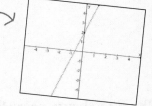

4. Figure out if the shading should be above the line or below the line.

   y > or ≥ Shade higher y values.

   y < or ≤ Shade lower y values.

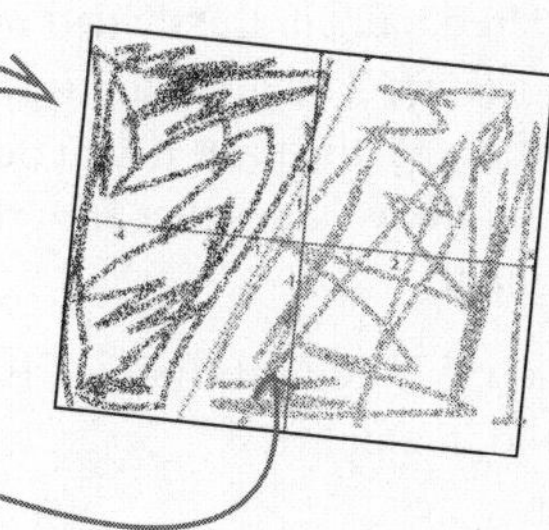

5. Shade the graph.

What does the shaded region actually represent?

Answers made in the shade

Look at our example from earlier, and plot one of the solutions. Does that tell you anything about what the shaded area represents?

$y > 3x + 2$

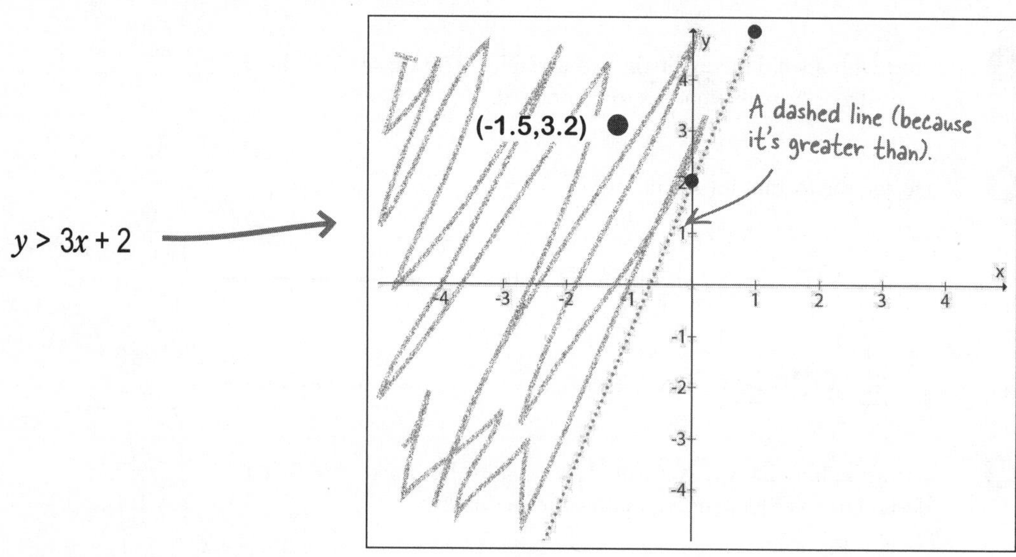

(-1.5,3.2)
A dashed line (because it's greater than).
y
x

Shading shows potential answers

The ordered pairs in the shaded region satisfy the inequality. *Every pair is a solution!* The great thing about this is that it makes trial and error a thing of the past. If you decide you want x to be -1.5, you can use all the y values that are greater than the line and still satisfy the inequality.

Give it a try! If you substitute in both numbers, you should always get a valid inequality:

It's plotted up on the graph, and it's in the shaded area. **Try (-1.5, 3.2)**

$$y > 3x + 2$$

$$3.2 > 3(-1.5) + 2$$

$$3.2 > -4.5 + 2$$

$$3.2 > -2.5$$

See, the inequality still holds!

All of the ordered pairs in the shaded region satisfy the inequality

Match each inequality to its graph.
Be careful, not every equation has a graph!

Inequality	Its graph
$t > 2(d - 0.5)$	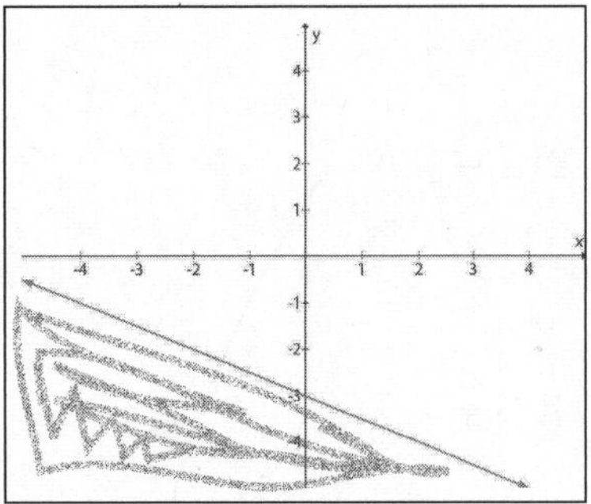
$y - 3 > x - 7$	
$y \geq -\frac{x}{2} - 3$	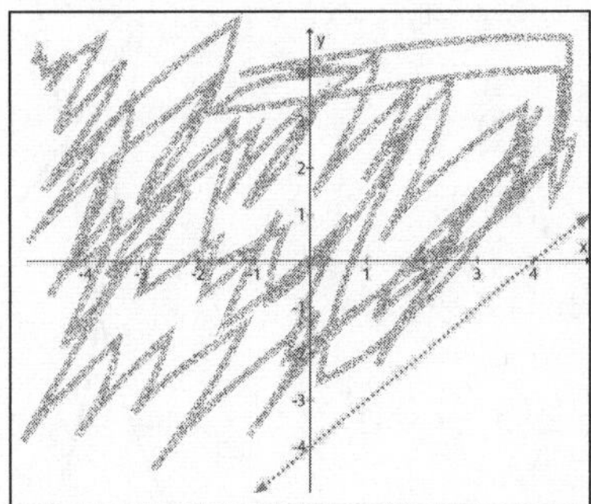
$y > 2(x - 0.5)$	
$y \leq -\frac{x}{2} - 3$	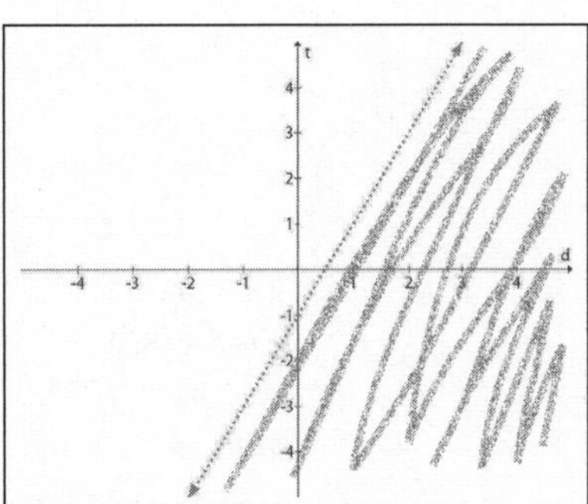
$y - 3 \geq x - 7$	

Who Does What? Solution

Match each inequality to its graph.
Be careful, not every equation has a graph!

Inequality

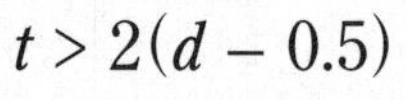

$y - 3 > x - 7$

The inequality is similar to the first graph, but the inequality is the wrong way

$y \geq -\frac{x}{2} - 3$

This one has the wrong variables (not t & d).

$y > 2(x - 0.5)$

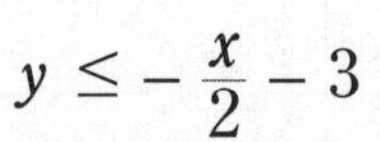

$y - 3 \geq x - 7$

This is similar to the second graph, but it would have to be a solid line.

Its graph

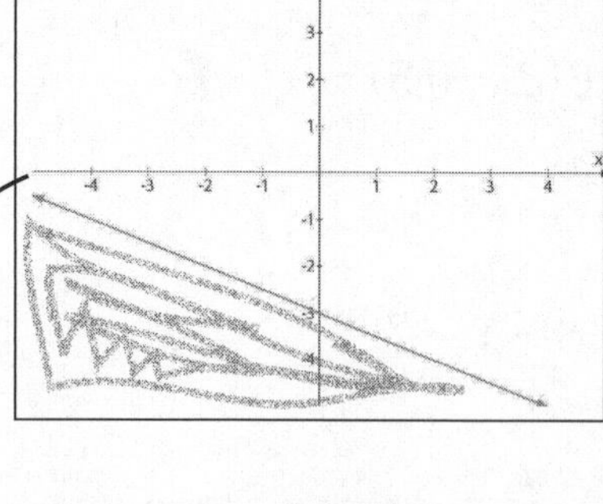

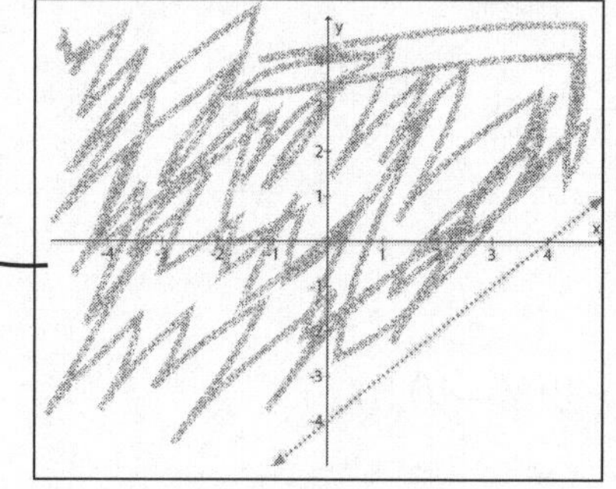

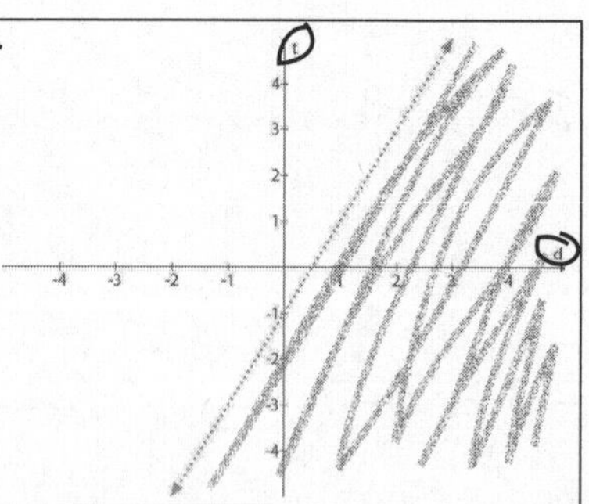

So I can graph the players I have left, then all I have to do is look at the rosters and the graph to figure out the options I've got.

Sharpen your pencil

This is from earlier...

$$d + q \leq 436,000$$

Graph Kathleen's inequality. You'll have to get ***d*** in terms of ***q*** first. We've setup the grid so it will work for the problem, so be sure to pay attention to the tick marks!

....

....

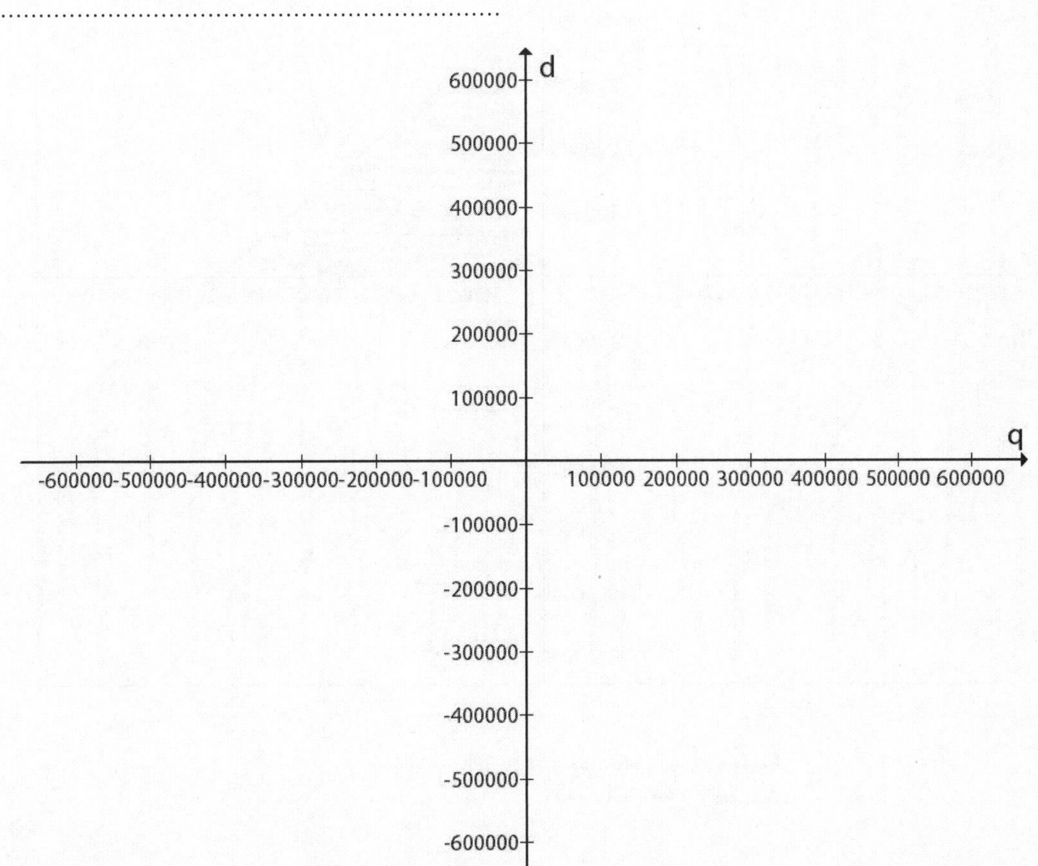

Sharpen your pencil Solution

Graph Kathleen's inequality. You'll have to get ***d*** in terms of ***q*** first. We've setup the grid so it will work for the problem, so be sure to pay attention to the tick marks!

$d + q \leq 436,000$

$-q + d + q \leq 436,000 - q$

$d \leq 436,000 - q$

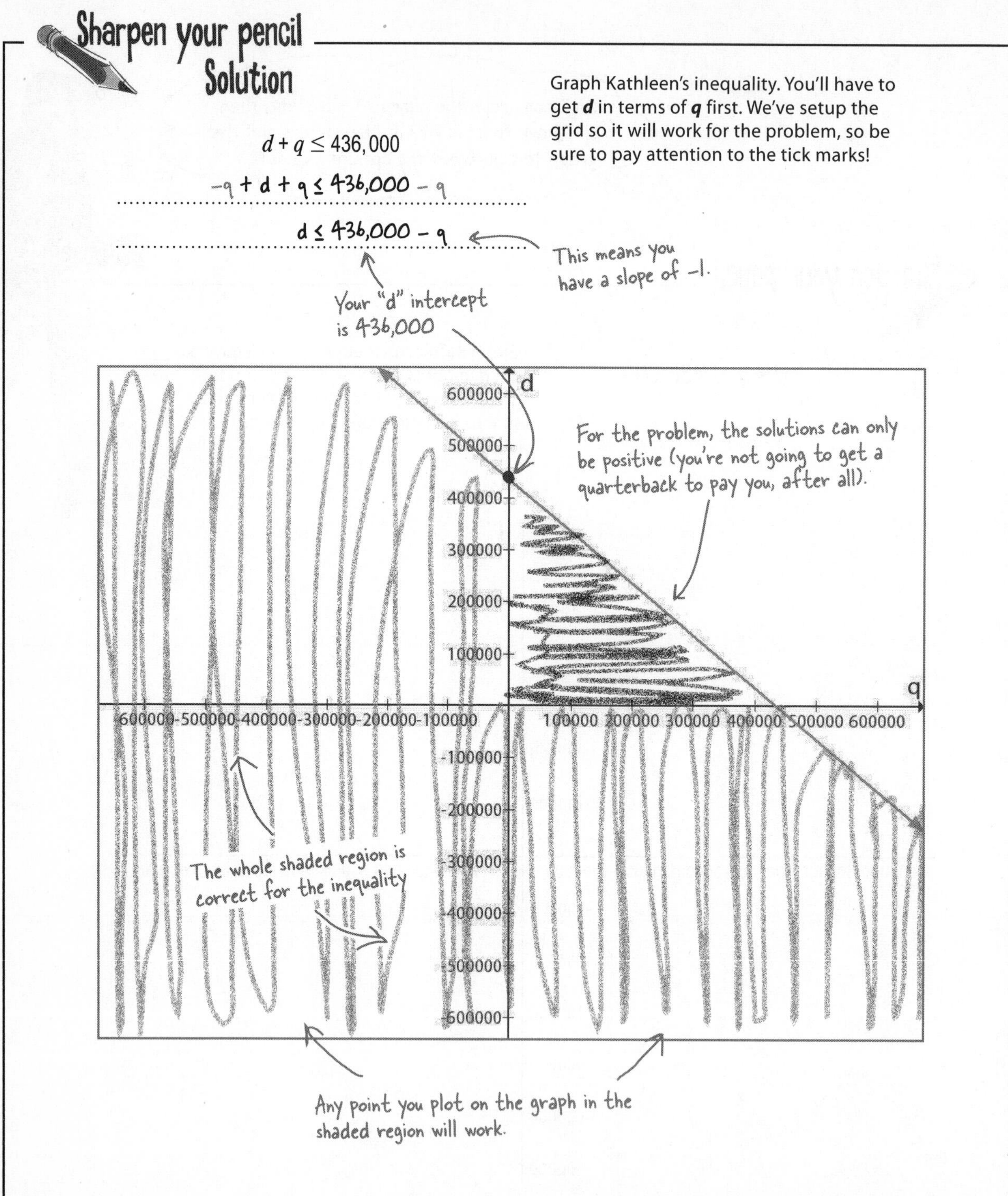

This means you have a slope of -1.
Your "d" intercept is 436,000
d
600000
500000
400000
300000
200000
100000
q
-600000 -500000 -400000 -300000 -200000 -100000
100000 200000 300000 400000 500000 600000
-100000
-200000
-300000
-400000
-500000
-600000
For the problem, the solutions can only be positive (you're not going to get a quarterback to pay you, after all).
The whole shaded region is correct for the inequality
Any point you plot on the graph in the shaded region will work.

Are you ready for some football?

Now, using your graph, help Kathleen make some picks. You need to pick a defensive team, and then see what quarterback you can afford. To use the graph, pick a defensive team, and find it on the ***d*** axis. Then, read across for ***q*** values that will work.

Try and get as close to using all Kathleen's available money as possible.

Defensive Teams

Team	Cost
Broncos	$300,000
Eagles	$200,000
Steelers	$333,000
Ravens	$250,000

Quarterbacks

Name	Cost
Tony Jaglen	$208,200
Eric Hemal	$175,000
Pat Brums	$199,950
Dan Dreter	$202,400

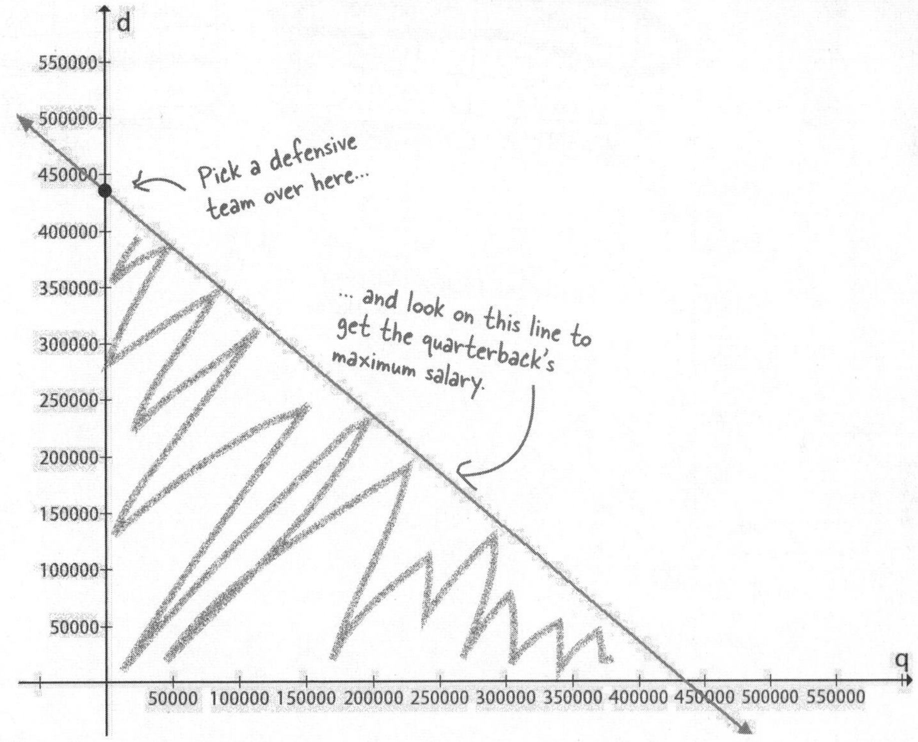

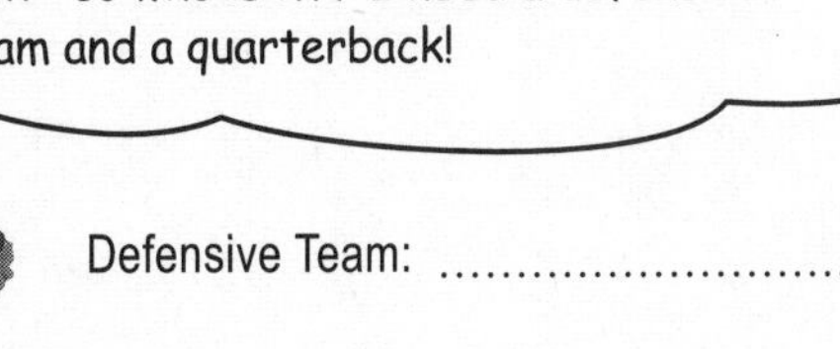

Ok - so who is it?? I need a defensive team and a quarterback!

Defensive Team: .... Quarterback: ....

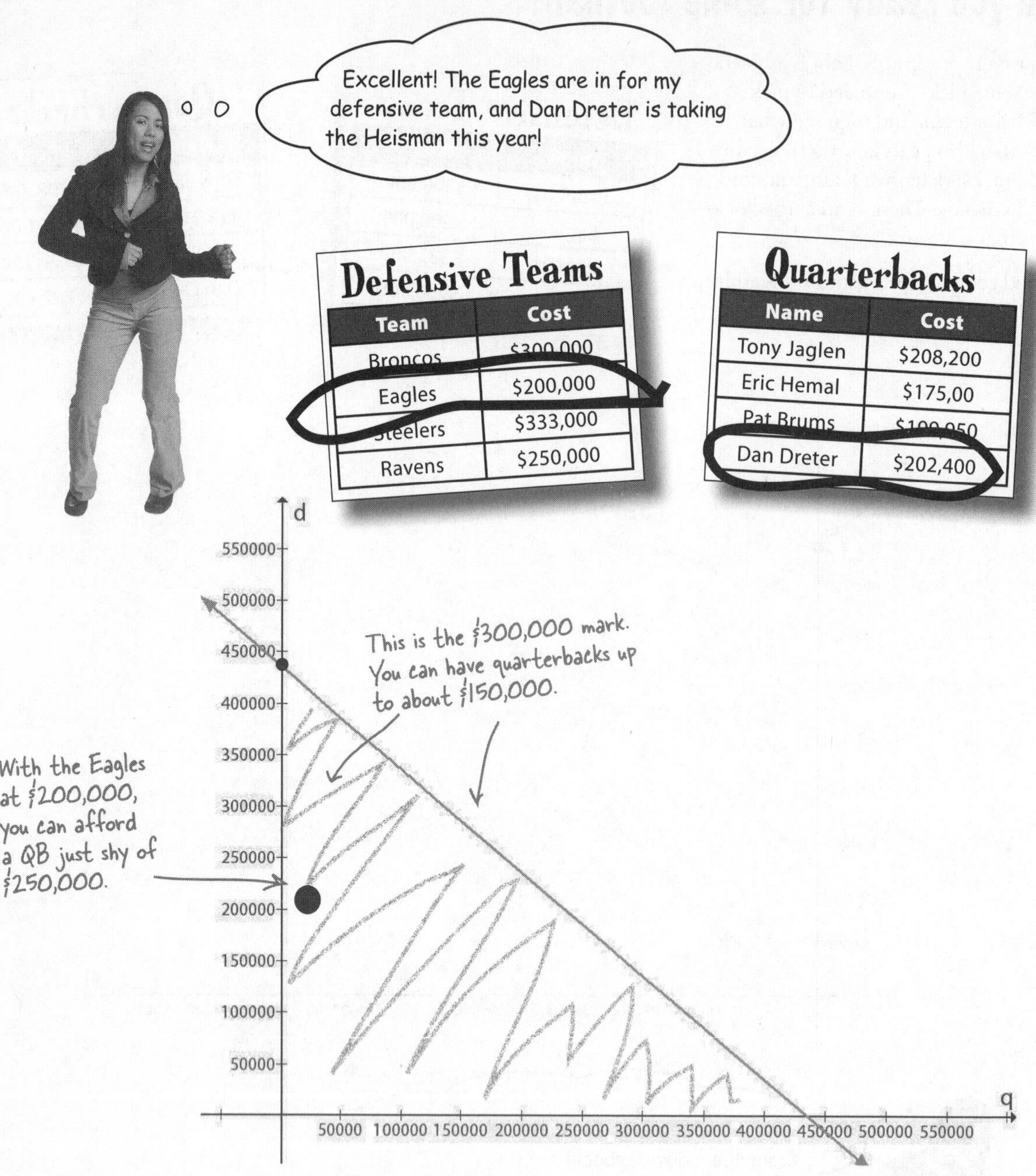

Team	Cost
Broncos	$300,000
Eagles	$200,000
Steelers	$333,000
Ravens	$250,000

Name	Cost
Tony Jaglen	$208,200
Eric Hemal	$175,00
Pat Brums	$199,950
Dan Dreter	$202,400

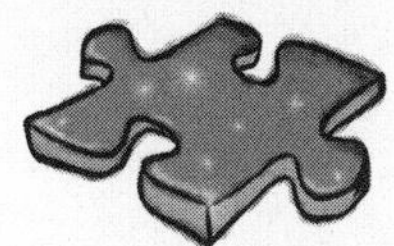

Inequalitycross

All things (not) being equal, while you're waiting for the season to play out, pass the time with a crossword.

Across

2. Inequalities have up to an __________ number of solutions.
4. Depending on how complex your inequality is, you might need to flip the sign ___________ _______.
5. When you graph an inequality, the answers are the ________ area
10. You can plot an inequality on a ________ _______ to help visualize the relationship.
11. The uppermost or lower most value that solves the inequality is called a
12. Negative numbers get ____________ as they get closer to zero.
13. Large negative numbers are _______ ______ small positive ones.

Down

1. An equality says two things are equal, but an inequality is a
3. If you multiply or divide by a negative number across an inequality, you need to __________ it.
5. The shaded part of a graph represents the _________ ________ for the inequality.
6. Solutions to inequalitiesshould be considered in the __________ of the problem.
7. If the boundary is not included in the solution set when you graph it you should use a _______ ______.
8. If the boundary is included in the solution (equal-to), you use a ________ circle.
9. If the inequality is greater than or greater than or equal to, then you shade the ________ part of the graph.

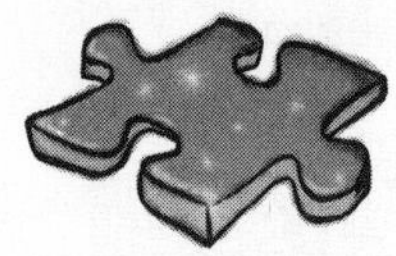

Inequalitycross Solution

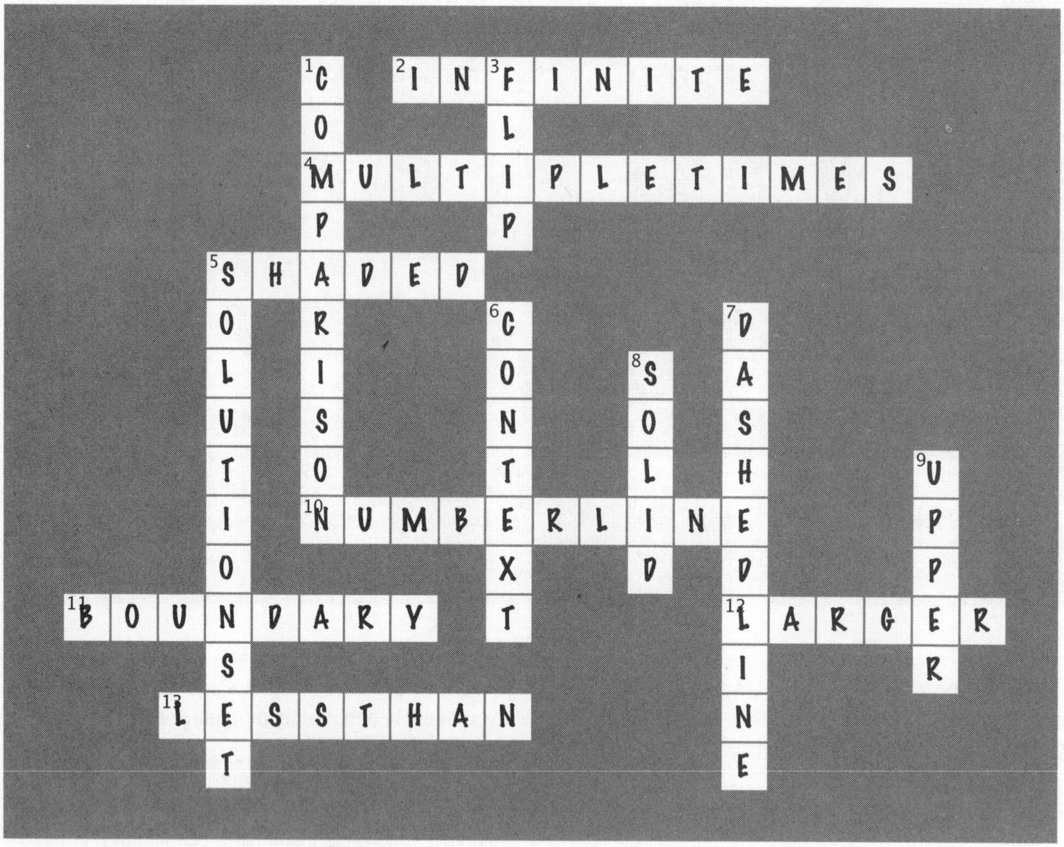

Across:

2. INFINITE
4. MULTIPLETIMES
5. SHADED
10. NUMBERLINE
11. BOUNDARY
12. LARGER
13. LESSTHAN

Down:

1. COMPARISON
3. FLIP
5. SOLUTIONSET
6. CONTEXT
7. DASHEDLINE
8. SOLID
9. UPPER

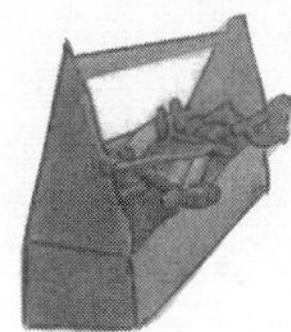

Tools for your Algebra Toolbox

This chapter was about algebraic inequalities.

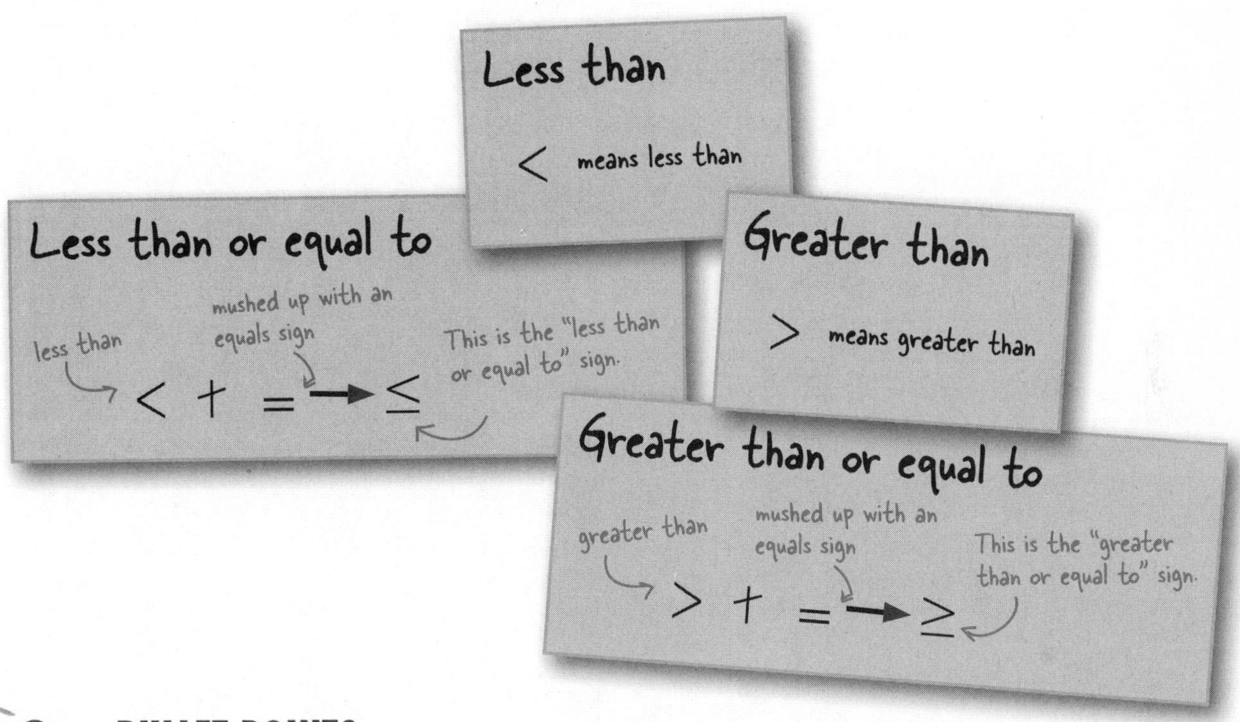
Less than
< means less than
Less than or equal to
less than
mushed up with an equals sign
< + = ≤
This is the "less than or equal to" sign.
Greater than
> means greater than
Greater than or equal to
greater than
mushed up with an equals sign
> + = ≥
This is the "greater than or equal to" sign.

CHAPTER 6

BULLET POINTS

- Number lines work for visualizing solutions for one variable inequalities.
- ● for $\leq$ or $\geq$
- ○ for. $<$ or $>$
- Indicate the direction of the range on the number line.
- The Cartesian Plane works for visualizing two variable inequalities.
- Use a solid line for graphing "less than or equal to" and "greater then or equal to" inequalities on a Cartesian Plane.
- Use a dashed line for graphing "less than" and "greater than" inequalities on a Cartesian Plane.
- Once you've drawn the line, shade the region above or below the graph where the valid solutions are.
- When you solve an inequality, you get a range of valid solutions called a solution set.
- To solve an inequality, you manipulate it like an equation, unless you have to multiply or divide both sides of the equation by a negative number.
- When you multiply and divide the equation by a negative number, you need to flip the inequality.

7 systems of equations

Know what you don't know

Well, I understand you don't know either... but what else can you tell me about what you don't know? Yes, dear... it's really important. I need to know what you know about what you don't know.

You can graph equations with two unknowns, but can you actually solve them? You've been graphing all kinds of expressions lately: ***C*** and ***t***, ***x***, and ***y***, and more. But what about actually ***solving*** equations with **two variables**? That's going to take more than one equation. In fact, you need an equation for every unknown you've got. But what then? Well, a little **substitution**, a few **lines**, and an **intersection** are all you need to solve two-variable equations.

New Year's
Eve Bash
9 pm - 1 am
Music!
Dancing!

It's New Year's Eve, and Zach has a bunch of people coming in just an hour. Problem is, he hasn't made drinks yet...

My girlfriend was supposed to do this, but she's still getting ready. I have no idea how to make punch. She said the punch has to be 52% sparkling, but that's all I know. What the heck am I supposed to do with these percentages?

Zach

100% + 40% = 52%

100% sparkling cider...

40% sparkling pineapple juice.

How does that make 52% sparkling punch?

5 gallon punch bowl

Help Zach out. He needs an equation to figure out how much cider and how much pineapple juice to use to get a 52% sparking punch.

Write the equation for the **amount** of punch to use: .... ← This is just the amount, in gallons, of each type of juice to use.

This should be in gallons.

....

What form of linear equation are you using? (circle one):

Standard form **Point slope form** **Slope-intercept form**

Graph the equation:

Don't forget to label the axis.

Don't forget to label the axis.

What are the intercepts? ....

What do they mean? ....

Help Zach out. He needs an equation to figure out how much cider and how much pineapple juice to use.

Write the equation for the amount of punch to use: The volume of the punch bowl is 5 gallons, so that's the total.

cider volume + pineapple juice volume = 5 gallons → $c + p = 5$ ← This is a linear equation. It's got slope, intercepts, and no exponents greater than 1.

What form of linear equation are you using? (circle one):

Standard form (circled) **Point slope form** **Slope-intercept form**

To graph this, you could solve for the intercepts or manipulate the equation to get it into slope-intercept. Then you'll know the slope, too.

$$c + p = 5$$

$$-p + c + p = 5 - p$$

$$c = 5 - p$$

$$c = -p + 5$$

Now this is of the form: $y = mx + b$

$m = -1 =$ slope

What are the intercepts? $c + p = 5$

Substitute 0 for c.

$$0 + p = 5$$

$p = 0$ So (5, 0) is a point.

Try $p = 0$

$c + 0 = 5$ (0, 5) is another point.

So they're (0, 5) and (5, 0)

What do they mean? If $c = 0$, then the punch is all pineapple juice, and if $p = 0$, then the punch is all cider.

Graph the equation:

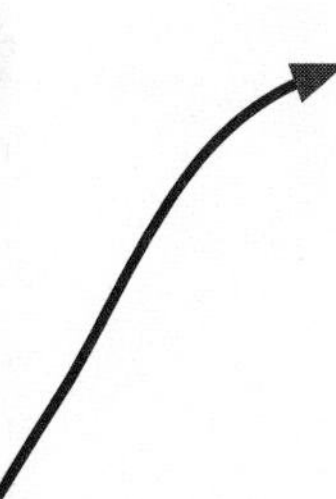

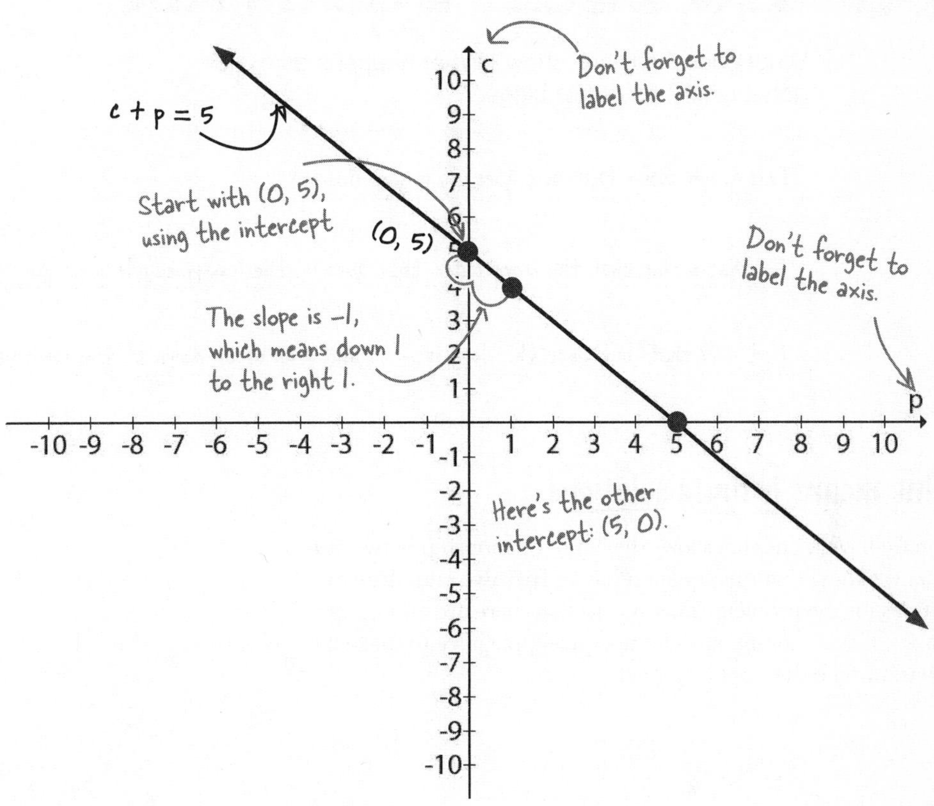

c + p = 5
Don't forget to label the axis.
Start with (0, 5), using the intercept
(0, 5)
The slope is –1, which means down 1 to the right 1.
Don't forget to label the axis.
Here's the other intercept: (5, 0).
c
p

BRAIN BARBELL

What does the graph show? What happens as c gets bigger? What about as ***p*** gets bigger?

....

....

....

Brain Barbell Solution

What does the graph show? What happens as c gets bigger? As ***p*** gets bigger?

This graph shows that as c goes up, p goes down.

So that means that the more cider that goes in, the less pineapple juice goes in.

Not only that, but since the slope is –1, c and p go up or down at the same rate..

A line means infinite solutions

Since we have a line that shows the entire relationship between ***p*** and ***c***, we know that this equation has an **infinite number** of solutions for the problem. That means that there are an infinite number of ways to mix up cider and pineapple juice to make 5 gallons... which does not help Zach.

A line represents an infinite number of points.

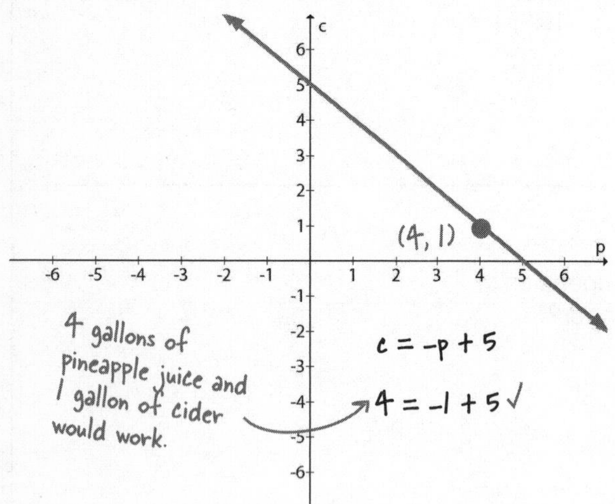

c
p
(4, 1)
4 gallons of pineapple juice and 1 gallon of cider would work.
c = –p + 5
4 = –1 + 5 ✓

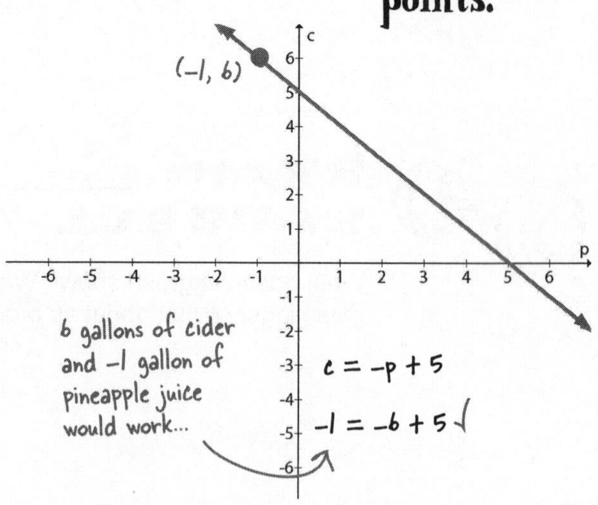

c
p
(–1, 6)
6 gallons of cider and –1 gallon of pineapple juice would work...
c = –p + 5
–1 = –6 + 5 ✓

Look closely at the graph. Do all of the solutions represented by the line make sense for THIS problem?

You can't have -1 gallons of liquid!

If you're working with a real-world problem, you have to keep in mind your problem's **context**. The values for c and p can't be less than zero because you can't have less than zero gallons of punch.

So we know that some of the answers on our graph don't work, but we still don't know what mixture Zach should use. So what do we do now?

So what was the point of all that graphing?

We have MANY solutions.

All Zach has is a bunch of ways to mix up cider and pineapple juice to get 5 gallons. But Zach needs more than just 5 gallons of mix. He needs 5 gallons of mix, and he needs a 52% sparkling punch at the end. So there's something we're missing...

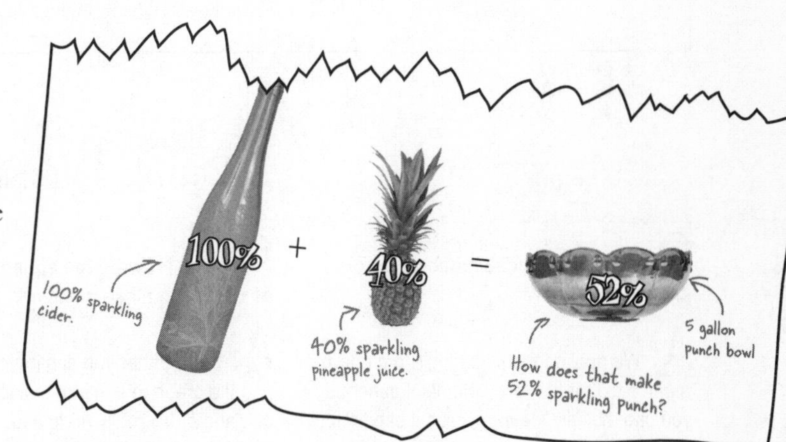

Sharpen your pencil

Zach is worried about how sparkling the punch is, not just what will fit in his bowl. Write another equation for the punch. This one has to be about how sparkling the punch is, not the overall amount.

....

....

....

....

Sharpen your pencil Solution

Your job was to come up with another equation, related to how sparkling Zach's punch is. Here's what we did:

To write an equation for this, figure out the amount of sparkling that would be in one mixture of punch.

The sparkling stuff is: 100% cider, 40% pineapple juice, and should create 52% punch.

$$1c + 0.4p = 0.52(5)$$

c stands for the volume of cider.

Cider is 100% (1.0).

The juice is 40% (0.4).

p stands for the volume of pineapple juice.

This is the total volume of punch.

This is our second equation

$$c + 0.4p = 2.6$$

We dropped the 1 from 1c.

there are no Dumb Questions

Q: Where did those numbers before c and p come from?

A: We applied the same logic to ***c*** and ***p*** that we did for the total volume of punch. If you can multiply the total volume of punch by the percentage of sparkling (converted to a decimal), you can do the same thing for the other two liquids.

100% sparkling cider = 1***c***
40% sparkling pineapple juice = 0.4***p***

Q: How can you just "figure out" the amount of sparkling punch?

A: This problem of Zach's is actually a classic mixture problem. Mixture problems like this are usually based on a proportion, in our case, how sparkling the punch will be. If 52% of the total punch needs to be sparkling, then the amount of punch that is ***just sparkling*** is 2.6 gallons (52% of 5 gallons).

Q: Can I rewrite the equation in terms of x and y to make it easier?

A: Sure. Either you can have the equation and the graph in terms of ***x*** and ***y*** or in terms of ***c*** and ***p***. It's really up to you.

Q: How much do we need to worry about negative solutions?

A: For this problem, you don't need to worry about negative solutions since you can't have negative punch.

We'll look at how to express those types of limitations as part of an equation when we learn about functions in a few chapters.

Q: How can we work with these two equations together?

A: That's what's coming up next. But the volume equation has an infinite number of solutions, and the sparkling equation has an infinite number of solutions. There's a way those solutions can work together...

How does the sparkling equation work?

The problem Zach has with just the first equation is that there are an infinite number of ways to mix the ingredients together and get 5 gallons of punch. But none of those solutions specifically deal with his girlfriend's request that the punch is 52% sparkling.

By coming up with another equation that deals with the mixing, we can figure out which solutions create the right mix. This second equation will help because we're using the ***same variables*** representing the ***same thing.***

c = The amount of sparkling cider, in gallons.

p = The amount of pineapple juice, in gallons.

To develop the second equation, we took the second piece of information that we had about each variable—how sparkling each liquid is—and applied it to the variables to come up with a new problem:.

Zach's girlfriend gave him this number.

52% carbonated punch = 5 gallons x 0.52

= 2.6 gallons

If you're hazy about how to work with decimals and percents – jump to the appendix and brush up.

This is how much pure sparkling liquid we need.

If you apply the same idea to the ***c*** and ***p*** variables, we've got a new equation. The sparkling cider is 100% sparkling, and the pineapple juice is 40% sparkling:

$$1c + (0.4)p = 2.6$$

This is the volume of cider that's sparkling (all of it).

The volume of Pineapple juice that's carbonated, just 40%

We've got another linear equation

This is another linear equation in terms of the same variables that we used in the first equation, ***c*** and ***p***. This equation is in standard form, and you can work with it in exactly the same way.

What can we do with the second equation to help find a solution to the punch problem?

BULLET POINTS

- The graph of a linear equation is the **infinite number of solutions** that solve the equation.
- You always need to remember the **context** of the problem.

Now we have TWO linear equations

So now we've got two equations with the same two variables. ***c*** is the volume of cider, and ***p*** is the volume of pineapple juice. Since both equations talk about the same thing, with the same variables, they can be looked at together:

$$c + p = 5$$

We know that this is a linear equation – which means that all the points on the solution line solve this equation.

$$c + 0.4p = 2.6$$

This is a linear equation, too. All the points on this equation's line solve this problem.

One equation is in terms of total volume of punch, and the other is in terms of the *sparkling* volume of punch, but ***both equations are about volume***. Each equation comes with an infinite number of ordered pairs that will solve the equation. Remember, solving an equation means coming up with values for the unknowns that will make that particular equation true.

What we need, though, is a set of numbers that will solve ***both*** equations. We want a set of numbers that will result in 5 gallons of total punch ***and*** that will also result in a punch that is 52% sparkling. So how can we find a solution that works for both equations?

Let's start out by graphing these two equations... on the ***same*** Cartesian plane.

A linear equation is represented by a LINE. Every point on that line is a solution for THAT PARTICULAR linear equation.

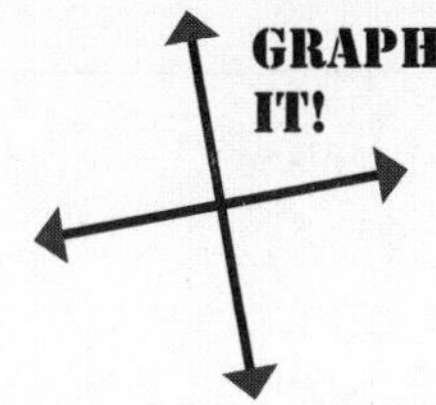

Graph both punch equations on the same Cartesian plane.

$c + p = 5$ $c + 0.4p = 2.6$

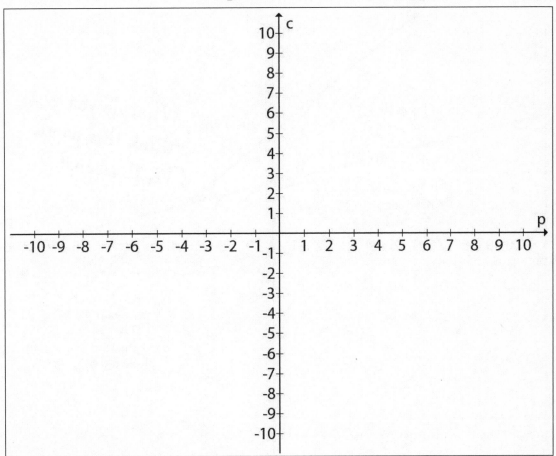

Use this space for any work you need to do to graph the equations.

Are there any points that solve both equations?

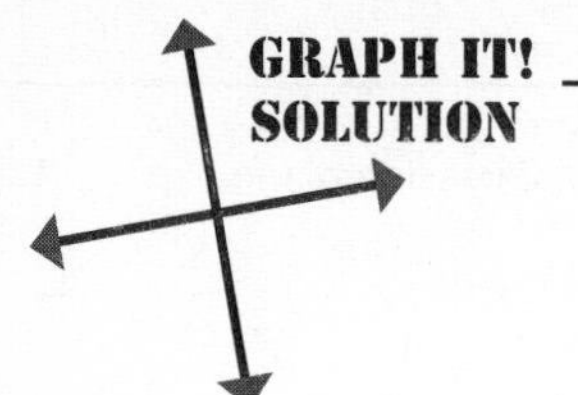

Graph both punch equations on the same Cartesian plane.

$c + p = 5$ $c + 0.4p = 2.6$

c + 0.4p = 2.6

c +p =5

What do you think this point, (4, 1), means?

When you plot (0, 2.6), put the point just past half-way between the 2 and the 3.

We already did c + p = 5, so this is the same line from earlier.

c + 0.4p = 2.6

-0.4p +c + 0.4p = 2.6 – 0.4p

c = –0.4p + 2.6

$c = -\frac{4}{10}p + 2.6$ ← The intercept is (0, 2.6)

Just convert that slope to a fraction and it's much easier to work with.

Are there any points that solve both equations?

It looks like (4,1) will work for BOTH equations.

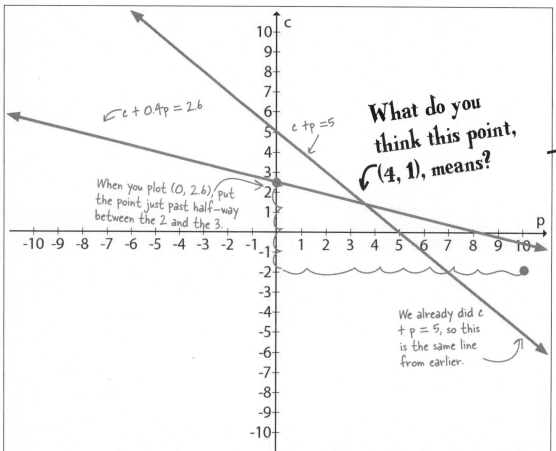

The INTERSECTION of the lines solves BOTH linear equations

The point where the lines meet is the point where both equations have a solution. So for Zach's punch problem, we need to find where the two lines meet. That's the amount of liquid we want to make the perfect party punch.

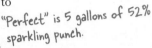

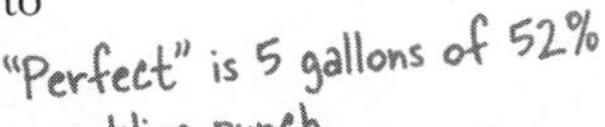
"Perfect" is 5 gallons of 52% sparkling punch.

To make the perfect punch:

The volume of pineapple juice, in gallons

p = 4 gallons

The volume of cider in gallons

c = 1 gallon

- **Both equations individually have an infinite number of solutions.**
  Each line is an infinite set of ordered pairs that satisfy that particular equation and make the equation true.

- **The point where the lines intersect solves the problem.**
  The intersection is the solution because it solves both equations **at the same time**. The intersection is a point on both solution lines.

Sharpen your pencil

Check the solution point (4, 1) in both equations to see if it works. Is our solution right?

$c + 0.4p = 2.6$ $\qquad\qquad$ $c + p = 5$

Sharpen your pencil Solution

Your job was to see if (4, 1) solved both punch equations.

The point we're working with is (4, 1).

$c + 0.4p = 2.6$	$c + p = 5$
$1 + 0.4(4) = 2.6$	$4 + 1 = 5$
$1 + 1.6 = 2.6$	$5 = 5$ ✓
$2.6 = 2.6$ ✓	

Both points work out – which proves that the intersection solves them both and solves the problem!

Solve multiple unknowns with a SYSTEM of EQUATIONS

You've just solved a system of equations! A **system of equations** is a group of equations that can be treated as one problem. The solution is the point that satisfies all of the equations at the same time.

A system of equations

$$c + 0.4p = 2.6$$
$$c + p = 5$$

Solution: (4, 1)

When it comes to a system of equations, though, ***you have to have two equations to find two unknowns***. Why? Because if you only have one equation with two unknowns, you have a line that goes on forever.

To know which point on the line to pick, you need more information. A second equation gives you a second line, and then you can find the intersection of the two lines to solve your system.

If you have <u>two</u> unknowns you need <u>two</u> independent relationships to find the values of the unknowns.

there are no Dumb Questions

Q: Do we have to check our work?

A: Yes! Really, we're going through this again? It takes like two seconds and then you know you got the right answer. Imagine, walking away from a problem knowing it's completely right.

It's also good practice for manipulating the equation and substitution.

Q: Is graphing the only way to figure the answer out?

A: Stay tuned, there are more options coming up! The big advantage of graphing, though, is that it helps you see what could happen if something changed in your system.

Q: How did you know you can work with both equations together?

A: Well, it goes back to understanding the situation, not just manipulating equations. ***Algebra is a tool to solve a problem you're dealing with, not the other way around.*** So when you're working with equations, you need to always keep the context of your problem in mind.

Zach was interested in a punch that was not only 5 gallons, but ***also*** had the perfect amount of sparkle. That means you need two equations, not just one.

Q: What if I only have one equation and two variables, can I get an answer?

A: Not without some kind of extra information. To figure out one unknown, you need one equation. To figure out two, you need two equations. To figure out three... well, you get the idea.

Q: What is a system of equations good for in real life?

A: Quite a lot, actually. Mixture problems (like we used here), supply and demand situations, area and perimeter, distance vs. time problems..... pretty much anything where you've got two related unknowns.

Q: If I have two equations with the same two variables, is there always a solution that solves both equations?

A: Not necessarily. Sometimes there isn't an answer to the two specific equations. This means the lines never intersect.

More trouble... Zach dropped some glasses

When Zach and his girlfriend were setting up for the party, Zach broke a few glasses. The glasses were rented, so Zach's got to pay to replace them. Now Zach's trying to figure out how many more glasses he needs, and he can't remember how many of each type he dropped. Here's what he knows:

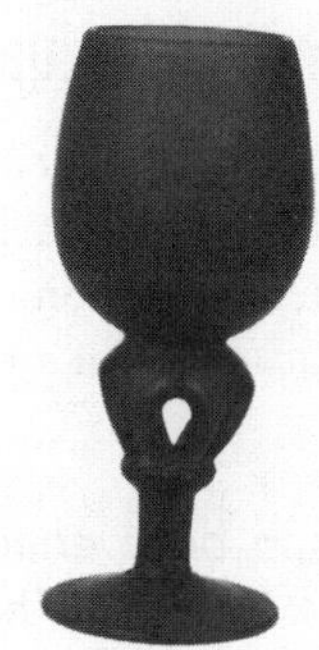

The wine glasses cost $6 to replace.

These tumblers cost $4 each to replace.

Kathleen said we paid $33 for broken glasses, and there were 7 broken glasses.

BRAIN BARBELL

Write the two equations that Zach can use to figure out what glasses he needs.

....

....

Two kinds of glasses... that's TWO unknowns

Now that you've worked with a system of equations, you know that's what we'll need to help Zach. He has two unknowns:

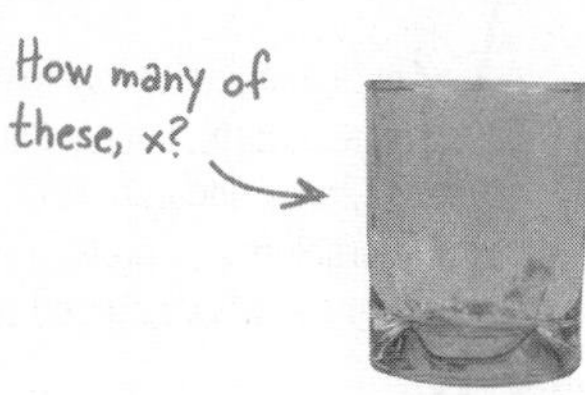

How many of these, x?

How many of these, y?

Since we have two unknowns, we need two equations:

- **A cost equation:** $4x + 6y = 33$
  The tumblers, x, are \$4 each, and the wine glasses, y, are \$6 each. Zach paid a total of \$33 in replacement fees.

- **A number of glasses equation:** $x + y = 7$
  Zach broke a total of 7 glasses.

Solve your system of equations using a graph

Graphing both equations is a way to find the values that satisfy both equations. We can graph the line that solves each equation, and then find the intersection point. Then we check our work in both equations, and Zach's in and out of the glasses store in time to party.

1. **Graph both equations.**
   Use whatever method you want to put both solution lines on the same graph.

2. **Determine the intersection point.**
   Using the gridlines on the graph, find the point where the two lines meet. That point satisfies both equations.

3. **Check your solution point.**
   Go back to both equations and substitute in your solution point. Make sure that the point is valid.

Let's solve the glasses problem

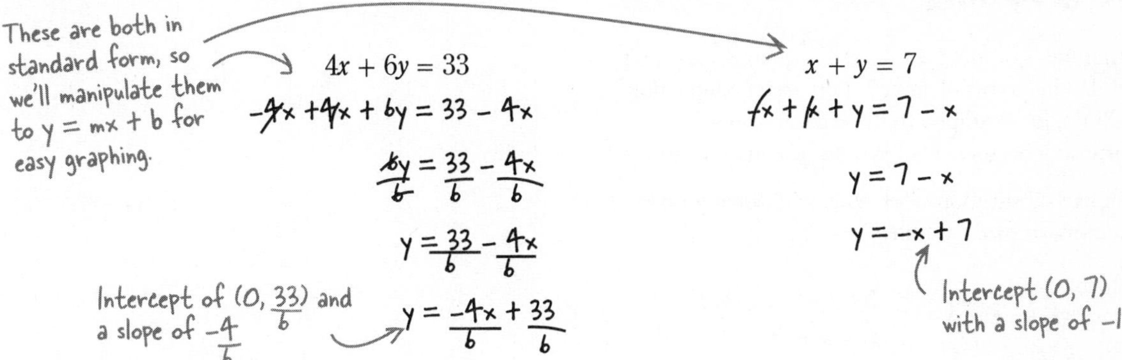
These are both in standard form, so we'll manipulate them to y = mx + b for easy graphing.
4x + 6y = 33
-4x + 4x + 6y = 33 - 4x
6y/6 = 33/6 - 4x/6
y = 33/6 - 4x/6
Intercept of (0, 33/6) and a slope of -4/6
y = -4x/6 + 33/6
x + y = 7
x + x + y = 7 - x
y = 7 - x
y = -x + 7
Intercept (0, 7) with a slope of -1

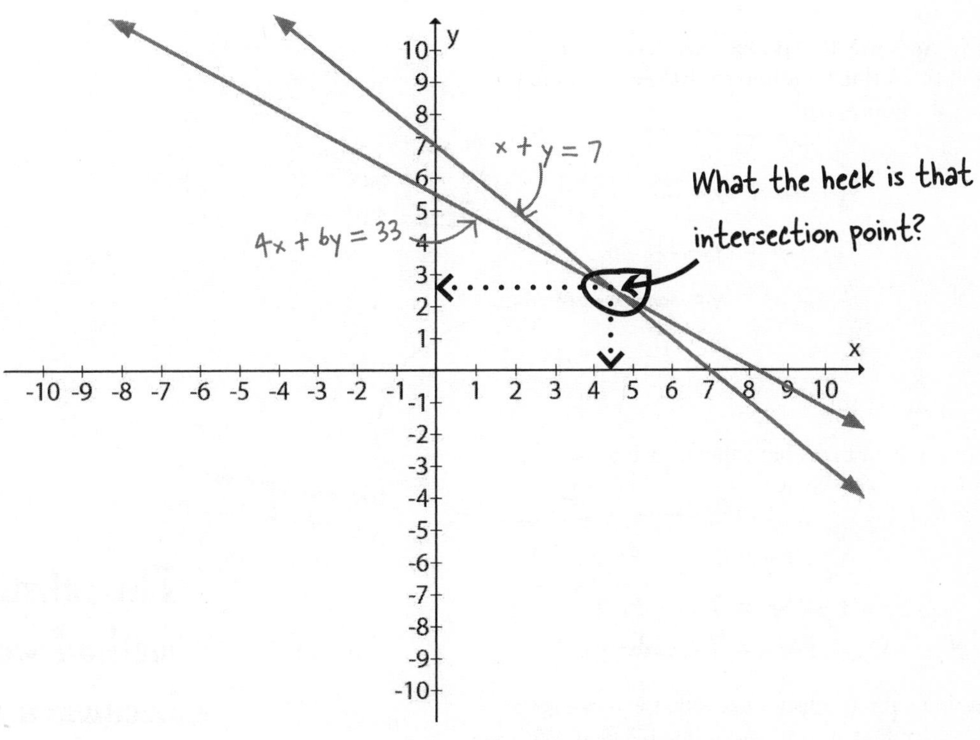
y
x
x + y = 7
4x + 6y = 33
What the heck is that intersection point?

We can't read the intersection off the graph. Now what do we do?

You can substitute substitution for graphing

Sometimes graphing a system of equations just isn't useful. The points of intersection aren't falling along gridlines, for example. In those cases, you can use substitution to solve your system of equations instead.

To make substitution work, start with solving ***one*** equation for ***one*** variable.

We'll solve this for x in terms of y.

$$x + y = 7$$

$$-y + x + y = 7 - y$$

$$x = 7 - y$$

Now you've got a way to represent x in terms of y. So now, put that value for x into the ***other*** equation. That's the substitution part:

We use this value for x, in the SECOND equation.

$$4x + 6y = 33$$

$$4(7 - y) + 6y = 33$$

$$y = \text{some number}$$

Now we've got all y, and can solve this equation.

Once you have y, just put that value back into your first equation and solve for x.

Use your value for y to solve for x.

$$x + y = 7$$

$$x + \text{number} = 7$$

$$x = 7 - \text{number}$$

The great thing about substitution is that you can get a fraction or a decimal answer, and it's no big deal. With the graphing method, that doesn't work well.

The substitution method works because a system of equations is a set of equations with the same variables. If you can solve for one variable in terms of another variable, you can get a valid equation in terms of one variable.

The substitution method works because a system of equations is a set of equations with the same unknowns.

BULLET POINTS

- To solve a system of equations with graphing, the **intersection** of the lines is the solution.
- To solve a system of equations with two variables using the **substitution** method, solve one equation **in terms of** one variable, substitute it into the second equation, and **solve for the only variable left**.
- A system of equations is a **group of equations** that can be treated as **one problem**.

Exercise

Using substitution, figure out how many of each type of glass Zach needs to replace.

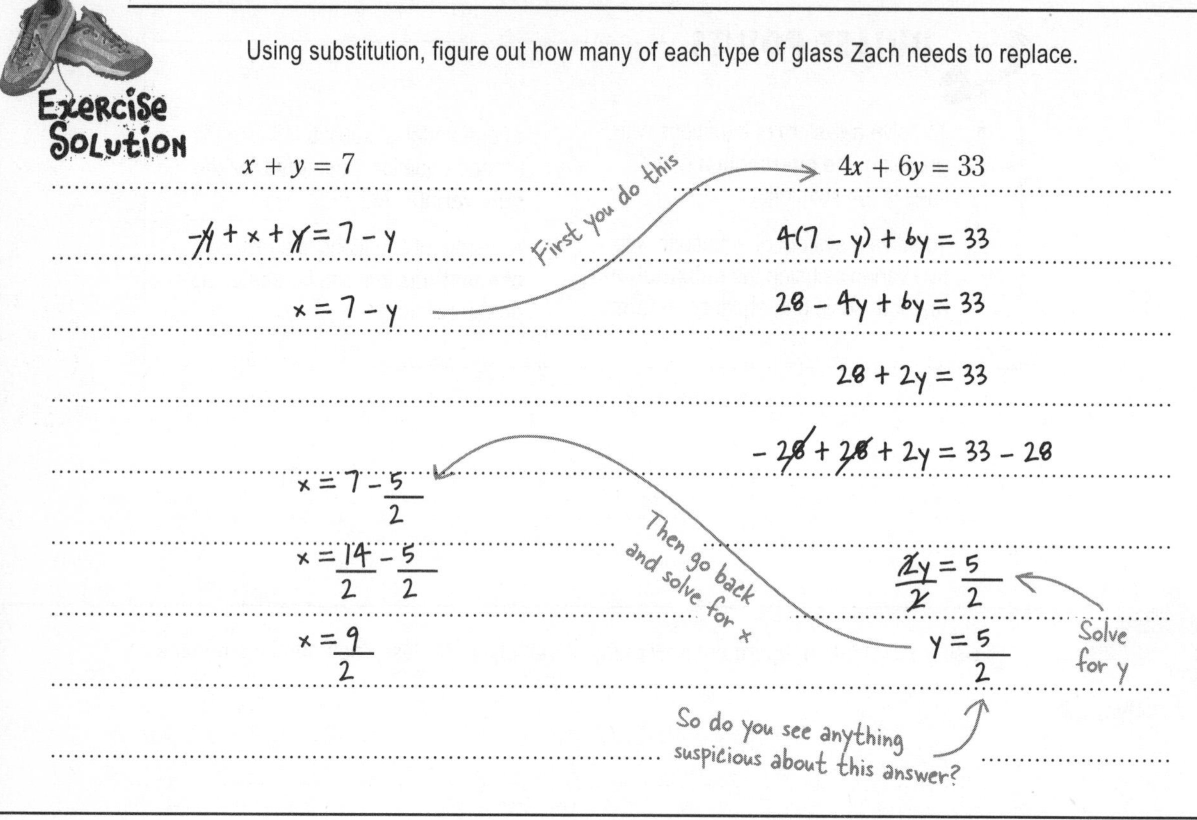

Exercise Solution
Using substitution, figure out how many of each type of glass Zach needs to replace.
x + y = 7
-y + x + y = 7 - y
x = 7 - y
First you do this
4x + 6y = 33
4(7 - y) + 6y = 33
28 - 4y + 6y = 33
28 + 2y = 33
-28 + 28 + 2y = 33 - 28
2y/2 = 5/2
y = 5/2
Solve for y
Then go back and solve for x
x = 7 - 5/2
x = 14/2 - 5/2
x = 9/2
So do you see anything suspicious about this answer?

there are no Dumb Questions

Q: Is the substitution method better than the graphing method?

A: Better isn't really the right word. Substitution has some advantages: you get an exact answer, and since you're not estimating on a graph, it may be more exact.

It has some disadvantages, too. There's a lot of equation manipulation to be done, and that can take a while. Substitution also doesn't allow you to *see* any of what's going on. If you are trying to figure out what happens anywhere else in your system of equations, substitution doesn't help much.

Q: Why is it ok to treat both equations as part of the same problem? Can you really exchange two variables in two different equations?

A: If the problem statement says that they are the same two variables, then they are! It can either be the problem situation (like Zach's punch) or two specific equations that are given together.

The reason you have two equations and two unknowns is that you know two different things about how these things are related.

Q: How do I decide which equation to use first? How about which variable to solve for first?

A: It's a bit of a judgement call, really. The best bet is to look at both equations and figure out which one is the easiest to work with. The important thing to remember is that you will always get to the right answer eventually, as long as you keep your algebraic manipulation within the rules.

We didn't break half a glass. Kathleen must have gotten how much we paid wrong.

Kathleen

Oh, sorry - I just checked again; it was $30.

Sharpen your pencil

Using Kathleen's new cost, figure out how many of each glass Zach needs to get. Graph the problem and then use substitution to check your answer.

y

10 9 8 7 6 5 4 3 2 1 -1 -2 -3 -4 -5 -6 -7 -8 -9 -10

x

-10 -9 -8 -7 -6 -5 -4 -3 -2 -1 1 2 3 4 5 6 7 8 9 10

Sharpen your pencil

Your job was to (re-)figure out how many of each glass Zach needs to get. Graph the problem and then use substitution to check your answer.

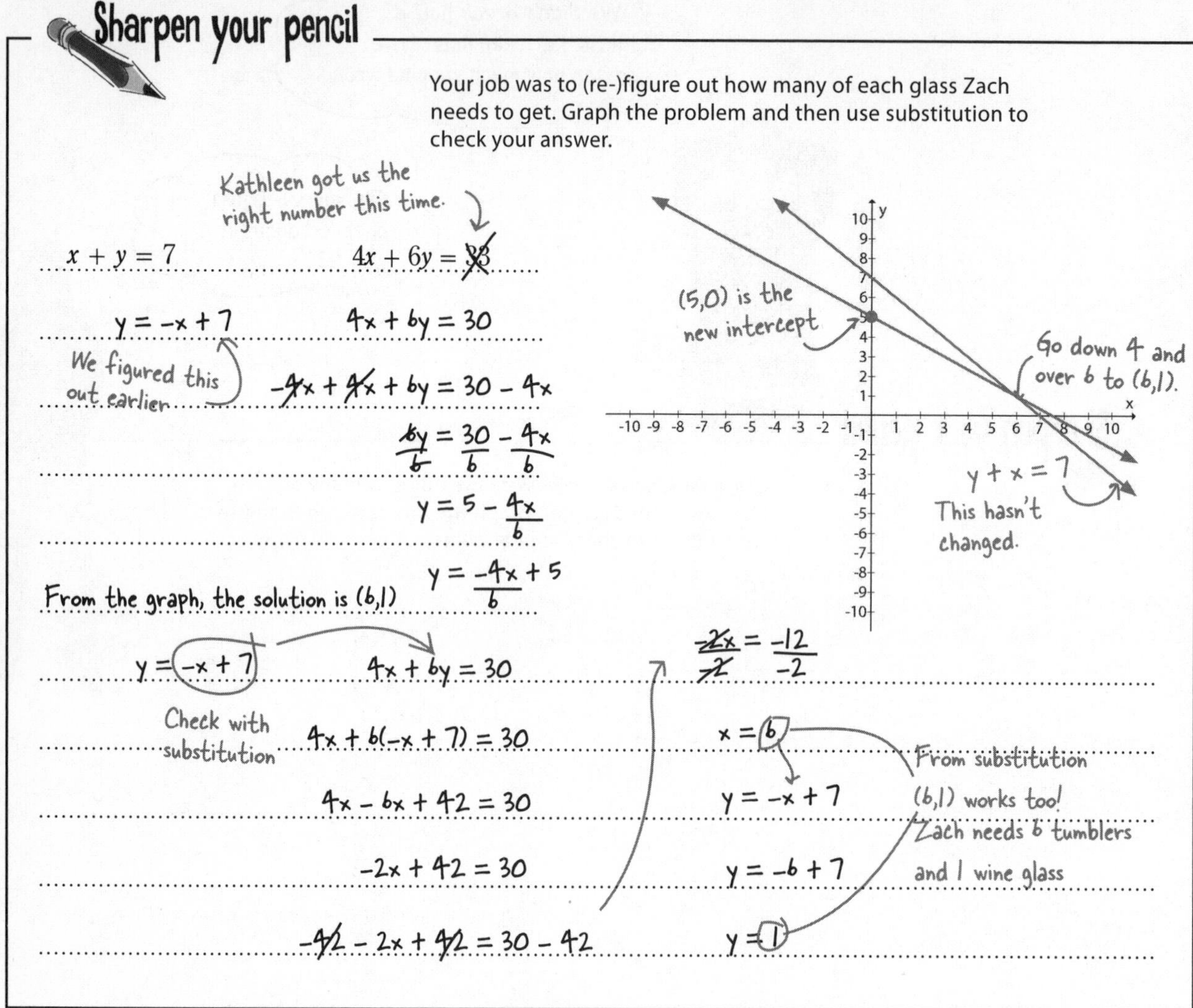

Kathleen got us the right number this time.
x + y = 7
4x + 6y = 33
y = -x + 7
4x + 6y = 30
We figured this out earlier
-4x + 4x + 6y = 30 - 4x
6y/6 = 30/6 - 4x/6
y = 5 - 4x/6
y = -4x/6 + 5
From the graph, the solution is (6,1)
(5,0) is the new intercept
Go down 4 and over 6 to (6,1).
y + x = 7
This hasn't changed.
y = -x + 7
4x + 6y = 30
Check with substitution
4x + 6(-x + 7) = 30
4x - 6x + 42 = 30
-2x + 42 = 30
-42 - 2x + 42 = 30 - 42
-2x/-2 = -12/-2
x = 6
y = -x + 7
y = -6 + 7
y = 1
From substitution (6,1) works too!
Zach needs 6 tumblers and 1 wine glass

Zach and Kathleen have enough glasses and the perfect punch mix. The party is well on its way. But now there's a new wrinkle...

The girls are going to leave if I don't get the music right. What do I need to do?

Everyone loves a little slow dancing.

Zach's figured out that if he increases the number of slow songs, then more girls will stay. He wants to play a slow song for every two fast songs.

But Zach's got to put a playlist together for the rest of the party, and he needs your help.

Each song is about four minutes long, and the party's four hours long.

Four hours of music
$= 4$ hours $\cdot$ 60 minutes an hour
$= 240$ minutes

$\frac{240 \text{ minutes}}{4 \text{ minutes a song}} = 60 \text{ songs}$

The total number of songs for the party.

Sharpen your pencil

Come up with two equations in standard form with two unknowns to figure out how many fast songs and how many slow songs Zach will need.

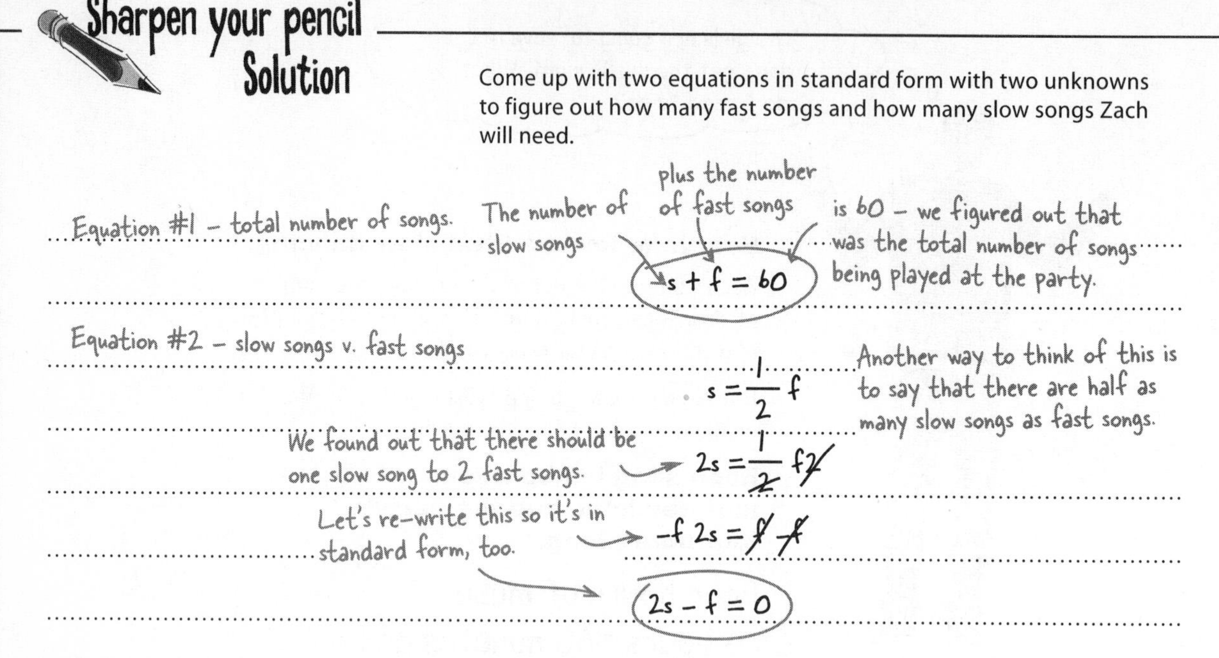

Sharpen your pencil
Solution
Come up with two equations in standard form with two unknowns to figure out how many fast songs and how many slow songs Zach will need.
Equation #1 - total number of songs.
The number of slow songs
plus the number of fast songs
is 60 - we figured out that was the total number of songs being played at the party.
s + f = 60
Equation #2 - slow songs v. fast songs
s = 1/2 f
Another way to think of this is to say that there are half as many slow songs as fast songs.
We found out that there should be one slow song to 2 fast songs.
2s = 1/2 f2
Let's re-write this so it's in standard form, too.
-f 2s = f -f
2s - f = 0

there are no Dumb Questions

Q: Does it matter which equation comes first?

A: When you're working with a system of equations, it doesn't matter what order you work with them. As long as you consistently follow the rules of manipulating an equation consistently, you'll be fine.

Q: Can I use any variables that I want?

A: Yes! We picked ***s*** and ***f*** to stand for slow and fast, but you can always stick with the basic ***x*** and ***y***. If you use something besides ***x*** and ***y*** when you're graphing, make sure to label your graph axes correctly.

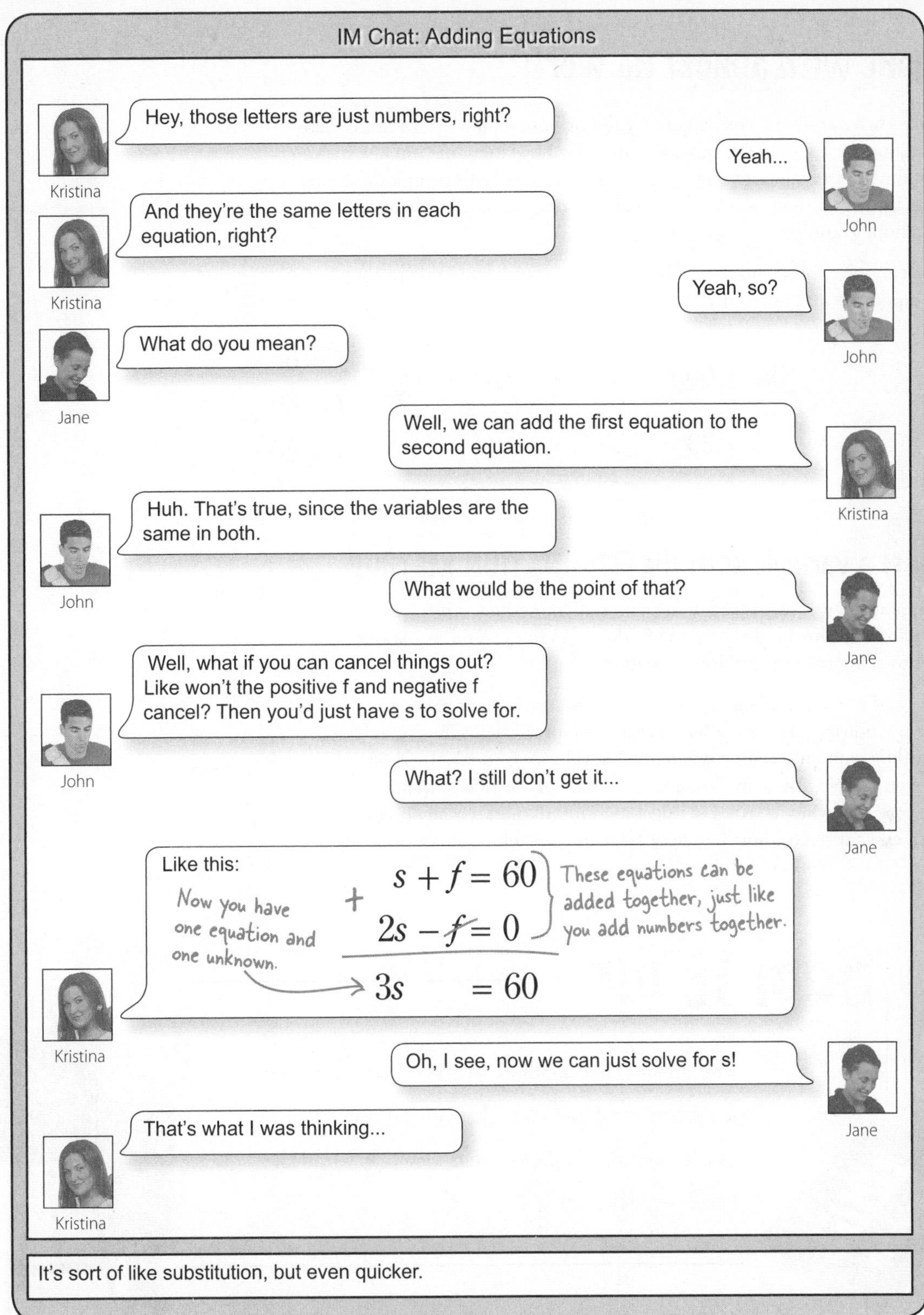
IM Chat: Adding Equations
Kristina
Hey, those letters are just numbers, right?
John
Yeah...
Kristina
And they're the same letters in each equation, right?
John
Yeah, so?
Jane
What do you mean?
Kristina
Well, we can add the first equation to the second equation.
John
Huh. That's true, since the variables are the same in both.
Jane
What would be the point of that?
John
Well, what if you can cancel things out? Like won't the positive f and negative f cancel? Then you'd just have s to solve for.
Jane
What? I still don't get it...
Kristina
Like this:
$s + f = 60$
$+ \quad 2s - f = 0$
$3s = 60$
Now you have one equation and one unknown.
These equations can be added together, just like you add numbers together.
Jane
Oh, I see, now we can just solve for s!
Kristina
That's what I was thinking...
It's sort of like substitution, but even quicker.

f is gone with almost no work

With some clever addition, you've gotten ***f*** out of both equations. Because these two equations are a system of equations, and they have the same unknowns, you can skip some steps by just adding up all of the terms of both equations. If the equations are set up right, you'll lose one of the variables and make life much easier, without graphing ***or*** substitution.

Now you can solve for **s**, and then substitute it back into either one of the original equations and solve for ***f***:

First this:

$$\frac{3s}{3} = \frac{60}{3}$$

$$s = 20$$

Then substitute this back in...

$$s + f = 60$$

$$-20 \quad 20 + f = 60 \quad -20$$

$$f = 40$$

... and solve for f

Eliminate a variable with the ELIMINATION METHOD

Solving equations this way is called the elimination method. The **elimination method** is the process of adding up both of the equations and then working with the resulting equation.

This is a valid way of working with equations because both equations have the same unknowns, just with different relationships. Specifically, since you're dealing with an equation where the left side, **s**+***f***, equals the right side, 60, you're doing the same thing to both sides of the other equation. So you're still following the rules of Algebra. This method is really useful because it can cut out a whole bunch of steps in your solution.

Sum it up

The elimination method – a method of solving a system of equations through adding the two equations together and working with the resulting equation.

there are no Dumb Questions

Q: Can we always just make variables disappear?

A: Yes, if your equations are set up the right way first. You need coefficients that will cancel each other out. We had -1***f*** and+1***f***, which add up to zero. We'll talk more about how you can set that up in a moment.

Q: Does it matter which equation we put the first solution into?

A: No. Once you solve for one of the unknowns, you can substitute your value into either equation to solve for the remaining unknown.

That makes sense, right? The solution for a system of equations is still the ONE point that will solve both equations. That means for that one point, ***x*** and ***y*** (or ***s*** or ***f***) are the same for both equations.

Q: Why is it ok to just add up two equations like this?

A: The elimination method is like high-speed substitution. The variables are the same thing at the solution point. So you can work with both equations because they are both describing the same variables. Each equation is just written a different way, so combining them gets you a single solution.

Adding the equations together is just another way of working with all of the variables. It's like manipulating both equations at once.

Q: Which method should I use: graphing, substitution, or elimination?

A: The good news is that any of them will work for any problem. As you saw, the limits on the graphing method is you have to be able to read the graph, so it's tough with decimal or fractional answers.

Substitution is the most straightforward, but it's a lot of back and forth. That can take a lot of time, and there's a greater chance for mistakes. Elimination is great if your equations are setup right, and they're usually pretty fast, too.

Q: What if the equations aren't set up to cancel each other out?

A: Then you'll need to manipulate one (or both) of the equations so you end up with variables that will cancel each other out.

Suppose you have two equations, one with a -4***f***, and one with a +***f***. You could multiply that entire second equation by 4 (all of the terms, both sides of the equation so it stays equal), to get 4***f***, and then add the equations up. That will give you a situation where the variables will cancel out.

BRAIN BARBELL

Go back to Zach's punch problem. Try solving it this time using the elimination method.

Total punch equation → $c + p = 5$

Sparkling equation → $c + 0.4p = 2.6$

BRAIN BARBELL SOLUTION

Your job was to solve the punch problem from earlier using the elimination method.

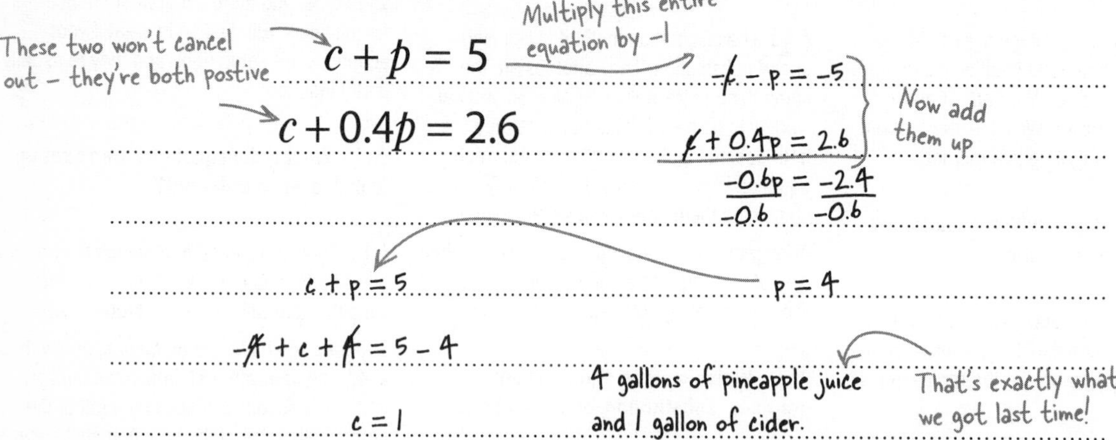

These two won't cancel out – they're both postive
$c + p = 5$
Multiply this entire equation by –1
$-c - p = -5$
$c + 0.4p = 2.6$
$c + 0.4p = 2.6$
Now add them up
$-0.6p = -2.4$
$-0.6 \quad -0.6$
$c + p = 5$
$p = 4$
$-4 + c + 4 = 5 - 4$
4 gallons of pineapple juice and 1 gallon of cider.
That's exactly what we got last time!
$c = 1$

The elimination method requires PLANNING

For elimination to work properly, ***you must have two variables cancel each other out.***

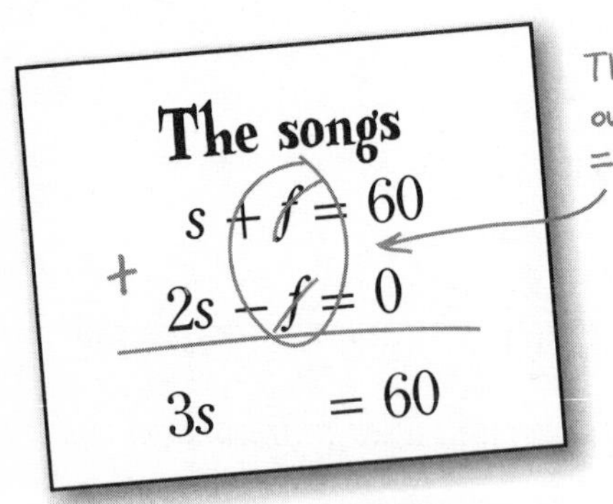

The songs
$s + f = 60$
+
$2s - f = 0$
$3s = 60$
These two canceled out because f – f = 0

The songs system of equations was already set up to easily cancel out.

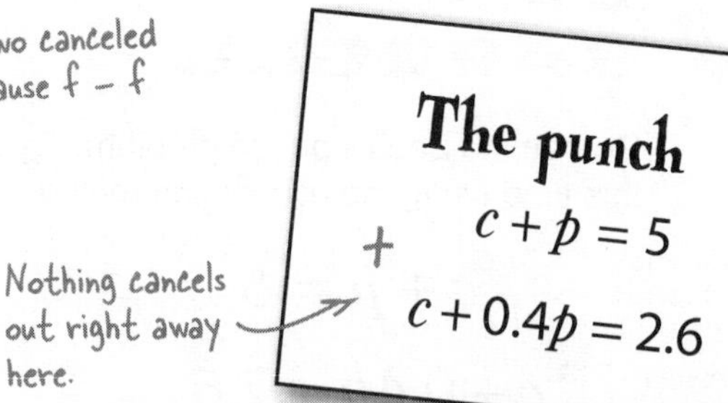

The punch
+
$c + p = 5$
$c + 0.4p = 2.6$
Nothing cancels out right away here.

The punch system of equations wasn't set up to cancel out.

Manipulate your equations for elimination

Add the two punch equations together and, you'll get an equation in two variables that you still can't solve. To eliminate a variable, you need to have the coefficient in front of one of your variables be the opposite of the coefficient in front of the same variable in your second equation.

After figuring out your system of equations, you need to look at the system and figure out a couple of things:

Do either of the variables cancel each other out when you add the equations together?
For the punch equation, we started with $+1c$ and $+1c$, and $+1p$ and $+0.4p$.
None of these cancel out right away.

If not, figure out what variable to eliminate.
Which variable are you going to try and get to be the opposite coefficient of the same variable in your other equation? Here's where some strategy comes in.

Which variable?

The options with the punch are to change one of the $+1c$s to a $-1c$, or change around the $+1p$ to a $-0.4p$, so the ps cancel out.

Let's try and work with c since there aren't any decimals. So how do we change a $+1c$ to a $-1c$? Multiply the entire equation by negative one. Then we can add the two equations and use elimination.

Determining the variable to cancel out, and manipulating the equation to get there, is the tricky part of the elimination method. Here are a few things you can look out for:

Look for variables with constants that are multiples of each other, like 2 and 4. And 1 works with anything.

To make the variable constants cancel each other out and keep your equation balanced, you need to multiply the entire equation by the right constant.

there are no Dumb Questions

Q: What if I pick the wrong variable to eliminate?

A: There's really no "wrong variable." As long as you properly apply the rules of manipulating an equation, you won't get the wrong answer.

There is usually an "easiest" variable, though. You'll get a better idea of which variables to choose as you have more experience solving equations, but if there's one that has a coefficient of 1, it's usually easiest to go for that variable. Then, it's easy to figure out what has to be done to the entire equation to make it work and eliminate the variable.

Q: Why is it ok to add up the two equations?

A: Because both equations are using the same variables, representing the same things. It's just like adding two numbers together. But you can't swap around sides or otherwise change the equations. Just add the two equations together to eliminate one variable, and solve.

Q: Does it matter which equation I put my solution value back into?

A: No, either one will work. But just like choosing which variable to eliminate, there's usually an easier equation to use. If one equation uses whole numbers, and the other uses fractions or decimals, it's probably easier to use the equation with whole numbers to get a quicker solution.

Q: So I should use substitution for solving all my systems of equations?

A: It really just depends on the equation. You can use graphing, substitution, or elimination, and all will work. It's really just a matter of figuring out what works best for you in a certain situation, and going with that.

Remember: as long as you follow the rules, you'll get the same answer no matter which method you use.

BULLET POINTS

- Always check your work.
- The toughest part of the elimination method is figuring out which variable to eliminate.
- Always apply the same multiplier to both sides of the equals signs.
- After you've eliminated and solved for one variable, use that value to solve for the other variable.

ELIMINATION CONSTRUCTION

Look at the following situations, write the system of equations, and then solve them using either the elimination method or the substitution method.

Zach has been working on figuring out his income from the party. He let people get tickets two ways—either in advance for $18, or at the door for $22. He had a total of 1,512 people come to the dance and collected $31,566. How many tickets were advance sales, and how many were purchased at the door?

Zach ordered 11 cakes for the party and paid for them, but the bakery called him up and said they couldn't find his order. He knows that the tiered cakes will serve 150 and the sheet cakes will serve 104. How many of each kind of cake did he order to feed everyone?

ELIMINATION CONSTRUCTION SOLUTION

Look at the following situations, write the system of equations, and then solve them using either the elimination method or the substitution method.

Zach has been working on figuring out his income from the party. He let people get tickets two ways—either in advance for $18, or at the door for $22. He had a total of 1,512 people come to the dance and collected $31,566. How many tickets were advance sales, and how many were bought at the door?

To have the as cancel out, the first equation needs to be multiplied by −18

$a + d = 1512 \rightarrow -18(a + d) = -18(1512) \rightarrow -18a - 18d = -27{,}216$

$18a + 22d = 31{,}568$

$$\begin{aligned} -18a - 18d &= -27{,}216 \\ + \quad 18a + 22d &= 31{,}568 \\ \hline 4d &= 4{,}352 \\ \frac{4d}{4} &= \frac{4{,}352}{4} \\ d &= 1088 \end{aligned}$$

$a + 1088 = 1512$

$-1088 + a + 1088 = 1512 - 1088$

$a = 424$

The advance sales were 424, and the tickets at the door were 1088 tickets.

Zach ordered 11 cakes for the party and paid for them, but the bakery called him up and said they couldn't find his order. He knows that the tiered cakes will serve 150 and the sheet cakes will serve 104. How many of each kind of cake did he order to feed everyone?

The total number of tiered cakes plus sheet cakes is 11.

$t + s = 11$

This is the total number of people from earlier.

$150t + 104s = 1512$

150 for each tiered cake.

104 pieces for each sheet cake.

$-150(t + s) = -150(11)$

$$\begin{aligned} -150t - 150s &= -1650 \\ + \quad 150t + 104s &= 1512 \\ \hline -46s &= -138 \\ \frac{-46s}{-46} &= \frac{-138}{-46} \\ s &= 3 \end{aligned}$$

$t + 3 = 11$

$-3 + t + 3 = 11 - 3$

$t = 8$

Zach needs 3 sheet cakes and 8 tiered cakes to feed everyone.

System of Equations Exposed

This week's interview:
One problem but two equations?

Head First: What's it like being a system of equations? Do you find you have identity problems?

System: Not at all—just because I have multiple equations, doesn't mean I have multiple personalities! All of my equations work for the same problem.

Head First: I didn't mean to upset you. I just meant that being made up of two completely different equations must be tricky.

System: It's all I've ever known. Really, I just think it's easier to have a couple of equations that describe me. The equations working together is what really makes me tick. Since I usually have two unknowns, I couldn't be solved uniquely without both of them.

Head First: There are a number of ways to solve you—let's talk about substitution, okay?

System: Ok. The substitution method is one of my favorites because it gives you an exact answer, but there's really not much planning involved. You just go ahead and get one variable in terms of the other, and then switch equations and substitute. It doesn't require any fancy thinking ahead.

Head First: Doesn't working that way occasionally get complicated?

System: That's true. You get one variable, let's say x, in terms of the other, which is probably y with some constants. As you start performing the substitution, there can be a lot of manipulating to work out a solution.

Head First: There's another option for finding an exact solution, right?

System: Certainly, that's the elimination method. With elimination, you have to figure out in advance what variable you need to cancel out. Once you do that, all of the manipulation is done in advance.

Head First: What do you mean?

System: Well, with the elimination method, you add up the two equations, all the terms on the left of both equals signs, and all of the terms on the right of both equals signs together. As long as one variable is eliminated, you will end up with one equation in one variable.

Head First: I see, that does sound easy. But speaking of seeing—what about those of us who like to see what's going on?

System: Well then there's the graphing method. To solve me that way, you just graph both equations and look for an intersection. Graphing is great for getting a sense of the trends of both lines and a good starting point for guessing what may be coming next, say if you add more cider.

Head First: System, it's been a pleasure. You're complicated, but nobody can say there aren't plenty of ways to work with you!

Zach's party rocks!

You've done a lot of work, so I want **you** to come out to the best New Year's party ever!

Using the graphing method and elimination, you figured out the perfect punch.

The punch

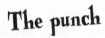

$$c + p = 5$$

$$+$$

$$c + 0.4p = 2.6$$

$c = 1$ 1 gallon of ginger cider

$p = 4$ 4 gallons of pineapple juice

Using the elimination method, you figured out how many slow songs and how many fast songs Zach needs to pick to keep the girls happy.

The songs

$$s + f = 60$$

$$2s - f = 0$$

$s = 20$ 20 slow songs

$f = 40$ 40 fast songs

We're ready to party!

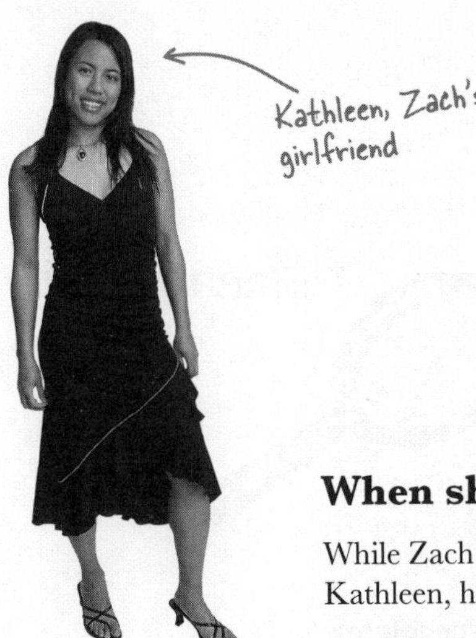

When shopping leads to relationship ruin!

While Zach has been busy planning his party, his girlfriend, Kathleen, has been getting ready too.

In addition to being a fantasy-football mogul, Kathleen loves to shop. She's been hitting the mall pretty hard lately, looking for something to wear to the party.

Fifteen Minute Mystery

Kathleen's shopping is not something that Zach is a big fan of, so when she gets a good deal, Kathleen makes sure Zach knows about it.

For the party, she needed dresses and shoes. She found an awesome sale: dresses for only \$16 and shoes for \$8 a pair. That means Kathleen only spent \$72 in total!

While showing off her purchases, Kathleen said that she bought twice as many dresses as shoes and got 6 things all together.

Using his mad Algebra skills, Zach did some quick math and then got pretty upset. They'd argued about this before. "How could you lie to me?" he asked her.

How did Zach know that Kathleen was lying?

When shopping leads to relationship ruin - solved!

How did Zach know Kathleen was lying?

To figure out that Kathleen was lying, Zach needed to to figure out how many pairs of shoes and how many dresses she actually bought. Let's start with figuring up some equations:

1. **Figure out the knowns from the problem:**

Dresses: We'll call them ***d***, and they're \$16 each.

Shoes: We'll call them ***s***, and they're \$8 a pair.

Totals: \$72 spent, and twice as many dresses as pairs of shoes for a total of 6.

2. **Let's write some equations and set up the system:**

Twice as many dresses → as shoes → for a total of 6 items.

$$2d + s = 6$$

$$16d + 8s = 72$$

Each dress is \$16. Shoes are \$8 a pair. Kathleen spent a total of \$72.

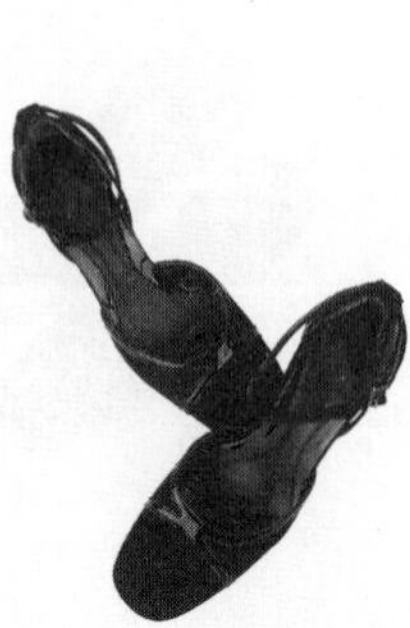

Now we just add them up...

3 **Use elimination to remove a variable.**

Let's eliminate s, so we need a −8s for the top equation. Multiply the entire top equation by −8.

$$-8 \cdot (2d + s = 6) \cdot -8$$
$$16d + 8s = 72$$

$$-16d - 8s = -48$$
$$+ \quad 16d + 8s = 72$$
$$0d + 0s = 24?$$

What the... Why the tough factors?

Equations that aren't true are INCORRECT.

0 times the number of dresses plus 0 times the number of pairs of shoes ***doesn't*** equal 24. This doesn't make a lot of sense, and it doesn't explain why Zach is mad at Kathleen. So elimination isn't working. Let's try graphing the lines and see what's going on.

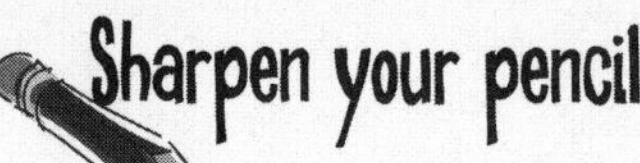

Rewrite the equations into a form that you can graph. Then plot the solution lines on the Cartesian plane. Does their intercept tell you anything about Kathleen's story?

$$16d + 8s = 72$$

$$2d + s = 6$$

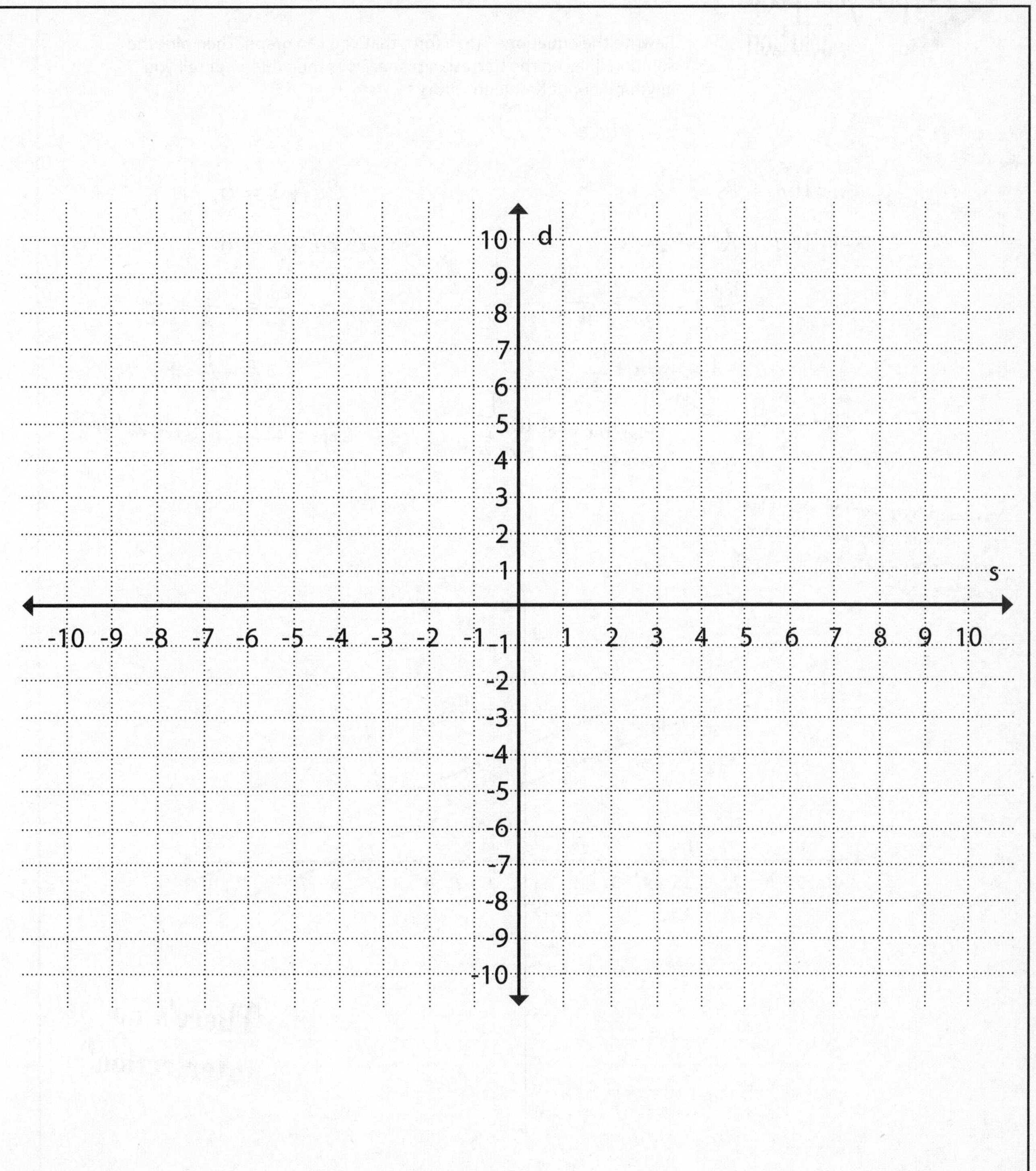
d
s
10
9
8
7
6
5
4
3
2
1
-1
-2
-3
-4
-5
-6
-7
-8
-9
-10
-10 -9 -8 -7 -6 -5 -4 -3 -2 -1 1 2 3 4 5 6 7 8 9 10

Rewrite the equations into a form that you can graph. Then plot the solution lines on the Cartesian plane. Does their intercept tell you anything about Kathleen's story?

Remember, d is the y - so if we want to get it in y = mx + b format, we have to isolate the d.

$$16d + 8s = 72$$

$$-8s\ 16d + 8s = 72 - 8s$$

$$\frac{16d}{16} = -\frac{8s}{16} + \frac{72}{16}$$

$$d = -\frac{1}{2}s + \frac{9}{2}$$

Slope = $-\frac{1}{2}$

Plot the point $(0, \frac{9}{2})$ and go down 1 over 2

$$2d + s = 6$$

$$-s\ 2d + s = 6 - s$$

$$\frac{2d}{2} = -\frac{s}{2} + \frac{6}{2}$$

$$d = -\frac{1}{2}s + 3$$

Slope = $-\frac{1}{2}$

Start at (0,3)

You could have worked with these equations any way you wanted to graph them, but the graph should look the same.

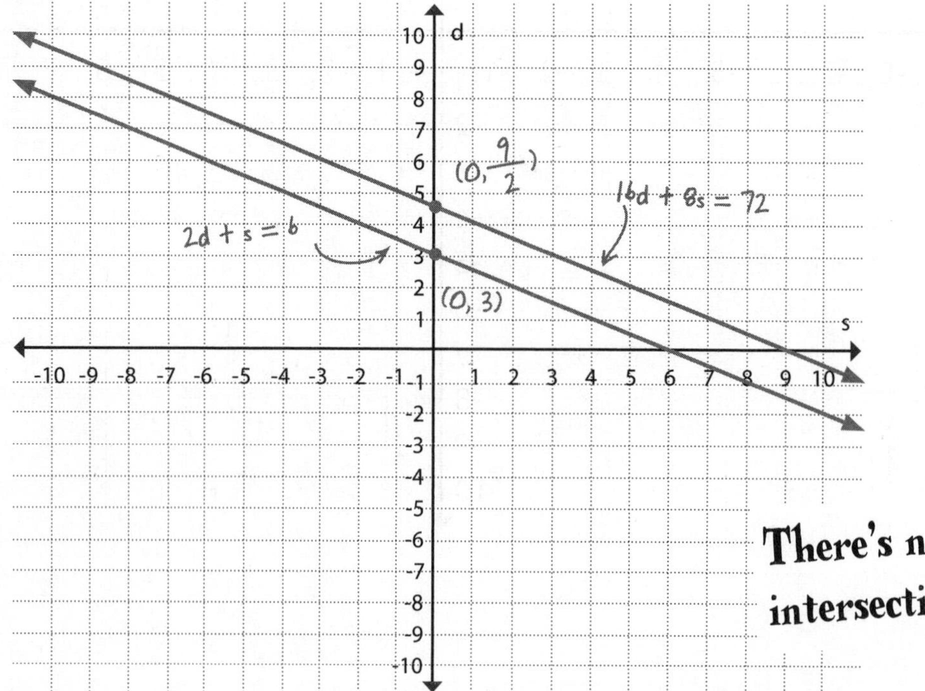

There's no intersection!

When shopping leads to relationship ruin - solved!

How did Zach know Kathleen was lying?

Zach knew from his quick math that there is no solution to Kathleen's equations, so ***she must be lying about something***.

Fifteen Minute Mystery Solved

The lines are parallel! That means that there isn't any one point that will satisfy both equations. Parallel lines go on forever without crossing, so there's no intersection.

That means that what Kathleen said can't be true because there's no combination of shoes and dresses that will satisfy both equations. That means she has an extra dress or something!

Hmm - I noticed that those lines have the same slope but different intercepts...

Parallel lines share a slope, but nothing else.

That's actually the key indicator that two lines are parallel. If you have two equations with the same slope, you don't need to graph it. ***The lines are parallel, and there's no solution***.

Well, except for one case...

Sometimes two equations aren't two lines

You know a lot about equations at this point: you can graph them a bunch of different ways, and if you have a system of equations, you can solve them three different ways. Not only that, but if you have two equations with the same slope, there's no answer, right?

Not quite.

You can have two equations that have the same slope, but they may look a little different. However, when you plot them, the lines look exactly the same:

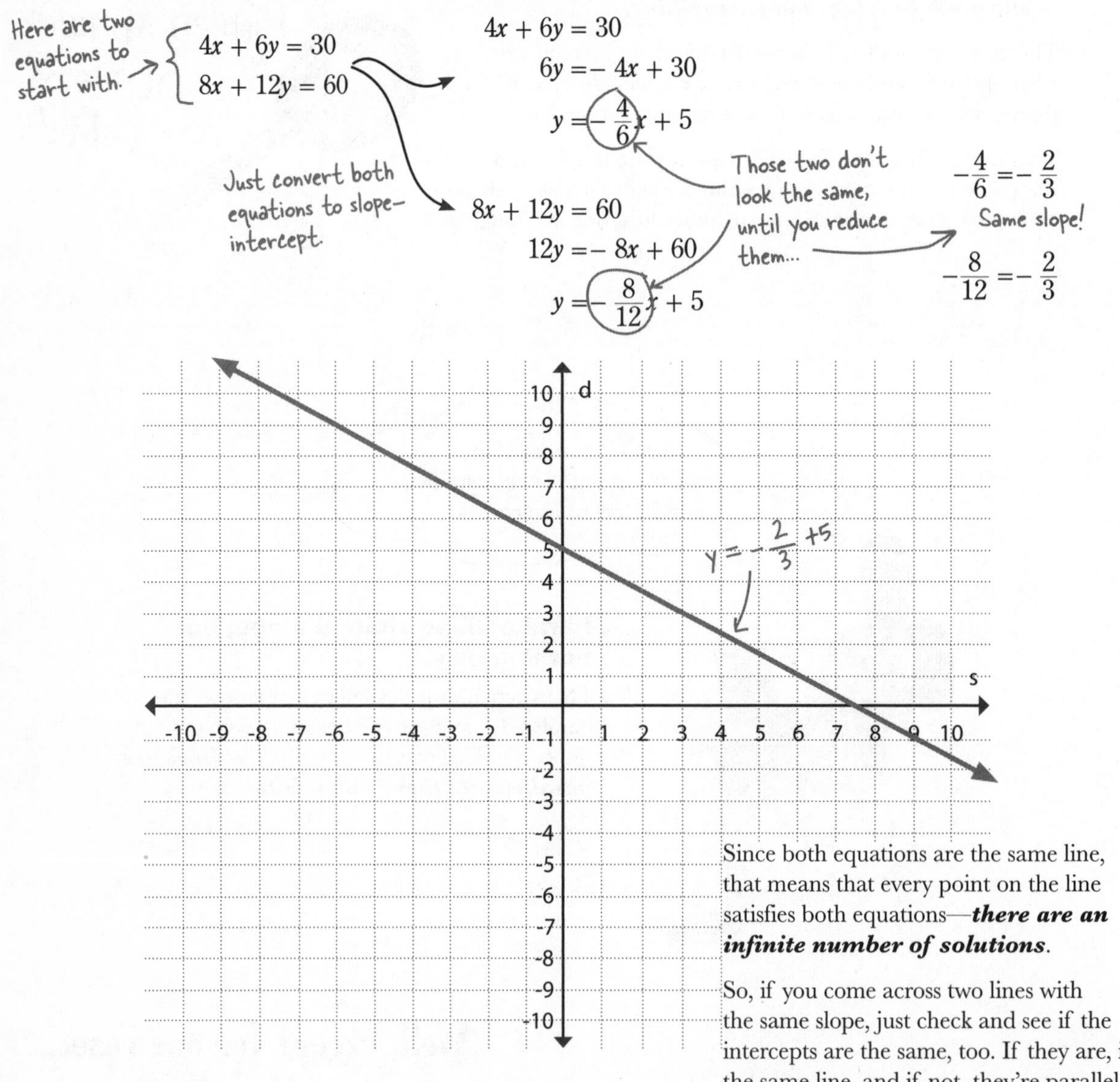
Here are two equations to start with.
4x + 6y = 30
8x + 12y = 60
Just convert both equations to slope-intercept.
4x + 6y = 30
6y =– 4x + 30
y =– 4/6 x + 5
8x + 12y = 60
12y =– 8x + 60
y =– 8/12 x + 5
Those two don't look the same, until you reduce them...
–4/6 =– 2/3
Same slope!
–8/12 =– 2/3
y = – 2/3 +5
d
s

Since both equations are the same line, that means that every point on the line satisfies both equations—***there are an infinite number of solutions***.

So, if you come across two lines with the same slope, just check and see if the intercepts are the same, too. If they are, it's the same line, and if not, they're parallel.

Look at the following systems of equations and answer the questions about them. Then solve them with whatever method you want..

$$2x + 3y = 100$$
$$-0.5x - .75y = -25$$

What describes these two lines? (circle one)

Intersecting **Parallel** **The same line**

Why ....

....

....

$$-2x + 2y = -8$$
$$-3x - 3y = -30$$

What describes these two lines? (circle one)

Intersecting **Parallel** **The same line**

Why ....

....

....

Solve the following systems of equations using the method shown.

$$2x + 3y = 100$$
$$-0.5x - .75y = -25$$

$-4(-0.5x - 0.75y) = -4(-25)$

These are the same! → $2x + 3y = 100$

What describes these two lines? (circle one)

Intersecting Parallel **The same line** (circled)

Why Because these lines are the same, there are an infinite number of solutions.

You can do this problem with substitution or graphing, too. No matter how you do that, you should get the same answer.

The best way to handle this is to actually multiply both equations.

Like when you find a common denominator for fractions.

$$-2x + 2y = -8$$
$$-3x - 3y = -30$$

$3(-2x + 2y) = 3(-8)$

$2(-3x - 3y) = 2(-30)$

$-6x + 6y = -24$
$+ \quad -6x - 6y = -60$

$\frac{-12x}{-12} = \frac{-84}{-12}$

$x = 7$

$-2x + 2y = -8$

$-2(7) + 2y = -8$

$+14 - 14 + 2y = -8 + 14$

$\frac{2y}{2} = \frac{6}{2}$

$y = 3$

What describes these two lines? (circle one)

Intersecting (circled) Parallel The same line

Why Because there's a point for the solution

Systemcross

Got a system for equations? Great. Now use your system for solving a crossword. Hint: it's your brain!

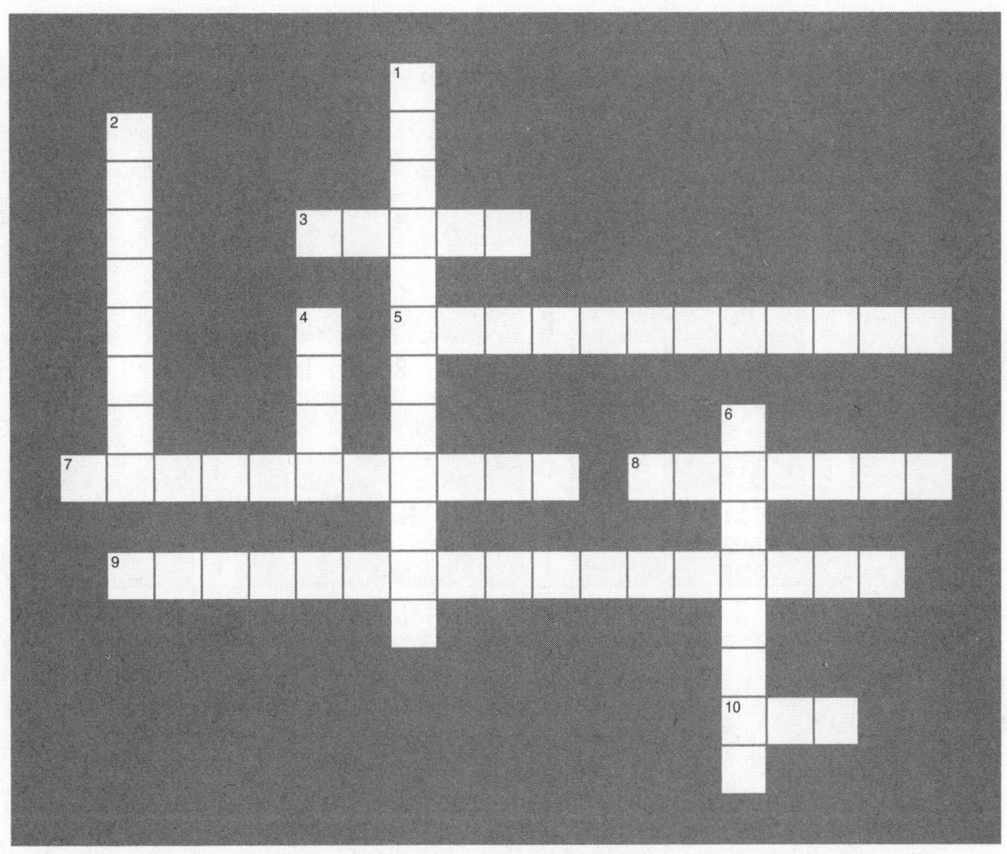

Across

3. Always _________ your work.
5. Solving one equation in terms of one variable, then putting it into the other is the _________ method.
7. Setting the system of equations up and adding them together is called the __________ method.
8. Always remember the _____________ of the problem.
9. A group of equations that can be treated as the same problem.
10. If you have two unknowns, you need ________ equations.

Down

1. The solution to a system of equations is the ____________ of their graphs.
2. Lines that have the same slope are ___________.
4. For the elimination method to work, you have to _____.
6. The number of solutions to an equation that a line represents

Systemcross Solution

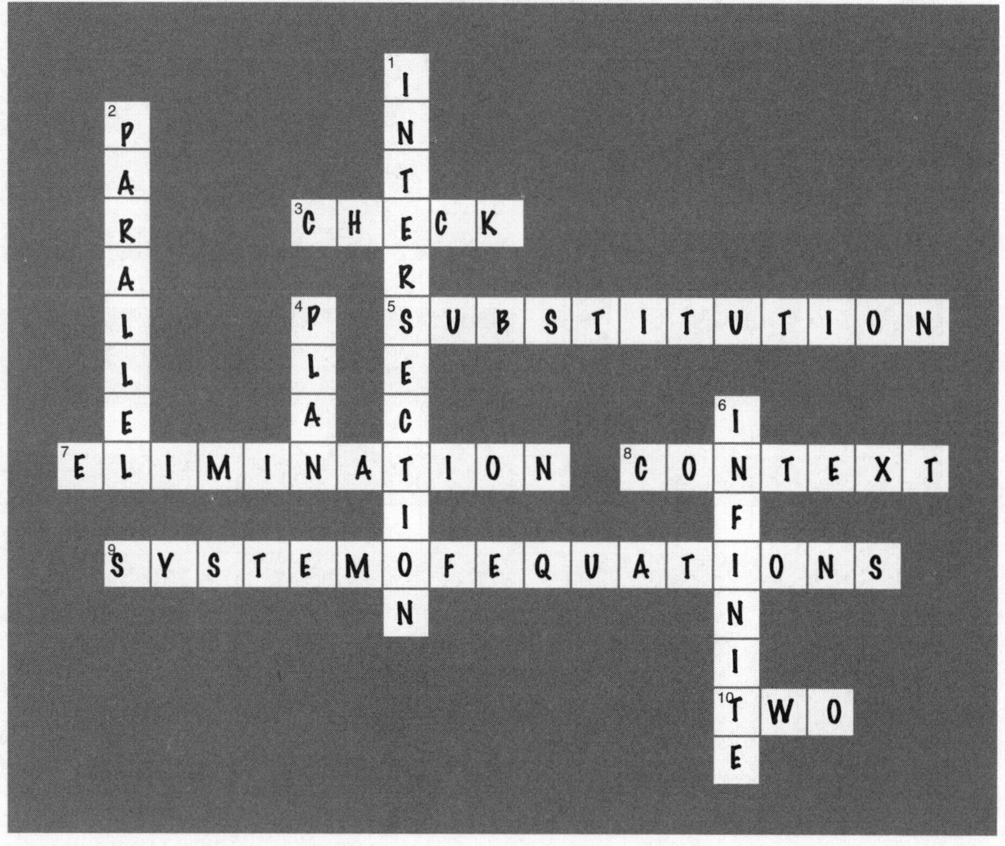

Tools for your Algebra Toolbox

This chapter was about solving systems of equations with three methods.

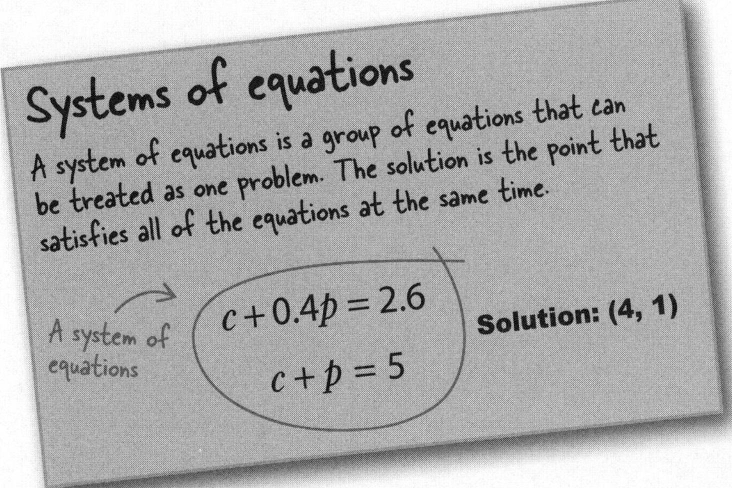

Systems of equations
A system of equations is a group of equations that can be treated as one problem. The solution is the point that satisfies all of the equations at the same time.
A system of equations
$c + 0.4p = 2.6$
$c + p = 5$
Solution: (4, 1)

BULLET POINTS

- Always check your work.
- The toughest part of the elimination method is figuring out which variable to eliminate.
- Always apply the same multiplier to both sides of the equals signs.
- After you've eliminated and solved for one variable, just go and finish it up with substitution.
- For 2 linear equations with 2 unknowns, you either have a single solution (intersect once), no solutions (don't intersect), or infinite solutions (same line).

CHAPTER 7

8 expanding binomials & factoring

Breaking up is hard to do

So she said, "I totally hate the new 90210," and I said, "Forget it. You're totally square. And we're breaking up." And that was it... we were over.

Sometimes being square is enough to give you fits.

So far, we've dealt with variables like x and y. But what happens when x is **squared** in your equations? It's time to find out—and you already have the tools to solve these problems! Remember the distributive rule? In this chapter, you're going to learn how to use **distribution** and a special technique called **FOIL**, to solve a *new* kind of equation: **binomials**. Get read—it's time to ***break down*** some really tough equations.

Math or No Math semi-regional masters final

Our champion, Kate, is back to defend her title against a new challenger, James. It's up to you to be the judge again... but this time, the problems are even harder.

Problem #1: Simplify this expression

$$(x + 3)(2x - 1)$$

Here's what they got:

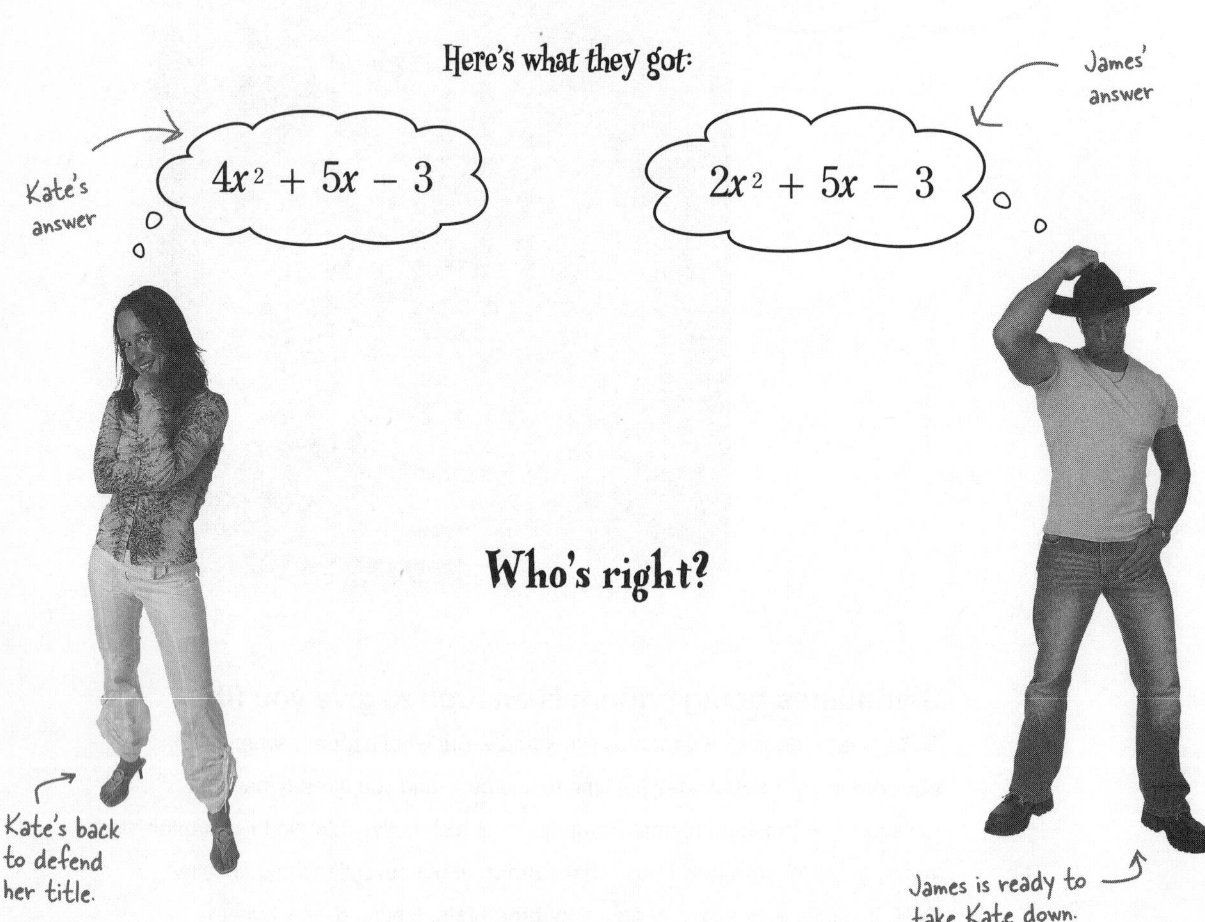
Kate's answer
$4x^2 + 5x - 3$
James' answer
$2x^2 + 5x - 3$
Kate's back to defend her title.
James is ready to take Kate down.

Who's right?

Who's right?

Below, Kate and James both showed their work. Kate did her work by multiplying the whole first expression by the second. James did his work by splitting up the first binomial.

original expression

$$(x + 3)(2x - 1)$$

$2x^2 - x + 2x^2 + 6x - 3$

Combine like terms. $4x^2 + 5x - 3$

$$(x + 3)(2x - 1)$$

This is distributing 2x–1 to both parts of the first binomial.

$x(2x - 1) + 3(2x - 1)$

$2x^2 - x + 6x - 3$

$2x^2 + 5x - 3$

How do we figure out who's right? Well, if you know a solution, substitution will work. So you can try plugging the right x value into each of their equations and see if the math works out. If the values don't come out right, then one of the simplifications that Kate and James ended up with isn't correct.

Sharpen your pencil

The producers told you the correct value for x- is 3 and the equation is equal to zero. Substitute -3 in the different equations for x, and see who's right. Show your work, and don't forget to circle the correct expression, too.

$x = -3$ into $4x^2 + 5x - 3$ and $2x^2 + 5x - 3$

The producers told you the correct value for x- is 3 and the equation is equal to zero. Substitute -3 in the different equations for x, and see who's right. Show your work, and don't forget to circle the correct expression, too.

$x = -3$ into $4x^2 + 5x - 3$ and $2x^2 + 5x - 3$

$4x^2 + 5x - 3$	$2x^2 + 5x - 3$
$4(-3)^2 + 5(-3) - 3$	$2(-3)^2 + 5(-3) - 3$
$4(9) - 15 - 3$	$2(9) - 15 - 3$
$36 - 15 - 3$	$18 - 15 - 3$
$21 - 3$	That works! 0 ✓
This is wrong! ~~18~~	

The equation was supposed to equal zero, so this one is correct.

Okay, but how did James actually come up with his answer? I mean, that's great that the... ahem... producers had the right answer... but we don't usually have people giving us answers, right? So what's really going on?

Those expresions are binomials.

To understand what James did, you need to know how to handle exponents in equations. That means learning about binomials and polynomials.

Binomials are groups of two algebraic terms

Kate and James are working with binomials on this round of Math or No Math. A **binomial** is an expression that contains ***two*** algebraic terms. Binomials are actually part of a larger family: **polynomials**. Poly just means many, so a polynomial is any expression with multiple terms.

So anytime you see an expression with more than one term, think ***polynomial***. And if there are just two terms, think ***binomial***. Look:

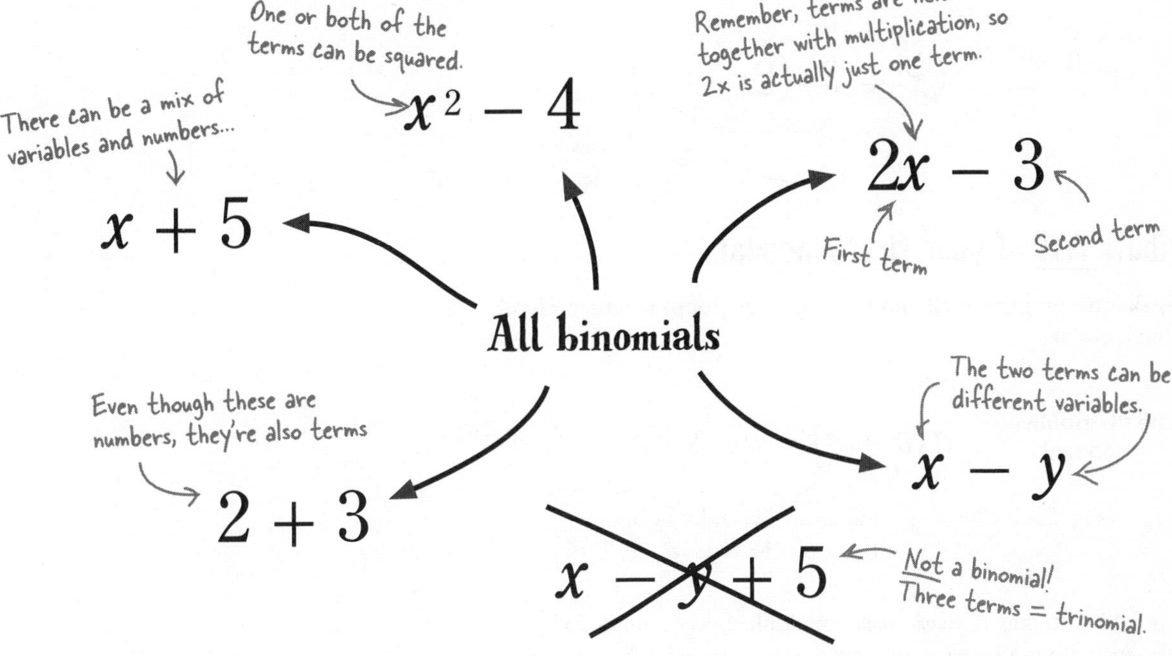
All binomials
There can be a mix of variables and numbers...
x + 5
One or both of the terms can be squared.
x² − 4
Remember, terms are held together with multiplication, so 2x is actually just one term.
2x − 3
First term
Second term
Even though these are numbers, they're also terms
2 + 3
The two terms can be different variables.
x − y
x − y + 5
Not a binomial! Three terms = trinomial.

Sum it up

Polynomial – an expression containing any number of algebraic terms with whole number exponents of at least 0.

Binomial – is a special case of a polynomial that is a group of two algebraic terms.

y=mx+b

The distributive property, revisited

$$(x + 3)(2x - 1)$$

Look back at the problem Kate and James were given in the contest. You can't simplify the algebraic terms within the parentheses. So how about just doing the multiplication? Since the expression can't be simplified, we're going to have to do multiplication with multiple terms—we're going to have to **distribute**. ***Both terms in the first binomial need to be distributed over both terms of the second binomial.***

We need to multiply x times 2x and −1...

$$(x + 3)(2x - 1)$$

...and 3 times 2x and −1, also.

Distribute ALL of your first binomial...

The distributive property is all about multiplying groups together. Here's what that looks like:

The distributive property $a(b + c) = ab + ac$ ← to get this...

The first term can be multiplied by both terms of the binomial

That's great if you only have *one number* multiplied over a binomial. But we're dealing with is a *binomial* multiplied over a binomial. So we've first got to distribute both terms of the first binomial over the entire second binomial.

Multiplying two binomials means you have to distribute both terms from the first binomial over both terms of the second binomial.

Simplify binomials with the distributive property

So now we just need to actually distribute our entire binomial. Before digging into Kate and James' problem, let's take a look at how we'd do this distribution in general:

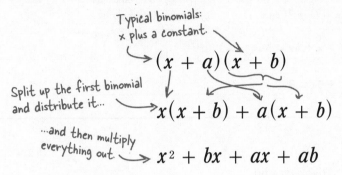

Typical binomials: x plus a constant.

$$(x + a)(x + b)$$

Split up the first binomial and distribute it...

$$x(x + b) + a(x + b)$$

...and then multiply everything out.

$$x^2 + bx + ax + ab$$

Now we can simplify a little bit more. We have two $\boldsymbol{x}$ terms, $\boldsymbol{bx}$ and $\boldsymbol{ax}$. Since $\mathbf{a}$ and $\mathbf{b}$ are constants, $\boldsymbol{bx}$ and $\boldsymbol{ax}$ are actually like terms. We can combine those:

a and b are constants, like 2 or 18. They can be added up.

These are the two constants multiplied together. This usually is another number, like 35 or 90.

$$x^2 + (a + b)x + ab$$

This is combining like terms, to simplify the equation even further.

Sharpen your pencil

Use the distributive property and simplify the Math or No Math challenge.

$$(x + 3)(2x - 1)$$

Sharpen your pencil Solution

Use the distributive property and simplify the Math or No Math challenge.

$(x + 3)(2x - 1)$

$x(2x - 1) + 3(2x - 1)$ ← Split the first binomial, so we can distribute it...

$2x^2 - x + 6x - 3$ ← Multiply things out, and look for like terms.

After combining like terms, we end up with the same thing James got. → $2x^2 + 5x - 3$

there are no Dumb Questions

Q: The distributive property says we have to spread out both terms?

A: Yes. The distributive property says that you have to distribute the first binomial over the entire second binomial. That means both terms of the first binomial need to be multiplied by both terms of the second.

Q: Expanding binomials seems complicated. Is there an easier way?

A: There are some tools you can use, but you've still got to know that multipying binomials is really about distribution. It's not that bad as long as you take the first terms and multiply them over the second binomial.

Q: So this works the same for numbers and variables?

A: It sure does. Of course, in cases where you've got all numbers, you can just work inside the parentheses first and avoid all this distribution. But you've noticed something important: any rule that applies to variables also applies to numbers.

Q: How often do you really have to multiply binomials?

A: Actually, you're most likely to use this the other way around. We'll talk about it in the next section, but when you have certain types of equations, you'll have to go from that equation to a couple of binomials. Sound confusing? Don't worry, we're going to spend a lot more time on that next.

BULLET POINTS

- Multiplying two binomials is a common Algebra problem.
- To multiply two binomials, you need to apply the distributive property.
- Each term of the first binomial needs to be multiplied by each term of the second binomial.

Sharpen your pencil

Simplify the following binomial expressions. Make sure to combine like terms!

$(y - 1)(y - 7)$

$(4 + x)(3 - x)$

Watch the signs!

$(a + 4)(a - 6)$

$(-x - 3)(x + 3)$

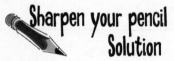

Sharpen your pencil Solution

Your job was to simplify the following binomial expressions. You should have combined like terms, too.

It will work with y, too!

$$(y - 1)(y - 7)$$

$y(y - 7) - 1(y - 7)$

$y^2 - 7y - 1y + 7$

$y^2 - 8y + 7$

You could also rewrite the binomials and multiply $(x + 4)(-x + 3) = 0$.

$$(4 + x)(3 - x)$$

$4(3 - x) + x(3 - x)$

$12 - 4x + 3x - x^2$

$-x^2 - x + 12$

$$(a + 4)(a - 6)$$

$a(a - 6) + 4(a - 6)$

$a^2 - 6a + 4a - 24$

$a^2 - 2 - 24$

Watch the signs!

Just carry through the sign for each operation.

$$(-x - 3)(x + 3)$$

$-x(x + 3) - 3(x + 3)$

$-x^2 - 3x - 3x - 9$

$-x^2 - 6x - 9$

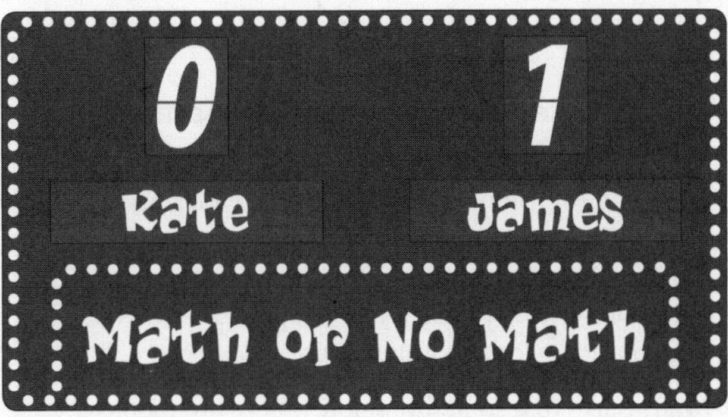

Problem #2: Simplify this expression - fast!

$$(x + 2)(x - 2)$$

Here's what the contestants got:

Kate's answer → $x^2 - 4$ ← This is a binomial.

James' answer:

$(x + 2)(x - 2)$

$x(x - 2) + 2(x - 2)$

$x^2 - 2x + 2x + 2(-2)$

$x^2 - 4$

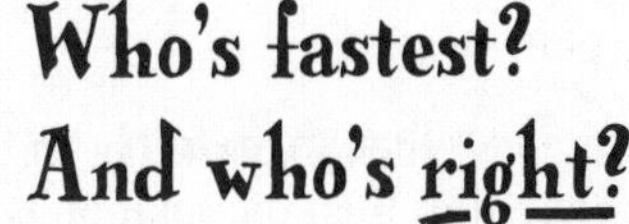

Kate's up to her old tricks, cranking through the speed round. But how?

Kate cheated, didn't she? She didn't even show her work!

Kate looked for patterns instead of working the problem out.

Have you ever noticed little patterns in a friend's phone number or how a football team runs their offense? A pattern lets you figure out what's going on based on some key things you recognize... and you don't have to work anything out.

Math is like that sometimes, too. There are patterns that let you avoid lots of extra work.

The SQUARE pattern

If you have two binomials that only differ in the sign of the second term, you're dealing with a **square pattern**. In the problem that Kate and James just worked on, the first term of both binomials was *x* and the second term was 2 and -2.

$$(x + 2)(x - 2)$$
$$x(x - 2) + 2(x - 2)$$
$$x^2 - 2x + 2x + 2(-2)$$
$$x^2 - 4$$

We can work this out the normal way to see what the answer is.

The first term in both binomials is x.

The second term in both binomials is 2.

The minus sign carried through to the final solution.

This is the first term, x, squared.

This is the second term squared.

What you've got here is the **difference of two squares**. You square the *first* term and subtract from it the square of the *second* term. Everything else gets canceled out and goes away.

So anytime you see two binomials that are ***exactly the same***, except for the *sign* before one of the constants, then the solution is the difference of the squares of the two terms. You can just skip the middle steps and jump right to the answer!

This is what Kate saw when she solved the last problem.

$$(x + a)(x - a) = x^2 - a^2$$

One of these signs must be positive, the other must be negative.

The answer always has a negative sign here.

What about when the signs are the SAME?

But what about when you've got two terms that are the same in both binomials, and the signs before the constants are the ***same***? Let's work one out:

The first term in both binomials is x.

The second term in both binomials is 5.

$$(x + 5)(x + 5)$$

$$x(x + 5) + 5(x + 5)$$

$$x^2 + 5x + 5x + 25$$

$$x^2 + 10x + 25$$

This is the first term, x, squared.

This is the second term, 5, squared.

This is 2 times the second term.

Another way to think about this is the square of a binomial. So this is really just a binomial multiplied by itself.

This is with the same signs...

Binomial squared: $(x + a)^2 = x^2 + 2ax + a^2$

...and this is with opposite signs.

Binomials with different signs: $(x + a)(x - a) = x^2 - a^2$

The solution is the difference of two squares.

Sharpen your pencil

Below are some binomial problems that you can test out the squares patterns on. No showing your work here... just look for the right pattern, and write down your answer. Good luck!

$(x + 3)(x - 3)$

....

$(x + 9)(x + 9)$

....

$(2x - 10)(2x + 10)$

....

Ok, this one's a little tricky. The signs are the same, but the sign is negative this time. What does that do to the solution?

$(x - 7)^2$

....

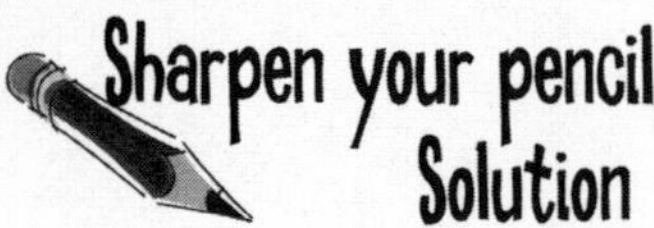

Your job was to solve the binomial multiplications below, without working through each problem step-by-step. How did you do?

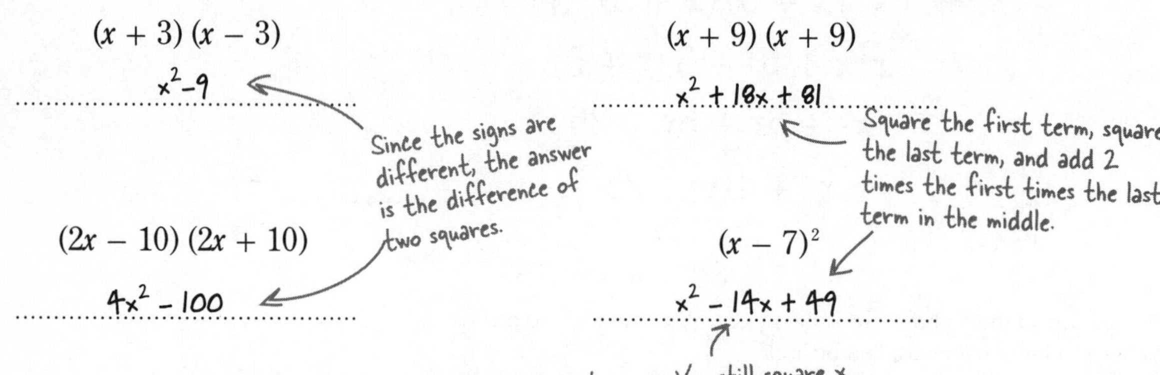

$(x + 3)(x - 3)$
x²-9
$(x + 9)(x + 9)$
x² + 18x + 81
Since the signs are different, the answer is the difference of two squares.
Square the first term, square the last term, and add 2 times the first times the last term in the middle.
$(2x - 10)(2x + 10)$
4x² - 100
$(x - 7)^2$
x² - 14x + 49

This isn't as hard as it seems. You still square x and 7, but the difference is in the middle term. You multiply 2 by x by the second term, which is negative 7. So you get negative 14x.

there are no Dumb Questions

Q: Those patterns are great. Will they always work?

A: As long as the binomials match one of the patterns, then you bet! Just make sure you watch your signs and coefficients.

Q: What if the binomials are close but don't exactly match one of those square patterns?

A: Then you're going to have to use distribution to figure out what the simplified equation really is. Multiply the terms through, collect your like terms, and then you have the answer. But these patterns only work if there's an exact match.

Binomial squared: $(x + a)^2 = x^2 + 2ax + a^2$

Binomials with different signs: $(x + a)(x - a) = x^2 - a^2$

Sometimes there's just not a pattern...

Suppose you're on a fast-paced, high-dollar game show, forced to simplify binomials at a moment's notice. When the square patterns aren't working, you've got to have another way to simplify binomials.

But distribution isn't that fast. Thankfully, there's another sort of pattern you can follow, called **FOIL**. That stands for **First**, **Outside**, **Inside**, **Last**. Here, let's take a closer look:

Simplify this:

$$(x + a)(x + b)$$

Let's suppose that a and b aren't the same, so our square patterns won't work.

F **First:**

Multiply together the ***first*** terms of both binomials.

$(x + a)(x + b) \rightarrow x^2$

1st term 1st term

Multiply x and x together.

\+

O **Outside:**

Multiply together the ***outside*** terms of both binomials.

$(x + a)(x + b) \rightarrow bx$

outside term outside term

Multiply x and b together.

\+

I **Inside:**

Multiply together the ***inside*** terms of both binomials.

$(x + a)(x + b) \rightarrow ax$

inside term inside term

Multiply a and x together.

\+

L **Last:**

Multiply together the ***last*** two terms of both binomials.

$(x + a)(x + b) \rightarrow ab$

last term last term

Multiply a and b together.

Then just add it all up: $x^2 + bx + ax + ab$

FOIL ALWAYS works

Let's use FOIL with some real binomials... ones that don't fit into one of our square patterns:

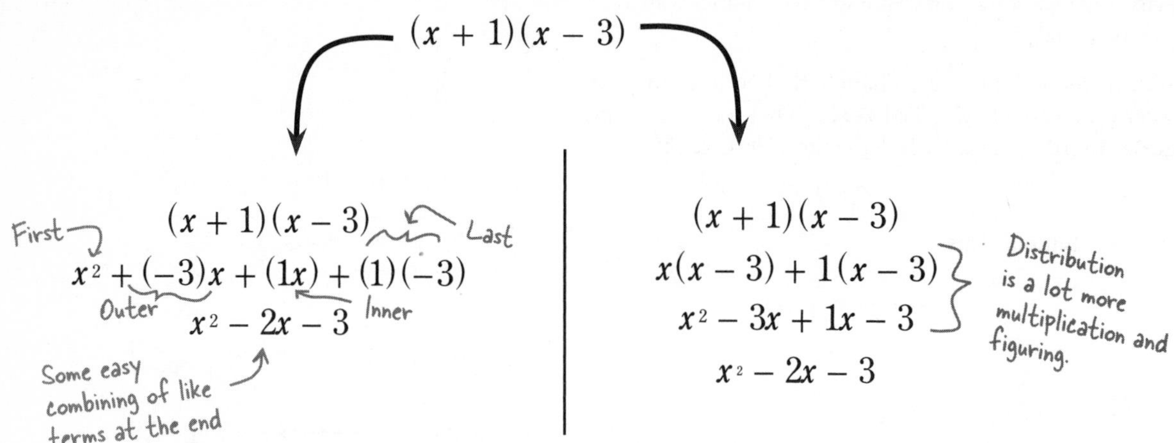

$(x + 1)(x - 3)$
First
$(x + 1)(x - 3)$
Last
$x^2 + (-3)x + (1x) + (1)(-3)$
Outer
Inner
$x^2 - 2x - 3$
Some easy combining of like terms at the end
$(x + 1)(x - 3)$
$x(x - 3) + 1(x - 3)$
$x^2 - 3x + 1x - 3$
$x^2 - 2x - 3$
Distribution is a lot more multiplication and figuring.

Solved using FOIL

Solved using distribution without FOIL

FOIL saves you an extra step, but it also makes the steps you do have to do a lot easier. You don't have to think about distribution as much, and you're usually left with an easy combining of like terms at the end.

And, best of all, FOIL always works, even if there's not a pattern to use!

BULLET POINTS

- Multiplying two binomials is a special case of **distribution**.
- FOIL stands for **First**, **Outside**, **Inside**, and **Last**.
- **FOIL** lets you apply distribution without lots of... well... distribution.
- FOIL is just a tool to help you apply the **distributive property** easily and consistently.
- Patterns like the **squares patterns** can help you solve a problem really quickly, but FOIL will always work.

Binomial Multiplication Magnets

Here are some problems from previous Math or No Math competitions.
See how you would have stacked up by filling in the missing pieces.

$(x - 3)(x + 4)$

$x^2 + \ldots - \ldots - 12$

$\ldots$

$(y - 10)(y + 2)$

$y^2 \ldots - 10y \ldots$

$y^2 \ldots$

Use FOIL here to solve these in just two steps!

$\left(x - \frac{1}{2}\right)\left(x + \frac{1}{2}\right)$

$\ldots$

$(c - 5)(c - 2)$

$c^2 \ldots + \ldots$

$c^2 \ldots + \ldots$

$\left(x + \frac{1}{8}\right) + \left(x + \frac{1}{8}\right)$

$\ldots \quad \ldots \quad \ldots$

$(3 + x)(7 - x)$

$\ldots - \ldots + \ldots - \ldots$

$\ldots$

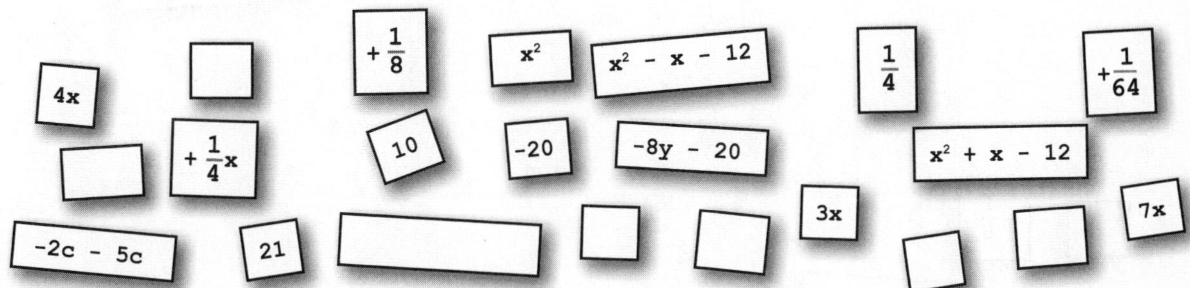

Binomial Multiplication Magnets Solution

Here are some problems from previous Math or No Math competitions. See how you would have stacked up by filling in the missing pieces.

$(x-3)(x+4)$

x^2 + 4x − 3x − 12

x² + x − 12

First, Outside, Inside, Last

$(y-10)(y+2)$

y^2 +2y − 10y -20

y^2 -8y − 20

$\left(x-\frac{1}{2}\right)\left(x+\frac{1}{2}\right)$

x² − $\frac{1}{4}$

$(c-5)(c-2)$

c^2 -2c − 5c + 10

c^2 -7c + 10

$\left(x+\frac{1}{8}\right)+\left(x+\frac{1}{8}\right)$

x² + $\frac{1}{4}$x + $\frac{1}{64}$

$(3+x)(7-x)$

21 − 3x + 7x − x²

21 + 4x − x²

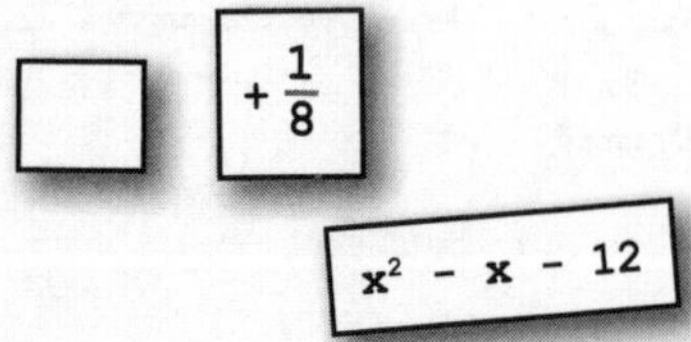

We're more than halfway through the Math or No Math competition, and here's where the contestants stand:

Kate needed some help keeping her terms straight in round 1:

$$(x + 3)(2x - 1)$$

$2x^2 - x + 2x^2 + 6x - 3$

$4x^2 + 5x - 3$

Kate did this one twice. FOIL would've helped her keep track of her distribution and prevented her mistake.

James used the distributive property, but not FOIL. Still, he got the right answer and took the points in round 1:

$$(x + 3)(2x - 1)$$

~~$x(2x - 1) + 3(2x - 1)$~~ You can skip this using FOIL.

$2x^2 - x + 6x - 3$

$2x^2 + 5x - 3$ ✓

He's still right because he manipulated the equation properly.

But Kate totally nailed the next round by using the squares pattern.

$$(x + 2)(x - 2)$$

$x^2 - 4$

Kate recognized identical terms and used the difference of two sqares.

Although he got the right answer, James got crushed in the speed round.

$$(x + 2)(x - 2)$$

$x(x - 2) + 2(x - 2)$

$x^2 - 2x + 2x + 2(-2)$

$x^2 - 4$

James got the right answer, but was too slow in the speed round.

The contestants are tied up... on to the next round.

Problem #3: Another speed round...

$$\frac{11x + \frac{11}{3}x - \frac{33}{4}x}{1 + \frac{1}{3} - \frac{3}{4}} = 1$$

Wow... this is the nastiest looking problem yet.

Kate wants to get rid of the fraction

$$\left(1 + \frac{1}{3} - \frac{3}{4}\right) \cdot \frac{11x + \frac{11}{3}x - \frac{33}{4}x}{\left(1 + \frac{1}{3} - \frac{3}{4}\right)} = 1 \cdot \left(1 + \frac{1}{3} - \frac{3}{4}\right)$$

Kate uses a common denominator to combine like terms in this step.

$$11x + \frac{11}{3}x - \frac{33}{4}x = 1 + \frac{1}{3} - \frac{3}{4}$$

$$\frac{132}{12}x + \frac{44}{12}x - \frac{99}{12}x = \frac{12}{12} + \frac{4}{12} - \frac{9}{12}$$

Kate multiplied everything by 12 to get rid of fractions.

$$12 \cdot \frac{77}{12}x = \frac{7}{12} \cdot 12$$

$$\frac{77x}{77} = \frac{7}{77}$$

All that's left is to reduce the fraction.

$$x = \frac{7}{77} = \frac{1}{11}$$

Time: 1 minute 35 s

$$\frac{11x + \frac{11}{3}x - \frac{33}{4}x}{1 + \frac{1}{3} - \frac{3}{4}} = 1$$

Whoa... what happened here? All these terms canceled out somehow?

$$\frac{11x\left(1 + \frac{1}{3} - \frac{3}{4}\right)}{\left(1 + \frac{1}{3} - \frac{3}{4}\right)} = 1$$

$$\frac{11x}{11} = \frac{1}{11}$$

$$x = \frac{1}{11}$$

Time: 35 s

Same answer, but James blew Kate away. How do you think he did that? ....

....

Un-distribution is called FACTORING

So far, we've talked a lot about distribution. That's when you take a number or term and multiply it over a group of terms. But James did just the opposite of that... he *un*-distributed, which is called **factoring** in Algebraland. Let's take a closer look. Here's what James started with:

A factor is a term that's multiplied over an ENTIRE expression.

Here's the original expression to solve.

$$\frac{11x + \frac{11}{3}x - \frac{33}{4}x}{1 + \frac{1}{3} - \frac{3}{4}} = 1$$

James actually saw this another way, though... he saw that everything on the top part of the fraction could be represented as being multiplied by 11. So here's how the same expression looks, with that 11 shown explicitly:

This was the tricky one... but James realized that 33/4 is really the same as 11 * 3/4.

James basically pulled an 11 out of each term... making the expression look like this.

$$\frac{11 \cdot x + 11 \cdot \frac{1}{3} \cdot x - 11 \cdot 3 \cdot \frac{1}{4} \cdot x}{1 + \frac{1}{3} - 3 \cdot \frac{1}{4}} = 1$$

Nothing's changed here... this is the same equation as before.

When you've got a number or term that everything else is multiplied by, that number or term is called a **factor**. So 11—and x, for that matter—is a factor of the top part of this equation:

A factor can be multiplied by your terms to get back to the original form of your equation or expression.

11x is a factor... so we can un-distribute this factor and write is separately, like this.

$$\frac{11x\left(1 + \frac{1}{3} - \frac{3}{4}\right)}{\left(1 + \frac{1}{3} - \frac{3}{4}\right)} = 1$$

But look... with that 11x pulled out, the remaining terms on the top match the fraction's denominator exactly. So we can cancel all that out!

$$\frac{11x\left(1 + \frac{1}{3} - \frac{3}{4}\right)}{\left(1 + \frac{1}{3} - \frac{3}{4}\right)} = 1$$

These two are the same quantity, so they cancel out.

Factoring is un-mulitplying

Pulling out common factors from a term or group is called **factoring**. (Clever, isn't it?) Once you've pulled out a common factor, you can work with the group or the factor, doing things like canceling terms (which is what James did).

Factoring is, basically, the opposite of multiplying. And when you're working with a group of terms, factoring is the opposite of distributing. Remember, the distributive property is all about multiplying groups together:

The distributive property

$$a(b + c) = ab + ac$$

This is the common term.

That term is multipled by this group...

...to get this answer.

So the opposite of this is to take the solution, and pull out the common factor:

$$ab + ac = a\,(b + c)$$

a is a factor in both terms...

...so we can turn that into a common factor multiplied over a group.

By taking advantage of the distributive property in reverse, you can pull out common terms. By manipulating the terms this way, you can set up an equation that is usually easier to work with. Sometimes you can cancel out a common factor, or even an entire group.

- **Factoring is the inverse of the distributive property.**
- **Factoring is pulling numbers and terms apart by finding terms that multiply together to make another term or expression.**

The distributive property means that you need to evenly distribute a factor over all the terms of a group.

Factor by looking for common terms

Factoring isn't a nice process like FOIL... it's more like looking for patterns. But these aren't patterns that are as easy as the square patterns. Instead, you've got to try and "see" common factors in an expression. Sound tricky? It is a bit, but you'll get better with practice.

Here are some ways to get started when you're solving an equation and you think factoring might help:

- [x] Look at your equation. If there are quite a few terms, look for numbers that repeat, or *multiples* of numbers that keep showing up.

$$\frac{11x + \frac{11}{3}x - \frac{33}{4}x}{1 + \frac{1}{3} - \frac{3}{4}} = 1$$

Start out by looking for repeating numbers. Here, you've got 11, 11/3, and 33/4... but 33 is 3 * 11, so that's another 11.

- [x] Next, figure out the ***greatest common factor*** (GCF) of the terms you're interested in. This is probably one of the numbers that kept showing up in the first step.

If you're rusty on the GCF, jump to the appendix and brush up.

$$\frac{11x + \frac{11}{3}x - \frac{33}{4}x}{1 + \frac{1}{3} - \frac{3}{4}} = 1$$

This is the part to factor - find the GCF of $11x$, $\frac{11}{3}x$ and $\frac{33}{4}x$.

- [x] Pull out the GCF and write it down, and then write in a set of parentheses. Inside the parentheses, write the new terms that are left after the GCF is pulled out. These are your original terms, each ***divided*** by the GCF you just pulled out.

The common factor for all of those terms is $11x$, so pull that out.

What's left inside is the original terms divided by the GCF.

$$\frac{11x\left(1 + \frac{1}{3} - \frac{3}{4}\right)}{1 + \frac{1}{3} - \frac{3}{4}} = 1$$

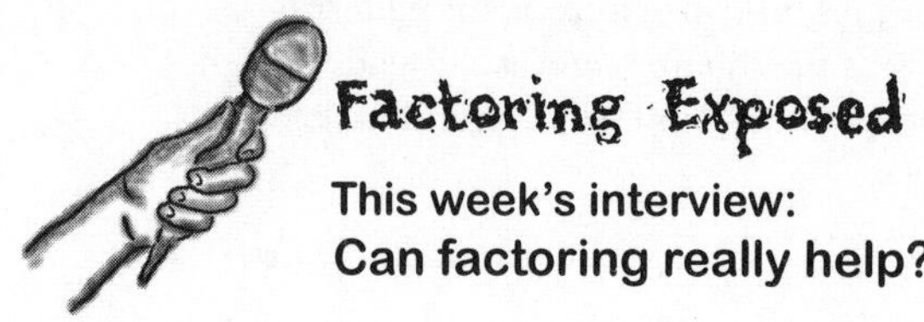

Factoring Exposed
This week's interview:
Can factoring really help?

Head First: Welcome, Factoring. Good to have you!

Factoring: Thanks! I'm really looking forward to explaining more about who I really am.

Head First: There does seem to be some confusion. What exactly are you?

Factoring: Any time a term, set of terms, or equation has a factor removed from it, I'm the one doing with work.

Head First: That seems pretty general; can you be more specific?

Factoring: Not really, I'm just flexible that way. I'm a general term, that's all.

Head First: Well then, what would you say your strengths are? How can you help?

Factoring: Well, when you use me to get your variable away from all of its coefficients, equations seem to get a lot easier to solve.

Head First: Ok, that makes sense. So it's sort of like high-speed variable isolation, right?

Factoring: Exactly! One easy step and you've got your variable alone. Then you can solve your equation. Who doesn't love that?

Head First: It does sound like you're pretty powerful.

Factoring: Well, yeah! I don't like to flex my muscles too much, but I do have some mad skills.

Head First: Do you work with any other properties?

Factoring: I'm the way you can undo a distribution or multiplication, since I identify factors and remove them from a group you're working with.

Head First: Hmm, interesting.

Factoring: And, although I don't work directly with them, the associative and commutative properties need to be involved most of the time, too.

Head First: Anything else you want to add?

Factoring: Don't forget implied numbers! If a variable is written alone, there's a one if you factor away the variable. People forget that all the time!

Head First: Thanks for your time... this was great.

Factoring: Thank you for letting me clear the air. I feel so separated from the group most of the time.

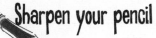

Sharpen your pencil

Use factoring to solve these Math or No Math bonus problems. Once you've factored, go ahead and solve these completely.

$$\frac{x}{2} + \frac{3x}{2} = 6$$

$$5y - 3 = \frac{4y - 3y}{y}$$

$$\frac{z}{3} - \frac{5}{3} = 5$$

$$\frac{8x + 24x - 16x}{100} = 24$$

Sharpen your pencil Solution

Your job was to factor and then solve all of the equations.

$$\frac{x}{2} + \frac{3x}{2} = 6$$

Here we only factored a half, but you could've factored out 1/2x if you wanted.

$\frac{1}{2}(x + 3x) = 6$

$(2)\frac{1}{2}(x + 3x) = 6(2)$

$x + 3x = 12$

$\frac{4x}{4} = \frac{12}{4}$

$x = 3$

$$5y - 3 = \frac{4y - 3y}{y}$$

$5y - 3 = \frac{y(4-3)}{y}$

$5y - 3 = 4 - 3$

$3 + 5y - 3 = 1 + 3$

$\frac{5y}{5} = \frac{4}{5}$

$y = \frac{4}{5}$

You could have solved these problems a number of ways, but these are how we chose to do it. As long as you end up with the same value for the unknown, you got it right!

$$\frac{z}{3} - \frac{5}{3} = 5$$

$\frac{1}{3}(p-5) = 5$

$(3)\frac{1}{3}(p-5) = 5(3)$

$p - 5 = 15$

$p - 5 + 5 = 15 + 5$

$p = 20$

$$\frac{8x + 24x - 16x}{100} = 24$$

$(100)\frac{(8x + 24x - 16x)}{100} = 24(100)$

$8x + 24x - 16x = 2400$

$8x(1 + 3 + 2) = 2400$

$8x(6) = 2400$

$\frac{48x}{48} = \frac{2400}{48}$

$x = 50$

The final round: Kate can pull even and force a tie, or James can ride off with the grand prize.

Problem #4: Solve this...

$$(x - 3)(x + 7) = 0$$

$(x - 3)(x + 7) = 0$ Kate expanded the binomial.

$x^2 + 7x - 3x - 21 = 0$

$x^2 + 4x - 21 = 0$

Kate doesn't know what to do next! ?

$(x - 3)(x + 7) = 0$

James split up the equation and set both expressions equal to zero.

$x - 3 = 0$ and / or $x + 7 = 0$

$x - 3 + 3 = 0 + 3$ $\quad$ $x + 7 - 7 = 0 - 7$

$x = 3$ and / or $x = -7$

Why did James get two answers? Which of his answers is correct? Is ***either*** answer correct?

Zero times anything is 0

Kate was working her problem out the normal way, but James noticed something... his two terms multiplied out to zero. Suppose the first grouping is **a**, and the second grouping is **b**:

$$(x-3)(x+7)=0$$

If: $a \cdot b = 0$

Both a and b can be quantities, numbers, variables, whatever.

Then: $a = 0$ **and / or** $b = 0$

There's no way two NON-zero numbers can multiply together to get zero.

Now, if **a** is (x-3) and **b** is (x+7), we just have to solve for x so that either (x-3) is 0, or (x+7) is 0.

If either of those groupings is zero, then it will cause the entire equation to go to zero. That's called the **Zero Product Rule**: any number or term multipled by zero is always zero.

Let's zero these things out...

$$\underbrace{(x-3)}\underbrace{(x+7)}=0$$

$$a \quad \cdot \quad b \quad = 0$$

We need either a or b to be zero to make this work out.

a could be 0 — b could be 0

$x - 3 = 0$ and/or $x + 7 = 0$

$x - 3 + 3 = 0 + 3$ $\qquad$ $x + 7 - 7 = 0 - 7$

$x = 3$ and $x = -7$

Both of these solutions will work because both make the entire expression work out to 0.

If either **a** *or* **b** is zero, then the entire equation goes to zero. So if x is 3 *or* -7, the equation works out. James was right; there are actually *two* right answers for this problem.

there are no Dumb Questions

Q: Does a common factor have to appear in all the terms in an equation?

A: You can factor any or all terms of the equation. But when you pull out the factor, it's only multiplied by the terms you pulled that factor out of.

In Algebra, all of the terms on one side of the equals sign are typically factored at once. But in more complicated math, you may have several factors in a single equation.

Q: What if I choose a right factor, but it doesn't help me solve my problem?

A: As long as you follow the rules and factor correctly, you can always go back to your original equation. Just like inverse operations and distributing, all of these tools can be tried out and undone if they don't help.

Q: How come James got to work on only part of his equation at a time?

A: That's thanks to the Zero Product Rule. Since only one part of the equation has to be zero, both of those pieces can be set equal to zero and then solved independently. By applying the ZPR, you can work with two equations separately, and both solutions will work.

Q: How can we have two solutions for a single equation?

A: Good question, and we'll talk about that a lot more in the next chapter. For now, you just need to know that if you substitute each value you found for ***x*** back into the equation, and you get the right solution, you're good to go.

Q: How can I tell when to factor?

A: Well, it depends. Usually you can get at the solution for a problem several different ways, factoring being just one of them.

If you try factoring something and it doesn't get you anywhere, don't worry; just try something else!

Q: What about equations in terms of multiple variables? What if we have x, y, z, and w?

A: They can be factored the same way. If the GCF has multiple variables, you can factor them all out. Another strategy with an equation of multiple variables is to factor out just one of the variables, if you can. And, of course, the Zero Product Rule is always a great option.

BULLET POINTS

- Factoring is a **tool** to help manipulate the equation.
- Any time you pull out a **common number or term**, you're factoring .
- The **Zero Product Rule** means that if terms multiplied together equal zero, then one or all of those terms must be zero.
- Factoring to get an equation in the form of two terms multiplied together to **equal zero** means that you can use the ZPR to solve.

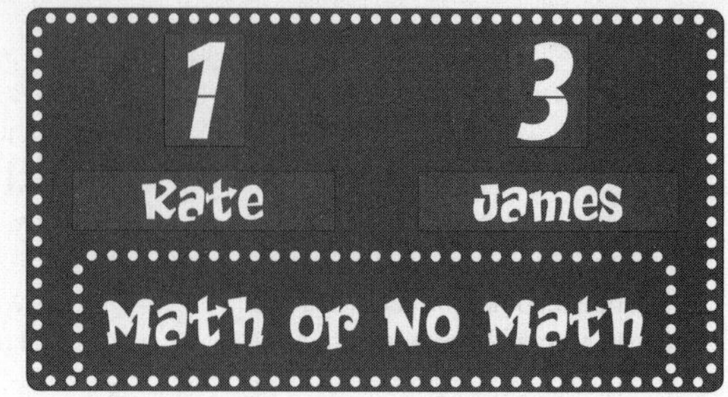

Sorry, Kate, but the Zero Product Rule put me over the top.

To recap the rulings:

1. **James and Kate expanded binomials**
   James was careful with distribution, but Kate wasn't, so she got her first problem wrong. Then, they both learned FOIL. Next time they'll both be faster and more reliable.

2. **Kate's squared pattern skills rocked**
   Kate knew the difference of two squares right off the top of her head and didn't need to do any work.

3. **James' factoring frenzy blew Kate away**
   James knew that factoring to make a manipulation easier worked great... and he proved it.

4. **The Zero Product Rule works.**
   James finished up with applying the ZPR to solve a tough equation. Since Kate didn't know about the ZPR, she was out of luck.

Pool Puzzle

Your **job** is to take terms from the pool and place them into the blank lines in the expressions. You may **not** use the same term more than once, and you won't need to use all the terms. Your **goal** is to complete the solutions for each problem.

$$6d + 4d - 18 = d$$

$$6d + 4d - 18 = d$$

$$6d + 4d = d + 18$$

$$\ldots\ldots = 18$$

$$\ldots\ldots = 18$$

$$d(\ldots\ldots) = 18$$

$$d = \ldots\ldots$$

$$-12a - 3ab + 9ab = 0$$

$$\ldots\ldots(\ldots\ldots) = 0$$

$$\ldots\ldots(\ldots\ldots) = 0$$

$$\ldots\ldots = 0 \quad \ldots\ldots = 0$$

$$\ldots\ldots = 12$$

$$b = \ldots\ldots$$

Note: each term from the pool can only be used once!

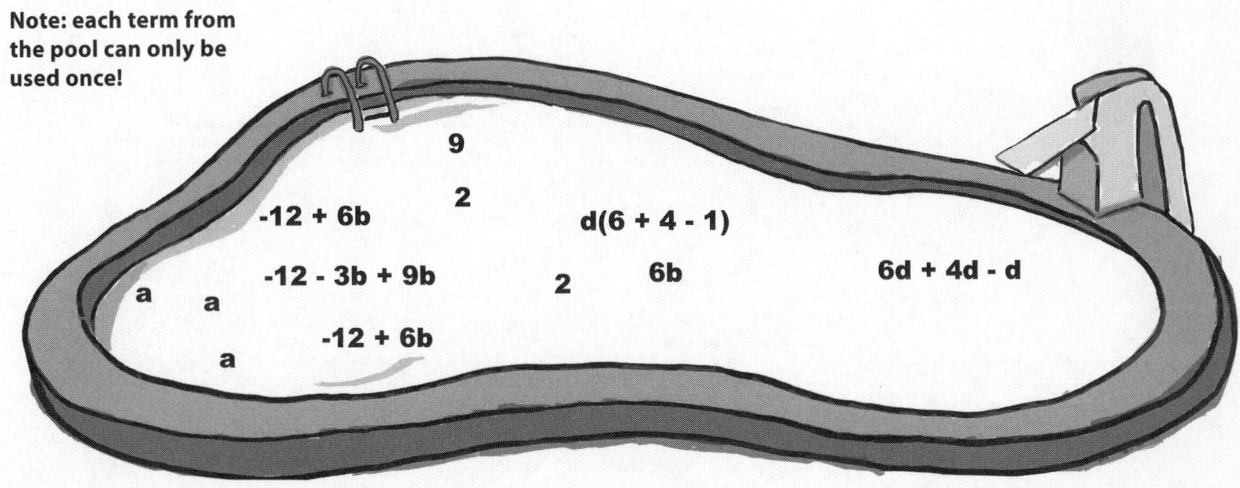

Pool Puzzle Solution

Your **job** is to take terms from the pool and place them into the blank lines in the expressions. You may **not** use the same term more than once, and you won't need to use all the terms. Your **goal** is to complete the solutions for each problem.

$$6d + 4d - 18 = d$$

$$6d + 4d - 18 + 18 = d + 18$$

$$6d + 4d - d = d + 18 - d$$

$$6d + 4d - d = 18$$

$$d(6 + 4 - 1) = 18$$

$$\frac{d(9)}{9} = \frac{18}{9}$$

$$d = 2$$

$$-12a - 3ab + 9ab = 0$$

$$a(-12 - 3b + 9b) = 0$$

$$a(-12 + 6b) = 0$$

$$a = 0 \quad \text{and/or} \quad -12 + 6b + 12 = 0 + 12$$

$$\frac{6b}{b} = \frac{12}{b}$$

$$b = 2$$

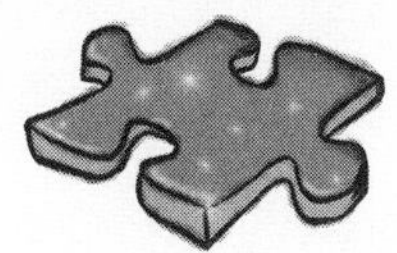

Binomialcross

First, use the inside of your brain to get at the clues that are outside your immediate recall to burn them into memory at last...

Across

4. A special case of polynomials made up of two terms
6. First, outside, inside, last
7. A quadratic equation will have _______ solutions.
8. Values held together by multiplication
9. When you factor, look for the ______ (abbr.)
10. Square the first term, twice the second, square the second
11. To help simplify solving quadratics, manipulate them to be equal to ______.

Down

1. The product of two binomials that only differ by sign is the difference of two _______.
2. When the equation equals zero and is a product of terms, one or more must be
3. FOIL really works because of the ______________ property.
5. The distributive property describes how to multiply _________.
6. Un-distributing

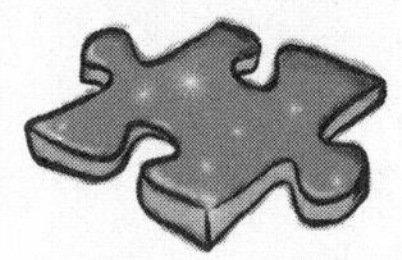

Binomialcross Solution

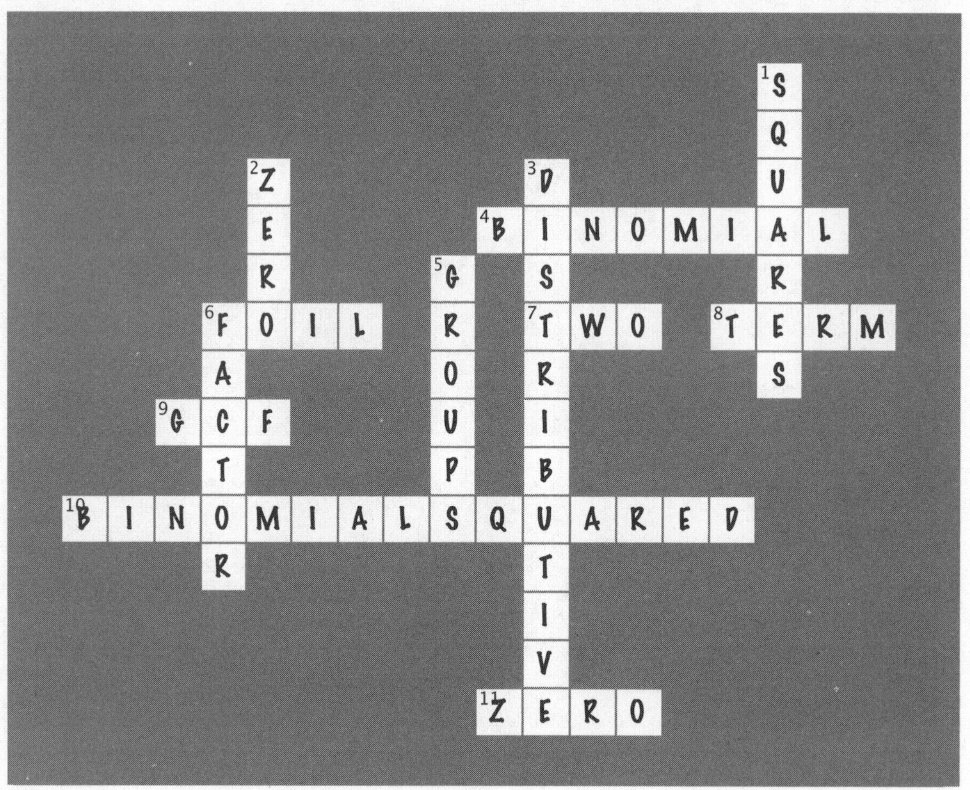

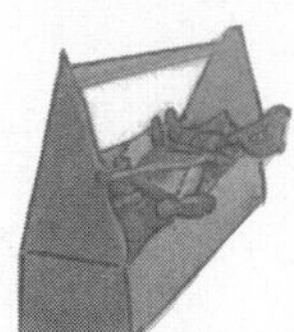

Tools for your Algebra Toolbox

This chapter was about expanding binomials and basic factoring.

BULLET POINTS

- Multiplying two binomials is a special case of distribution.
- FOIL is just an acronym to help you apply the distributive property correctly.
- Patterns like the squares patterns can help you solve a problem really quickly, but FOIL will always work.
- Factoring is a tool to help manipulate the equation.
- Any time you pull out a common factor, it's factoring.
- The Zero Product Rule means that if terms multiplied together equal zero, then one or all of those terms must be zero.
- Factoring to get an equation in the form of two terms multiplied together to equal zero means that you can use the ZPR to solve.

FOIL

To multiply these two binomials: $(x + a)(x + b)$

F First: Multiply together the first terms of both binomials.

$(x + a)(x + b)$ (1st term, 1st term)

O Outside: Multiply together the outside terms of both binomials.

$(x + a)(x + b)$ (outside term, outside term)

I Inside: Multiply together the inside terms of both binomials.

$(x + a)(x + b)$ (inside term, inside term)

L Last: Multiply together the last two terms of both binomials.

$(x + a)(x + b)$ (last term, last term)

$$x^2 + bx + ax + ab$$

Binomial squared: $(x + a)^2 = x^2 + 2ax + a^2$

Binomials with different signs: $(x + a)(x - a) = x^2 - a^2$

The Zero Product Rule

If: $a \cdot b = 0$

Then: $a = 0$ and / or $b = 0$

CHAPTER 8

9 quadratic equations

Getting out of line

This is stupid. I'm just supposed to wait in this line all day? Look, it even curves around the block!
Cinema

Not everything in life is linear. But just because an equation doesn't graph as a **straight line**, doesn't mean it's unimportant. In fact, some of the most important **unknowns** you'll have to work with in life end up being **non-linear**. Sometimes you've got to deal with terms that have **exponents greater than 1**. In fact, some equations with **squared terms** graph as **curves**! How's that work? Well, there's only one way to find out...

Head First U is at war!

HFU has a tradition of frat wars during pledge week.

The pledges have to nail the opposing fraternity's president with water balloons. Jon's a pledge and has to hit the president at least three times during pledge week, or he's got no hope of getting into Theta Theta Pi.

And, guess who he's asked for a little help... ***you!***

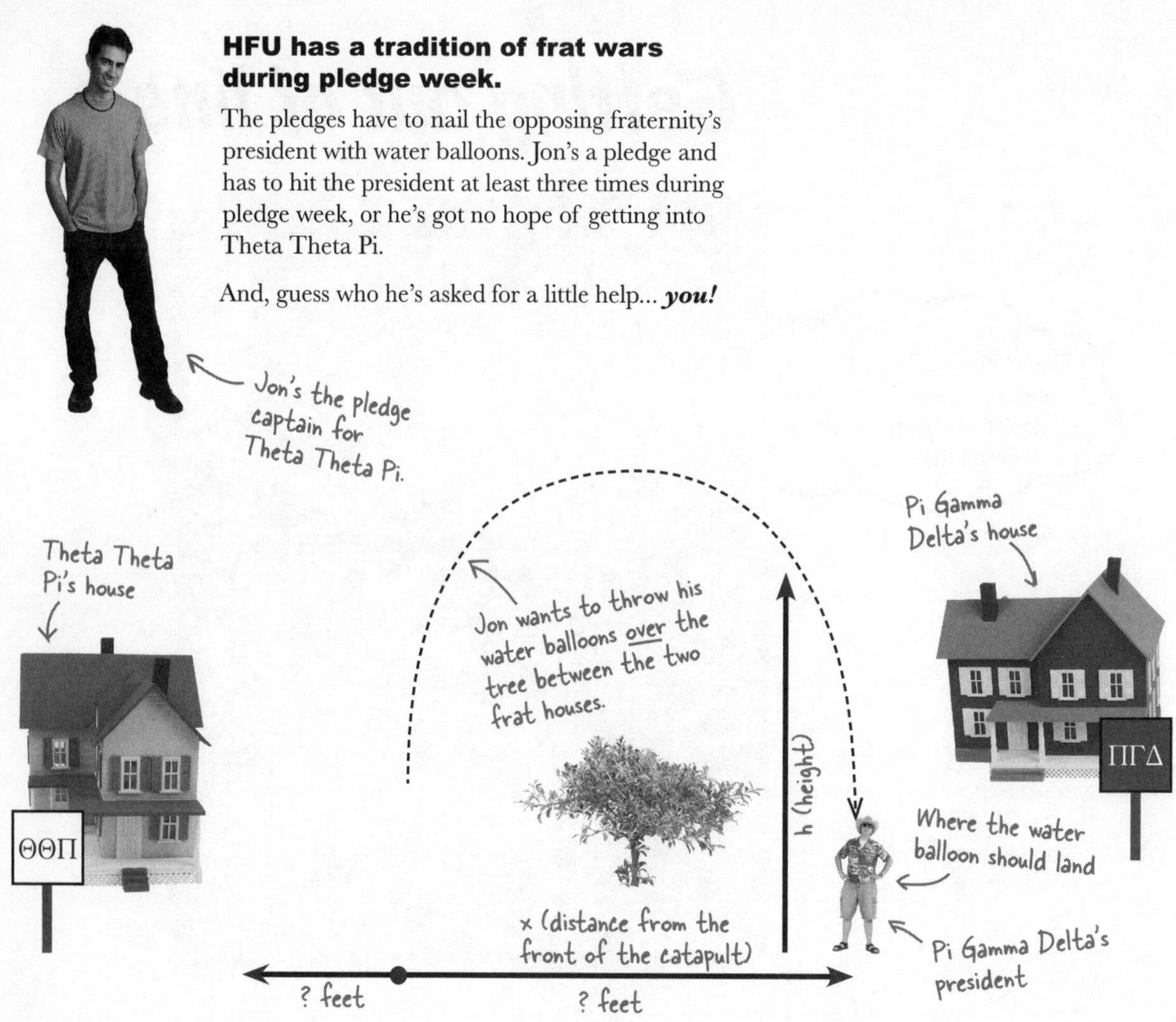

Traditionally, the water balloons have been launched by hand, or maybe with a slingshot. Jon is out to make Theta Theta Pi history, though...

Jon's upgrading his technology

Jon's bringing in a catapult. With a catapult, he can be more accurate and hit Pi Gamma Delta's president even harder, and that means he'll get into the frat for sure!

The catapult comes with a range equation in terms of height and distance. If you know how high you want a balloon to go, you can figure out where to place the catapult so it plummets right down on the Pi Gamma Delta president's head.

Only trouble is, Jon doesn't know Algebra... that's where you come in.

This is the website where Jon ordered his catapult. The site lists the range of each catapult in terms of an equation

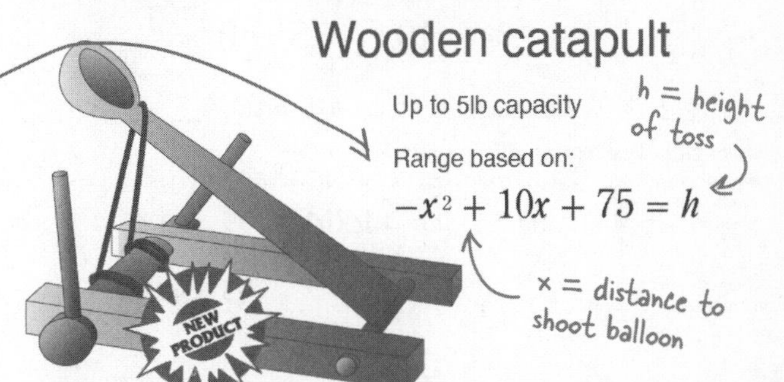

Here's the range equation.

h = height of toss

x = distance to shoot balloon

BRAIN BARBELL

For the first shot, the Pi Gamma Delta president is sitting on the lawn, so the height is zero, $h = 0$. Figure out x so Jon can place the catapult in the right spot.

We'll get you started...

$$-x^2 + 10x + 75 = h$$

h = 0 when the president is on the ground.

$$-x^2 + 10x + 75 = 0$$

To make this easier to work with, multiply both sides by -1.

$$-1 \cdot (-x^2 + 10x + 75) = -1 \cdot (0)$$

$$x^2 - 10x - 75 = 0$$

Now, what's x? Substitute in a couple of values, and see what you come up with.

For the first shot, the Pi Gamma Delta president is sitting on the lawn, so $h = 0$. Figure out x so they can place the catapult in the right spot.

$$-x^2 + 10x + 75 = h$$

$$-x^2 + 10x + 75 = 0$$

To make this easier to work with, multiply both sides by -1.

$$-1 \cdot (-x^2 + 10x + 75) = -1 \cdot (0)$$

$$x^2 - 10x - 75 = 0$$

Now, what's x? Try a couple of values out with substitution and see what you come up with.

try $x = 20$ $\quad 20^2 - 10(20) - 75 = 0\,?$

$125 \neq 0$

try $x = 0$ $\quad 0^2 - 10(0) - 75 = 0\,?$

$-75 \neq 0$

try $x = 10$ $\quad 10^2 - 10(10) - 75 = 0\,?$

$-75 \neq 0$

You may have picked different values. That's ok, but it's pretty obvious that trial and error isn't the way to go. There must be a way to solve these things algebraically.

Introducing a new type of equation: quadratic

A **quadratic equation** is any *polynomial equation* that has a degree of *two*; that just means that the largest exponent for any variable is two. The degree of the equation also dictates how many solutions there can be, and since a quadratic equation has a degree of two, these equations can have up to two solutions.

We saw with linear equations, like $3 = 4x - 2$ (which is a degree of one) that there is one solution per variable.

So Jon's equation needs to be manipulated to solve for x twice. The Zero Product Rule (ZPR) can also help. If we can manipulate the quadratic equation so it's a product of two quantities equal to zero, these equations will be a lot easier to solve.

We actually saw several quadratics last chapter... we were turning them into two binomials, but they started out as quadratics.

Expanding two binomials made a quadratic equation, so maybe factoring the quadratic *back* to two binomials is the way to go. Let's try that and see what happens.

Any equation with a degree of 2 is a quadratic equation.

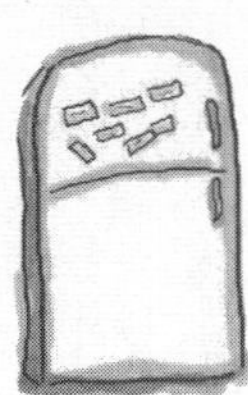

Factoring Magnets

Use the magnets at the bottom of the page to fill in the blanks, and see if you can solve this problem on your own.

Remember, this is 0 because the height is 0... the president is standing at the same level as the catapult.

$$x^2 - 10x - 75 = 0$$

The last two numbers need to be multiplied together to be −75.

To use the ZPR, we'll need two quantities that when multiplied together, equal zero.

$$(\_\_\_\_ - \_\_\_\_)(\_\_\_\_ + \_\_\_\_) = 0$$

Two of these values will be x, and two will be numbers.

We'll explain where this sign came from later.

We'll explain where this sign came from later.

$$(\_\_\_\_ - \_\_\_\_) = 0 \qquad (\_\_\_\_ + \_\_\_\_) = 0$$

$$+ \not{\_\_\_\_} + \_\_\_\_ - \not{\_\_\_\_} = 0 + \_\_\_\_ \qquad - \not{\_\_\_\_} + \_\_\_\_ + \not{\_\_\_\_} = 0 - \_\_\_\_$$

$$x = \_\_\_\_ \qquad x = - \_\_\_\_$$

Use this space to check your work by expanding back out the binomials:

....

....

....

....

x x 15 3 5 1 15

h 5 25 75 1 5 5

x

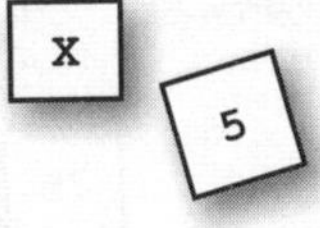

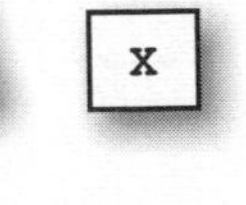

15

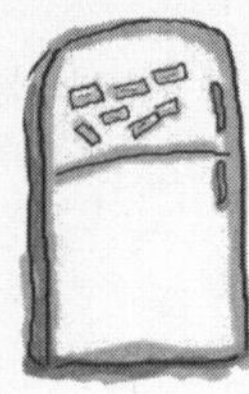

Factoring Magnets Solution

Your job was to use the magnets to solve the quadratic and help Jon hit the opposing frat's president.

$$x^2 - 10x - 75 = 0$$

This is the tricky part. To figure out these two numbers, you have to have two things: two numbers that multiply out to –75 and that, when added togeher, come out to –10, the coefficient of x in the original equation.

The first term of each is x, since x * x = x^2

$$(x - 15)(x + 5) = 0$$

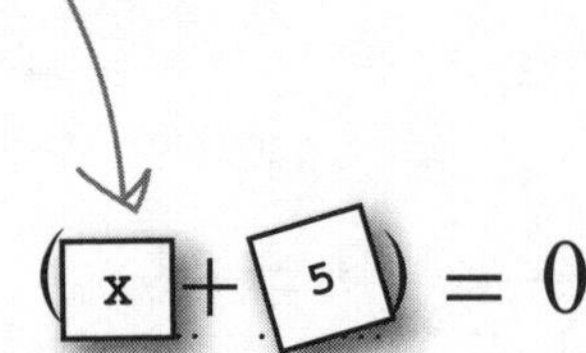

$$(x - 15) = 0 \qquad (x + 5) = 0$$

$$+15 + x - 15 = 0 + 15 \qquad -5 + x + 5 = 0 - 5$$

$$x = 15 \qquad x = -5$$

Use this space to check your work by expanding back out the binomials:

$(x - 15)(x + 5) = 0$

This is where we apply FOIL to make sure the factoring was right.

$x^2 + 5x - 15x - 75 = 0$

$x^2 - 10x - 75 = 0$

h 3 25 75 1 1 75 25

Where does Jon put the catapult?

So x is -5 or 15. But what does that actually mean? How can a catapult fire a balloon a distance of -5? Keep in mind the context of this problem. How can a water balloon travel -5 anything?

A *negative* distance means that the balloon goes *behind* the front of the catapult. That doesn't make sense in this context, so we can ignore that answer. What we needed was the 15. The balloon will travel 15 feet. So let's shoot the catapult from 15 feet away from the president...

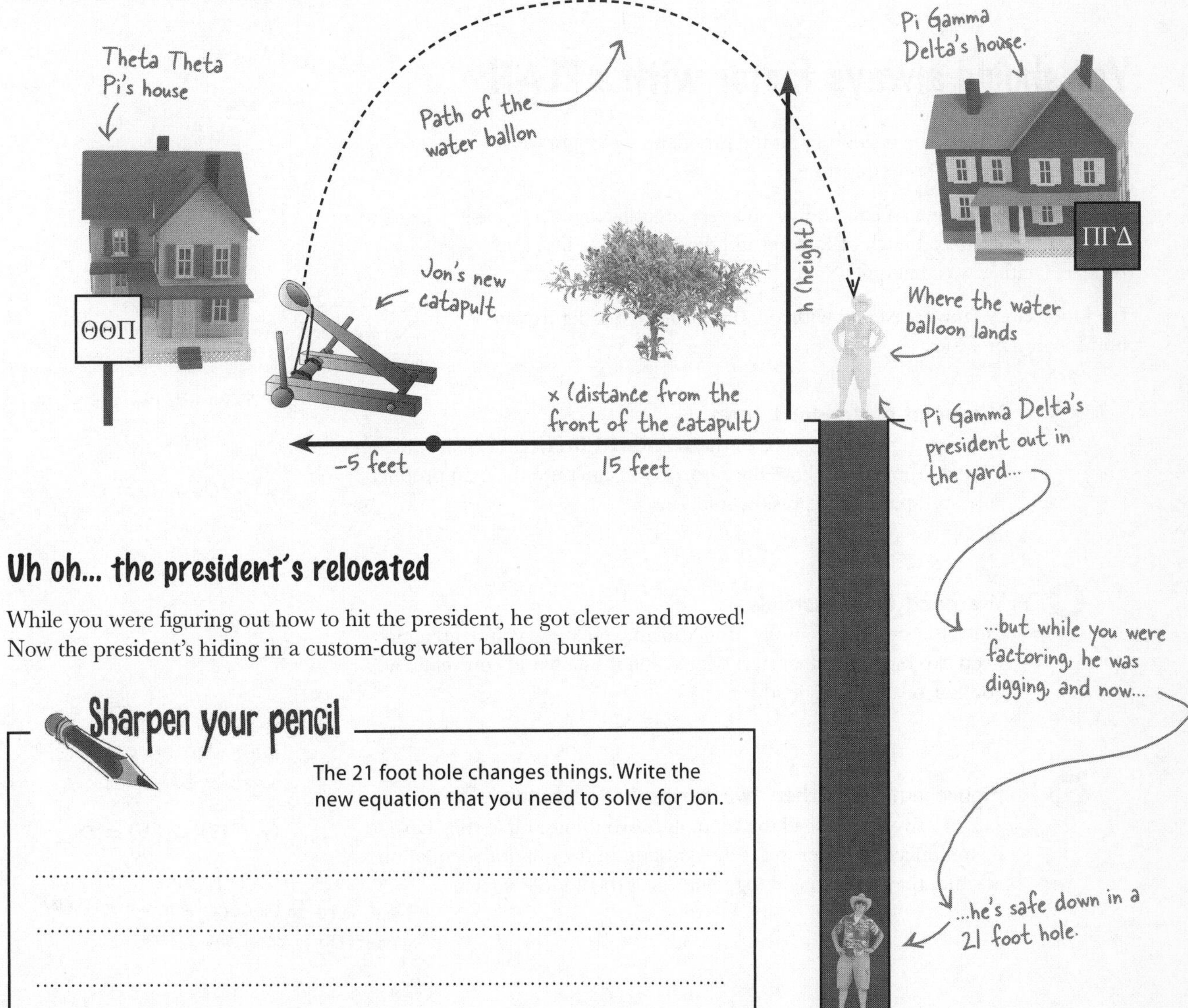

Uh oh... the president's relocated

While you were figuring out how to hit the president, he got clever and moved! Now the president's hiding in a custom-dug water balloon bunker.

Sharpen your pencil

The 21 foot hole changes things. Write the new equation that you need to solve for Jon.

....

....

....

Sharpen your pencil Solution

The 21 foot hole changes things. Write the new equation that you need to solve for Jon.

$h = -x^2 + 10x + 75$ — You have to go back to the original equation, so you can substitute the new h back in.

The height is a negative height because it's a a hole and it's below the ground.

$-21 = -x^2 + 10x + 75$

$+21 - 21 = -x^2 + 10x + 75 + 21$ → $0 = -x^2 + 10x + 96$

You should always factor with a PLAN

If we're going to get the balloon to hit the president, we've got to solve our quadratics faster this time.

What we did with the first equation wasn't very organized, or consistent. We tried a bunch of factors, and then had to do FOIL in reverse. Neither was that quick.

Let's look closer at what we did with $x^2 - 10x + 75 = 0$ and figure out how to get a bit faster.

1. **We need a standard form.**
   The equation needes to be in the **standard form** of a quadratic, set to zero. You have to have the zero, or you can't use the zero product rule and split up the possible solutions.

The standard form of a quadratic equation is three terms: an x^2 term, an x term, and a constant.

$x^2 - 10x - 75 = 0$

2. **We need two binomials.**
   Once the equation is in the proper form, you know you're going to need two binomials that start with x. Fill those in and you've already got half of the terms ready.

$(x \quad)(x \quad) = 0$

3. **Figure out the other two terms in the binomials.**
   The last two terms need to accomplish two things. First, they have to be multiplied together to get the constant in the quadratic equation. Second, they need to add together to get the middle x term.

These need to be multiplied together to get 75.

$(x \quad 15)(x \quad 5) = 0$

These need to be added (or subtracted) to get the x term the $-10x$.

Factoring Up Close

Picking the last two terms to fill in for the binomials is the trickiest part of factoring. Sometimes it helps to get even further into the details.

$(x \;\bigcirc)(x \;\bigcirc) = 0$

Picking these two

The easiest way to find out what numbers you can use is to list all of the possible factors of the constant term of the quadratic equation. Why? Because when you multiply this thing back out using FOIL, you know that those two numbers multiplied together will end up as the constant in your equation.

$$75 = 75 \bullet 1$$
$$= 25 \bullet 3$$
$$= 15 \bullet 5$$

Once you know that these are the options, then you move on to look at the x term, 10, in our case. There needs to be a set of these that added (or subtracted) together gets to 10.

$15 - 5 = 10$

Look over these factors closely. Which pair will let you get to the middle x term? Since our middle x term is 10, we can take 15 and 5, which have a difference of 10.

$(x \quad 15)(x \quad 5) = 0$

4 **Fill in the signs and check your work.**

To finish up your binomials, fill in the signs. Your binomial constants need to be multiplied to get the same sign in front of the constant (75), and added to get the right x term ($-10x$). Then, expand the binomials you came up with, using FOIL, and make sure that it matches what you started with.

$$(x - 15)(x + 5) = 0$$
$$x^2 + 5x - 15x - 75 = 0$$
$$x^2 - 10x - 75 = 0$$

Exercise

Use your new factoring skills and the equation you just wrote for the subterranean president to figure out where the catapult needs to go now.

....

....

....

....

Use your new factoring skills and the equation you just wrote for the subterranean president to figure out where the catapult needs to go now.

$$0 = -x^2 + 10x + 96$$

This is not quite the right form... but you can swap the zero to the other side.

$$-x^2 + 10x + 96 = 0$$

Then get rid of that -1.

$$-1(-x^2 + 10x + 96) = -1(0)$$

$$x^2 - 10x - 96 = 0$$

Now we're in standard form for a quadratic equation.

$$(x \quad)(x \quad) = 0$$

The options for the last two terms come from

$$96 = 96 \bullet 1$$
$$= 48 \bullet 2$$
$$= 32 \bullet 3$$
$$= 24 \bullet 4$$
$$= 16 \bullet 6$$

We also know that we need $-10x$ in the middle, so the two terms need to work out to 10.

$$(x \quad 16)(x \quad 6) = 0$$

For the signs to get a -96, one has to be positive, and one negative. Since we want a $-10x$, the bigger number is the negative, -16

$$(x + 6)(x - 16) = 0$$

$$x + 6 = 0 \qquad\qquad x - 16 = 0$$

$$-6 + x + 6 = 0 - 6 \qquad\qquad +16 + x - 16 = 0 + 16$$

$$x = -6 \qquad\qquad x = 16$$

Use FOIL to check this

$$(x + 6)(x - 16) = 0$$
$$x^2 - 16x + 6x - 96 = 0$$
$$x^2 - 10x - 96 = 0 \checkmark$$

Jon needs to stick the catapult back 16 feet to hit Pi Gamma Delta where it hurts!

Yeah! Direct hit! He can run, but he can't hide...

BULLET POINTS

- Quadratic equations have up to **two solutions**.
- Factoring a quadratic equation means finding the product of **two binomials**.
- You need to check that your factoring is correct using **FOIL**.
- Finding the **constant terms** for the binomials is the hardest part of factoring a quadratic.
- Quadratic equations need to be in **standard form** before you factor.

there are no Dumb Questions

Q: Do I have to write all of those teeny steps?

A: Not forever. Whenever you learn something new, it's good to write down as much as you can so that you can check your work. We actually wrote down a few middle steps that make it easier to understand. The binomial only needs to be written once, not 3 times.

The bottom line is that you should write down as many steps as you need to so that you can understand and complete your work.

Q: Is this just FOIL backward?

A: Yes! We're working on figuring out what two binomials will get you back to where you started. Why? Because if you apply the Zero Product Rule, you can solve the quadratic!

Q: What are quadratics good for in the real world?

A: Quadratic equations can be used to address a whole new set of real world scenarios. For instance, Jon's equation is a simplified form of a projectile motion equation that describes how objects will travel in the air.

Quadratic equations are also used to design parabolic microphones, satellite dishes, and suspension bridges. They are even used to design water fountains like those outside fancy Las Vegas hotels.

Q: How can one equation have two solutions?

A: When you have an equation with a degree of two, there are two solutions. Remember, a solution means a number that makes the equation true; that's it. As you check your solutions, you'll see that there really are two ways to make these equations work.

Q: What's standard form for a quadratic equation?

A: The general way to write it is $ax^2 + bx + c = 0$.

Q: What if there's an ax^2? We've only done just x^2 terms.

A: Things get more complicated. The process doesn't change, but if you think back to FOIL, that means two things. First, the first terms of the binomials need to come up to ax^2, not just x^2, so now you're looking at a coefficient in front of one or both of the x terms.

Second, that means when you expand the binomials back out you have to keep that coefficient in mind when you're getting the x term. It means you're looking at a lot more trial and error to get it done.

Q: Can you always factor an equation?

A: Yes and no. There are methods, like completing the square, that allow you to work with fractions to factor any equation. But it's tricky and really more of an Algebra 2 topic, anyway.

So what does a quadratic that doesn't easily factor look like? Good question. Turn the page and see...

Pi Gamma Delta built a wall!

Over night, Pi Gamma Delta got some of their engineering majors to supervise their pledges and build a wall between the frats. Now the president figures he's safe because there's a lot more than just a couple of bushes protecting him.

Jon has no idea how to figure this out.

But you do. Now you need to figure out another ***h*** value: 9 feet. No problem! Just do what you've been doing, and you can find out if ***h*** of 9 feet works.

Sharpen your pencil

Can Jon clear the wall? Start back with the original equation and put 9 in for ***h***, then solve.

Something weird is going to happen. Just flip the page when you get to a part where you feel you're stuck...

Sharpen your pencil Solution

Can Jon clear the wall? Start back with the original equation and put 9 in for ***h***, then solve.

We know we want it to go 9 feet.

$$-x^2 + 10x + 75 = h$$

$$-x^2 + 10x + 75 = 9$$

To get the equation in standard form, we need to get a zero over here.

$$-9 - x^2 + 10x + 75 = 9 - 9$$

Remove the -1 again

$$-1(-x^2 + 10x + 66) = -1(0)$$

$$x^2 - 10x - 66 = 0$$

Now we're in standard form for a quadratic equation.

$$(x \quad)(x \quad) = 0$$

The options for the last two terms come from $66 = 66 \bullet 1$
$= 22 \bullet 3$
$= 33 \bullet 2$
$= 11 \bullet 6$

We need two constants that can be added up to -10, but none of these work!

And that's the weird thing we warned you about. Factoring doesn't always work.

Now what?

Got any ideas where we should put the catapult, Oh Great and Wonderous Pledge Captain?
Tim
Jon
Scott

Jon: We can't figure it out. This equation won't work out.

Scott: Jon, if we don't hit these guys—

Tim: Hey, you guys, what's this thing in the small print on the website?

Scott: I didn't see that earlier. Let me see.

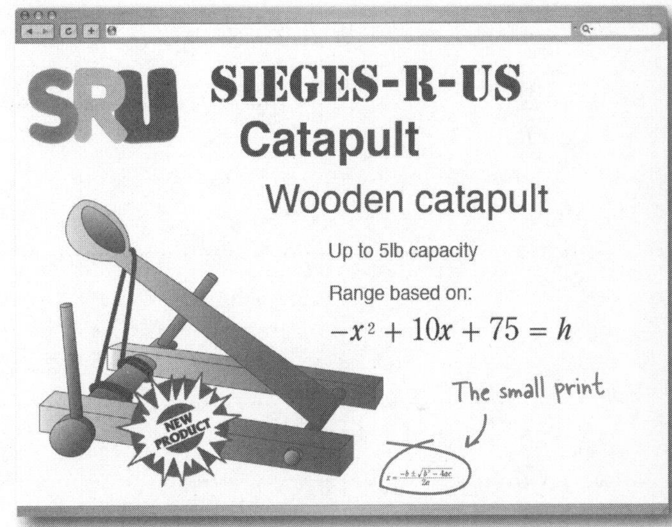

Scott: Oh yeah, there it is. It says: $x = \dfrac{-b \pm \sqrt{b^2 - 4ac}}{2a}$

Jon: What the heck is that?

Tim: It looks a little like the range equation that it came with. It uses x, right?

Scott: What about all those other letters: a, b, and c. Where are they coming from?

Jon: I'm not sure, but let's figure it out.

Scott: Quick, or the president's going to move again!

Try out that weird equation that Scott found and see if you can find **x**. Then Jon will know if he can whack the Pi Gamma Delta president on his head, even with a wall in the way.

The equation we need to solve

$x^2 - 10x - 66 = 0$

This is the equation that Jon found.

$$x = \frac{-b \pm \sqrt{b^2 - 4ac}}{2a}$$

This sign means plus or minus.

This is standard form of a quadratic equation.

Standard form is:

$ax^2 + bx + c = 0$

We left you plenty of room to work things out!

Try out that weird equation that Scott found and see if you can find **x**. Then Jon will know if he can whack the Pi Gamma Delta president on his head, even with a wall.

The equation we need to solve.

$x^2 - 10x - 66 = 0$

This is the equation that Jon found.

$$x = \frac{-b \pm \sqrt{b^2 - 4ac}}{2a}$$

This sign means plus or minus.

This is standard form of a quadratic equation.

Standard form is :

$ax^2 + bx + c = 0$

This is all the information you need to solve this problem. The tricky part is splitting up the formula for x into two pieces: one with minus and one with plus.

$b = -10$

$x^2 - 10x - 66 = 0$

$a = 1$

$c = -66$

Make sure you hold onto these negative signs...

Use these a, b, and c values...

... in here

$$x = \frac{-b \pm \sqrt{b^2 - 4ac}}{2a}$$

Watch the signs here!

After substitution, simplify as much as you can.

$$x = \frac{-(-10) \pm \sqrt{(-10)^2 - 4(1)(-66)}}{2(1)}$$

$$x = \frac{10 \pm \sqrt{100 + 264}}{2}$$

$$x = \frac{10 \pm \sqrt{364}}{2}$$

At this point, we need to split up the formula to deal with that "plus or minus thing." That means we'll just have one equation with addition, and one with subtraction.

This is how you get two solutions to the equation. Same equation, but one has a plus sign, and the other has a minus sign.

$$x_1 = \frac{10 + \sqrt{364}}{2}$$

$$x_2 = \frac{10 - \sqrt{364}}{2}$$

$$x_1 = \frac{10 + 19.0788}{2}$$

$$x_2 = \frac{10 - 19.0788}{2}$$

This is a rounded solution.

$$x_1 = \frac{29.0788}{2}$$

$$x_2 = \frac{-9.0788}{2}$$

$$x_1 = 14.5394$$

$$x_2 = -4.5394$$

Now, check your work and make sure these solutions are ok.

$$x^2 - 10x - 66 = 0$$

$$(14.5394)^2 - 10(14.5394) - 66 = 0$$

$$211.394 - 145.394 - 66 = 0$$

$$0 = 0$$

$$x^2 - 10x - 66 = 0$$

$$(-4.5394)^2 - 10(-4.5394) - 66 = 0$$

$$20.6061 + 45.394.54 - 66 = 0$$

$$0.0001 = 0$$

We rounded some, so this is close enough to be considered the same.

It works! Now Jon knows where to put his catapult!

9 feet is not a problem

After using that formula from the side of the box, we figured out that 9 feet works for the catapult at two locations: -4.54 feet and 14.54 feet.

Both of these numbers are the distance from the front of the catapult to the location where the water balloon will be 9 feet off the ground. That's good news for Jon because it means that the water balloon is above 9 feet before it passes the front of the catapult and until 14.54 feet away. As long as the catapult is less than 14.54 feet from the wall, it should clear it!

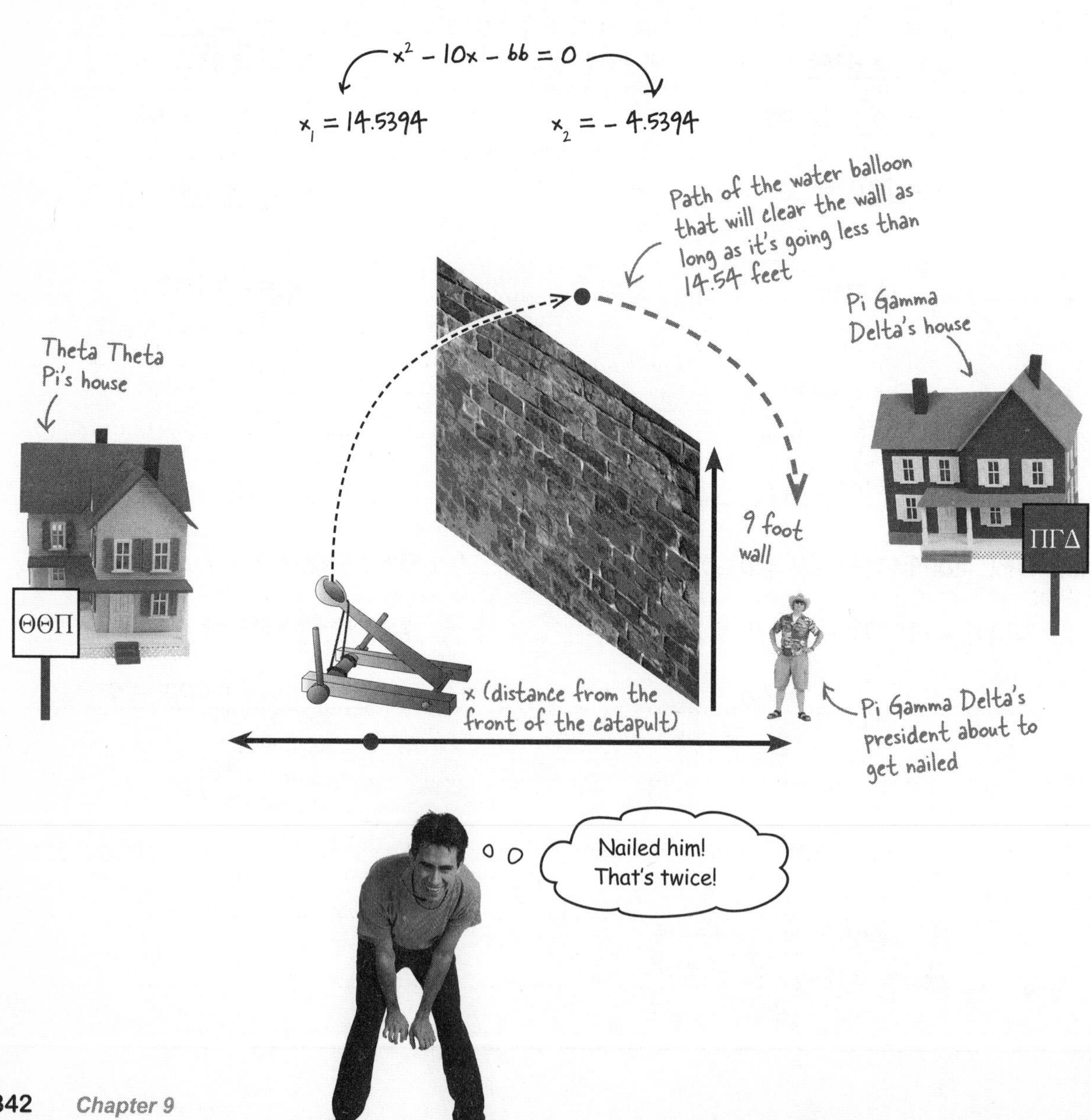

What was that thing that you used? Jon just found it in the small print on the website!

It's the quadratic formula.

There's a formula out there that can give you the solution to any quadratic equation—no factoring required— and it's what was printed on the side of the box.

The quadratic formula

The other way to solve a quadratic equation besides factoring is by using the **quadratic formula**. The quadratic formula is written for the standard form of a quadratic equation that allows you to solve any quadratic equation, whether or not it can be factored.

$$x = \frac{-b \pm \sqrt{b^2 - 4ac}}{2a}$$

This is the x in the quadratic equation.

The a, b, and c in this equation are the a, b, and c in the standard form of a quadratic.

This symbol means "plus or minus."

Standard form is :

$ax^2 + bx + c = 0$

The quadratic formula is great because you can use it to solve any equation, but it is a bit cumbersome. To get both solutions for the quadratic (remember, a quadratic has two solutions), you have to work with the "plus or minus" symbol. To handle that, you simplify the expression once for addition of the quantity under the root, and a second time for subtraction of the quantity under the root, like this:

$$x_1 = \frac{-b + \sqrt{b^2 - 4ac}}{2a} \quad \text{and} \quad x_2 = \frac{-b - \sqrt{b^2 - 4ac}}{2a}$$

The subscript is just a way to tell the answers apart.

there are no Dumb Questions

Q: Why didn't we use the quadratic formula first? Isn't factoring a waste of time?

A: We didn't do this first because we knew that you would say that! The quadratic formula is fantastic if you can't factor, but if you can, factoring quickly gives you the whole number solutions.

Another issue with the quadratic formula is that it's easy to mess up. You need to watch the order of operations and the square root. If you get either of those wrong, then your answers will be wrong.

Q: That square root was really long. How much of it do we have write down?

A: It's a judgement call. There is a standard for how many decimals you should use, called scientific notation. For now, let's say that two to four decimal places is enough.

Q: What if there are fractions for a, b, or c?

A: That's no problem. You can do two things:

First, just go ahead and put the fractions in the quadratic equation and simplify. It can sometimes be pretty difficult to do that, but if you watch your order of operations, there shouldn't be a problem.

The other option is to multiply the equation by a number that will allow you to clear the fractions, like 4 if you have 1/4 or 3/4, then use the resulting coefficients in the quadratic formula.

Q: Will the square root always be a decimal?

A: Most of the time it will be, so you'll need a calculator. There are, of course, perfect squares (9, 16, 25, etc.), but it's not very often you'll get one under the radical.

Q: Will real world equations have decimals?

A: Most of the time you'll have decimals. Unfortunately, the real world tends to be messy and not easily quantified. A lot of equations deal with real materials (like water or steel) or real phenomenon (like speed) are based on measurement, which typically are written in decimals.

Q: Where did the quadratic formula come from?

A: The quadratic formula can be derived using a special factoring technique called "completing the square." If you use the ***a, b***, and ***c*** general constants when you complete the square, you'll come up with the quadratic formula.

You'll learn how to complete the square for any equation later on in math... it's nothing to worry about right now.

Q: What if the value under the root is negative?

A: You mean, what if $b^2 - 4ac$ is negative? Well, you can't take the root without learning about a whole new class of numbers... and that's another book altogether.

So what do you do if you come across this now? Keep reading...

Watch it!

The quadratic formula requires precision.

You have to get the order of operations right when you're working with the quadratic formula. Everything under the radical has to be simplified before you can take the root.

You also have to watch the signs! It's easy to lose track, so if you need to write all your steps out, that's a good way to keep from making mistakes.

Sharpen your pencil

Tim and Scott pulled out the slingshot from last year to try and ping the vice president of Pi Gamma Delta while Jon's setting up the catapult again. Help them figure out where they need to shoot from.

x is the distance the slingshot needs to be from the wall to make it over.

$x^2 - 8x = -13$

Make sure you check your work!

Sharpen your pencil Solution

Tim and Scott pulled out the slingshot from last year to try and ping the vice president of Pi Gamma Delta while Jon's at it. Help them solve the new problem to figure out where they need to be.

x is the distance the slingshot needs to be from the wall to make it over.

$x^2 - 8x = -13$

$x^2 - 8x = -13$ ← This needs to be in standard form

$+13 + x^2 - 8x = -13 + 13$

$x^2 - 8x + 13 = 0$

$a = 1$, $b = -8$, $c = 13$

$$x = \frac{-b \pm \sqrt{b^2 - 4ac}}{2a}$$

$$x = \frac{-(-8) \pm \sqrt{(-8)^2 - 4(1)(13)}}{2(1)}$$ Watch the signs!

$$x = \frac{8 \pm \sqrt{64 - 52}}{2}$$

$$x = \frac{8 \pm \sqrt{12}}{2}$$

$$x_1 = \frac{8 + \sqrt{12}}{2} \qquad x_2 = \frac{8 - \sqrt{12}}{2}$$

$$x_1 = \frac{8 + 3.464}{2} \qquad x_2 = \frac{8 - 3.464}{2}$$

$$x_1 = \frac{11.464}{2} \qquad x_2 = \frac{4.536}{2}$$

$x_1 = 5.732$ $\qquad$ $x_2 = 2.268$

These are the two values that will work, so between 2.268 feet and 5.732 feet the slingshot will get the balloon over the wall.

Make sure you check your work!

$x^2 - 8x + 13 = 0$ $\qquad$ $x^2 - 8x + 13 = 0$

$(5.732)^2 - 8(5.732) + 13 = 0$ $\qquad$ $(2.268)^2 - 8(2.268) + 13 = 0$

$32.856 - 45.856 + 13 = 0$ $\qquad$ $5.1348 - 18.144 + 13 = 0$

$0 = 0$ ✓ $\qquad$ $-0.0092 = 0$ ✓

There was some rounding in there, so this is basically 0.

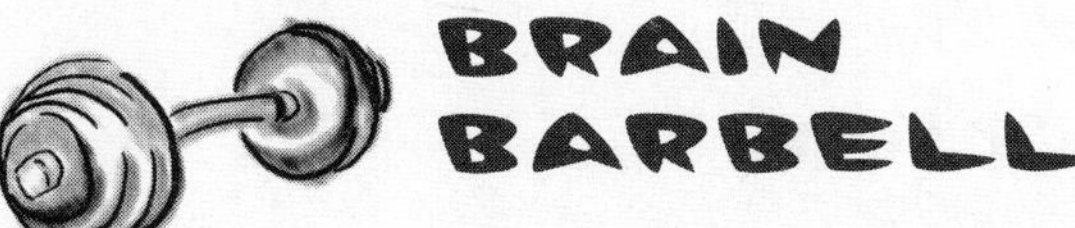

Here are two quadratic equation for you to work on—be careful—some weird things are going to happen.

$$x^2 + x + 7 = 0$$

You can choose between factoring or using the quadratic formula.

What's the weird thing that happened here?

$$x^2 + 10x + 25 = 0$$

Try to factor it here:

Use the quadratic formula here:

What's the weird thing that happened here?

BRAIN BARBELL SOLUTION

Here are two quadratic equation for you to solve. Be careful - some weird things are going to happen.

$$x^2 + x + 7 = 0$$

You can choose between factoring or using the quadratic formula.

Let's try factoring first:

$$(x + \quad)(x + \quad) = 0$$

$7 = 7 \cdot 1$ — The only factors add up to 6 or 8... no good.

This equation can't be factored, so we'll use the quadratic formula.

$$x = \frac{-b \pm \sqrt{b^2 - 4ac}}{2a}$$

$$x = \frac{-1 \pm \sqrt{1^2 - 4(1)(7)}}{2(1)}$$

To follow the order of operations, work under the radical first.

$$x = \frac{-1 \pm \sqrt{1 - 28}}{2}$$

That's a negative number! What does that mean?

What's the weird thing that happened here? There's a negative number under the radical, so you can't get the square root.

$$x^2 + 10x + 25 = 0$$

Try to factor this:

$$(x + 5)(x + 5) = 0$$

$$x + 5 - 5 = 0 - 5 \qquad x + 5 - 5 = 0 - 5$$

$$x = -5 \qquad x = -5$$

But there's only one answer!

Use the quadratic formula here:

$$x = \frac{-b \pm \sqrt{b^2 - 4ac}}{2a}$$

$$x = \frac{-10 \pm \sqrt{10^2 - 4(1)(25)}}{2(1)}$$

$$x = \frac{-10 \pm \sqrt{100 - 100}}{2}$$

$$x = \frac{-10 \pm \sqrt{0}}{2} = -\frac{10}{2} = -5$$

This goes away.

What's the weird thing that happened here? The numbers under the radical cancel out, so there's only one answer.

Wait a second! I thought that the quadratic formula was always supposed to work. With that first one, we're totally stuck before we even finish under the radical! In the second one, the root just goes away. What's that about?

The quadratic formula always works... but it may give you some surprising answers.

A quadratic can have two solutions... but it can also have one solution. And, to really throw you, sometimes the solutions are undefined. An undefined solution is what we call a solution that forces you to take the square root of a negative number.

So how do you know what to expect? It's all about the discriminant.

What the heck is a discriminant?

The **discriminant** is the portion of the quadratic formula that is under the radical, that square root sign:

$$x = \frac{-b \pm \sqrt{b^2 - 4ac}}{2a}$$

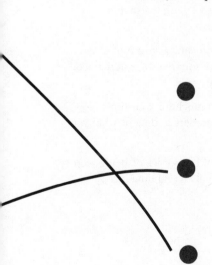

This piece is the discriminant.

- **IF $b^2 - 4ac > 0$, then there are 2 real solutions.**
  This is the typical situation we've been dealing with so far: two solutions that are independent and real.

- **IF $b^2 - 4ac = 0$, then there is only one unique solution.**
  In this case, there's just a single solution that makes the quadratic true.

- **IF $b^2 - 4ac < 0$, then the solutions are undefined.**
  Here, there aren't any real ***x*** values that will make the equation true. This is because you'd have to take the root of a negative number.

The Discriminant Exposed

This week's interview:
Are you hard to work with?

Head First: Hi Discriminant. It seems you haven't gotten much publicity lately.

Discriminant: So true. Most of the time, people are required to learn about me, but they just don't appreciate my usefulness.

Head First: Umm, yeah. So, what exactly are you good for?

Discriminant: I'm a short cut! Using the quick $b^2 - 4ac$ check can save you from doing lots of work.

Head First: How so?

Discriminant: If you just run that formula and compare it with zero, you can figure out how many answers you're looking for with a quadratic equation.

Head First: I see. So if you're less than zero, the solutions are undefined, right?

Discriminant: That's true. There aren't any real numbers that can be squared to get a negative number.

Head First: What about if you're equal to zero?

Discriminant: There's just one number that works - so if you're going to solve the equation, you know that you're only looking for one answer.

Head First: And if you're greater than zero?

Discriminant: Then there are two solutions, like you'd expect for a quadratic equation.

Head First: So it's a neat little trick to figure that stuff out, but how does it really help to figure you out first, instead of just factoring?

Discriminant: It's almost like checking your work in advance. If you know how many solutions you're after, you'll know when you get some solutions you're on the right track.

Head First: That seems helpful. Do you have any other tips?

Discriminant: If you figure me out and I turn out to be a perfect square, the solutions to the equation may be a round number.

Head First: Just to review, why are the solutions undefined if you're negative?

Discriminant: Because you can't take a root of a negative number. A negative square root is undefined. That's a property of a square root.

Head First: Thank you so much for your time. Now I think we all have a better understanding of what you do.

Discriminant: Thank you! I'm so sick of being seen as a waste of time. I'm actually a real time saver.

Simplify each discriminant and then match it to the corresponding possible solution to the quadratic equation.

Discriminant Value

$b^2 - 4ac = ?$

Possible solutions to the quadratic equation that created the discriminant:

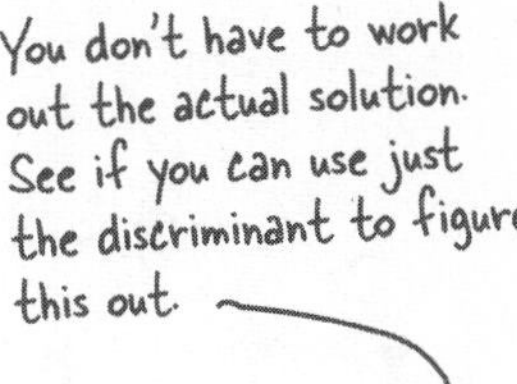

You don't have to work out the actual solution. See if you can use just the discriminant to figure this out.

$(6)^2 - 4(3)(3)$

....

-1

$(5)^2 - 4(1)(-14)$

....

- 2, 7

$(1)^2 - 4(7)(1)$

....

Undefined

$(4)^2 - 4(-2)(-2)$

....

$(2)^2 - 4(3)(3)$

....

Simplify each discriminant and then match it to the corresponding possible solution to the quadratic equation.

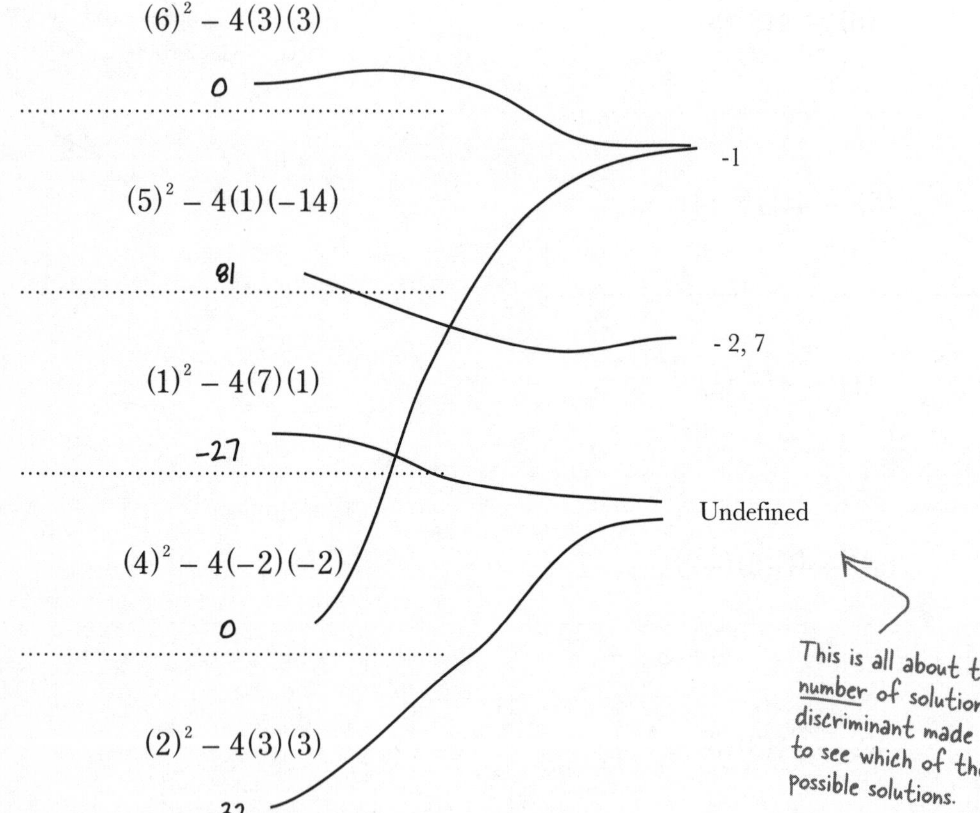

Make it Stick
The quadratic formula
$x = \frac{-b \pm \sqrt{b^2 - 4ac}}{2a}$
The standard form of a quadratic equation
$ax^2 + bx + c = 0$

there are no Dumb Questions

Q: Checking my work with the quadratic formula is a pain...

A: Yeah, it kinda is, but it's worth it. The quadratic formula is pretty easy to mess up... those signs can be a real killer.

The decimals make checking your work tricky, too, and they make solving the original equation difficult. That's why checking your work is so important!

Q: Why can't you take a negative square root?

A: Because a negative times a positive is a negative. A square root is supposed to be two things that are exactly the same multiplied together, and you can't do that and get a negative.

You can, though, take a *cube* root of a negative number. Since a cube root is *three* numbers multiplied together, a negative times a negative is a positive, times another negative is a negative.

Q: If I want to start with the discriminant, do I have to get the actual solution for it to be useful?

A: You can totally eyeball it! If you get 1 minus some giant number, it's obviously negative, and that's all you need. Just use your best judgement!

Q: Which is better, factoring or the quadratic formula?

A: It depends. Usually, if an equation's factorable, then factoring is easier. If you start with the discriminant and it comes out as a perfect square, it's probably a good idea to try factoring first. If you factor, you get nice whole number answers, which makes life so much easier.

On the other hand, if you just want to solve an equation, go quadratic, and you'll get the answer every time. The downside is that it's easy to mess up.

Frat Wars, part deux

After their unsuccessful attempt to hide their president in a hole and behind a wall, Pi Gamma Delta has figured out that Jon can be really exact with his water balloons, if he has time to calculate. So now the president has decided he's going to just keep running around since we can't work that fast.

To aim the catapult properly, we need to be able to work out two things:

1. **What heights can we hit?**
   Jon's water balloon can only go up so high, and there's a different x for every h.

2. **For every height, where does the catapult need to be?**
   Jon needs the position for aiming, which is x. Then he can nail the president and get himself and his fellow pledges into Theta Theta Pi.

The problem is that we've got to solve the quadratic over and over. We really need to just be able to look these things up somehow...

$-x^2 + 10x + 75 = h$
I'm just going to keep moving... then they're hosed.
Pi Gamma Delta's president

Pi Gamma Delta's house
ΠΓΔ

A graph lets us SEE values...

Just like when we were working with linear equations and wanted to skip calculations, a graph would help us see all possible solutions.

We need to figure out the relationship between distance and height, so for any distance, we know the height to shoot balloons, and vice versa. So we're looking for the relationship between two variables, x and h.

If we had a graph, we could just read off points and tell Jon without having to run a bunch of calculations. The thing is, this equation has an x^2 term in it, so how does that work? It's probably not a line, is it?

What shape do you think this graph will take?

How can you graph x^2?

We know lots of stuff about an equation in the degree of one: it has a slope, some intercepts, and it's a straight line. But a second-degree equation may have some other shape that we don't know about yet. We do know that it can have two solutions, not just one, which must mean something about the graph...

The simplest way to plot a graph of an equation is to plot a few points, and then connect the dots. Since we don't know the shape of this graph, let's start by picking a bunch of points and plotting them to see what we come up with.

Fill in this chart with the right ***h*** for each ***x***.

Here are some x-values. You figure out the h-values.

x	$-x^2 + 10x + 75$	h
5	$-(5)^2 + 10(5) + 75$	100
8		
10		
3		
0		

Use this space for extra work:

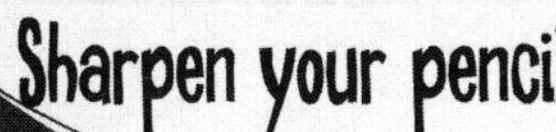

Plot using the points that you just found on the facing page, and draw the resulting curve.

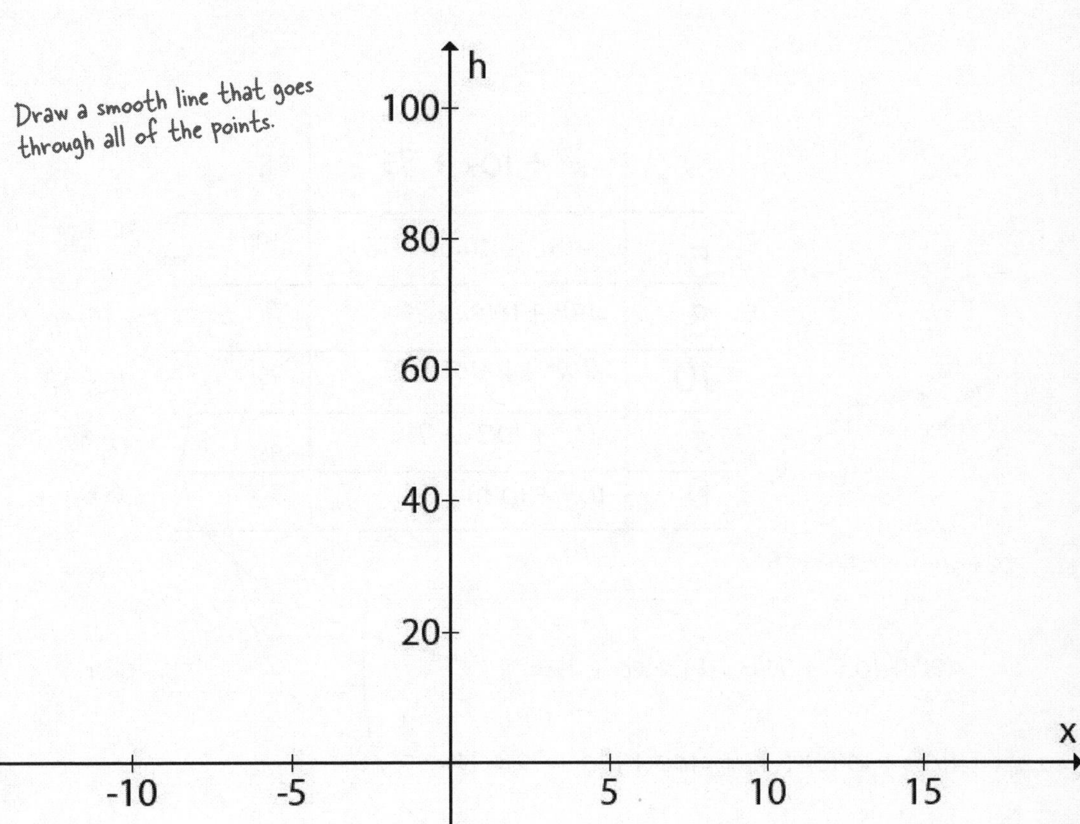

$$-x^2 + 10x + 75 = h$$

Sharpen your pencil Solution

Fill in this chart with the right ***h*** for each ***x***.

x	$-x^2 + 10x + 75$	h
5	$-(5)^2 + 10(5) + 75$	100
8	$-(8)^2 + 10(8) + 75$	91
10	$-(10)^2 + 10(10) + 75$	75
3	$-(3)^2 + 10(3) + 75$	96
0	$(0)^2 + 10(0) + 75$	75

This point is (5, 100)

(8, 91)

(10, 75)

(3, 96)

(0, 75)

Plot these

This one was actually nice and easy.

Use this space for extra work:

$-(8)^2 + 10(8) + 75 = -64 + 80 + 75 = 91$

$-(10)^2 + 10(10) + 75 = -100 + 100 + 75 = 75$

$-(3)^2 + 10(3) + 75 = -9 + 30 + 75 = 96$

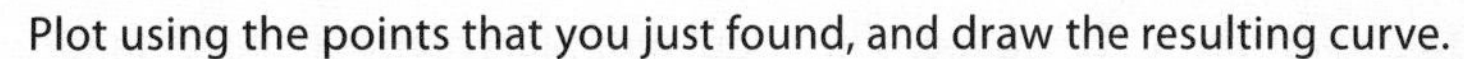

Plot using the points that you just found, and draw the resulting curve.

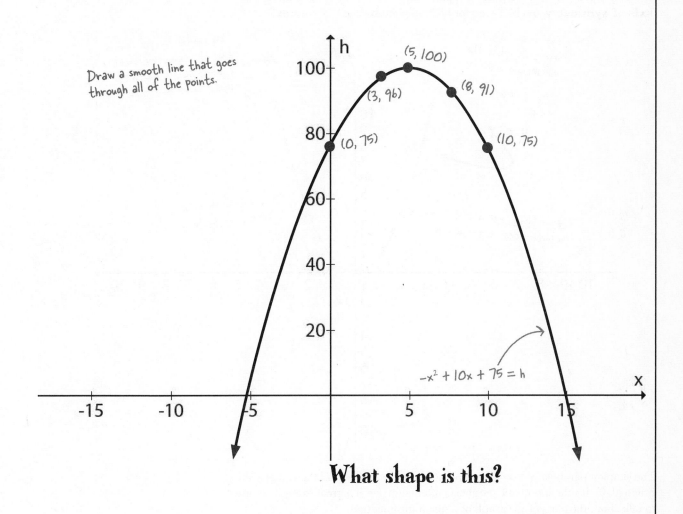

Draw a smooth line that goes through all of the points.

h

100 — 80 — 60 — 40 — 20

(5, 100)

(3, 96)

(8, 91)

(0, 75)

(10, 75)

$-x^2 + 10x + 75 = h$

x

-15 — -10 — -5 — 5 — 10 — 15

What shape is this?

Relax

Don't worry if your curve isn't exactly a smooth shape.

The point of the exercise was to have you figure out the general shape of the equation, not draw a curving masterpiece.

A parabola is the shape of a quadratic equation

A **parabola** is basically a "U" shape. The width and placement of the U-shape changes, depending on the equation. A parabola is also symmetrical around an **axis of symmetry** and it's lowest (or highest) point is called the **vertex**.

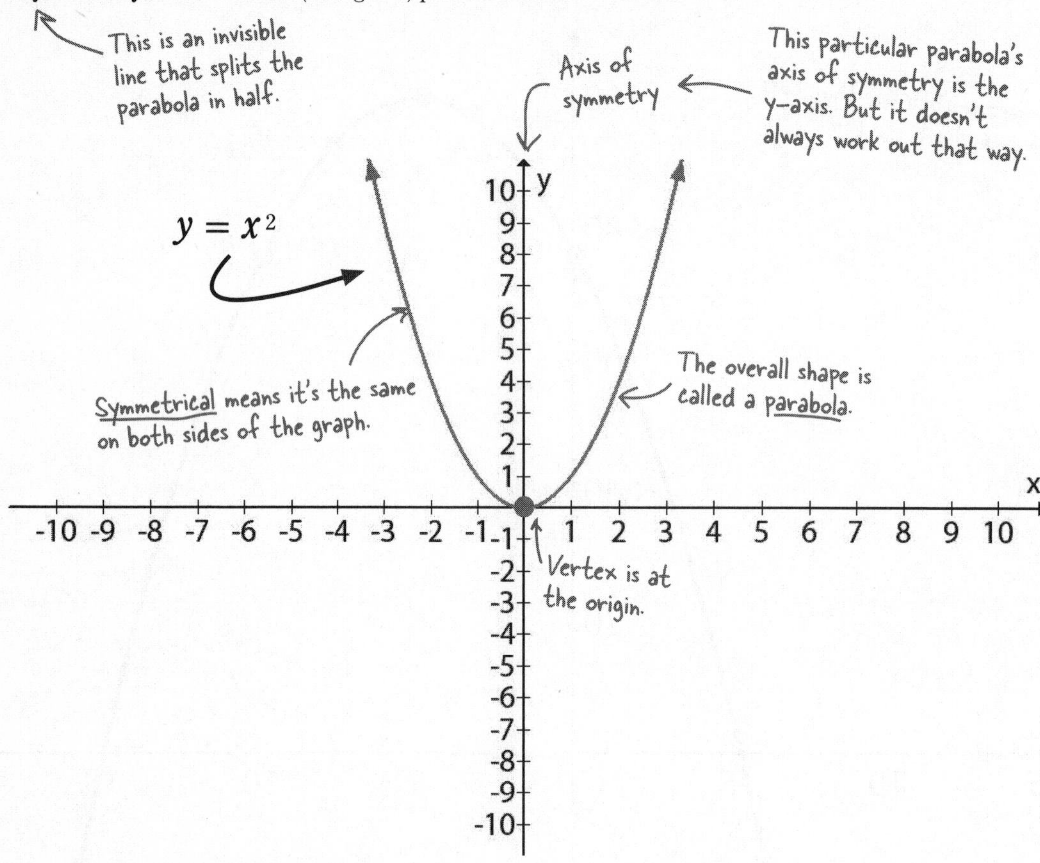
This is an invisible line that splits the parabola in half.
Axis of symmetry
This particular parabola's axis of symmetry is the y-axis. But it doesn't always work out that way.
$y = x^2$
Symmetrical means it's the same on both sides of the graph.
The overall shape is called a parabola.
Vertex is at the origin.
y
x
10
9
8
7
6
5
4
3
2
1
-1
-2
-3
-4
-5
-6
-7
-8
-9
-10
-10 -9 -8 -7 -6 -5 -4 -3 -2 -1
1 2 3 4 5 6 7 8 9 10

The simplest parabola, $y = x^2$, is symmetric about the y-axis, and the vertex is the origin (0, 0). It's the most basic parabola, and that makes it a great example to use to talk about the parts of the graph of a quadratic equation.

Once you start adding x terms, coefficients, and constants to a quadratic equation, the graph changes. Jon's catapult graph was upside down because our quadratic had a $-x^2$ term. Other coefficients in front of the x^2 affect how wide of narrow the U-shape is. The x term and the constant term in a quadratic equation move the basic parabola around on the plane: up, down, or side to side.

The $y = x^2$ parabola is the basic shape of all quadratic equations.

Graphing a parabola depends upon the vertex

Once you know what the vertex of a quadratic equation is, you can pick some points on the left and right of the vertex to draw a nice curve. That's all you need!

Finding the x-coordinate for the vertex is easy, it's: $x = -\frac{b}{2a}$

To find our catapult's vertex, start with the equation in standard form and use the formula:

$$-x^2 + 10x + 75 = h$$

a = −1 b = 10 c = 75

$$x = -\frac{b}{2a}$$
$$= -\frac{10}{2(-1)}$$
$$= 5$$

Substitute the x-coordinate for the vertex back into the equation to get the vertical coordinate (h, in our case).

$$-(5)^2 + 10(5) + 75 = y$$
$$-25 + 50 + 75 = y$$
$$100 = y$$

The vertex is (5, 100)

It may seem like a lot to remember. The more you work with quadratics, the more all of this will become second nature. Just keep at it, and don't give up.

Use and understand the vertex

The vertex is either the top or the bottom of the parabola. How can you tell? If the quadratic starts with a positive x^2 term, it's the bottom of the parabola; if the quadratic starts with a negative x^2 term, it's the top of the parabola.

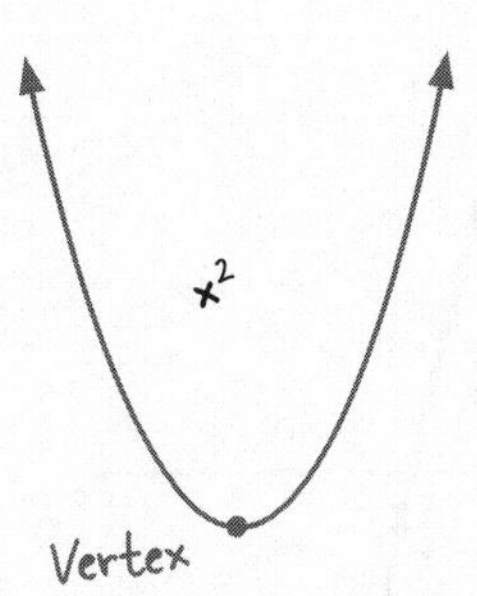

To finish the graph after you know the coordinates for the vertex, you just need a couple of points along the two sides of the parabola to finish the graph.

To pick the rest of the points to graph, just go to the left and the right of the vertex, and you'll have the basic shape.

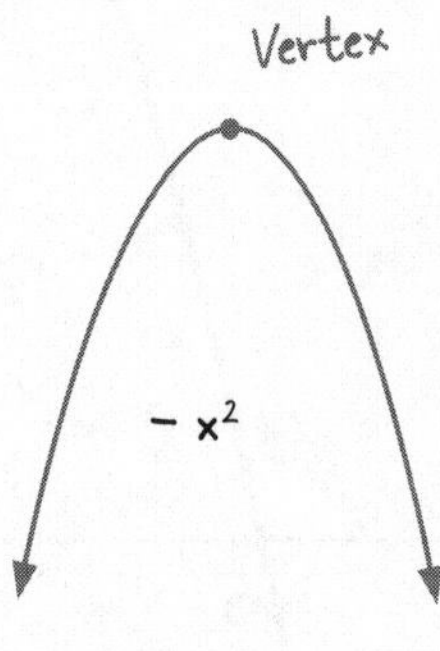

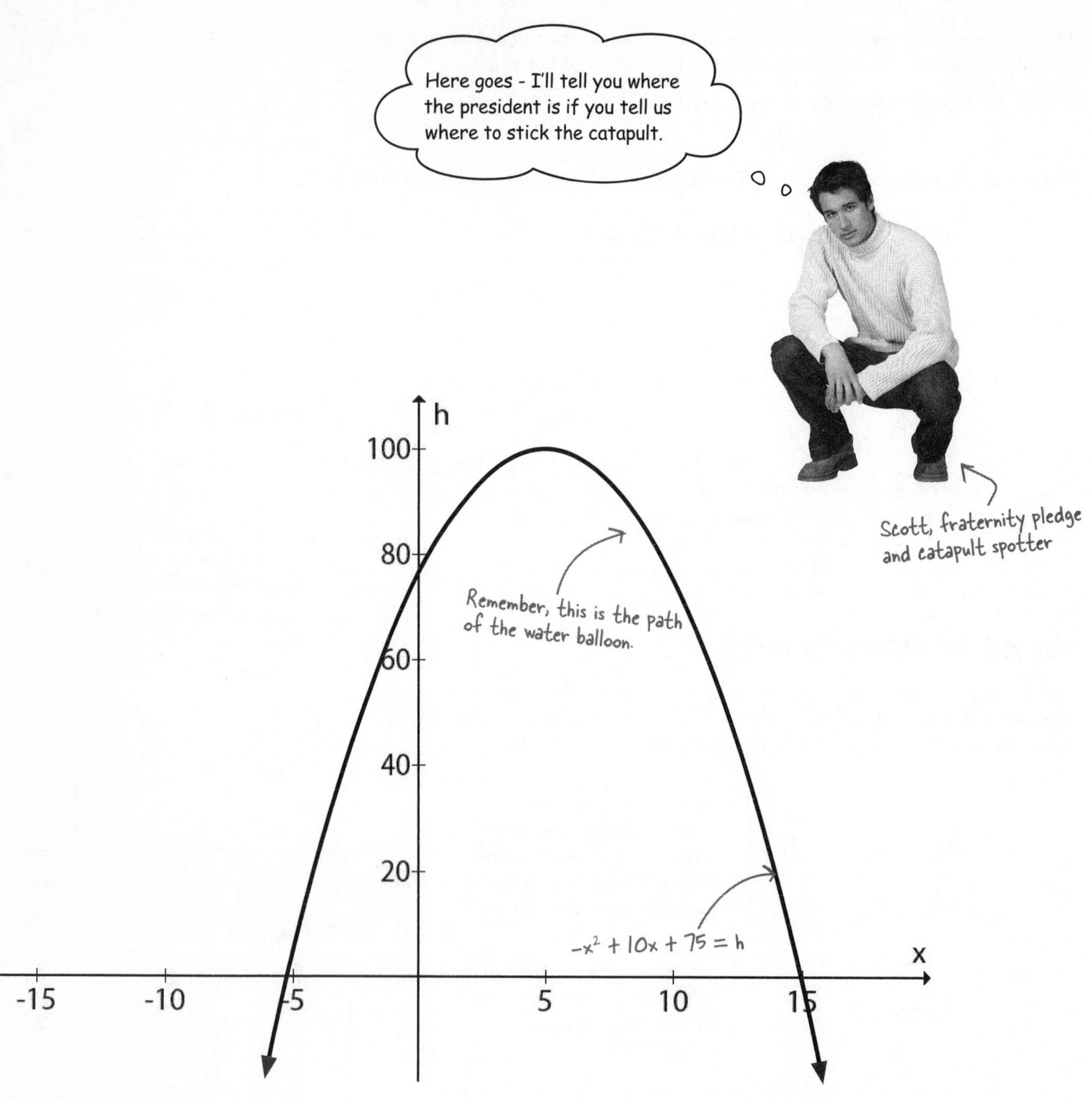
Here goes - I'll tell you where the president is if you tell us where to stick the catapult.
Scott, fraternity pledge and catapult spotter
h
100
80
60
40
20
x
-15
-10
-5
5
10
15
Remember, this is the path of the water balloon.
$-x^2 + 10x + 75 = h$

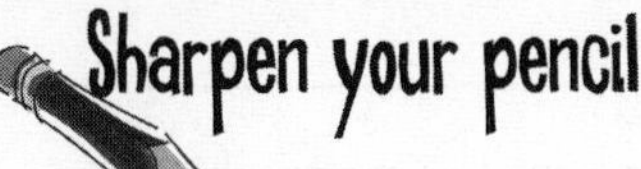

Here are the locations that Scott's calling out. Use the graph that you drew to figure out where to put the catapult.

The president climbed a flag pole, 30 feet up ....

Climbed down to a second-story porch, 15 feet up ....

Hot air balloon! He's 120 feet in the air! ....

Down the steps of the basement at -10 feet ....

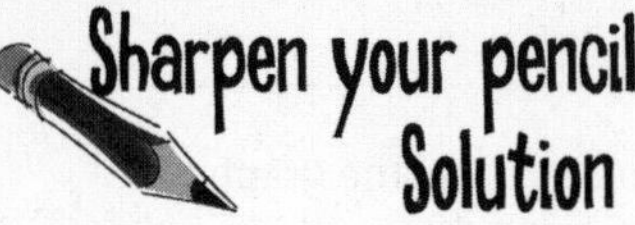

Sharpen your pencil Solution

Here are the locations that Scott's calling out. Use the graph that you drew to figure out where to put the catapult.

Direct hit!

The president climbed a flag pole, 30 feet up .... *If you go up 30 feet, the options are around 13 feet and -3 feet. Put the catapult at 13 feet from the flag pole.*

Climbed down to a second-story porch, 15 feet up .... *Nailed him!*

Up 15 feet, it's either -4 or 14.5 feet, so put the catapult 14.5 feet from the porch.

Hot air balloon! He's 120 feet in the air! ....

The catapult won't go that high! It's more than the vertex, at 100 feet, so we'll miss him.

Down the steps of the basement at -10 feet .... *One more right on the head!*

Down ten feet is -10, which is at about 16 feet and -5 feet.

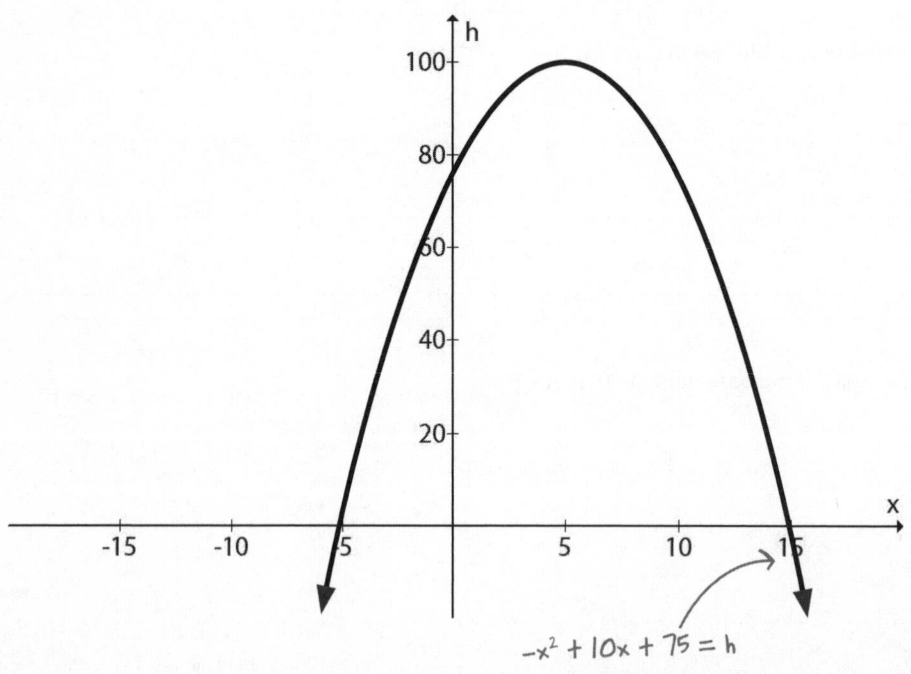

Work with the parabola, the SMART way

Now that you know all about the shape of a parabola and how it relates to the equation and the discriminant, we can put it all together and see how everything works.

Our original equation → $-x^2 + 10x + 75 = h$

1. We know that the parabola is pointing down because the x^2-coefficient is -1.

2. You can easily figure out the vertex with $x = -\frac{b}{2a}$

$$x = \frac{-(10)}{2(-1)} = 5$$

$$-(5^2) + 10(5) + 75 = h$$
$$-(25) + 50 + 75 = h$$
$$100 = h$$

Substitute this back into the original equation to find the other vertex coordinate.

That means the vertex is at (5, 100) when you plot the graph.

3. Evaluate the discriminant ($b^2 - 4ac$) to find out how many solutions there will be for the catapult's quadratic equation.

$$\text{discriminant} = 10^2 - 4(-1)(75)$$
$$= 100 + 300$$
$$= 400$$

Since this is greater than zero, that means there are two real solutions.

Two real solutions to the equation indicates two x values where the vertical value (h in our case) is zero. We already know those values:

(15, 0) and (–5, 0)

This was the first time you solved the equation, h = 0.

$$(x - 15)(x + 5) = 0$$

$$x - 15 = 0 \qquad x + 5 = 0$$

$$+15 + x - 15 = 0 + 15 \qquad -5 + x + 5 = 0 - 5$$

$$x = 15 \qquad x = -5$$

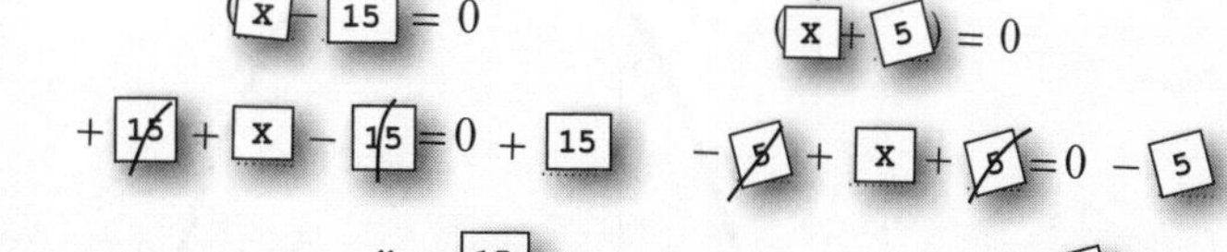

What if the discriminant was 0 or less than 0? What would that mean for the graph?

The parabola can sit two other ways on the Cartesian plane with respect to the x-axis.

The discriminant can help with our graph, too

Discriminant values fall into three categories: greater than zero, less than zero, or zero. We've talked about what those values mean for the number of solutions to a quadratic equation, but what about for the graph of a quadratic?

- IF $b^2 - 4ac > 0$ **There are 2 real solutions, 2 places that the parabola crosses the x-axis.**

  The solutions for the quadratic equation are the two values that work when the equation is set to zero and where it crosses the x-axis.

- IF $b^2 - 4ac = 0$ **There is only one unique solution, and the parabola only touches the x-axis.**

  Like the picture of $y = x^2$, the graph sits on the x-axis but won't cross the axis.

- IF $b^2 - 4ac < 0$ **The solutions are undefined and the graph sits above the x-axis and never crosses.**

  An undefined solution means that the quadratic equation has no number that will work if the equation is set to zero. On a graph, it'll look like it's floating...

So here's our final graph:

Take all that information, and you'll end up with the same graph as before. But, this time, we've got some built-in checks for our work. We knew to expect an upside-down parabola, with two real solutions, that crossed the x-axis in two places.

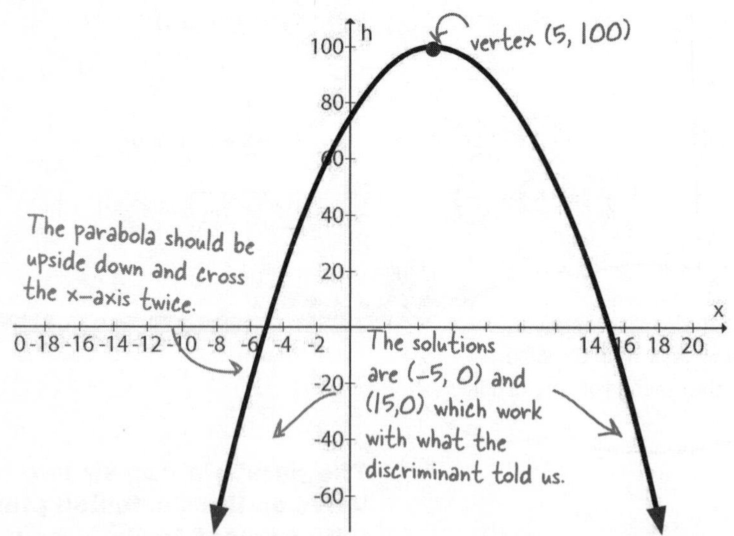

The solutions for a quadratic equation are the x values when the other variable (usually y) is 0.

there are no Dumb Questions

Q: What's the deal with always having two answers?

A: When we talked about the discriminant, there were three different options for answers. One answer, two answers, or no answers. The basic parabola, $y = x^2$, only touches the **x**-axis at one point, so that's one solution.

The typical parabola, like the catapult's, has two solutions and two places it crosses the **x**-axis. When we used the quadratic formula to solve this equation, the answers we got were the two places that the parabola crosses the **x**-axis.

Q: How do I find the vertex again?

A: The vertex has a consistent **x** coordinate, $x = -\frac{b}{2a}$ .It is a fantastic place to start plotting points because it's the top (or bottom) of the parabola, and you know that you'll have symmetrical points on either side of the vertex.

Q: Is there another way to graph a parabola besides plotting points?

A: Yes, but it's complicated. We're just scratching the surface on parabolas, and in order to learn all the deep dark secrets of parabolas, there's a lot more math involved. Algebra 2 will explain all of that! For now, we'll stick with plotting points, usually starting with the vertex.

Q: Which kind of quadratic equations have a graph that is an upside down parabola?

A: If an equation has a negative x^2 term, the parabola is upside down, or pointed down. If the x^2 term is positive, then the parabola is upward.

Q: Is there a way to find the axis of symmetry?

A: The axis of symmetry is the vertical line that goes through the vertex of the parabola.

So if you think back to the standard form of a vertical line, it's just **x** = the **x** - coordinate of the vertex, which is: $x = -\frac{b}{2a}$

BULLET POINTS

- The basic shape of a quadratic equation is called a **parabola**.
- The uppermost (or lowest) point of the parabola is the **vertex**.
- The **x**- coordinate of the vertex is $-\frac{b}{2a}$
- A quadratic equation has 0, 1, or 2 **solutions**.

Woo hoo! We're in! Even with the president moving and climbing, we nailed him.
ΘΘΠ
ΠΓΔ
I don't understand how they got me *every* time. Crazy pledges!

Let's put it all together. Work with the following quadratics using all of the techniques you've learned.

$x^2 - 4 = 0$

How many solutions are there? **0** **1** **2**

How will you solve it? **Factoring** **Quadratic Formula**

Which direction will the parabola be facing? **Up** **Down**

$5x^2 + 4x - 11 = 0$

How many solutions are there? **0** **1** **2**

How will you solve it? **Factoring** **Quadratic Formula**

Which direction will the parabola be facing? **Up** **Down**

$3x^2 - x + 13 = 0$

How many solutions are there? **0** **1** **2**

How will you solve it? **Factoring** **Quadratic Formula**

Which direction will the parabola be facing? **Up** **Down**

Exercise

Work on one more, but graph it this time.

$x^2 - 11x + 28 = 0$

How many solutions are there? **0** **1** **2**

How will you solve it? **Factoring** **Quadratic Formula**

Which direction will the parabola be facing? **Up** **Down**

Graph it here

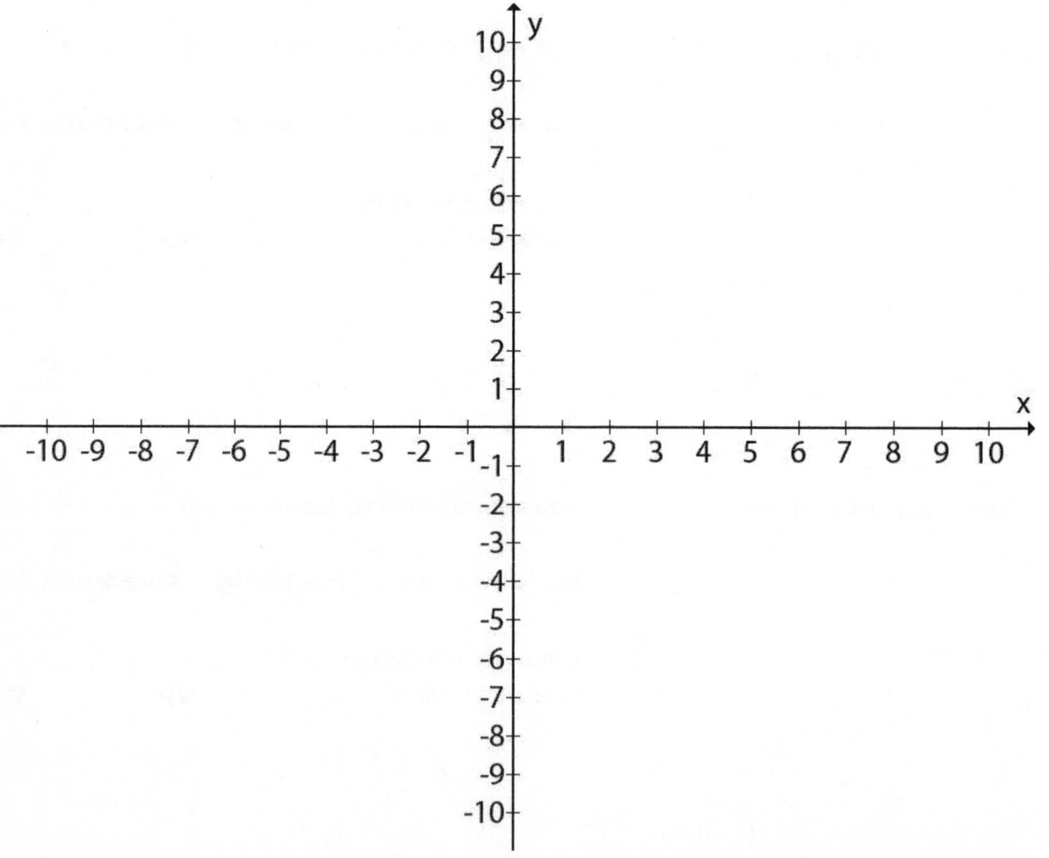

Let's put it all together. Work with the following quadratics using all of the techniques you've learned.

$x^2 - 4 = 0$

This is a difference of two squares.

$(x + 2)(x - 2) = 0$

$x + 2 = 0$ $\quad$ $x - 2 = 0$

$-2 + x + 2 = 0 - 2$ $\quad$ $+2 + x - 2 = 0 + 2$

$x = -2$ $\quad$ $x = 2$

How many solutions are there? **0** **1** **2** (circled: 2)

How will you solve it? **Factoring** **Quadratic Formula** (circled: Factoring)

Either one of these will work, but factoring is easier.

Which direction will the parabola be facing? **Up** **Down** (circled: Up)

You know this because the coefficient in front of the x^2 term is positive.

$5x^2 + 4x - 11 = 0$

$b^2 - 4ac = 16 - 4(5)(-11) = 236$

Greater than zero – there are 2 solutions.

$$x = \frac{-b \pm \sqrt{b^2 - 4ac}}{2a}$$

$$x = \frac{-4 \pm \sqrt{(4)^2 - 4(5)(-11)}}{2(5)}$$

$$x = \frac{-4 \pm \sqrt{236}}{10}$$

$$x = \frac{-4 + \sqrt{236}}{10} \qquad x = \frac{-4 - \sqrt{236}}{10}$$

$x = 1.136$ $\quad$ $x = -1.936$

How many solutions are there? **0** **1** **2** (circled: 2)

How will you solve it? **Factoring** **Quadratic Formula** (circled: Quadratic Formula)

Which direction will the parabola be facing? **Up** **Down** (circled: Up)

$3x^2 - x + 13 = 0$

$b^2 - 4ac = 1 - 4(3)13 = -155$

That's less than 0. There aren't any real solutions!

How many solutions are there? **0** **1** **2** (circled: 0)

How will you solve it? **Factoring** **Quadratic Formula**

Which direction will the parabola be facing? **Up** **Down**

There aren't any solutions, the discriminant is less than 0

Work on one more, but graph it this time.

Exercise Solution

$b^2 - 4ac = (-11)^2 - 4(1)28 = 9$

$x^2 - 11x + 28 = 0$

How many solutions are there? **0** **1** **(2)**

$(x - 7)(x - 4) = 0$

How will you solve it? **(Factoring)** **Quadratic Formula**

$x - 7 = 0$ $x - 4 = 0$

Which direction will the parabola be facing? **(Up)** **Down**

$+7 + x - 7 = 0 + 7$ $+4 + x - 4 = 0 + 4$

$x = 7$ $x = 4$

Vertex (x) $= -b/2a$

$x = \frac{-(-11)}{2(1)}$

$x = \frac{11}{2}$

$5.5^2 - 60.5 + 28 = y$

$30.25 - 60.5 + 28 = y$

$-2.25 = y$

This means you know two points, (7, 0) and (4, 0).

Graph it here

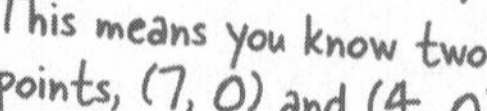

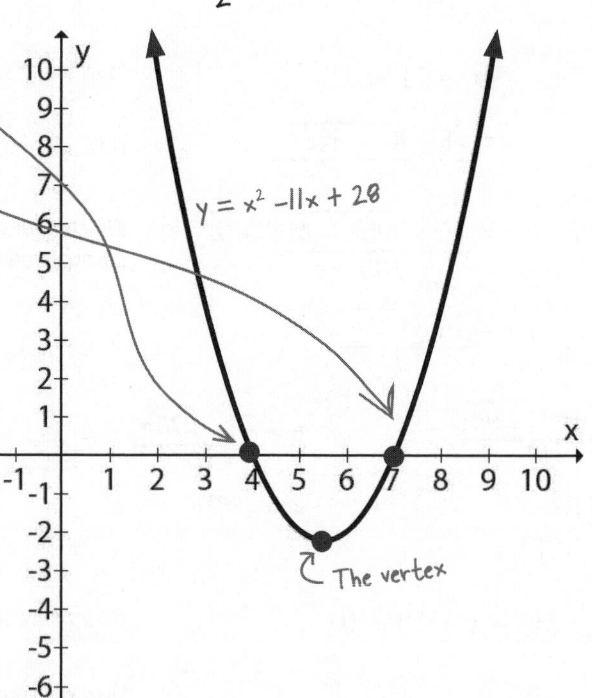

Quadraticcross

Make sure both sides of your brain balance. Here's a crossword to get the right side working.

Across

4. A quadratic is centered on the _____ __ ________
6. The shape of a quadratic equation is a ____________
9. If you can't factor an equation, use the _________ _________
10. Breaking a quadratic into two binomials

Down

1. The top (or bottom) of a parabola is a _____________
2. You can't take a ___________ root.
3. The terms under the root are the ________________
5. An equation with a squared variable
7. Factoring quadratics is ___________ backwards
8. A quadratic equation has _______ solutions

Answers on page 376.

CHAPTER 9

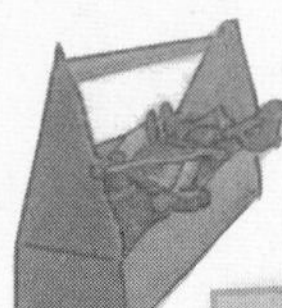

Tools for your Algebra Toolbox

Factoring quadratics:

The standard form of a quadratic equation

$$ax^2 + bx + c = 0$$

Form matters.

The equation needs to be in standard form of a quadratic, set to zero. You have to have the zero or you can't use the zero product rule and split up the possible solutions.

$x^2 - 10x - 75 = 0$

Set up the binomials.

Once the equation is in the proper form, you know you're going to need two binomials that start with x. Fill them in and you've already got half of the terms done!

$(x \quad)(x \quad) = 0$

Find out the other two terms in the binomials.

The last two terms need to accomplish two things. They have to be multiplied together to get the constant in the quadratic equation (75). They also need to add together to get the x term (–10x).

These need to be multiplied together to get 75.

$(x \quad 15)(x \quad 5) = 0$

These need to be added (or subtracted) to get the x term the –10x.

Fill in the signs and check your work.

To finish up the factor, fill in the signs. They need to be multiplied to get the same sign in front of the constant (75) and added to get the right x term (–10x). Then, expand the binomials you came up with, using FOIL, and make sure that it matches what you started with.

$(x - 15)(x + 5) = 0$

$x^2 + 5x - 15x - 75 = 0$

$x^2 - 10x - 75 = 0$

BULLET POINTS

- Quadratic equations have up to two solutions.
- Factoring a quadratic equation means finding the product of two binomials.
- You need to check that your factoring is correct, using FOIL, before you solve.
- Finding the constant terms for the binomials is the hardest part of factoring a quadratic.
- Quadratic equations need to be in standard form before you factor.

The quadratic formula

the discriminant

$$x = \frac{-b \pm \sqrt{b^2 - 4ac}}{2a}$$

Using the quadratic formula

The quadratic formula can be used to solve a quadratic equation of the form:

$$ax^2 + bx + c = 0$$

The discriminant $b^2 - 4ac$

The discriminant is a term that used to describe the portion of the quadratic formula that is under the radical. Greater than zero, there are two real solutions to the quadratic equation. If it equals zero, there is one real solution, and if it's less than zero, there are no real solutions to the equation.

Quadratic equations shapes on graphs

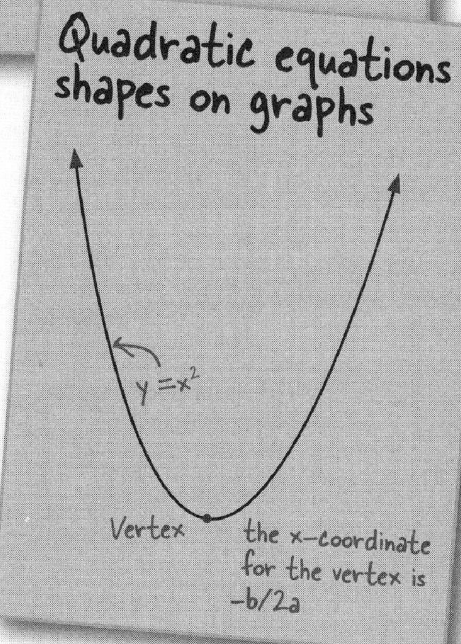

BULLET POINTS

- The basic shape of a quadratic equation is called a parabola.
- The uppermost (or lowest) point of the parabola is the vertex.
- The ***x***- coordinate of the vertex is $-\frac{b}{2a}$
- A quadratic equation has 0, 1 or 2 real solutions.

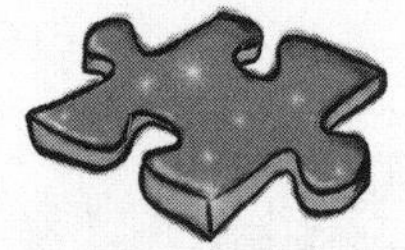

Quadraticcross Solution

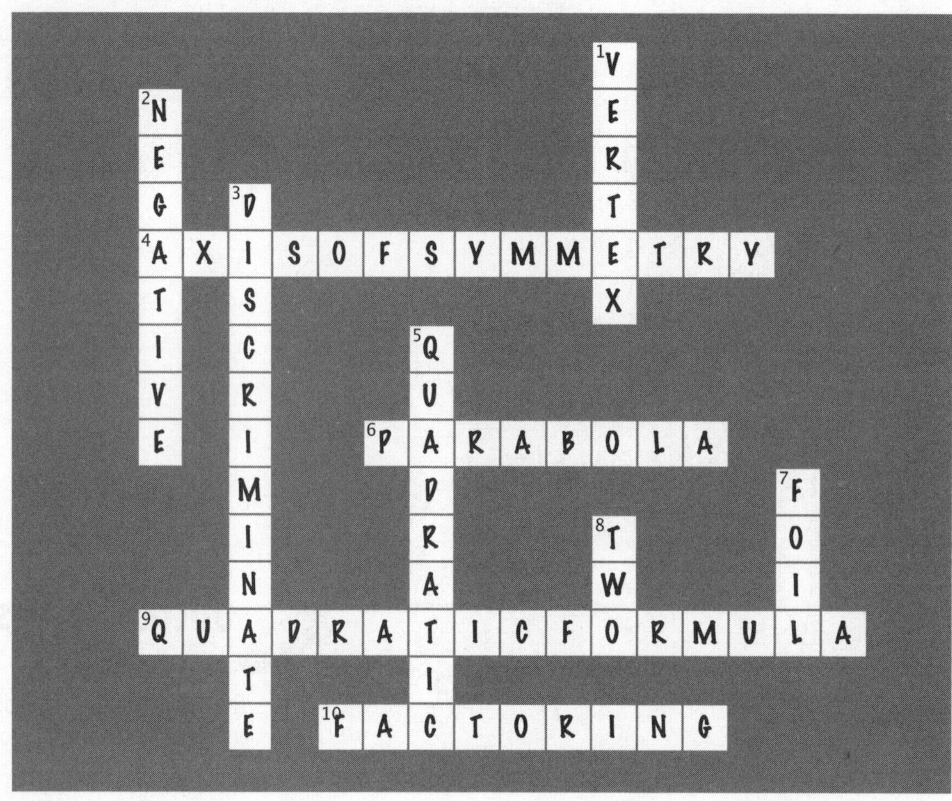

10 functions

Everyone has limits

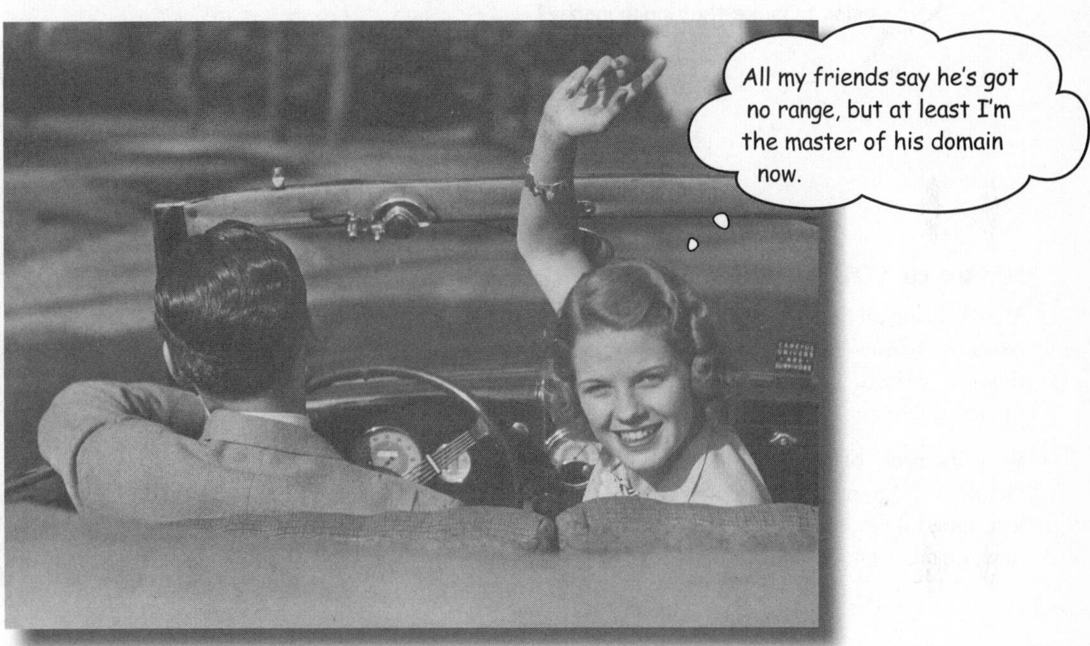

All my friends say he's got no range, but at least I'm the master of his domain now.

Some equations are like suburban neighborhoods...
...they're fenced in.

You'll find that in the real world, many equations are **limited**. There are only certain values that an equation is good for. For instance, you can't drive a car -5 miles or dig a hole 13 feet up. In those cases, you need to set **boundaries** on your equations. And when it comes to putting some limits on your equations, there's nothing better than a **function**. A function? What the heck is that? Well, turn the page, and find out... through the lens of reality TV.

Dude, it's time to go big time: bigger venues, more fans... But to get to the next level, we need $52,375 for new equipment. How are we going to make that much money?

Pajama Death

It's up to YOU to raise cash for Pajama Death.

After hearing all that you went through to get Paul to their concert, Pajama Death figures you're a real fan... and a financial wizard. They need you to take their future to a higher cash plane.

Your first order of business? Figure out how much Pajama Death stands to make off the new reality TV show their label just signed them up for. Nobody likes to sell out... but some new touring gear would sure be sweet..

Pajama Death TV

Here's the deal: the network wants an 11-show deal, and Pajama Death gets 5% of the proceeds. That includes advertising revenue, plus ticket sales to live tapings each week. The show's 90 minutes each week, and the network guarantees 20 ads per show.

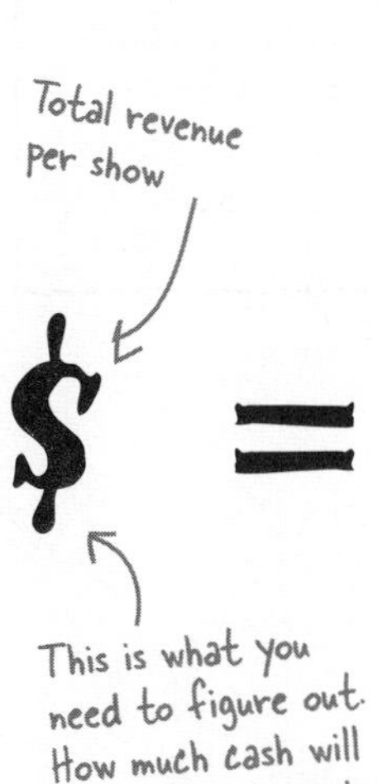

Total revenue per show
This is what you need to figure out. How much cash will Pajama Death make?

Ads on TV are one source of revenue.

Admission to the tapings make money, too.
ADMIT ONE
151332

$1\frac{1}{2}$ **hour show** $= 20$ **ads**

Ads are $1000 each.

Tickets are $100 a seat.

The number of seats sold varies per show.

Sharpen your pencil

Write up the equation for the per-show revenue and figure out how much Pajama Death will make if they sell 1,515 tickets to the season premiere.

....

....

....

....

Sharpen your pencil Solution

Your job was to figure out the revenue from doing the TV show.

Total revenue → $R = 1000(20) + 100x$ ← The number of ads per show times the cost per ad; x is the number of tickets; The cost per ticket

$R = 20{,}000 + 100x$

For 1515 seats, $R = 20{,}000 + 100(1515) = 171{,}500$

PJ Death's cut 5% $= (0.05)(171{,}500) = \$8{,}575$ ← Pretty good for one show!

An equation to model the revenue is based on the ads per show, the cost per ad, and cost of a ticket, and how many tickets are sold for a show.

Uh oh... there's a change in venue

At the last minute, the taping location changed. Now only 1,511 seats are available, not 1,515. Sound hokey? In the real world, limits exist all the time. An equation on paper doesn't have limits, but problems in the real world do.

Our current equation doesn't say enough about the problem it models. Sure, we can solve for ***R***, but there's nothing to prevent us from accidentally putting in too many seats for ***x***... and overestimating what Pajama Death will actually make.

The equation is valid for all R's. → $R = 20,000 + 100x$ ← The equation is valid for all x's, too

Limits for this equation will come from the ***x*** vales, not ***R***. Since ***R*** is totally dependent upon ***x***, if you limit ***x***, which is the number of tickets, then ***R*** will also be limited.

So what do we know about the number of tickets (***x***)?

Worst case?
The worst case scenario for Pajama Death and ticket sales? Nobody comes. That's zero tickets sold.

Best case?
Pajama Death sells out all 1,511 seats. If you sell more than that, people are going to be mad because they'll have nowhere to sit... so that's the maximum ***x*** value.

How can we turn the these facts into limits for our equation?

Equations have limits (most of the time)

The real world is full of limitations like tickets, time available for a TV show, the number of songs that Pajama Death can sing before they lose their voices... and each of these limits may need to be modeled with math.

Fortunately, Algebra just happens to have the perfect thing for this situation: a **function**. Among other things, a function can limit certain values for its variables. These variables are called **inputs** in a function.

In the Pajama Death equation, rewritten as a function, the input would be x, the number of seats sold.

For Pajama Death, the limits will be the maximum and minimum number of tickets, x, that can be sold. Those limits will make sure that we don't go figuring numbers for more seats than there are, and in turn, limits the amount of money that Pajama Death can make in a show.

A function can be expressed as an equation

A function is really just a special type of equation, and usually carries along some extra information. And since equations can also be functions, there needs to be different notations for functions, so we can tell what we're dealing with. A function isn't written in terms of a variable like y, but is instead written like this: $f(x)$.

So we can use the powers of a function to limit the revenue equation for Pajama Death, we need to rewrite the equation as a function. To limit the number of tickets, we're looking to work with the x variable, right? That means the expression needs to contain x and be set equal to $f(x)$. So we can dump R, and rewrite this equation as a function:

Just like regular variables, a function doesn't have to be f(x), it can be c(d) or r(x). But f(x) is the most common function you'll see.

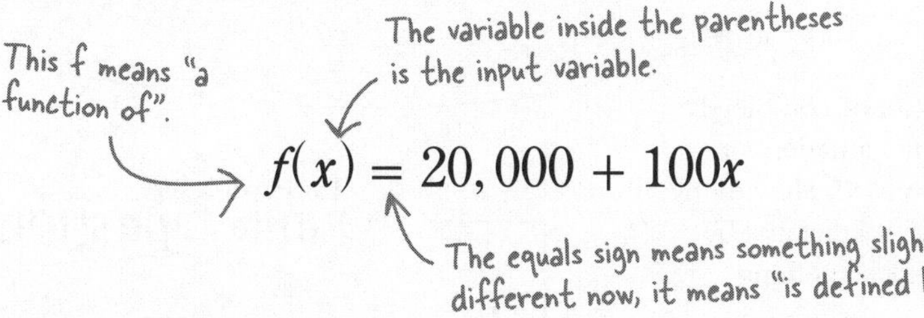
This f means "a function of".
The variable inside the parentheses is the input variable.
$f(x) = 20,000 + 100x$
The equals sign means something slightly different now, it means "is defined by".

BRAIN BARBELL

Fill in the limits for the Pajama Death revenue function.

$$f(x) = 20,000 + 100x \qquad \dots\dots \leq x \leq \dots\dots$$

Fill in the limits for the revenue function.

$$f(x) = 20,000 + 100x \qquad 0 \leq x \leq 1511$$

This is the worst case – nobody comes

The show sells out!

The input limits are the domain of the function

$f(x)$

Input variable

All functions have a **domain**, which is the set of valid inputs for the function. The group of values which is valid for the function is usually written as an inequality. The domain inequality is in terms of the input variable, which is the variable inside the parentheses of the ***f()***.

So for our function, ***x*** is the input, and the domain is all the numbers that ***x*** can take on: 0 to 1,511.

For the function:

$$f(x) = 20,000 + 100x$$

The domain is:

$$0 \leq x \leq 1511$$

x is valid from 0 through 1511 (that's why you needed less than or equal to, not just less than)

In general, the domain of a function can be completely **arbitrary** (based on the problem or it's situation) or defined **because of the expression** itself. How many tickets you can sell is arbitrary, because it's dictated by how many seats are in the arena. It's not something related to the math expression.

All functions have a domain.

there are no Dumb Questions

Q: What do you mean, limited because of the expression itself?

A: Many mathematical expressions don't go on forever, and are self-limiting. If you have an expression with an ***x*** denominator, a parabola that only covers part of the Cartesian plane, or a place where you may end up with a negative square root, the domain is limited due to the expression, not necessarily because of the real world problem the function models.

Only 473 people came to the first show. That was lame.

x is within the domain.

Only 473? It looks Pajama Death fell a lot short of what it was hoping for, at least for the first show.

Still, we knew that 473 was a possible value, because it's greater than (or equal to) 0 and less than (or equal to) 1,511. That's what the domain does: gives us all possible input values.

Sharpen your pencil

How much did the band make for the first show? If the numbers stay at 473 for the remaining 10 shows, will they make enough to buy the new equipment they want?

The function for revenue per show is over here.

Pajama Death needs $52,375 by the end of the season, 10 more shows

Sharpen your pencil Solution

How much did the band make for the first show? If the numbers stay at 473 for the remaining 10 shows, will they make enough to buy the new equipment they want?

$f(x) = 20000 + 100x \quad 0 \leq x \leq 1511$

473 is within the domain, so it's ok to use.

$f(473) = 20000 + 100(473)$

You know the input value – it's 473

$f(473) = 67300$

This is the total show revenue

5% of the revenue = $(0.05)(67300) = \$3{,}365$

That's a lot less than the band needs.

11 – that's 10 more shows plus the one that already happened

Total at 473 = 11 (3365)

Total at 473 = $37, 015

Pajama Death needs $52, 375 by the end of the season, 10 more shows.

$37, 015 is way less than $ 52,375

Pajama Death is not going to make enough for the new equipment they need!

What a terrible turnout. We're never going to sell out stadiums! We can't sell out and still not get our gear! That's a total zero-input situation.

Functions have minimum and maximum outputs

Function inputs make up the domain, we know that. By limiting the numbers that can go into a function, we're limiting the numbers that can come out of a function. If only 1,511 tickets can be sold, we can figure out the most that ***f(x)*** can equal... and we can do the same for the minimum value of the function. These minimum and maximum values are the limits for the **output** of the function.

The process of determining a function's outputs is called **evaluating the function**. So every time you plug in ***x***, and solve for ***f(x)***, you're evaluating the function for a certain input value. That's really just like solving the equation, which you've done a ton of times.

So what's the MAXIMUM we can make? What's the MINIMUM?

Using the domain as the starting point, we can find the minimum and maximum outputs of a function. That's basically saying, how small and how big can ***f(x)*** get?

$$f(x) = 20,000 + 100x \qquad 0 \leq x \leq 1511$$

We can evaluate the functions at these extremes to get the minimum and maximum outputs of THIS function (not all functions).

Exercise

Evaluate the revenue function to determine the maximum amount of money that Pajama Death can make for the rest of the season. Suppose Pajama Death sold out every remaining show after the poor season premiere sales. What would they make over the entire season?

....

....

....

....

Evaluate the revenue function to determine the maximum amount of money that Pajama Death can make for the rest of the season. Suppose Pajama Death sold out every remaining show after the poor season premiere sales. What would they make over the entire season?

$f(x) = 20000 + 100x \qquad 0 \leq x \leq 1511$ ← Back to the original function

$f(1511) = 20000 + 100(1511)$ ← Evaluate for a sold out show

$f(1511) = 171100$ ← The way to write this set of input output would be as an ordered pair (1511, 171100)

5% of the revenue = (0.05)(171100) = $8,555

This is how much the band would make for one sold out show.

10(8555) + 3365 = 88915

Suppose there are 10 more sold out shows with the one lousy show...

That's much more than the $52,375 they need!

All of the valid outputs are called the range

Just as there is a set of inputs over which the equation is valid (the domain) for every function - there is also a **range**. The **range** of a function is the set of numbers over which there are valid outputs. The range actually provides the minimum of the function (the lowest possible output) and the maximum (the highest possible output) of the function.

You write the range like the domain: as an inequality. Also like the domain, functions are written with their range so you can know the limits of the function's outputs. And, of course, the range for certain functions will be limited because the graph may not cover the entire Cartesian plane.

Watch it!

The limits of the domain don't always give you the maximum and minimum outputs of a function.

In the case of the revenue function, the least value of the domain results in the minimum value of the range, and the highest value for the domain results in the maximum. That's because we're dealing with a line that has a positive slope. If the line had a negative slope, though, the values would be reversed. If we had a curve, the maximum could be in the middle of the domain!

Ok, so functions have a domain and a range. Is there anything else I've got to worry about? This stuff seems pretty useful...

Functions have a very specific definition.

The thing about functions is that they aren't always defined in terms of equations. Functions are even more general...

Functions Way Up Close

What we've learned about functions so far are some characteristics of functions: domain, range, inputs, and outputs. But what's a function, really?

The official definition of a function is:

A function is a relation where one input results in a single output. It can be represented as an equation or as a set of ordered pairs, and has both a domain and a range.

Algebra is really about relations

A **relation** is the general way in which two sets of numbers go together, and they are usually represented as ordered pairs. The difference between a relation and an equation is that there does not need to be a pattern in a relation: it can be totally random. A relation does have a domain and a range, which you can determine directly from reading the ordered pairs.

Let's look at a simple relation that's not based on an equation to get an idea of what we're talking about:

All you need for a relation is a set of ordered pairs.

This is a relation. The curly braces are the beginning and the end of the relation.

$$\{(4,1),(4,-1),(2,0)\}$$

No function in sight, but this is still a relation.

The domain and range come from reading the points in the relation.

Domain: $\{2,4\}$

Range: $\{-1,0,1\}$

These are just the possible values that this relation can take on.

A function is just a specific type of relation. It's a relation that has ***one output*** for ***one input***. That means that the relation above is ***not a function***, because for the input 4, there are two possible outputs, 1, and -1.

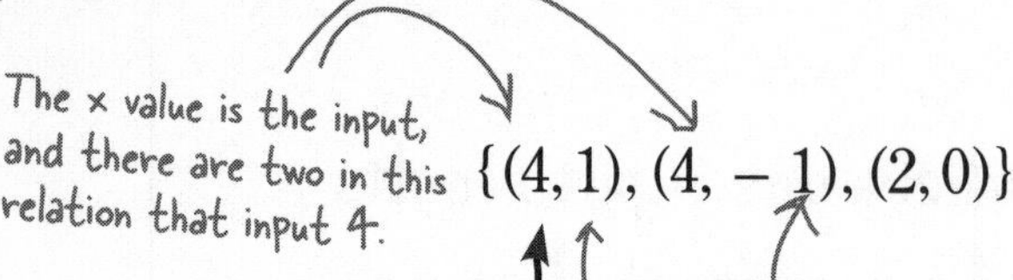

The x value is the input, and there are two in this relation that input 4.
{(4, 1), (4, – 1), (2, 0)}

Domain: $\{2,4\}$

Range: $\{-1,0,1\}$

There are two possible outputs for the input 4, -1 and 1

NOT a function

Sum it up

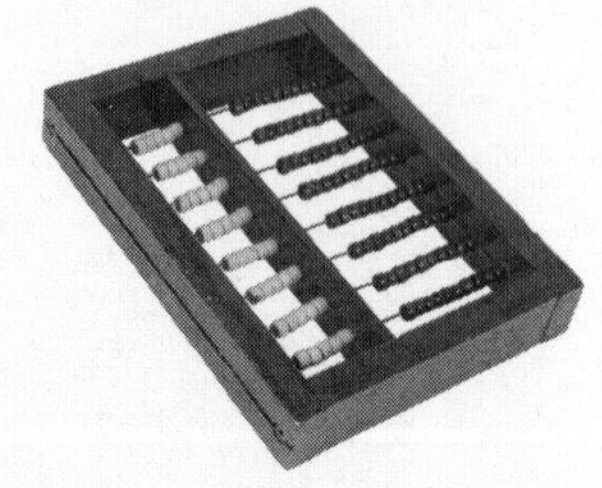

Relation – A set of ordered pairs with a domain and a range.

Who am I?

A bunch of function terminology, in full costume, are playing a party game, "Who am I?" They'll give you a clue — you try to guess who they are based on what they say. Assume they always tell the truth about themselves. Fill in the blanks to the right to identify the attendees.

Tonight's attendees:

FUNCTION, DOMAIN, RANGE, RELATION, EQUATION, INPUT, OUTPUT, f(X)

	Name
I'm the values that come out of a function.	....
I can be an equation or a set of ordered pairs, but either way I'm fun!	....
I represent minimum and maximum outputs of the function, but be careful, you can't just use the lowest and highest inputs to get me!	....
I might be a function, but I'm not as organized as an equation, I'm just a set of ordered pairs.	....
I limit the inputs for the function.	....
The easy way to tell if ordered pairs or an equation are a function is if you see me around, I'm a dead giveaway.	....
I might be a function, or I might not, but I define one or more variables in terms of numbers.	....
I'm a value that goes into a function.	....

A bunch of function terminology, in full costume, are playing a party game, "Who am I?" They'll give you a clue — you try to guess who they are based on what they say. Assume they always tell the truth about themselves. Fill in the blanks to the right to identify the attendees.

Tonight's attendees:

FUNCTION, DOMAIN, RANGE, RELATION, EQUATION, INPUT, OUTPUT, f(X)

	Name
I'm the values that come out of a function.	Output
I can be an equation or a set of ordered pairs, but either way I'm fun!	Function
I represent minimum and maximum outputs of the function, but be careful, you can't just use the lowest and highest inputs to get me!	Range
I might be a function, but I'm not as organized as an equation, I'm just a set of ordered pairs.	Relation
I limit the inputs for the function.	Domain
The easy way to tell if ordered pairs or an equation are a function is if you see me around, I'm a dead giveaway.	f(x)
I might be a function, or I might not, but I define one or more variables in terms of numbers.	Equation
I'm a value that goes into a function.	Input

there are no Dumb Questions

Q: A function doesn't have to be an equation?

A: Nope. A function can be an equation, but it can also be a set of ordered pairs. In that case, the ordered pairs are going to be (input, output). Ordered pairs probably seem weird, but it's much easier to figure out the domain and the range - it's given!

And remember, an equation for a line defines an infinite set of ordered pairs.

Q: Are range and domains given, or do I have to figure them out?

A: It depends upon the situation. Sometimes the domain or the range will be given, sometimes it will be part of the problem statement (like ticket sales), and sometimes you'll need to figure it out. But more about that later.

Q: One input and one output... what's the big deal about that?

A: It means that in a function you won't get two ***f(x)*** values for one ***x***, ever.

Q: Are all equations functions?

A: Nope. That's something that you'll need to figure out soon enough, too. If an equation has multiple output values for one input, it's not a function. It's still a perfectly good equation, though.

Q: The ordered pairs look a lot like ordered pairs for graphing. Is that a coincidence?

A: No, you can graph functions just like you graph points or equations. We're coming up on the details of how it works, but graphing helps out a lot with functions.

Q: Does the one input, one output thing mean that you can't have the same f(x) value as an output for different inputs?

A: You're saying, can you have a function like this **f(x) = {(1,4), (-1,4)}**? The function can have the same outputs for different inputs, that's fine. You just can't go the other way... so **{(1,4), (1,3)}** is NOT a function. You can't have ***different*** outputs for the ***same*** input.

Q: Is the maximum of the function the maximum of the range?

A: Yes. But the tricky thing is you can't necessarily get those values from plugging in the low and high values for the domain to find the range.

Q: Why not?

A: Because if your function has any kind of a curve in the middle of your domain limits, you may miss a maximum or a minimum for the function.

For example, if the vertex of a parabola was between the two points for your domain, the maximum range for that function would be based on a value between your domain's limits, not at one extreme or the other.

Q: This function thing has a lot of rules...

A: The definition of a function was created by mathematicians, to make sure everybody's consistent with what they're talking about. That's important so that everyone gets the same results.

Q: Are all functions relations?

A: Yes. Since functions have a domain and a range and represent a set of ordered pairs, they're a relation.

It doesn't work the other way, though. A relation does not have to have a one output to one input requirement, so it isn't necessarily a function.

Q: Why would you use a relation?

A: It's a math way to tie numbers together that may not have an easy pattern. Think about statistics; if you were keeping track of housing data, you'd just need an address and a price. It really wouldn't matter what order they were in or how much they each cost.

A function is always a relation, but relations are not always functions.

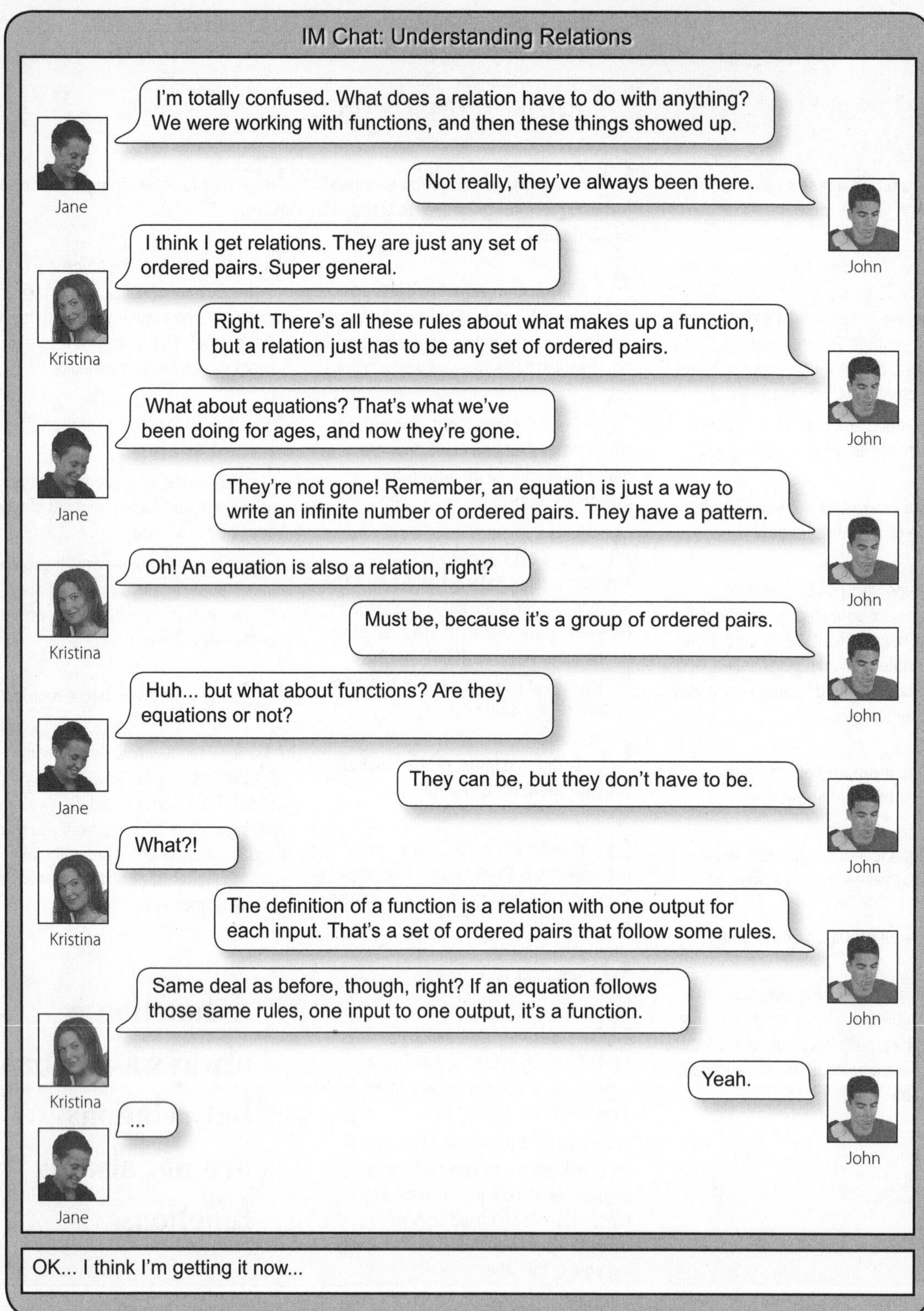
IM Chat: Understanding Relations
I'm totally confused. What does a relation have to do with anything? We were working with functions, and then these things showed up.
Jane
Not really, they've always been there.
John
I think I get relations. They are just any set of ordered pairs. Super general.
Kristina
Right. There's all these rules about what makes up a function, but a relation just has to be any set of ordered pairs.
John
What about equations? That's what we've been doing for ages, and now they're gone.
Jane
They're not gone! Remember, an equation is just a way to write an infinite number of ordered pairs. They have a pattern.
John
Oh! An equation is also a relation, right?
Kristina
Must be, because it's a group of ordered pairs.
John
Huh... but what about functions? Are they equations or not?
Jane
They can be, but they don't have to be.
John
What?!
Kristina
The definition of a function is a relation with one output for each input. That's a set of ordered pairs that follow some rules.
John
Same deal as before, though, right? If an equation follows those same rules, one input to one output, it's a function.
Kristina
Yeah.
John
...
Jane
OK... I think I'm getting it now...

Relations, equations, and functions all go TOGETHER

These three terms are just different ways to characterize a group of ordered pairs. That's a lot to keep up with, so let's take a step back and look at all three terms. How do they relate? How are they the same? When are they different?

- **Relations, linear equations, and functions are all sets of ordered pairs.**
  An equation is just a set of (x, y) points if you graph it. The same is true for functions and relations, although the ordered pairs don't have to connect for functions and relations.

- **Relations, linear equations, and functions all have a domain and a range.**
  The domain is the valid **input** values, and the range is the valid **output** values. The domain and range may be infinite, but they always exist.

- **An equation in two variables MUST be a relation and MAY be a function.**
  Since an equation is a list of an infinite number of points, that means all equations are relations. But functions still have that special one-input-to-one-output rule, something not required by equations. So not all equations are functions.

- **A function MUST be a relation and MAY be an equation.**
  Functions are very specific. They have to be a set of ordered pairs, and that makes them a relation, too. But not all sets of ordered pairs can be expressed as an equation.

Since some functions can be written as an equation, how do you think you work with functions? Can you graph them and solve them?

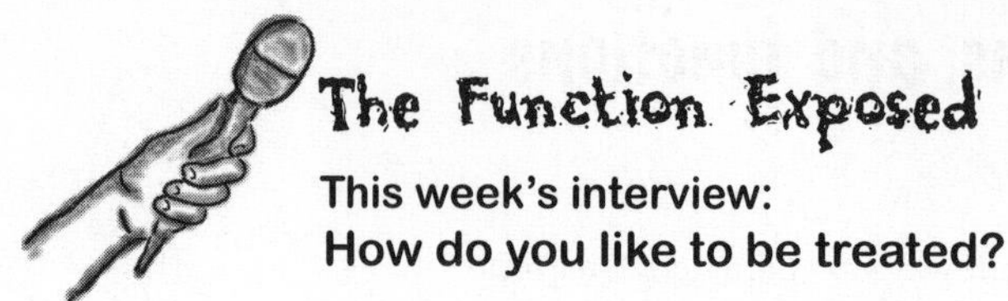

This week's interview:
How do you like to be treated?

Head First: Hi function! We've been learning a lot about you lately.

Function: Thanks, that's very flattering.

Head First: But what everyone wants to hear about you, is, how do people work with you? Just like a plain equation?

Function: In some ways, yes. If I happen to also be an equation, you can solve me for values in the same way you do with an equation.

Head First: So if we set $f(x) = 0$ that's ok?

Function: Yes. That way you can solve for the zeroes of the function.

Head First: What about graphing?

Function: Again, if I'm written as an equation with a domain and range, you can graph me in the much the same way as you would a regular equation. But you have to show the constraints on the graph.

Head First: Oh, I see, so we should only graph the values where you're valid, over your domain, for example.

Function: Right. So if my domain goes from -1 to 10, then the graph for me should also only go from -1 to 10.

Head First: So really, if you're written as an equation, we can just treat you the same as equations?

Function: Yes and no. You can, as long as whatever you're doing is within my domain and range. Don't go drawing a graph that goes too far, or solving me for values that aren't in my domain. If you do that, it won't work.

Head First: So you're more constrained than an equation?

Function: Yes, but I prefer to call it realistic. The world has limits, and so do I. It means that with me, you can work with more realistic situations.

Head First: Thanks. Now we have a much better idea of how to work with you...

Constraints on functions allow you to be more <u>realistic</u> in the way math represents the world.

Does that mean we can graph a function?

You can always graph a function.

Not only that, but you know enough from graphing linear equations to know the general shape of most functions. You just have to show the domain and range on your graph in some way.

BRAIN BARBELL

Find the range for the Pajama Death revenue function by graphing the function.

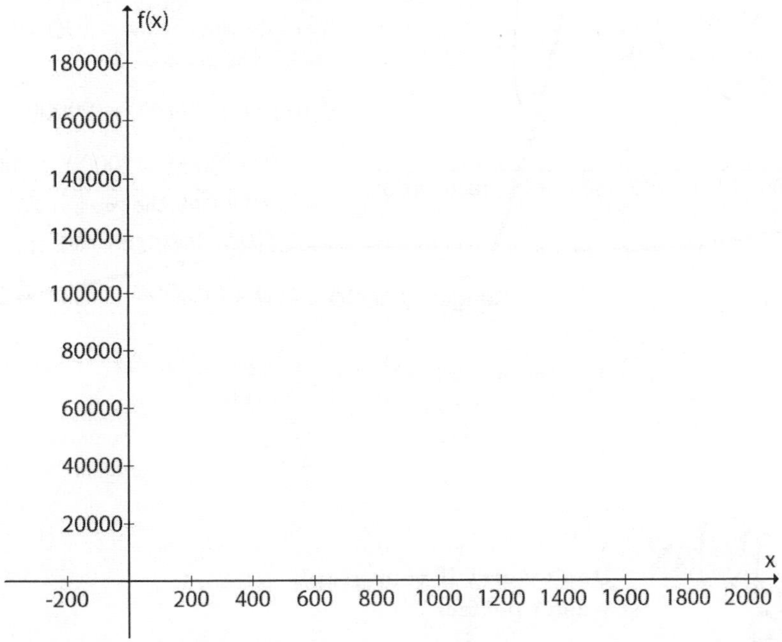

f(x)

180000
160000
140000
120000
100000
80000
60000
40000
20000

-200 200 400 600 800 1000 1200 1400 1600 1800 2000

x

$f(x) = 20,000 + 100x$

$0 \leq x \leq 1511$

Range:

Brain Barbell Solution

Find the range for the revenue function by reading it off the graph.

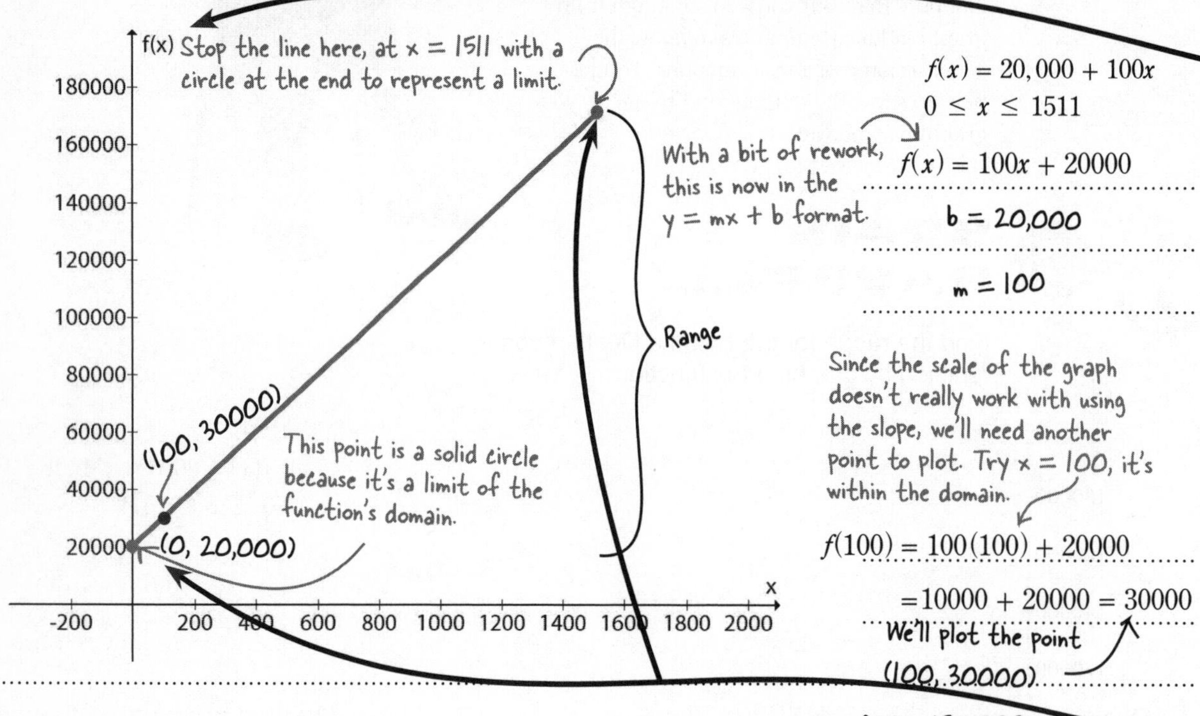

f(x)
Stop the line here, at x = 1511 with a circle at the end to represent a limit.
180000
160000
140000
120000
100000
80000
60000
40000
20000
(100, 30000)
(0, 20,000)
This point is a solid circle because it's a limit of the function's domain.
Range
-200
200
400
600
800
1000
1200
1400
1600
1800
2000
x
With a bit of rework, this is now in the y = mx + b format.
Since the scale of the graph doesn't really work with using the slope, we'll need another point to plot. Try x = 100, it's within the domain.
We'll plot the point (100, 30000)

$$f(x) = 20,000 + 100x$$
$$0 \le x \le 1511$$

$$f(x) = 100x + 20000$$

b = 20,000

m = 100

$$f(100) = 100(100) + 20000$$
$$= 10000 + 20000 = 30000$$

Range: $20000 \le f(x) \le 170{,}000$

This is the possible range of revenue for the function, read right off the graph.

Relax

Don't worry if your graph wasn't perfect.

Graphs aren't always super-exact. Focus on getting the basic shape, and finding the points at the domain limits.

Function graphs have LIMITS

The graph of a function doesn't look that much different from the graph of an equation. You just have to represent the limits of the function on the graph. Even marking the endpoints of the domain with filled-in dots is familiar: we did that when we graphed inequalities.

Just think of graphing a function as graphing an equation, with a twist.

1. **Graph the basic equation on a scale of f(x) vs. x scale.**
   Just substitute the ***f(x)*** values for ***y*** on a typical Cartesian plane, and go ahead and graph your function. So you'll usually have an ***x*** axis, and an ***f(x)*** axis.

2. **Look at the domain for the function and remove the graph outside of the domain.**
   Erase, scratch out, don't graph that part, but remove the pieces that don't matter.

   This is the domain. Just cut the graph off before 0 and after 1511.

   $0 \leq x \leq 1511$

3. **Draw the termination points.**
   When you draw an inequality on the number line, the termination points are solid or empty circles, depending upon the inequality. Functions are the same way.

   Since our graph is less than or equal to for both domain values, we have two solid points.

$\leq$ & $\geq$	Draw a solid point.	●	This indicates that the graph includes the boundary point
$<$ & $>$	Draw an open circle	○	This indicates that the graph goes up to the boundary, but does not include the boundary point

4. **Read all the values you need directly off the graph.**
   Just like when you graphed a linear equation, now you can get any value you need. This includes the range, even when it's not a linear equation.

BULLET POINTS

- **Graphing functions** is similar to graphing equations.
- To show the limits of a function, use a **solid point** for an "or equal to" inequality, and an **open point** for just a less than or greater than symbol.
- The best way to get the **range** of a function is to **read it** from a graph.
- To find the zeros of a function, set ***f(x)*** equal to 0.

there are no Dumb Questions

Q: How can I figure out the domain of my function?

A: It depends upon how the function is presented. If it's given as an equation in a problem, analyze the problem situation and see what limits there are. If the function is a set of ordered pairs, the domain can be read directly from the points presented.

Q: How do I find the range of my function?

A: The best, easiest way is to graph the function and then read the range points off your graph. Why? Because if anything weird happens to the equation (or with the points) between the boundaries of the domain, you'll see it.

Q: Can I just evaluate the function for the upper and lower values for the domain and get the range?

A: Sometimes, but not always. This really only works if you have a line with a positive slope. If you have a quadratic, with the vertex between the boundary points, that won't work at all.

Unless you are positive that you know exactly what is going on with an equation or a relation presented, it's best to graph your relation and then determine the range.

Q: What's the difference between the range and the maximum and minimum values?

A: Good question. The range is ***all*** the ***f(x)*** values the equation can take on. The maximum and minimum values for the function are the boundary values... but *just* those values.

Q: How can you tell if an equation is a function?

A: An equation is a function ***if and only if*** for each input there is a *single* output. That means for each ***x***, there is only one ***f(x)*** value. If you think about it, there's a way to see it on a graph, too... but more on that later.

The best way to find the range of a function is to read it off a graph.

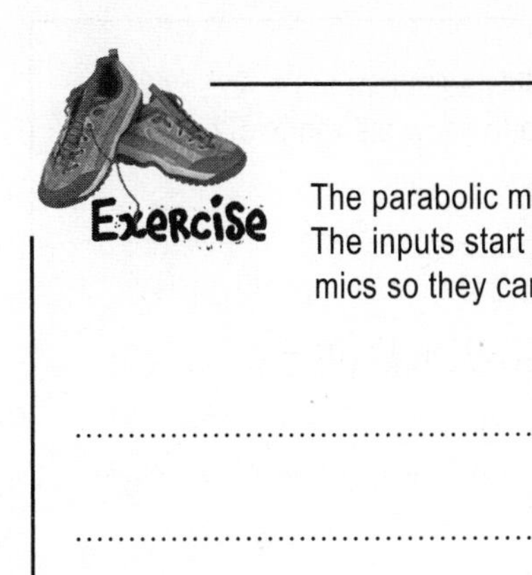

The parabolic mics that Pajama Death wants to buy conform to this equation: $f(x) = -2x^2 + 12x - 9$. The inputs start just above zero, and go through 5. They need to know the range of outputs for the mics so they can buy the right PA for their stage rig.

Range:

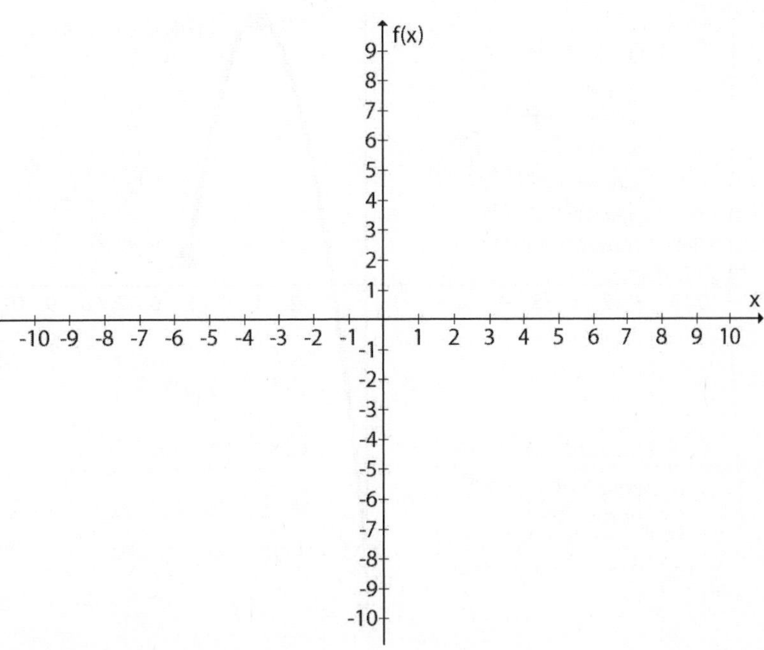

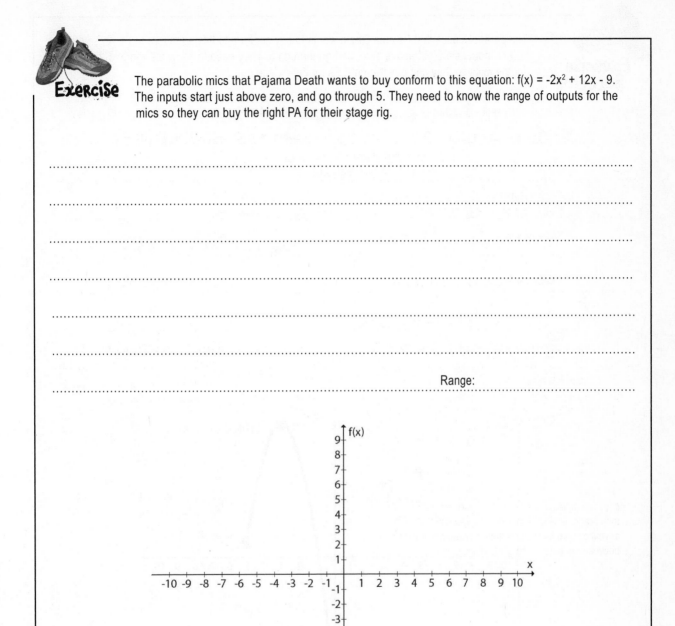

Exercise Solution

The parabolic mics that Pajama Death wants to buy conform to this equation: $f(x) = -2x^2 + 12x - 9$. The inputs start just above zero, and go through 5. They need to know the range of outputs for the mics so they can buy the right PA for their stage rig.

This negative means the parabola is upside down.

$$f(x) = -2x^2 + 12x - 9$$

$$0 < x \le 5$$

This is the domain: greater than zero, up to and including 5.

Since this is a quadratic, first find the vertex.

$$\text{Vertex } x = -\frac{b}{2a} = -\frac{12}{2(-2)} = 3$$

$$\text{Vertex } y = -2(3)^2 + 12(3) - 9$$

$$= -18 + 36 - 9 = 9$$

Plot the vertex at (3, 9)

This is one of our domain boundaries.

$$\text{Subst. } x = 0 \quad -2(0)^2 + 12(0) - 9$$

$$f(x) = -9$$

Here's the other domain boundary.

$$\text{Subst. } x = 5 \quad -2(5)^2 + 12(5) - 9$$

$$= -50 + 60 - 9 = 1$$

Range: $-9 < f(x) \le 9$

Now, just read the range right off the graph

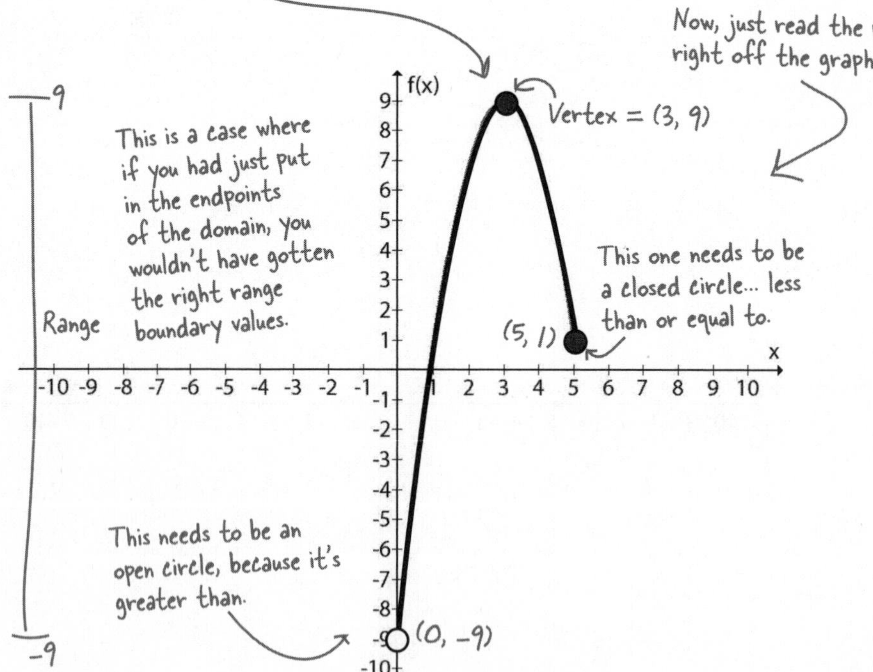

f(x)
x
Vertex = (3, 9)
(5, 1)
(0, -9)
9
Range
-9
This is a case where if you had just put in the endpoints of the domain, you wouldn't have gotten the right range boundary values.
This one needs to be a closed circle... less than or equal to.
This needs to be an open circle, because it's greater than.

Just before the second episode of Pajama Death TV...

So is this a general admission seating thing? Or do we have to worry about assigning seats?
ROCK

Assigned seating SHOULD BE like a function!

Assigned seating is really a lot like a function. There's an input - the seat number - and an output - the customer ID who's sitting in that seat.

Take a look:

Input (seat #, s)	Output (Customer ID, c)
Seat 110	75
Seat 112	63
Seat 125	85
Seat 110	40
Seat 75	56

There should be a one-to-one mapping here. One seat produces a SINGLE customer ID, no more.

In a function, we can only have one ***f(x)*** for every ***x***, and that's what we want here. We don't want any duplicate ***f(x)*** values, because that would mean two customer IDs are associated with the same seat. Not good!

BRAIN BARBELL

Jot down some ideas of how you could check to see if the relation between seats (***s***) and customer IDs (***c***) a function.

....

....

....

Graphing reveals the nature of a relation

The difference between having the seat assignments right and wrong is ***exactly*** the same as the difference between a relation and a function. If the seating assignments are one input to one output, then everybody has somewhere to go... and that's a function. If not, there are going to be fist fights over a relation that should have been a function, but wasn't!

Let's downshift and examine the difference in the graphs of two sets of ordered pairs; one that's just a relation and one that is actually a function.

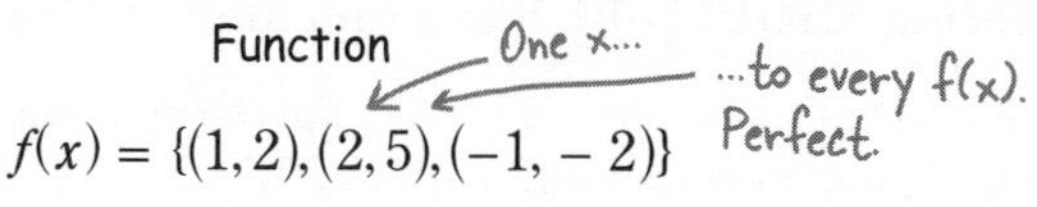

Function
One x...
...to every f(x).
Perfect.
$f(x) = \{(1,2),(2,5),(-1,-2)\}$

Just a relation

$\{(1,2),(2,5),(1,4)\}$

For one x, there are multiple f(x) values. So this isn't a function.

Two f(x) values for one x... that's like two y values for the same x, right? Sort of like a vertical line?

A vertical line is not a function!

A function can't have more than ***f(x)*** for any ***x***. That's like saying for a given ***x*** value, there can't be two points lined up vertically:

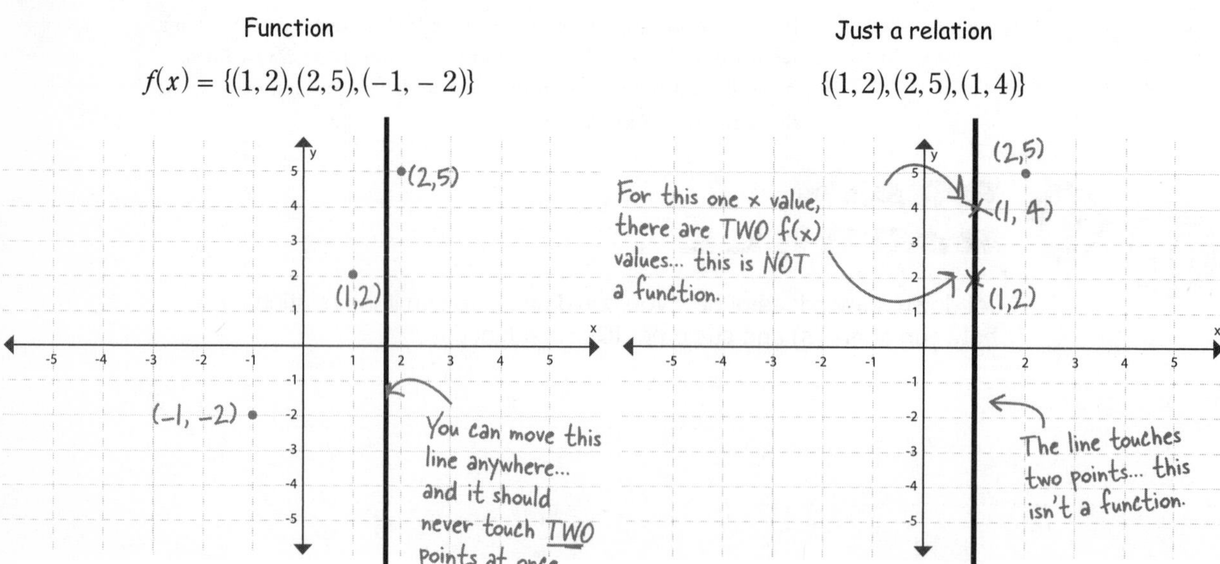

Function
$f(x) = \{(1,2),(2,5),(-1,-2)\}$
(2,5)
(1,2)
(–1, –2)
You can move this line anywhere... and it should never touch TWO points at once.
Just a relation
$\{(1,2),(2,5),(1,4)\}$
(2,5)
(1, 4)
(1,2)
For this one x value, there are TWO f(x) values... this is NOT a function.
The line touches two points... this isn't a function.

Functions pass the vertical line test

This whole "looking at the graph" thing is called the ***vertical line test***. It's based on the fact that if you have two points that are vertically stacked, anywhere on a graph, then that graph is **NOT a function**.

So if a vertical line passes through two points on the graph, it's NOT a function. That's it. To test a graph, you can use a ruler, your eyeball, or run the edge of a piece of paper over the graph to see if it's a function.

Sharpen your pencil

Are the seat assignments a function?

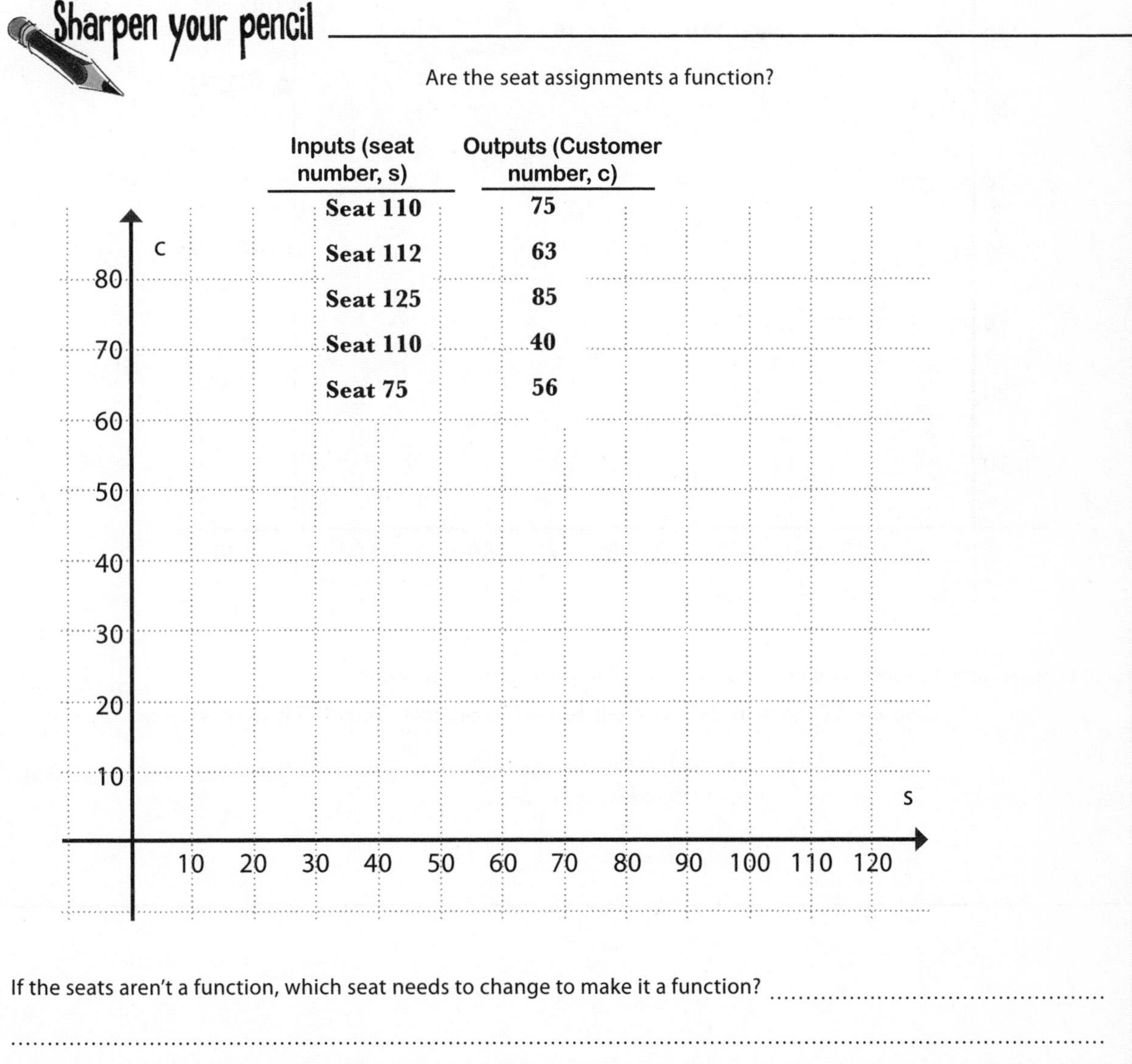

Inputs (seat number, s)	Outputs (Customer number, c)
Seat 110	75
Seat 112	63
Seat 125	85
Seat 110	40
Seat 75	56

If the seats aren't a function, which seat needs to change to make it a function? ....

....

Sharpen your pencil Solution

Are the seat assignments a function?

Inputs (seat number, s)	Outputs (Customer number, c)
Seat 110	75
Seat 112	63
Seat 125	85
Seat 110	40
Seat 75	56

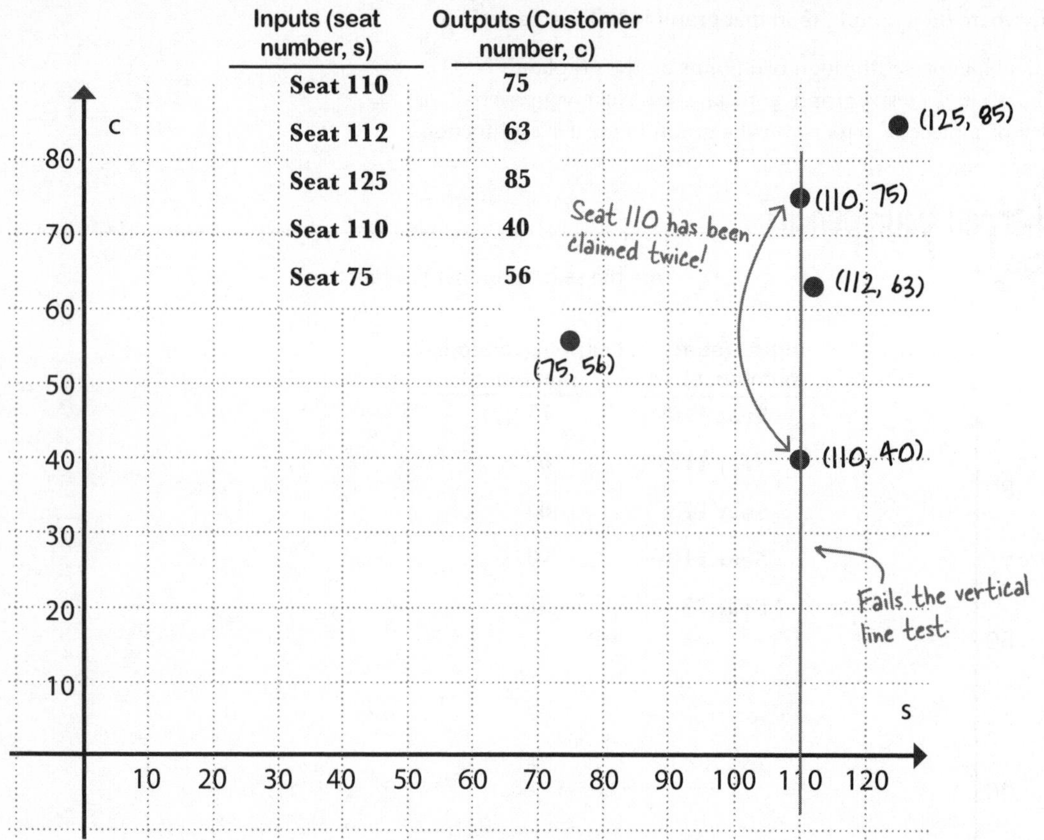

If the seats aren't a function, which seat needs to change to make it a function?

Customer 40 needs to be reassigned to a different seat. Seat 110 is already taken.

This relation didn't pass the vertical line test at s = 110, so that's the s value that need to generate different c values.

BULLET POINTS

- A graph allows you to use the **vertical line test** and see if it's a function.
- A graph shows the **domain** of a function or relation.
- A graph also shows the **range** of a function or relation.
- A graph can provide the **zeroes** of an equation or a function.

there are no Dumb Questions

Q: So, should I always graph a function?

A: It depends upon what you're after. A graph of a function, equation, or relation, is the best way to visually look at the situation and it does give you valuable information, but there are some limits.

You are limited to what you can actually read from the graph. If there are decimals involved, or the scale of your graph is large, it can be difficult to read values.

Creating a graph takes time. If you are in a test situation and you're only after a specific value, then just evaluate the function for your specific value, and move on.

Q: Can you really just treat functions like equations?

A: It's a really good starting point. If a function is represented as an equation, you can solve for the zeroes of the function, and then graph the function.

The only things to watch are that you limit the graph properly (keeping the domain and range in mind) and substitute 0 for ***f(x)***.

Q: Is it always f(x)? Can it be, like, f(t) or something?

A: Sure! It's typically ***f(x)***, but that's not required. The format is really what you're looking for; it can be ***r(q)*** or ***g(t)*** too. ***r(q)*** indicates ***r*** is a function of ***q***, for example. The axes on any graph will need to be adjusted, but other than that, everything else would be the same.

Q: Functions seem like more work than just equations. You've got to keep up with domains and ranges. So why are they so much better?

A: Better is a pretty subjective word. The thing is, functions are just more ***realistic*** in most cases. The real world has limits, and with functions you can easily communicate what those limits are and see them on a graph. That's a helpful thing.

What's going on? During the sound check we had VIPs showing up with free tickets. What's the deal with them?

The second show is starting, and now there are free tickets? How does that change things?

After running back to check the fine print, there was a clause in the contract that you (and Pajama Death) missed!

Apparently, the network can issue 350 free tickets per show to VIPs. It makes the audience look cooler... or so they think.

What does that mean for Pajama Death's income?

The FIRST 350 tickets are always free!

Wow, this complicates things. So now, there are still 1,511 possible seats. But for each show, the first 350 seats are given away, no charge (and no revenue for Pajama Death). Then, the rest of the seats are sold at the normal rate. So we've got something like this:

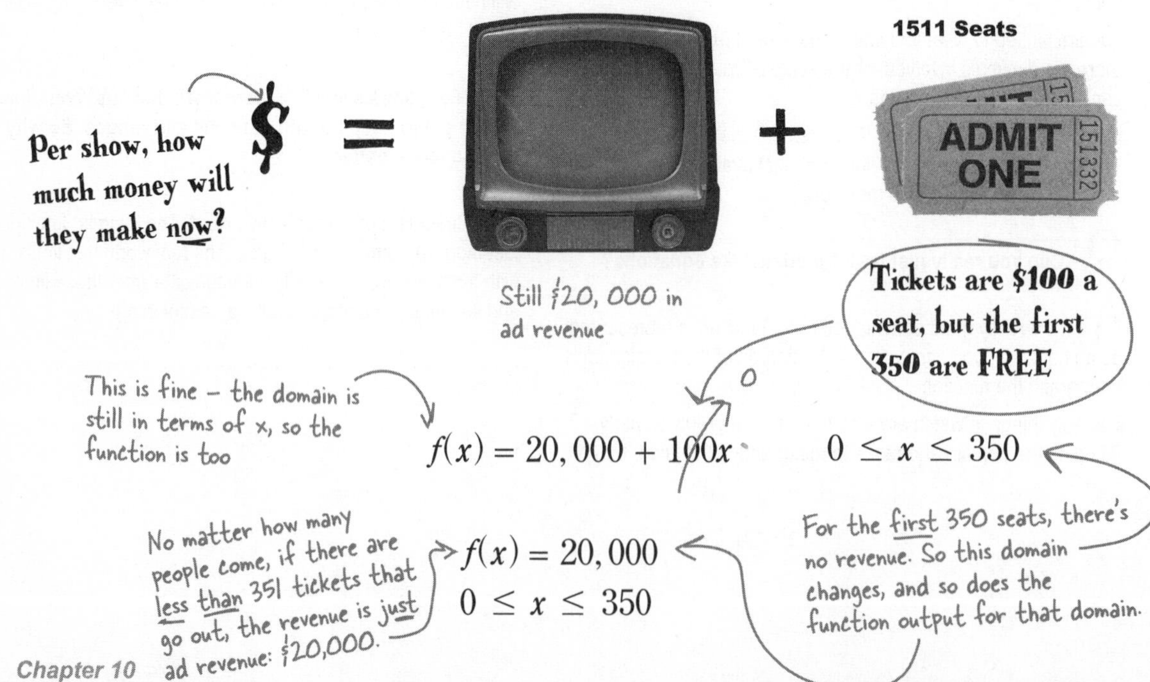

$$f(x) = 20,000 + 100x \qquad 0 \leq x \leq 350$$

$$f(x) = 20,000$$
$$0 \leq x \leq 350$$

But... what about the REST of the tickets?

The tickets that are sold after the VIP's are handled are still $100 a piece. That means the equation is almost the same as before... but only after those first 350 tickets are gone.

So that throws in a new wrinkle. Revenue is $100 per ticket, but only for tickets 351 and up. We've got to account for that:

This is the old equation. → $f(x) = 20,000 + 100x$

It's not 100 * x anymore... We have to pull out the first 350 tickets, since those all go for free now.

$f(x) = 20,000 + 100(x - 350)$

This is the updated equation.

$351 \leq x \leq 1511$

The domain changes, too. This whole equation ONLY works for tickets 351 to 1511.

So now we've got two different functions, each with a separate equation and domain. So how do we show that mathematically?

One function, two parts = real life

These functions can be treated together, as long as you keep track of ***where things are happening***. When you have a situation like this, where the function needs to have different values over different parts of the domain, is called a **piecewise defined function**. We show this as one big function, with different domains, like this:

This bracket means that you have to consider the equations together.

Each domain is written next to the equation the domain applies to.

$$f(x) = \begin{cases} 20,000 & 0 \leq x \leq 350 \\ 20,000 + 100(x - 350) & 351 \leq x \leq 1511 \end{cases}$$

So this function behaves differently based on the input domain.

How do you evaluate a piecewise defined function? You can't figure out Pajama Death's income until you know how to do that...

Use the function piece you NEED

This is another case where it's pretty important to keep in mind what your function represents. In the real world, you have to consider an entire situation, and with piecewise functions, the domain is the key. Since you have different equations applying to different inputs (in this case, x values), you just need to determine which equation applies to the situation you're working on..

To evaluate a piecewise function

What happens if you need to evaluate the function for a number? You just see what domain that number falls in, and use the matching equation to evaluate for that value. It goes like this:

1. **After you have a value to evaluate, determine where the value falls in your function's domain.**

2. **Evaluate only the portion of the function that applies to the domain.**
   Use the correct equation, evaluate, and use or plot your answer. Any other pieces of the function ***don't apply***!

Graph piecewise functions just like "normal" functions!

Graphing is much like it was for regular functions. You just graph the equation that applies, over the domain where it applies. So you'll have a different section of your graph for each piece of the function... and that's okay. That's why the function's called piecewise in the first place.

The conventions for the points at the end of the domain are the same: solid circles and open circles, depending on the type of inequality.

BULLET POINTS

- **Piecewise functions** are just a series of functions grouped together.
- The domains typically **don't overlap**.
- The graphing rules for functions **don't change**.
- To evaluate a piecewise function, just **determine the domain** that applies and **evaluate the portion of the function** that applies.
- Piecewise functions allow you to express that **different things** are happening at **different times**.

Graph Pajama Death's revenue function. Show both sections of the domain in one graph to see the entire picture of what's going on. The new graph will show you all the possible revenue values for each type of ticket sales.

First, graph just this piece of the revenue function.

$$f(x) = \begin{cases} 20{,}000 & 0 \le x \le 350 \\ 20{,}000 + 100(x - 350) & 351 \le x \le 1511 \end{cases}$$

Then graph the second piece of the function.

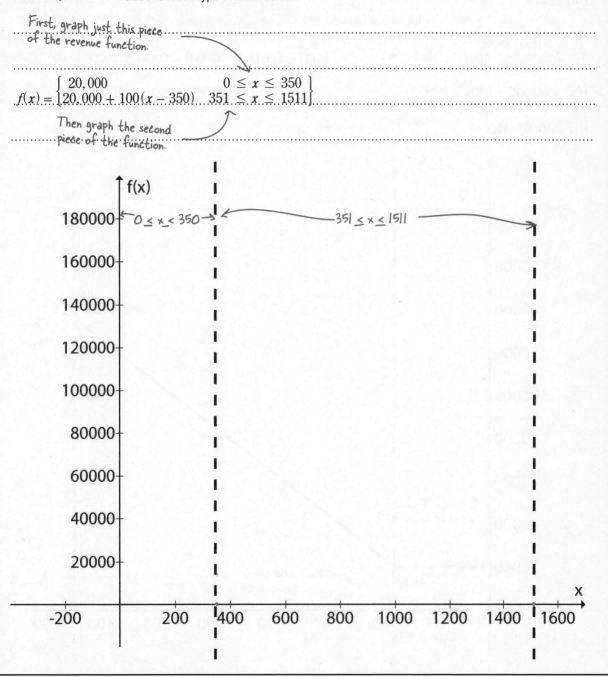

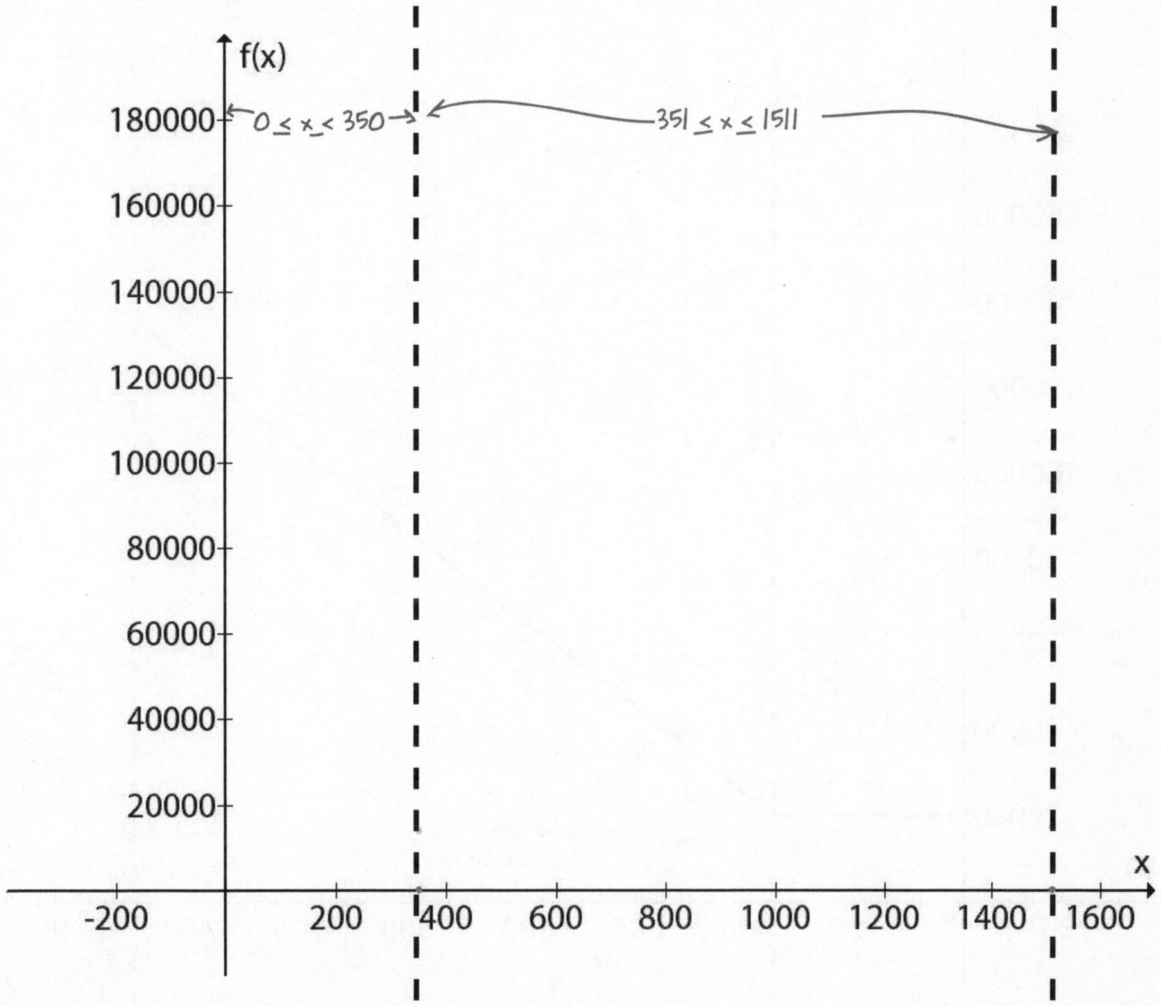

f(x)

180000 ← 0 ≤ x ≤ 350 → ← 351 ≤ x ≤ 1511 →

160000

140000

120000

100000

80000

60000

40000

20000

x

-200 200 400 600 800 1000 1200 1400 1600

Graph Pajama Death's new function. Graph both sections of the domain in one graph to see the entire picture of what's going on. The new graph will show you all the possible revenue values for each type of ticket sales.

This part is just a straight line!

$$f(x) = \begin{cases} 20,000 & 0 \le x \le 350 \\ 20,000 + 100(x - 350) & 351 \le x \le 1511 \end{cases}$$

$f(x) = 20000 + 100x - 35000$

$f(x) = 100x - 15000$

Simplify the expression and get it into an easier format.

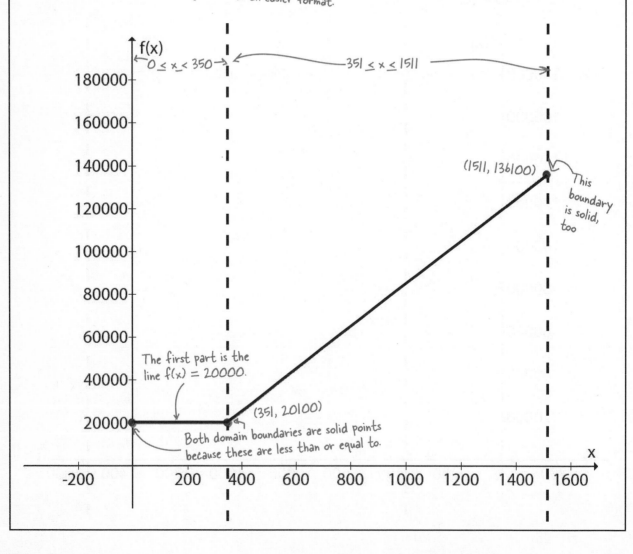

We can use the boundary points to graph this one because it's linear – if it was a parabola we'd need to use the vertex

$f(351) = 100(351) - 15000 = 20100$ (351, 20100)

$f(1511) = 100(1511) - 15000 = 136100$ (1511, 136100)

f(x)
180000
160000
140000
120000
100000
80000
60000
40000
20000
x
-200
200
400
600
800
1000
1200
1400
1600
0 ≤ x ≤ 350
351 ≤ x ≤ 1511
(1511, 136100)
This boundary is solid, too
The first part is the line f(x) = 20000.
(351, 20100)
Both domain boundaries are solid points because these are less than or equal to.

The numbers are in... and?

The season has finished up and it's time to figure out how much Pajama Death cleared. You need to go back and re-figure the first show revenue with the VIP tickets in mind, and then crunch the rest of the season numbers.

REVENUE CONSTRUCTION

Using the graph for the revenue, fill out the chart to find out if Pajama Death made enough money to go shopping!

Show Number	Attendance	Total Revenue	Pajama Death's 5%
1	473		
2	123		
3	789		
4	974		
5	1246		
6	1234		
7	1499		
8	1412		
9	1461		
10	1511		
11	1503		
		Total Pajama Death Revenue	

Is it enough for the new equipment, costing $52,375?

YES **NO**

REVENUE CONSTRUCTION SOLUTION

Using the graph for the revenue, fill out the chart to find out if Pajama Death made enough money to go shopping!

Show Number	Attendance	Total Revenue	Pajama Death's 5%
1	473	32, 300	1,615
2	123	20,000	1000
3	789	63,900	3195
4	974	82,400	4120
5	1246	109,600	5480
6	1234	108,400	5420
7	1499	134,900	6745
8	1412	126,200	6310
9	1461	131,100	6555
10	1511	136,100	6805
11	1503	135,300	6765
			$ 54, 010

Don't worry if the values you read off the graph aren't exactly the same. Since the graph is large, you have to estimate some.

Total Pajama Death Revenue

Is it enough for the new equipment at $52,375?

NO

Pajama Death's show was a hit!

The show got way more popular and Pajama Death made enough money to buy everything they wanted.

Now they're going to make that new album and play huge concerts, all thanks to you! Now that you helped them understand how much money they can make, they might take this music thing seriously!

Long Exercise

Look at the piecewise function and graph it to determine the range.

$$f(x) = \begin{cases} 2x^2 + 8x - 1 & -5 \le x < 0 \\ x & 0 < x < 3 \\ -\dfrac{x}{3} + 5 & 3 < x \le 8 \end{cases}$$

Use this space for your work:

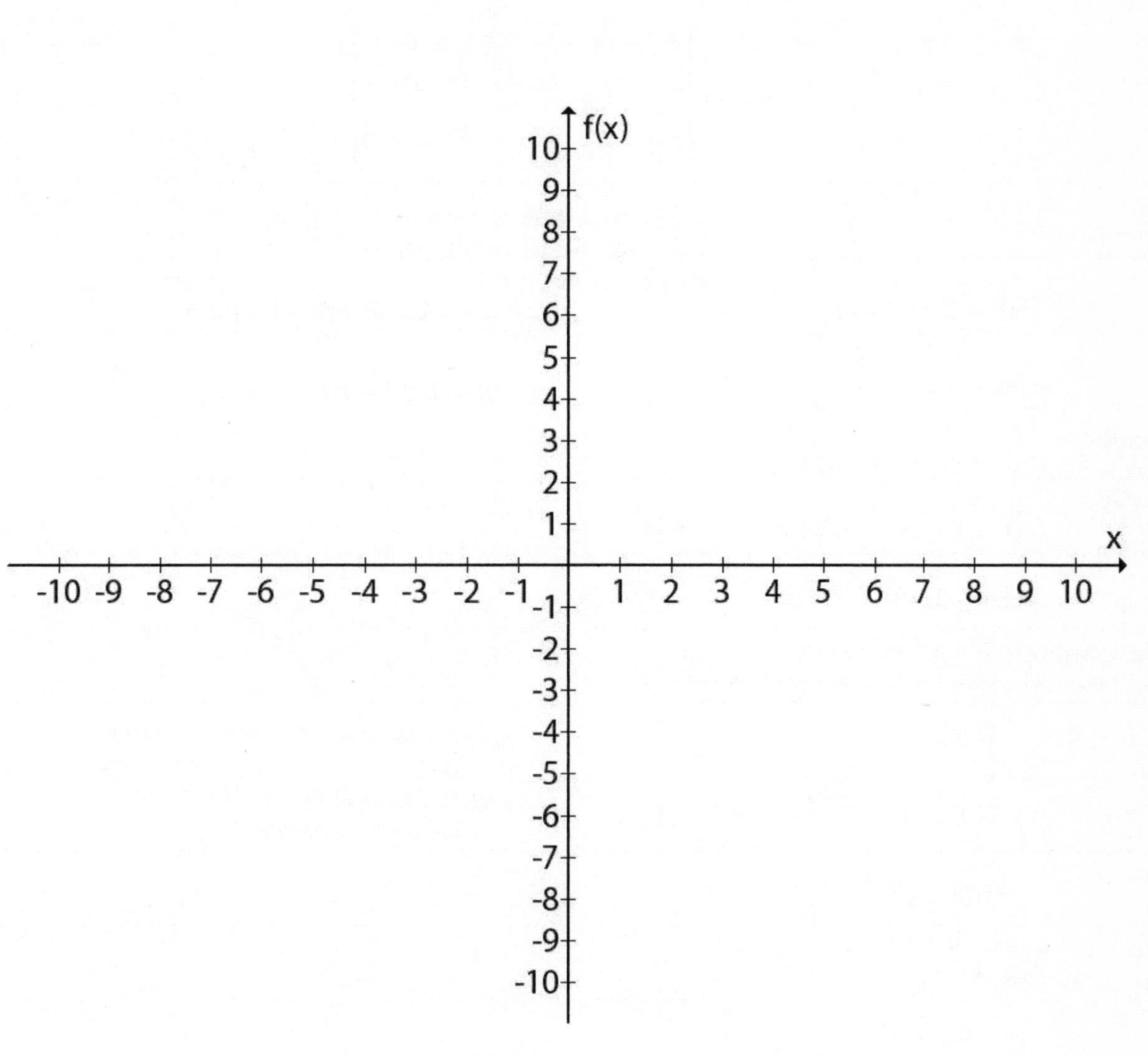

Range: ....

Look at the piecewise function and graph it to determine the range.

$$f(x) = \begin{cases} 2x^2 + 8x - 1 & -5 \leq x < 0 \\ x & 0 < x < 3 \\ -\frac{x}{3} + 5 & 3 < x \leq 8 \end{cases}$$

Since the boundaries of the domain are on the right and left of the vertex, we'll use them for points

$f(x) = 2x^2 + 8x - 1$

$f(-5) = 2(-5)^2 + 8(-5) - 1 = 9$ ← (-5, 9)

vertex $x = \frac{-b}{2a} = \frac{-8}{2(2)} = -2$

$f(0) = 2(0)^2 + 8(0) - 1 = -1$ (0, -1)

Substitute back in for the f(x)

$f(x) = 2(-2)^2 + 8(-2) - 1$

$f(x) = 8 - 16 - 1 = -9$

vertex (-2, -9)

The second piece of the function is a linear equation, so just using the boundary points is fine

$f(x) = x$

Now, just plot these points. But be careful about what type of points you're plotting... solid for "or equal to" inequalities, and open for regular inequalities

$f(0) = 0$ ← (0,0)

$f(3) = 3$ ← (3,3)

Same deal with the last piece of the function, it's linear.

$f(x) = \frac{-x}{3} + 5$

$f(3) = \frac{-3}{3} + 5 = 4$ (3,4)

$f(8) = \frac{-8}{3} + 5 = \frac{7}{3}$ (8, $2\frac{1}{3}$)

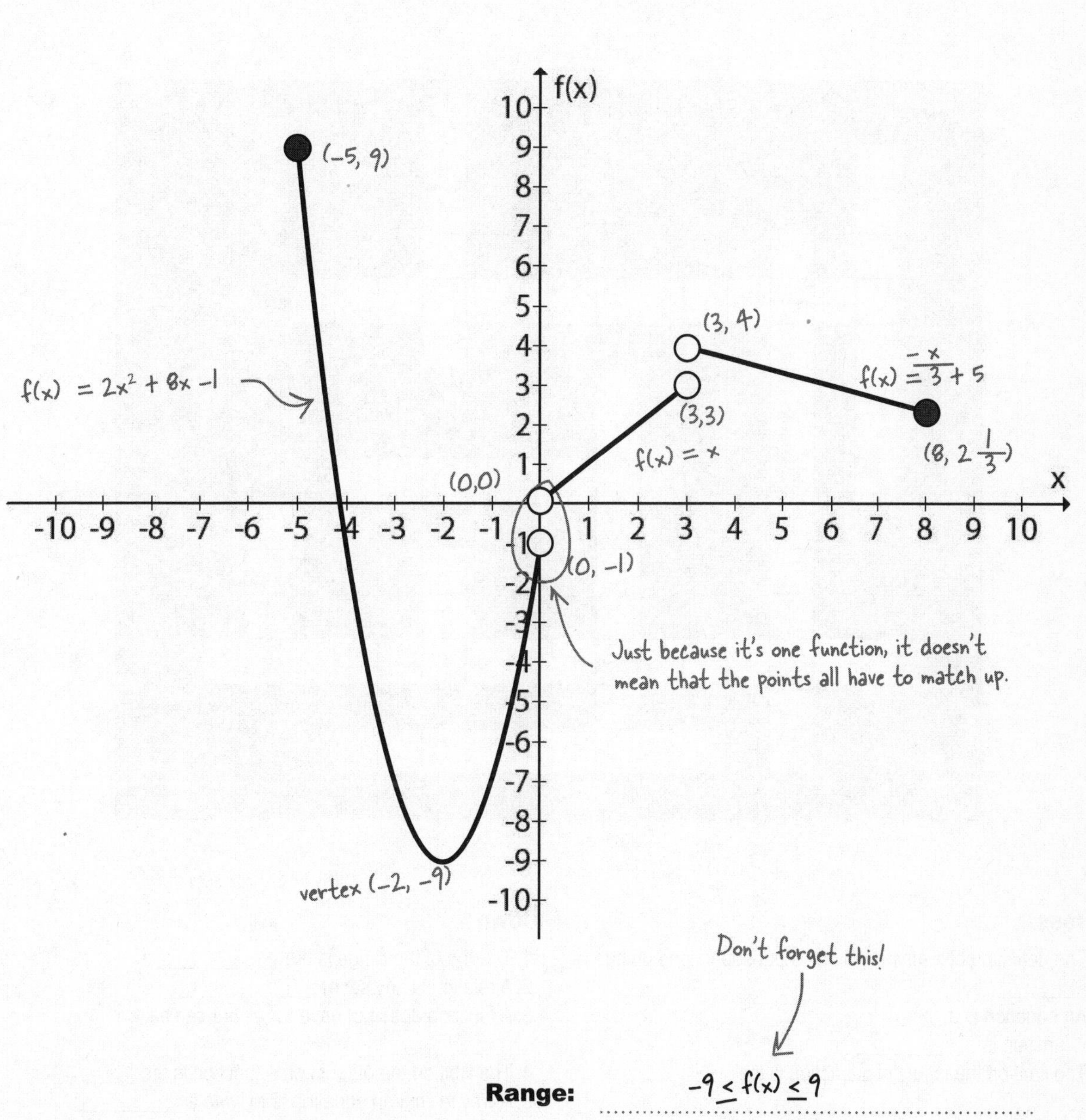

Range: $-9 \leq f(x) \leq 9$

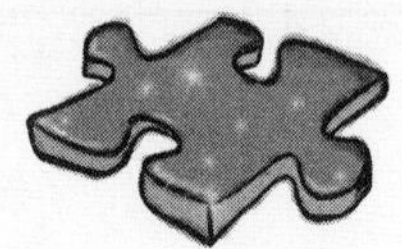

Functioncross

Get your right brain functioning. It's your standard crossword; all of the solution words are from this chapter.

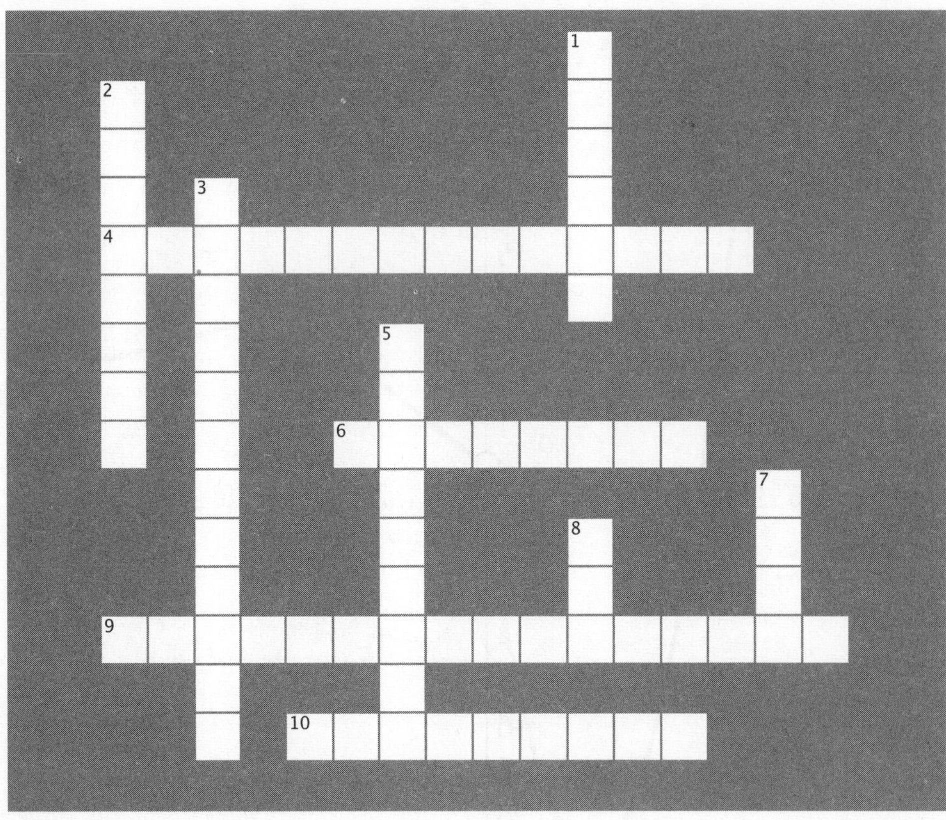

Across

6. The definition of a function is one input equals one unique _________
7. An equation is a ___________
8. A function is a __________
9. The limit on the inputs of a function is the

Down

1. The top of the range is the ____________
2. A relation is any set of _____ _______
3. A function does not have to be, but can be, an ____________
4. The limit on the outputs on a function is the ________
5. A way to limit an equation is to write a __________

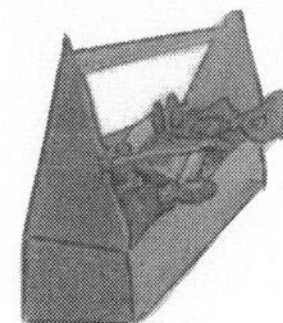

Tools for your Algebra Toolbox

This chapter was all about functions.

CHAPTER 10

Find the domain of a function.

When presented with an equation that's also a function, you can figure out the domain by understanding the equation. Watch for division by 0, or cases where a negative root could happen. If it's a line, it goes on forever; if it's a parabola it only covers part of the Cartesian plane and that can limit the domain. Graphing may help.

Find the range of a function/graph the function.

The best way to go about finding the range is to graph. There are some cases (like linear equations) where you can easily tell what the range will be – but typically graphing is the way to go. Once you know the domain, it's just a matter of cutting off the graph and interpreting what's left.

Solve for the zeroes of the function.

You already know how to do that! Simply set f(x)=0 and solve the remaining expression for x. It's exactly the same rules as when working with equations; inverse operations, FOIL, the quadratic equation, the whole thing. The solution is the value for x that makes the function true.

If you have a graph and you can read the value where $f(x) = 0$, you can even read it straight off the graph!

Evaluate the function.

This just means that you are given an input and they are looking for the output. The hardest part about this question is understanding what it's asking!

Working with piecewise functions.

Piecewise functions are just different functions grouped together that apply over different parts of the domain. The important thing to remember is to work with each piece separately.

Functioncross Solution

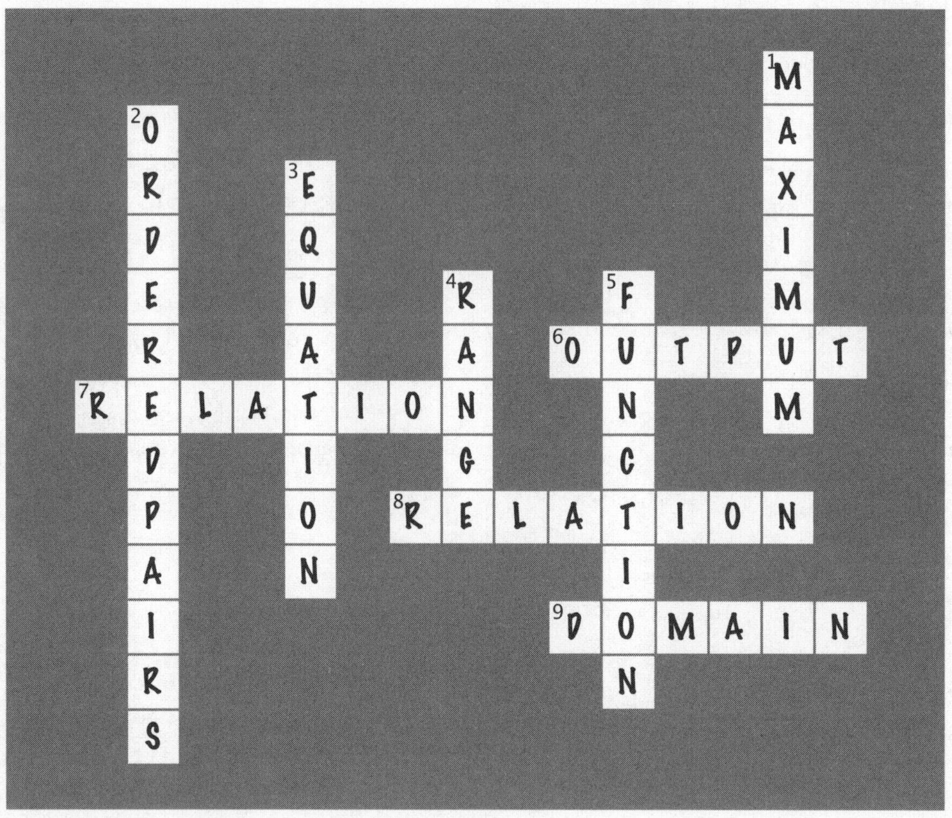

11 real-world algebra

Solve the world's problems

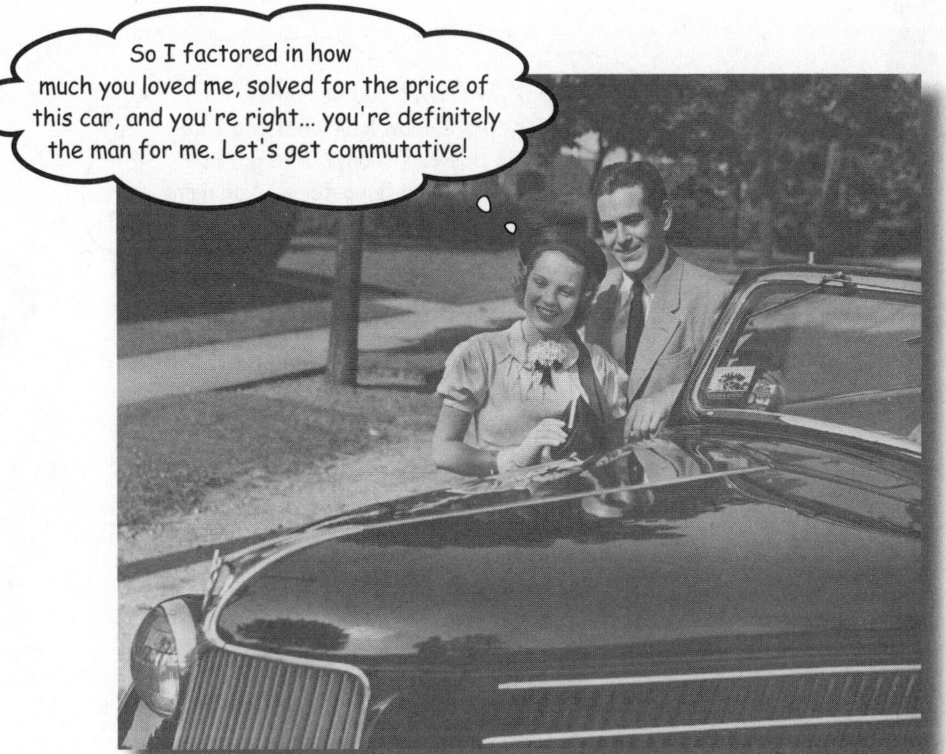

The world's got big problems... you've got big answers.

Hundreds of pages of math, and what do you really have? A bunch of ***x***'s and ***y***'s, **a**'s and **b**'s? Nope... you've got **skills** to **solve for an unknown**, even in the most difficult situations. So what's that good for? Well, in this chapter, it's all about the **real world**: you're going to use your Algebra skills to ***solve some real problems***. By the time you're done, you'll have won friends, influenced people, and saved yourself a whole bucket full of cash. Interested? Let's get started.

I've got some money saved up, and I'm going to get a great job when I graduate in 9 months. I definitely need some wheels... I just need a good loan and some financing terms I can handle.

Financing? My friend, you've come to the right place. Now look what can I do to get you in this car, today? I've got no-interest, low-interest, long-term... you name it, we can make it happen. Budget? No worries, we'll get you way more car than you thought you could afford.

Max is a few months away from a new job and needs a sweet car to go with his new gig.

Wow, can you say used-car salesman? Better watch out for this guy...

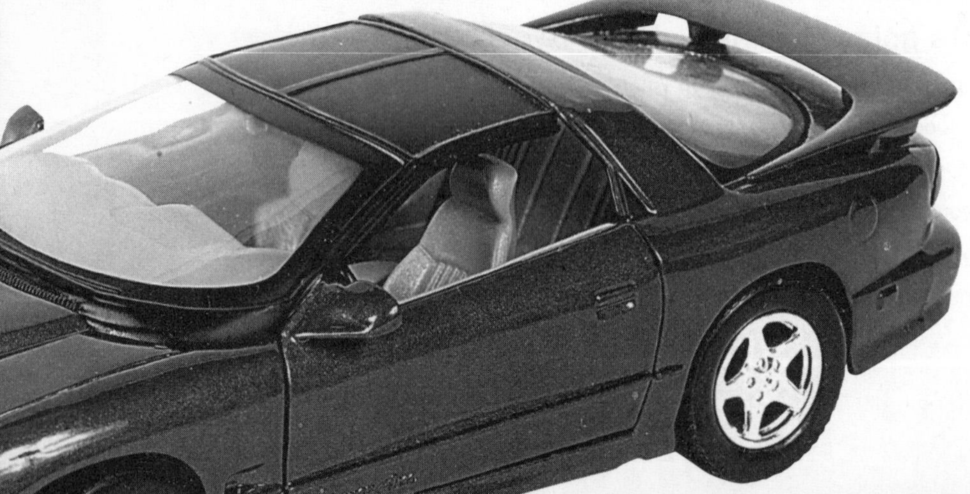

The object of Max's auto-affection, a $25,000 sports car.

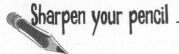

Sharpen your pencil

Max has a lot to think about to get into the car of his dreams. Here are just a few of the pieces of paperwork Max has to consider... write down a few details to get Max started.

What are the algebraic unknowns you'll need to deal with? ....

....

What kinds of things are going to affect the costs of Max's car? (What are things going to be "in terms of"?)

....

....

Earl's Autos

VEHICLE DESCRIPTION

FORMULAS

1992 HF 5.0L SALOON
4-PASSENGER SPECIALTY
5.0L H1 HF V- ENGINE
AUTO OVERDRIVE TRANSMISSION

VIN 1HFACALG4UISCOOL

EXTERIOR
BLUE AD SHINY
INTERIOR
GRAY LEATHER

...RE INCLUDED AT NO EXTRA CHARGE
...VEHICLE SHOWN AT RIGHT:

- TINTED GLASS
- ELECTRONIC AM/FM STEREO RADIO WITH CLOCK
- LEATHER-WRAPPED STRG.WHEEL
- POWER WINDOWS
- INTERVAL WINDSHIELD WIPERS
- FULL INSTRUMENTATION
  - TACHOMETER
  - TEMPERATURE GUAGE

PRICE INFORMATION

STANDARD VEHICLE PRICE	$18540.00
OPTIONAL EQUIPMENT	
PREFERRED EQUIPMENT PKG.4560	
.SPEED CONTROL	
.ELEC AM/FM RADIO W/CASS/CLOCK	
AUTO OVERDRIVE TRANSMISSION	1641.00
P233/H323F4778 HFA PERFORMANCE	1190.00
CONVENIENCE GROUP	NC
FRONT LICENSE PLATE BRACKET	198.00
8-WAY POWER HEATED DRIVER'S SEAT	NC
LIMITED EDITION	366.00
OPTIONAL TRACTION-LOCK BUTTON	1700.00
POLAR AIR CONDITIONING	NC
LEATHER STAIN REPELLANT SEAT SPRAY	1634.00
GRAPHIC EQUALIZER	NC
	278.00
TOTAL VEHICLE & OPTIONS	28230.00
DESTINATION AND DELIVERY	440.00
TOTAL BEFORE DISCOUNTS	28670.00.

$3670.00

$25000.00

Shady Dealings Insurance Broker

CAR INSURANCE

Full Name: John...
Address: Kappa...
City, State, Zip:
Email Address:
Home Phone:
Length of Ti...
yrs.
Rent or Ow...
Social Sec...

1st National Savings

AUTO LOAN APPLICATION

Vehicle Make: Formulas

1st National Savings
1234 SQL St. PO Box 1000
Dataville DV 26849

CHECKING ACCOUNT STATEMENT

PAGE: 1 of 1

...ent Period	Account No.
...to 2009-03-31	00004-323-3477-8

...ITHDRAWALS	DEPOSITS	BALANCE
		$7,267.00
...00		$7,164.00
		$7,104.00
		$7,074.00
	$500.00	$7,574.00
		$7,374.00
		$7,264.00
	$36.00	$7,300.00
		$7,300.00

9 months to go! Yippee!

Calendar 2009

Sharpen your pencil Solution

Your job was to write down some of the things Max has to think about when it comes to buying a car. Here's what we wrote down...

What are the algebraic unknowns you'll need to deal with? We're going to have to deal with the cost of the car, car insurance, car loan rates, how much money Max can put down, and how long he wants to take a loan for.

What kinds of things are going to control the costs? (What are things going to be "in terms of"?)
Some of the costs will depend on how much the car is (insurance, how much he pays in interest on the loan, etc.), and some cost will depend on time, like how long he takes his loan for, and how fast he can pay things off.

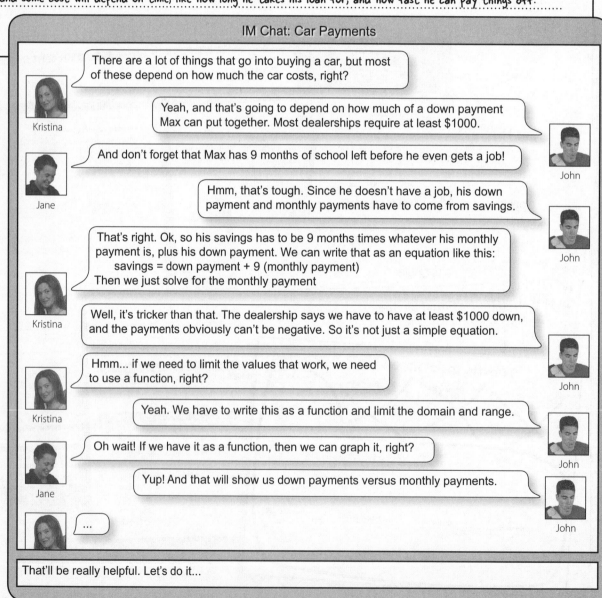

IM Chat: Car Payments

Kristina: There are a lot of things that go into buying a car, but most of these depend on how much the car costs, right?

John: Yeah, and that's going to depend on how much of a down payment Max can put together. Most dealerships require at least $1000.

Jane: And don't forget that Max has 9 months of school left before he even gets a job!

John: Hmm, that's tough. Since he doesn't have a job, his down payment and monthly payments have to come from savings.

Kristina: That's right. Ok, so his savings has to be 9 months times whatever his monthly payment is, plus his down payment. We can write that as an equation like this:
savings = down payment + 9 (monthly payment)
Then we just solve for the monthly payment

John: Well, it's tricker than that. The dealership says we have to have at least $1000 down, and the payments obviously can't be negative. So it's not just a simple equation.

Kristina: Hmm... if we need to limit the values that work, we need to use a function, right?

John: Yeah. We have to write this as a function and limit the domain and range.

Jane: Oh wait! If we have it as a function, then we can graph it, right?

John: Yup! And that will show us down payments versus monthly payments.

Kristina: ...

That'll be really helpful. Let's do it...

Sharpen your pencil

Write Max's equation as a function, find the domain and the range, and graph it. Suppose Max has $7300 saved. How much can he afford per month for a car?

$7,300 → *savings balance* = *down payment* + 9(*montly pament*)

Call this d(m). (down payment)

Call this m. (montly pament)

Domain: ....

Range: ....

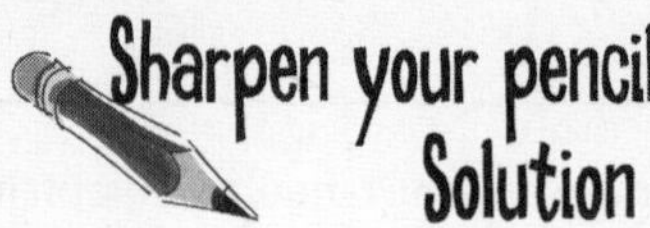

Write Max's equation as a function, find the domain and the range, and graph it. Using that information, how much can he afford per month for this car?

$7,300 — Call this d(m). — Call this m.

savings balance = down payment + 9(montly pament)

7300 = d(m) + 9(m) ← We want d(m) in terms of months, so let's isolate d(m) and think in terms of the down payment.

- 7300 + 7300 - d(m) = 9(m) -7300 ← You could also subtract 9m from both sides to get d(m) by itself. Either approach works just fine... it's up to you.

-1(- d(m)) = (-7300 + 9(m) -1

Multiply by -1 to make d(m) positive.

d(m) = 7300 - 9m

Domain: $0 \leq m \leq 700$ ← The monthly payment has to be 0 or more.

Range: $1000 \leq d(m) \leq 7300$ ← The down payment has to be at least $1000, and less than Max's total savings.

For the upper boundary, the most that Max will have available is $ 7300- $1000, the minimum down payment. So that's $6300. Take that and divide over 9 months, and his maximum monthly payment is $700/month.

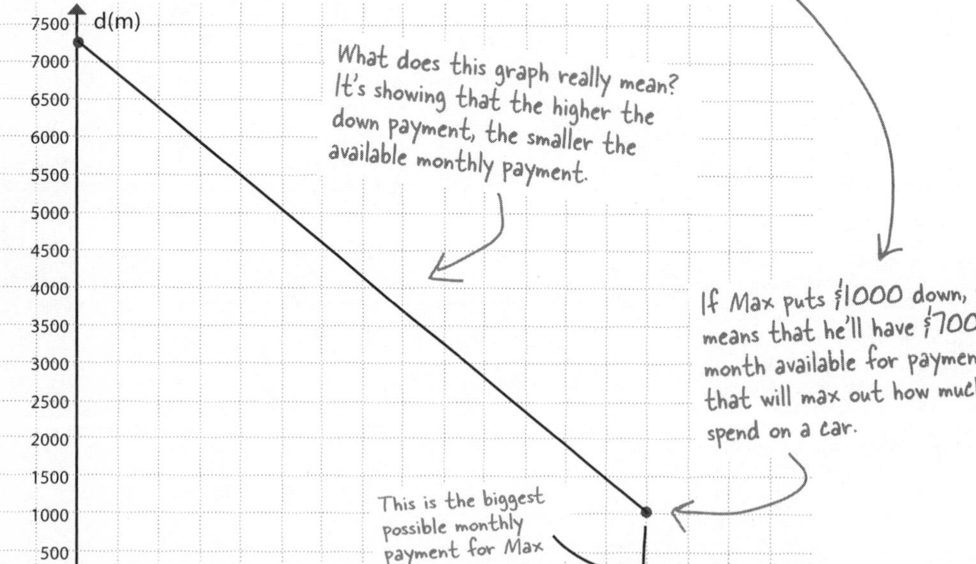

If Max puts $1000 down, that means that he'll have $700 a month available for payments. And that will max out how much he can spend on a car.

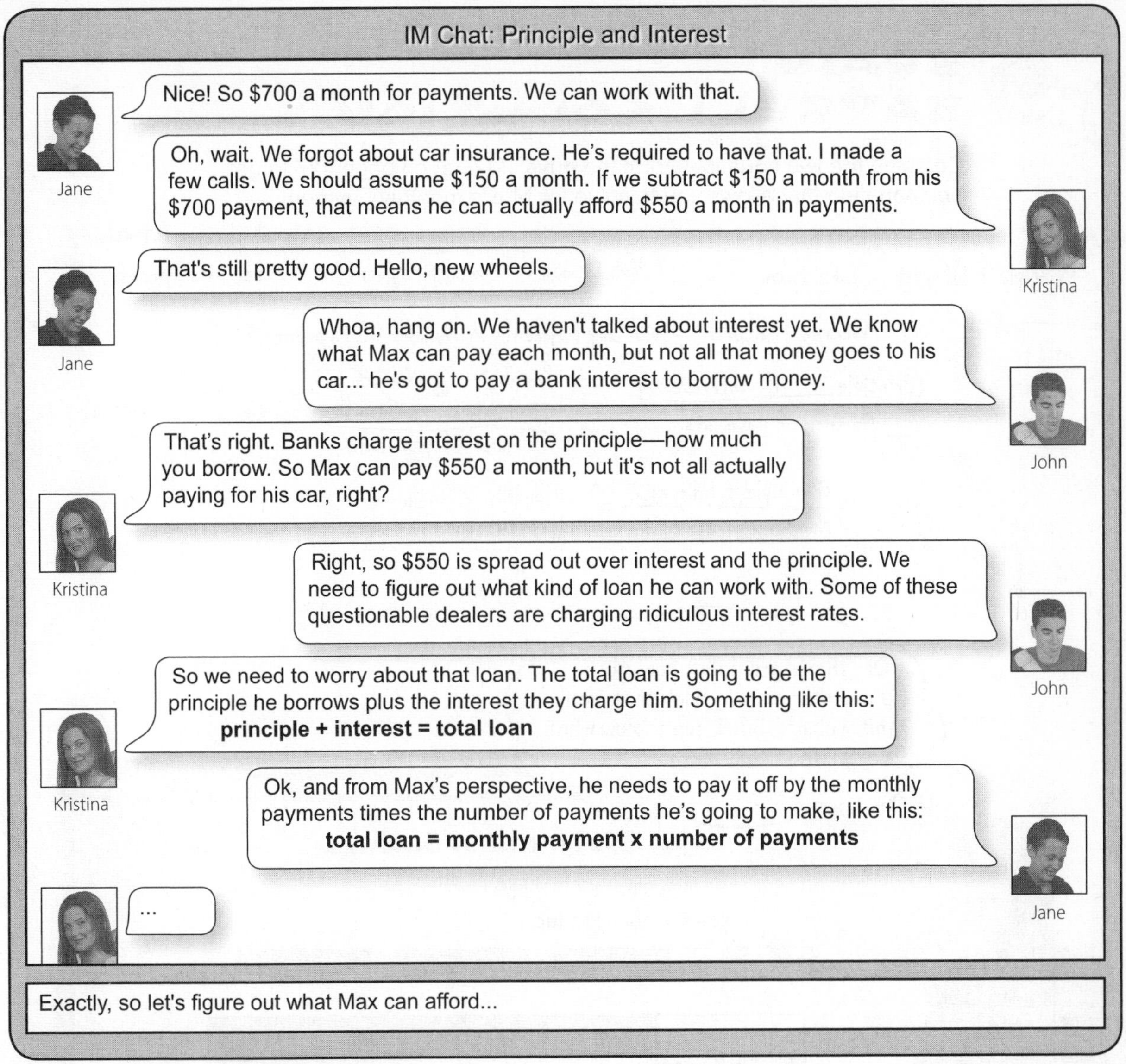

IM Chat: Principle and Interest

Jane: Nice! So $700 a month for payments. We can work with that.

Kristina: Oh, wait. We forgot about car insurance. He's required to have that. I made a few calls. We should assume $150 a month. If we subtract $150 a month from his $700 payment, that means he can actually afford $550 a month in payments.

Jane: That's still pretty good. Hello, new wheels.

John: Whoa, hang on. We haven't talked about interest yet. We know what Max can pay each month, but not all that money goes to his car... he's got to pay a bank interest to borrow money.

Kristina: That's right. Banks charge interest on the principle—how much you borrow. So Max can pay $550 a month, but it's not all actually paying for his car, right?

John: Right, so $550 is spread out over interest and the principle. We need to figure out what kind of loan he can work with. Some of these questionable dealers are charging ridiculous interest rates.

Kristina: So we need to worry about that loan. The total loan is going to be the principle he borrows plus the interest they charge him. Something like this:
principle + interest = total loan

Jane: Ok, and from Max's perspective, he needs to pay it off by the monthly payments times the number of payments he's going to make, like this:
total loan = monthly payment x number of payments

Kristina: ...

Exactly, so let's figure out what Max can afford...

BRAIN BARBELL

Combine the two equations from the guys' IM chat to get *one* equation that allows us to solve for Max's monthly payment.

....

....

....

BRAIN BARBELL SOLUTION

Combine the two equations form the guys' IM chat on page 427 to get *one* equation that allows you to solve for Max's monthly payment.

Equation 1.

Principle + Interest = Total Loan

Equation 2

Total Loan = Monthly Payments * Number of Payments

Principle + Interest = Monthly Payments * Number of Payments

We can set the left side of equation 1 equal to the right side of equation 2.

$$\frac{(\text{Principle} + \text{Interest})}{\text{Number of payments}} = \frac{\text{Monthly payment} \cdot \text{Number of payments}}{\text{Number of payments}}$$

Now we just need to isolate the monthly payment.

$$\frac{(\text{Principle} + \text{Interest})}{\text{Number of payments}} = \text{Monthly payment}$$

This is what we want... Max's monthly payment.

Ok, that's great. But how much is my interest? My bank sent me to this website, but I don't know what any of this really means...

1st National Savings

1st National Bank Home > 1st NB Rates > Consumer Loan Rates

New Car Loan Products

New Vehicles | Used Vehicles | Hybrid Vehicles | Leisure Vehicles

	3 Years	4 years	5 years
Term	3.0%	3.5%	4.0%

$$\frac{(\text{Principle} + \text{Interest})}{\text{Number of payments}} = \text{Monthly payment}$$

\$24,000

Determined by the term of the loan Max chooses

\$550 max

Max's dream car costs \$25,000.

We can fill in the things we know and then solve for the rest. We know that Max can afford \$550 in a monthly payment, and that the principle of the car he wants is \$24,000. With interest rates and terms, we should be able to fill in the rest of that equation, too.

Remember, the car costs \$25,000, but Max is putting \$1000 down. So that leaves \$24,000 to borrow.

Calculate interest from your interest rate and the principle amount you're borrowing

We know what terms the bank offers: 3 years (36 months), 4 years (48 months), and 5 years (60 months). And we know the interest rates from each of those terms. What we need for our equation, though, is the amount of interest. So how can we get that?

Well, you can Google "simple interest," but we've done that for you. Here's an equation to give you the interest based on a term and principle:

$$I_{simp} = (r \cdot B_0)n$$

I means total interest.

Interest rate, as a decimal

If you look up a simple interest equation, you will find an equation like this.

The number of payments (in YEARS, not months)

This means simple interest – there's also compound interest, but you don't need to worry about that here.

Initial amount borrowed, the principle

The bank offers three different terms, and the interest will be different for each term. So we need to figure out the total interest for each term, and then plug that back into our original equation to get Max's monthly payment:

$$I_{simp} = (r \cdot B_0)n$$

$$\frac{\text{(Principle+ Interest)}}{\text{Number of payments}} = \text{Monthly payment}$$

Sharpen your pencil

Figure out the interest for the first option: 3 years at 3.0% interest.

$$I_{simp} = (r \cdot B_0)n$$

$$I_{simp} = (\underline{\quad\quad} \cdot 24000)\underline{\quad\quad}$$

Watch that your numbers are in the right form (decimal form of the percent).

....

....

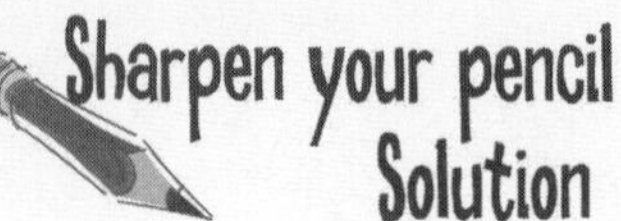

Figure out the interest for the first option: 3 years at 3.0% interest.

$$I_{simp} = (r \cdot B_0)n$$

$$I_{simp} = (0.03 \cdot 24000)\ 3$$

$$I_{simp} = (720)3$$

$$I_{simp} = 2160$$

3.0% is 0.03 in decimal form.

Remember, this term is in YEARS, not months.

25,000 – 1000 down payment

That means that the total interest for the 3-year term will be $2,160.

Now SUBSTITUTE to figure out Max's monthly payment

Now that we know the total amount of interest in the first option, we can figure out the monthly payment for Max's dream car on a 3-year term:

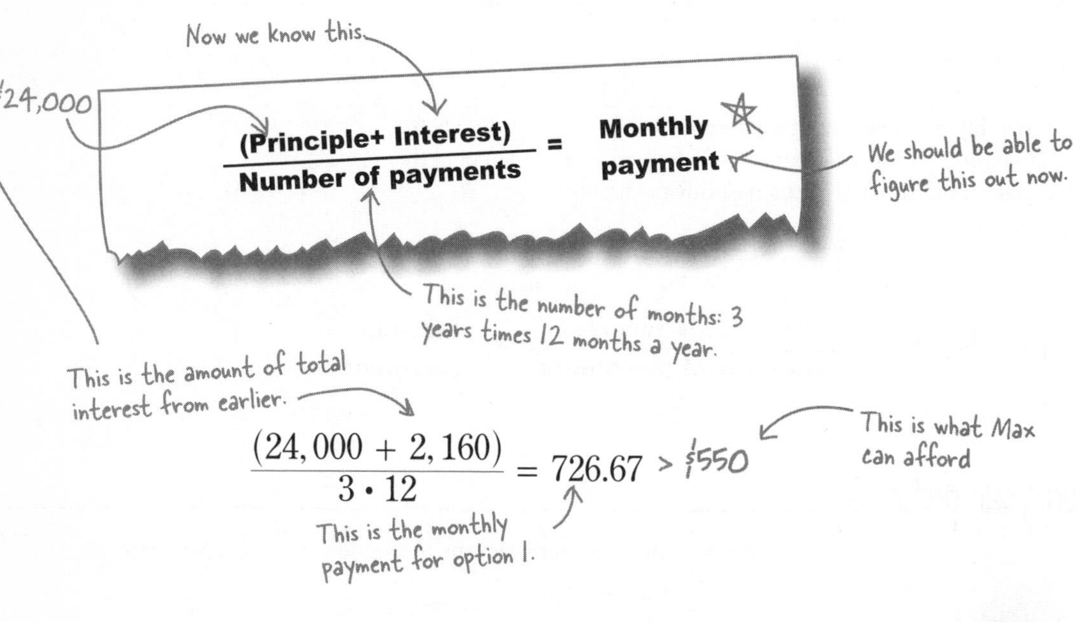

$$\frac{(24,000 + 2,160)}{3 \cdot 12} = 726.67 > \$550$$

This is what Max can afford

This is the monthly payment for option 1.

I can't afford that! What about the longer terms? I really want that car!

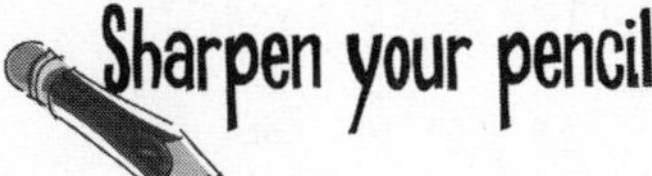

Figure out if Max can afford either of the other two loan options, the 4-year term or the 5-year term.

Option #2: 4 years at 3.5% interest

Option #3: 5 years at 4.0% interest

Your job was to figure out if Max can afford a 4-year or 5-year loan term.

Option #2: 4 years at 3.5% interest

$I_{simp} = (r \cdot B_0)n$ — Watch the decimal here.

$I_{simp} = (0.035 \cdot 24000)4$

$I_{simp} = (840)4$

$I_{simp} = 3360$ — This is the total interest to be paid over the life of the loan

Substitution back into the equation that you created earlier

$$\frac{(24{,}000 + 3{,}360)}{4 \cdot 12} = 570$$

$\$570 > \$550$

This monthly payment is still too much for Max. Here's hoping that the next one works!

Option #3: 5 years at 4.0% interest

$I_{simp} = (r \cdot B_0)n$

$I_{simp} = (0.04 \cdot 24000)5$

$I_{simp} = (960)5$

$I_{simp} = 4800$

Max will end up owing a total of 28,800... the principle plus $4400 in interest.

$$\frac{(24{,}000 + 4{,}800)}{5 \cdot 12} = 480$$

$\$480 < \$550$

Sweet! It works! Max can afford his car if he takes the longest loan. He'll even have $70 left over each month.

BRAIN POWER

What is the ***real*** price of the car when you include interest for the different loans? What if Max could make a larger down payment? What if he waited until closer to graduation so he could pay more per month when he gets his job?

there are no Dumb Questions

Q: Why is the interest equation so complicated?

A: The equation that we're using for interest is a standard equation for figuring out simple interest. It's actually not that complicated, although the unusual letters and terms do make it look a bit trickier at first glance.

Q: How does that interest equation really work?

A: This equation: $I_{simp} = (r \cdot B_0)n$ is actually fairly simple. It states that the total interest of the loan is the interest rate times the amount borrowed, multiplied by the time the money is borrowed over.

The lesson here? Two things really drive up the cost of a loan: how much your interest rate is, and how long you borrow the money for.

Q: When I figured out the interest, it was more for the five year loan, but the monthly payment was lower. Why is that?

A: With the way the loan works, the time factor is a bigger influence than the interest rate. Adding a year onto the loan adds 12 more payments. So there's a lot more interest, but it's spread out over 12 extra months. The end result is a smaller monthly payment.

It's worth noting in the long term, a longer loan will make the car cost quite a bit more. A shorter loan would mean less interest and would cost Max less overall. Shorter is better, if you can afford it.

BULLET POINTS

- In the real world, equations often need to be **limited** and turned into **functions**.
- Calculating interest is just solving an equation for an unknown.
- The interest equation states that the **length of a loan** and the **interest rate** affect the monthly payment.
- Car insurance is a **constant**, not a variable. It's a fixed amount.

I can afford the five year loan. That's awesome! Now I can buy the car!

Max doesn't own that car just yet...

Max is ready to go. He can afford a $1,000 down payment, and the monthly payments on a $24,000 loan if he pays the loan back over five years at an interest rate of 4%. He can even still cover insurance!

But now there's something else to consider...

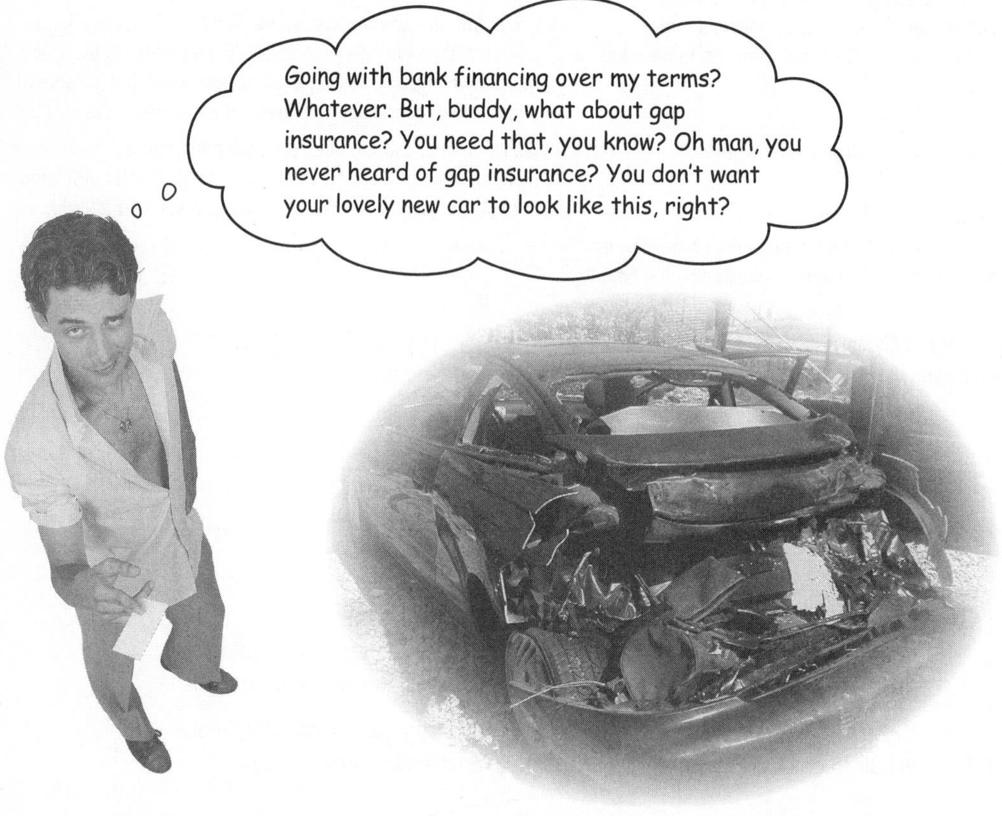

Going with bank financing over my terms? Whatever. But, buddy, what about gap insurance? You need that, you know? Oh man, you never heard of gap insurance? You don't want your lovely new car to look like this, right?

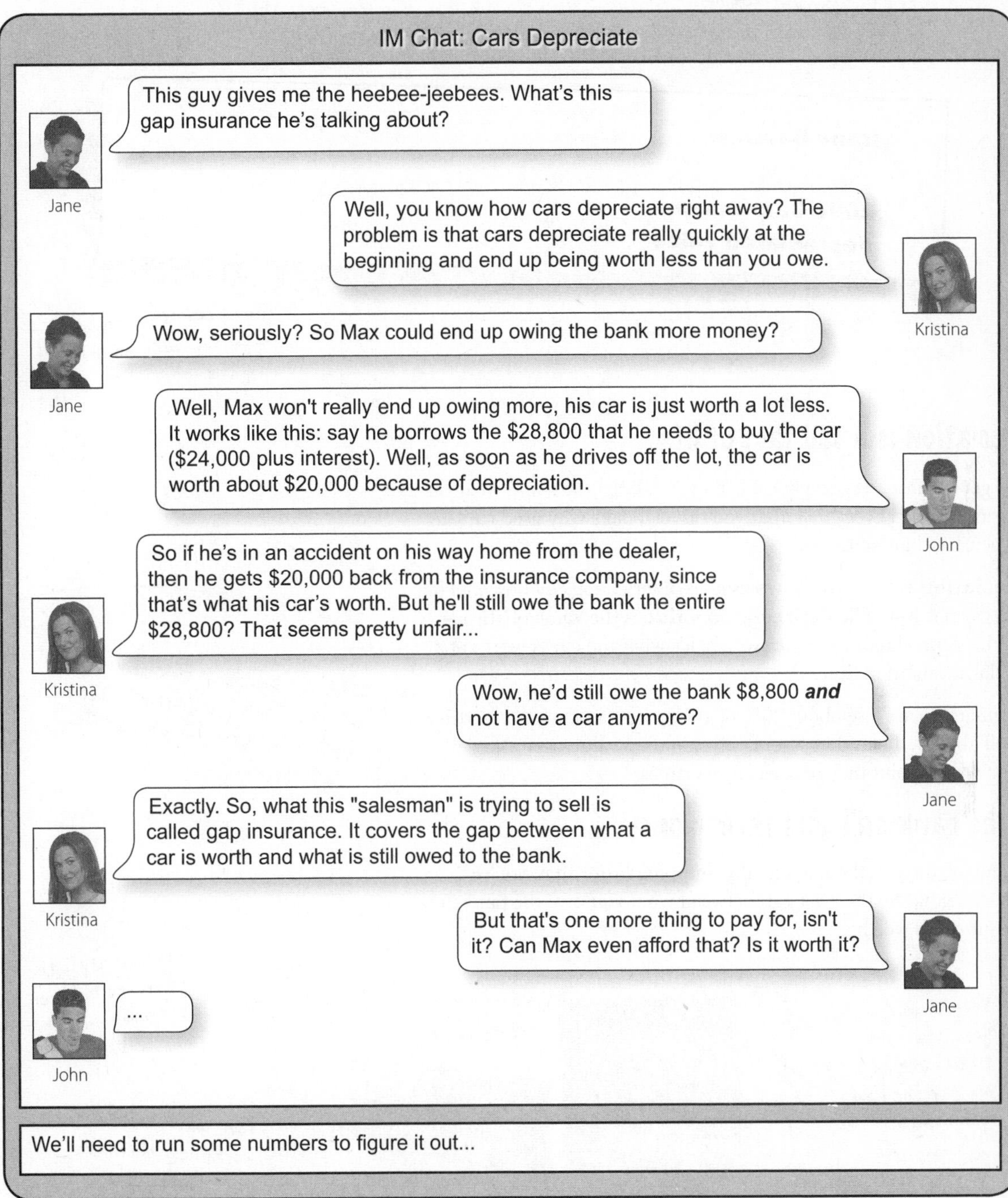
IM Chat: Cars Depreciate
This guy gives me the heebee-jeebees. What's this gap insurance he's talking about?
Jane
Well, you know how cars depreciate right away? The problem is that cars depreciate really quickly at the beginning and end up being worth less than you owe.
Kristina
Wow, seriously? So Max could end up owing the bank more money?
Jane
Well, Max won't really end up owing more, his car is just worth a lot less. It works like this: say he borrows the $28,800 that he needs to buy the car ($24,000 plus interest). Well, as soon as he drives off the lot, the car is worth about $20,000 because of depreciation.
John
So if he's in an accident on his way home from the dealer, then he gets $20,000 back from the insurance company, since that's what his car's worth. But he'll still owe the bank the entire $28,800? That seems pretty unfair...
Kristina
Wow, he'd still owe the bank $8,800 ***and*** not have a car anymore?
Jane
Exactly. So, what this "salesman" is trying to sell is called gap insurance. It covers the gap between what a car is worth and what is still owed to the bank.
Kristina
But that's one more thing to pay for, isn't it? Can Max even afford that? Is it worth it?
Jane
...
John
We'll need to run some numbers to figure it out...

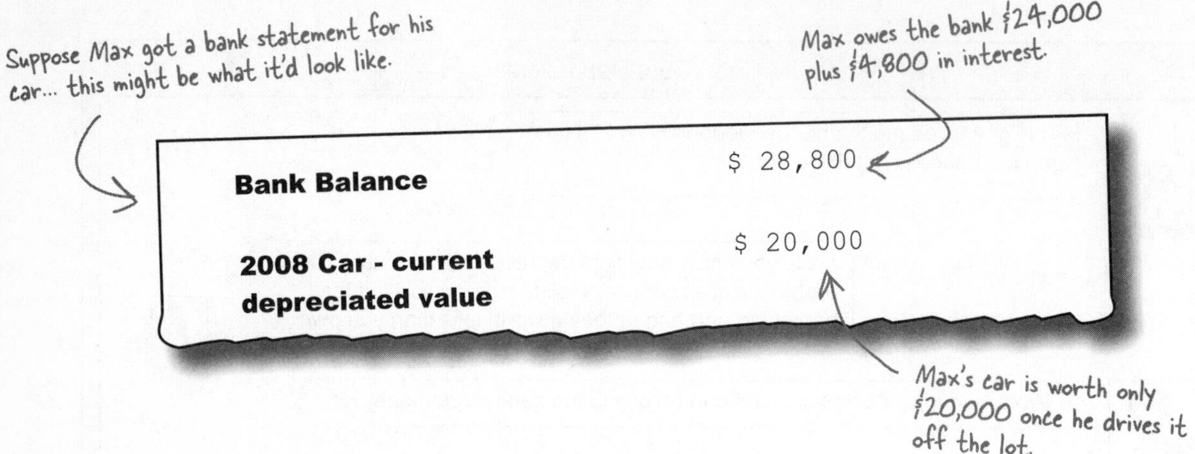

Depreciation is a sad fact of life

Things get old fast... especially cars. Tires, brakes, fluids, and wear on the engine happen every mile that you drive. That's why used cars are less expensive than new cars.

Depreciation is the term that is used to describe exactly how much value a car has lost. The **depreciated value** is the value of the car minus the depreciation. In other words, it's what the car is worth at a particular moment in time.

Unfortunately, cars lose about 20% of their value as soon as you drive them off the lot. Then they lose their remaining value over about 10 years, at which point they're basically worthless.

This is a piecewise function! Forgot what that means? Check back in Chapter 10 for details.

But the bank still gets their money

The value of the car, then, is actually dropping faster than you're paying the bank. So there's a gap between what you owe the bank, and what your car is worth:

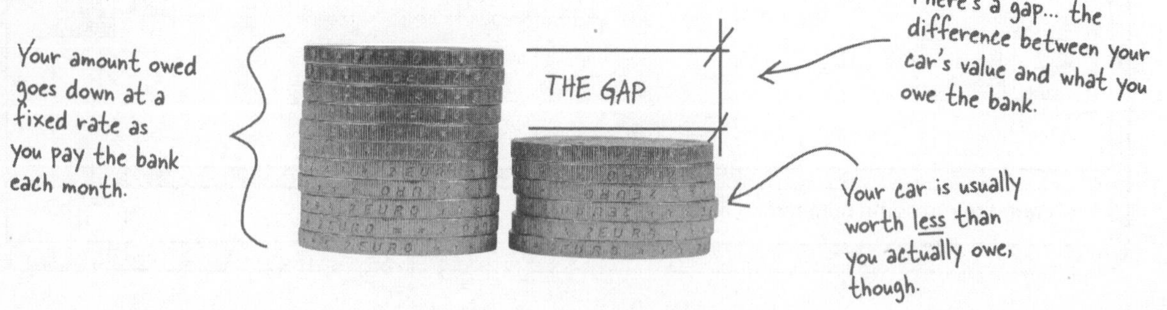

Your amount owed goes down at a fixed rate as you pay the bank each month.
THE GAP
There's a gap... the difference between your car's value and what you owe the bank.
Your car is usually worth less than you actually owe, though.

You need to figure out some functions to model the gap. Write a function for Max's loan balance, and the piecewise function for depreciation. Don't forget your domains!

These functions should both be v(t), values related to time.

Loan Balance: ....

....

....

....

....

Domain: ....

Assume the car depreciates evenly over 10 years... except for that first 20% drop.

Depreciated Car Value: ....

....

....

....

....

Domain: ....

Exercise Solution

You need to figure out some functions to model the gap. Write a function for Max's loan balance, and the piecewise function for depreciation. Don't forget your domains!

This means loan value as a function of time

Use minus because each payment is lowering what you owe.

Loan Balance: $v(t) = 28800 - 480t$

480 a month, times t, the number of months.

This number is the total cost of the loan: principle plus the interest we figured out earlier.

Domain: $0 \le t \le 60$

The domain for this function starts when Max buys his car, and goes for 60 months (five years), the term of his loan.

Piece one of the function

The initial value of the car

less the depreciation

Depreciated Car Value: $v(t) = 25000 - 0.2(25000)t$

This is 20% of the initial car value, times t.

$v(t) = 25000 - 5000t$ $\quad 0 \le t < 1$

This is only true for the first month, piece 1 of the function.

Piece two of the function

First, figure out what the car is worth after the initial drop. $= 25000 - 0.2(25000) = 20000$

The new initial value

$v(t) = 20000 - \frac{0.1}{12}(20000)t$

We know the car's going to drop over 10 years. Average that out as a 10% per year drop

Divide by 12 to get from a yearly rate to a monthly rate.

This is still the value of the car over time, just for t > 1.

$v(t) = 20000 - 166t$

Domain: $2 \le t \le 60$

This is true for the rest of the life of the loan: 60 months.

Put it all together, and here is the total gap situation:

$$v(t) = \begin{cases} 25000 - 5000t & 0 \le t \le 1 \\ 20000 - 166t & 1 < t \le 60 \end{cases}$$

$$v(t) = 28800 - 480t \quad 0 \le t \le 60$$

Systems of equations and systems of functions can be handled exactly the same way.

It's a system of ~~equations!~~ Functions

Well, technically equations are functions, so it's both.

Hey, that's just a system of equations. If we graph them, we'll "see" the gap, and we can solve for where the gap stops.

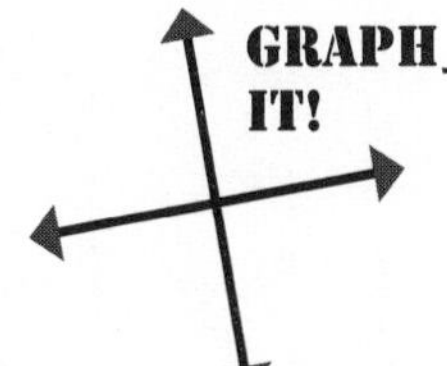

Graph both functions to "see" the gap.We got you started by graphing the first part of the depreciation graph.

Depreciation

$$v(t) = \begin{cases} 25000 - 5000t & 0 \le t \le 1 \\ 20000 - 166t & 1 < t \le 60 \end{cases}$$

Loan Balance

$$v(t) = 28800 - 480t \quad 0 \le t \le 60$$

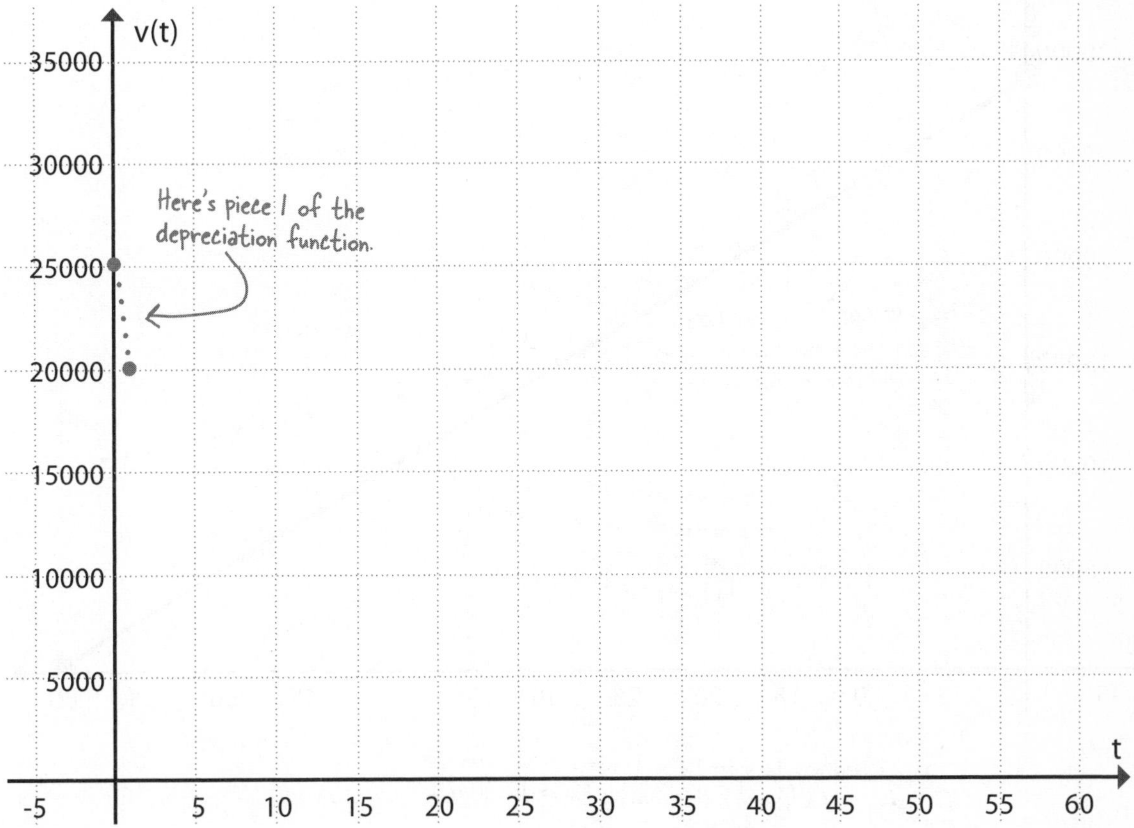

v(t)

35000

30000

25000

20000

15000

10000

5000

-5 5 10 15 20 25 30 35 40 45 50 55 60

t

Here's piece 1 of the depreciation function.

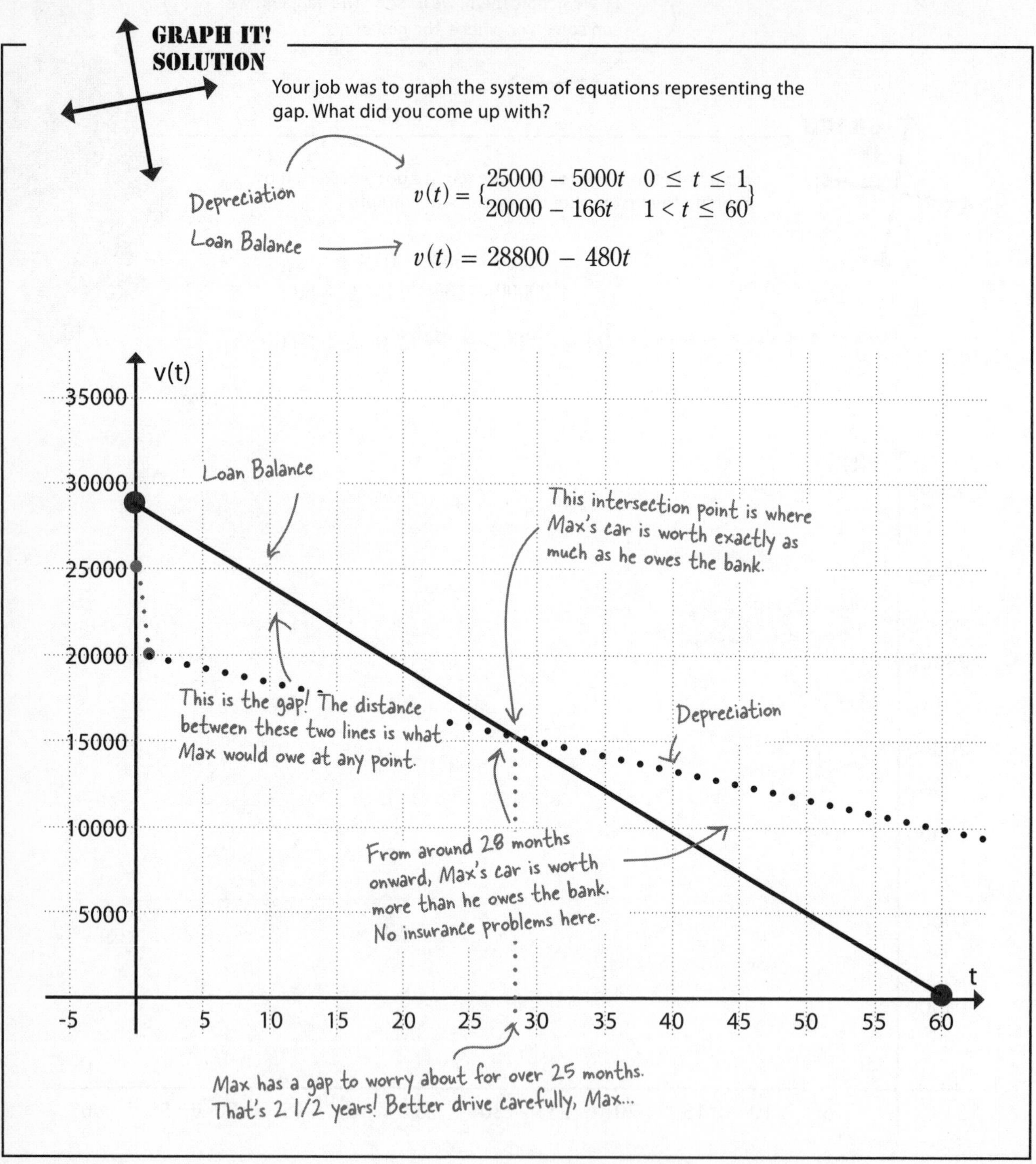

GRAPH IT! SOLUTION
Your job was to graph the system of equations representing the gap. What did you come up with?
Depreciation
$v(t) = \{ \begin{matrix} 25000 - 5000t & 0 \leq t \leq 1 \\ 20000 - 166t & 1 < t \leq 60 \end{matrix} \}$
Loan Balance
$v(t) = 28800 - 480t$
v(t)
35000
30000
25000
20000
15000
10000
5000
-5
5
10
15
20
25
30
35
40
45
50
55
60
t
Loan Balance
This intersection point is where Max's car is worth exactly as much as he owes the bank.
This is the gap! The distance between these two lines is what Max would owe at any point.
Depreciation
From around 28 months onward, Max's car is worth more than he owes the bank. No insurance problems here.
Max has a gap to worry about for over 25 months. That's 2 1/2 years! Better drive carefully, Max...

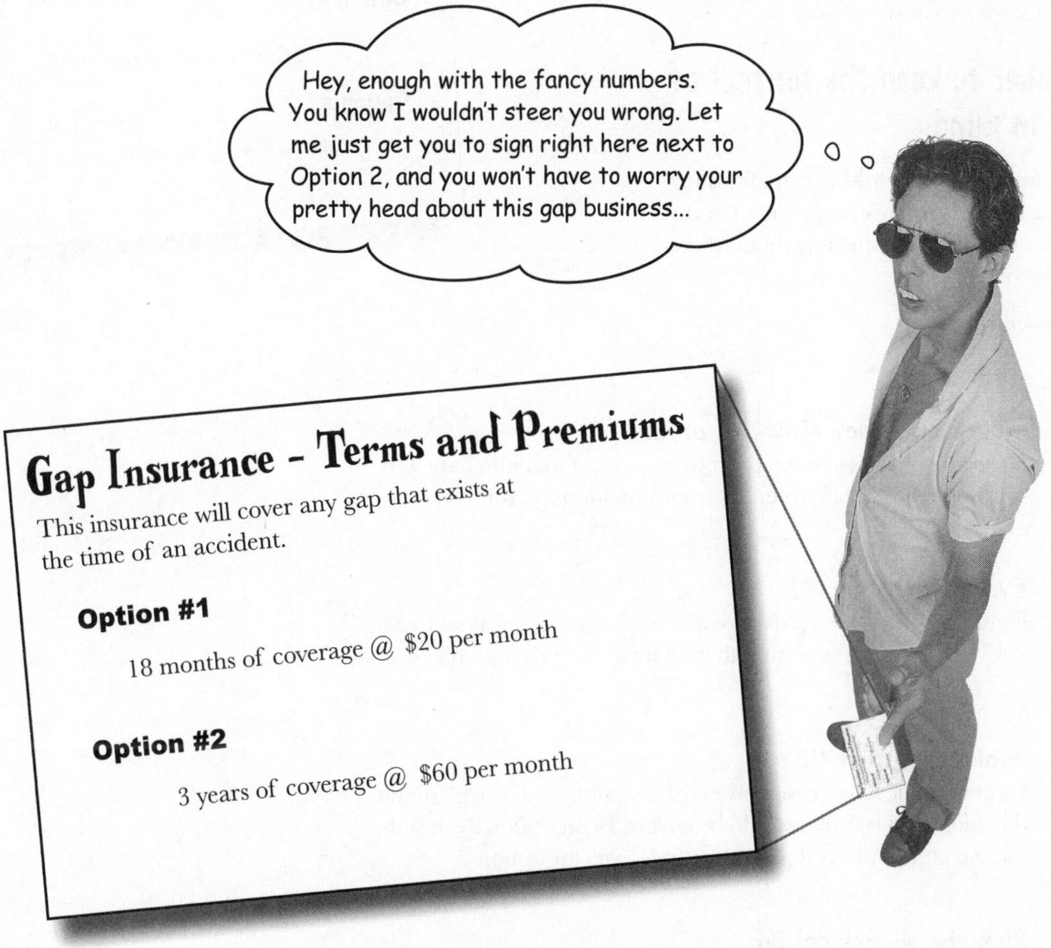

Hey, enough with the fancy numbers. You know I wouldn't steer you wrong. Let me just get you to sign right here next to Option 2, and you won't have to worry your pretty head about this gap business...

Gap Insurance - Terms and Premiums

This insurance will cover any gap that exists at the time of an accident.

Option #1

18 months of coverage @ $20 per month

Option #2

3 years of coverage @ $60 per month

What should Max do? Should he buy gap insurance? If so, which option is better?

You don't need to GUESS with Algebra

Max definitely needs some type of gap insurance to cover him while his car is worth less than he owes. But which option? 1 or 2?

Between the graph, the functions, and your mad Algebra skills, you can figure this out. You need to determine Max's maximum potential risk at any time, and how much the different gap insurance options cost. You also need to think about his monthly budget and the overall cost of each gap insurance option.

Gap Insurance - Terms and Premiums

This insurance will cover any gap that exists at the time of an accident.

Option #1

18 months of coverage @ $20 per month

Option #2

3 years of coverage @ $60 per month

... but remember to keep the context of the problem in mind

First, we need to know how much Max can afford in additional premiums on top of his car payment and existing insurance. But that's not the only thing Max needs to worry about.

Here's what we need to do:

1. **Figure out what Max can afford.**
   Determine Max's leftover balance from his initial purchase and make sure he can afford either premium for gap insurance.

2. **Evaluate Option #1.**
   Figure out the worst case gap over the first 18 months (that's Max's risk), and the total premium that he'll pay for this option's coverage.

3. **Evaluate option #2.**
   Figure out the worst case between 18 months and 3 years (that's the additional risk that would be covered with option 2), and the total in premiums he'd pay over three years for option 2.

4. **Pick the wisest option.**
   Using the new information and the graph of the gap, what's the smart plan? Which option should Max choose?

Sharpen your pencil

Use this space to figure out Max's gap problem.

1. Figure out what Max can afford.

2. Evaluate Option #1.

3. Evaluate option #2.

4. Pick!

Sharpen your pencil Solution

Your job was to choose the smartest insurance for Max.

1. Find out what Max can afford.

$$\frac{(24{,}000 + 4{,}800)}{5 \cdot 12} = 480$$

$480 < $550

Sweet! It works! Max can afford his car if he takes the longest loan and have $70 leftover..

We figured this out way back on page 432 when Max was picking a loan.

2. Evaluate Option #1.

The worst gap in the first 18 months comes at the initial dip – it looks like about $8,000. ← Maximum risk for Max

Premiums paid = 18(20) = $360 ← Total option 1 premiums (Months; Monthly Premium)

The premium is $20 a month, which Max can definitely afford.

So, for $360, Max covers a maximum possible risk of $8000.

3. Evaluate option #2. The worst gap between 18 months and 3 years is about $2000.

The overall maximum risk is still $8000 in the first few months. But the maximum risk not covered by option 1 is $2000, around month 18 or 19.

Premiums paid = 36(60) = $2160 ← Total option 2 premiums (Months; Monthly Premium)

The premium is $60 a month, so Max can afford that, too.

For a total of $2160, Max would cover the same $8000 risk as option 1, but he'd also cover a risk of $2000 that would occur after option 1's plan expired.

4. Pick! Option #2 is not worth it! Max would have to pay over $1440 extra (option 2 cost – option 1 cost) to cover an additional risk of $2000. So his extra premiums end up only potentially saving him $600! Better to take option 1 and drive carefully around month 19 and 20!

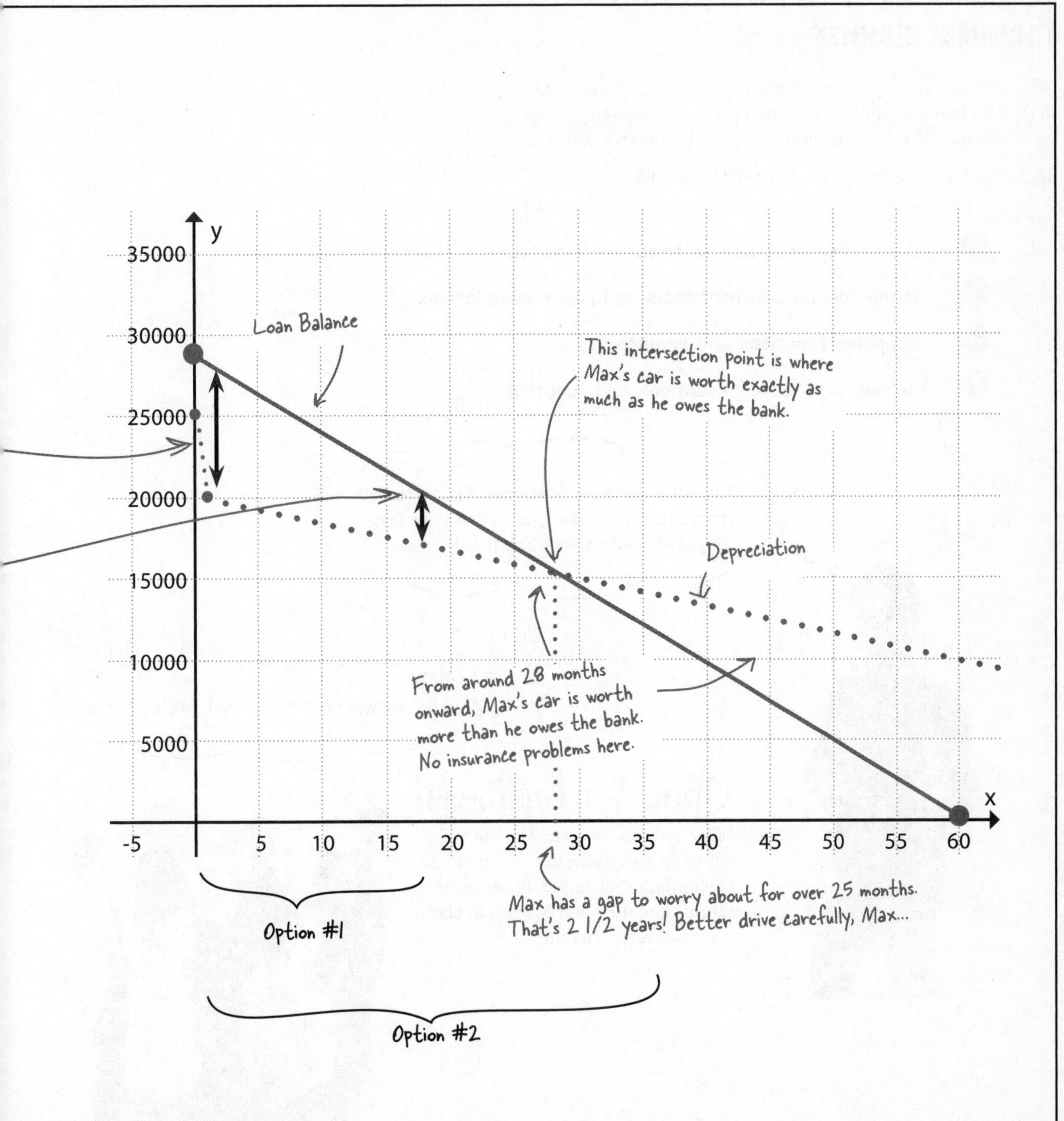

y
35000
30000
25000
20000
15000
10000
5000
x
-5
5
10
15
20
25
30
35
40
45
50
55
60
Loan Balance
This intersection point is where Max's car is worth exactly as much as he owes the bank.
Depreciation
From around 28 months onward, Max's car is worth more than he owes the bank. No insurance problems here.
Max has a gap to worry about for over 25 months. That's 2 1/2 years! Better drive carefully, Max...
Option #1
Option #2

Max plans to pay you to be his financial planner

Max is thrilled you saved him so much money and still managed to get him into the car of his dreams. He's even promised to keep using you for all his financial decisions.... and tell his friends!

To get here, you used a lot of advanced Algebra skills:

- **Expressing equations in terms of variables**
- **Using functions with domain and range restrictions**
- **Graphing functions and reading results**
- **Solving systems of equations and functions**

I'm out of here in my new car! I'm protected from an early accident, and I can make my payments, no problem. You totally rule!

Open for business!

Since your success with Max, there's a whole line of people just waiting to pay you for financial advice. Better open an office... and use some Algebra to plan your own financial future!

Hey, we want our turn. What do you charge?

Wrapupcross

You did it, you made it through the whole book! But don't relax just yet, there's a crossword to do first...

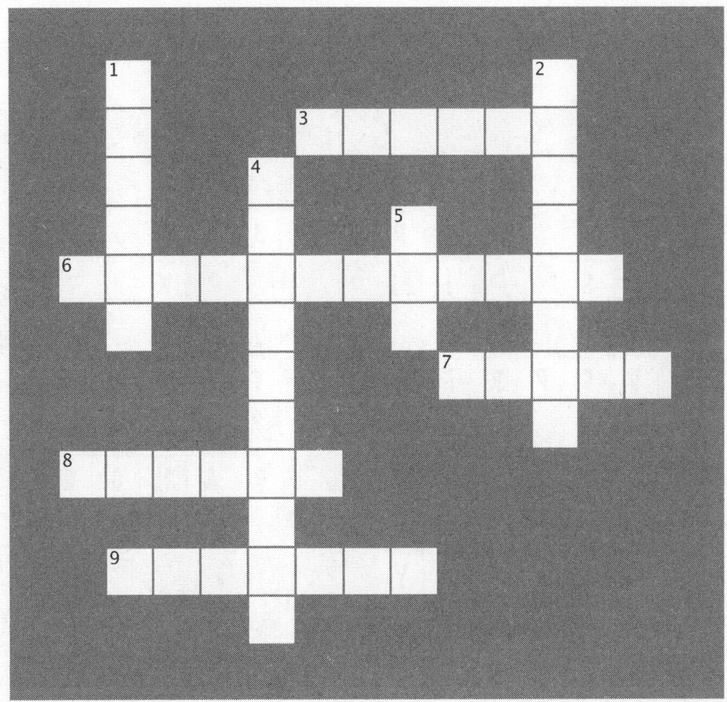

Across

3. Legal inputs to a function are the function's
6. A car loses its value when it leaves the dealer because of
7. The legal outputs from a function is the
8. You should always drive
9. Always evaluate the solution in the __________ of the problem.

Down

1. Using more than one function with the same variable is solving a ____________ of functions.
2. To use Algebra to solve real world problems you need to look for
4. Represents a comparison between values rather than equality
5. The difference between what a vehicle is worth and how much is still owed on a loan.

Wrapupcross Solution

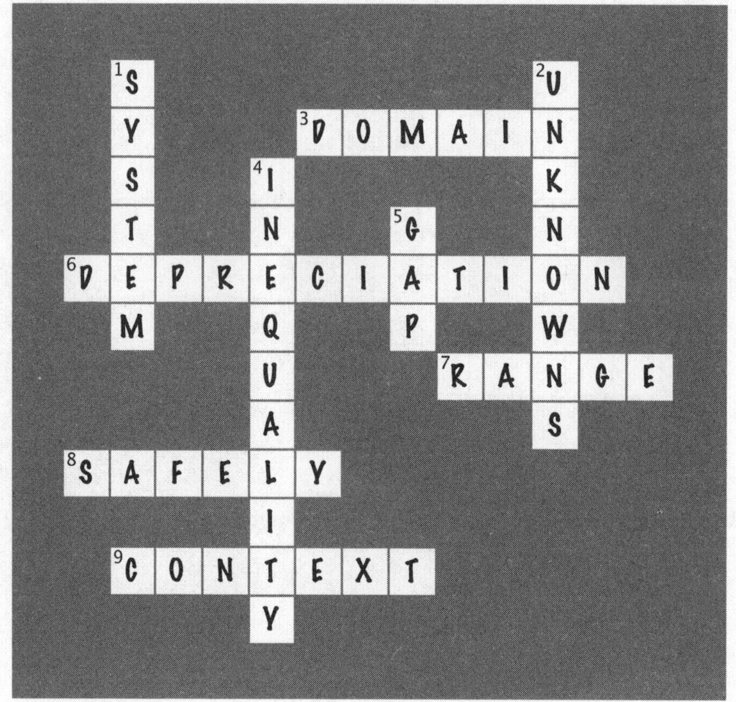

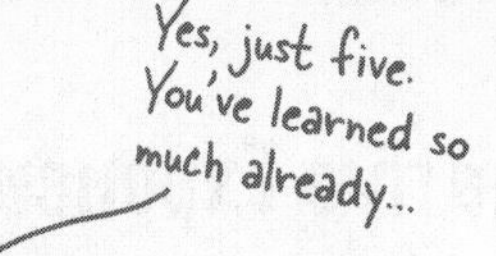

appendix i: leftovers

The Top Five Things (we didn't cover)

Even after eating all of this, we're going to have some left.

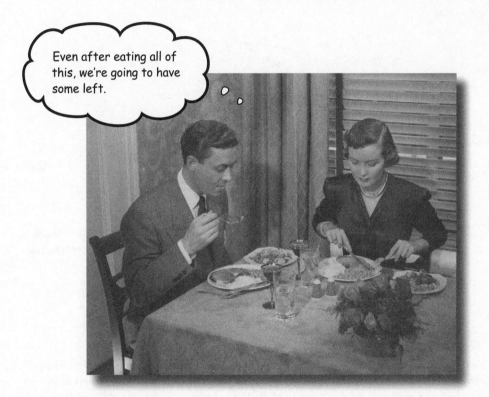

You've learned a lot in this book, but Algebra has even more to offer. Don't worry, we're almost done! Before we go, there are a just few gaps we want to fill in. Then you'll be onto Algebra 2, and that's a whole additional book...

#1 Negative Exponents

We touched on them in chapter 3, but as a quick refresher, here's what a negative exponent looks like:

$$x^{-a} = \frac{1}{x^a}$$

Combine that rule with multiplying exponential terms, and you can divide exponential terms too, like this:

$$\frac{x^a}{x^b} = x^{a-b}$$

What's going on here? It's really a matter of cancelling things out... here's a simple example, where you can see the cancelling in action:

Distributing the exponent →

$$\frac{2^2}{2^4} = \frac{(2 \cdot 2)}{(2 \cdot 2 \cdot 2 \cdot 2)} = \frac{1}{2^2}$$

Since both approaches are both valid, the answers are the same.

Using the division of exponential term rules →

$$\frac{2^2}{2^4} = 2^{2-4} = 2^{-2}$$

A negative exponent means that you divide one by the exponential term, with its exponent stripped of its negative sign. Negative exponents are a great way to get rid of fractions. Any fraction can be rewritten as a negative exponent, and you can work with it the same way you would work with any other exponent.

Negative exponents mean you can get rid of fractions.

Take this exponential term...

In general: $x^{-a} = \frac{1}{x^a}$

...and put it as a denominator, without the negative sign.

Working with negative exponents

The only difference in working with negative exponential terms and regular exponential terms is keeping track of the sign.

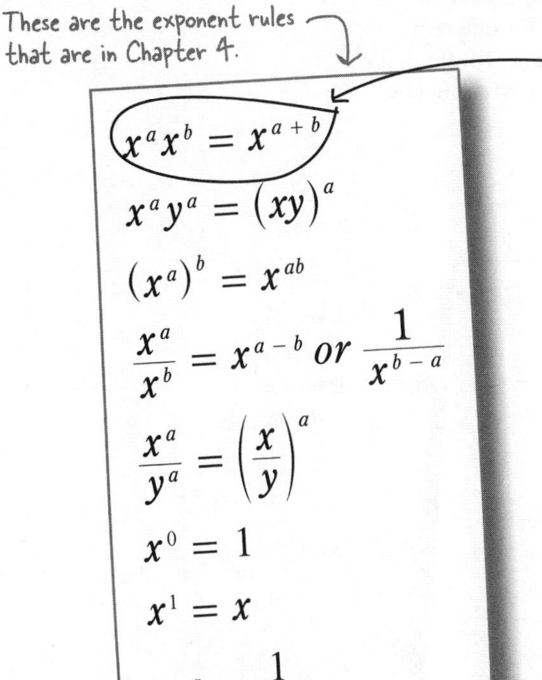

These are the exponent rules that are in Chapter 4.
$x^a x^b = x^{a+b}$
$x^a y^a = (xy)^a$
$(x^a)^b = x^{ab}$
$\frac{x^a}{x^b} = x^{a-b}$ or $\frac{1}{x^{b-a}}$
$\frac{x^a}{y^a} = \left(\frac{x}{y}\right)^a$
$x^0 = 1$
$x^1 = x$
$x^{-a} = \frac{1}{x^a}$
$x^{(1/a)} = \sqrt[a]{x}$

Take this one for example...

$$x^a x^{-b} = x^{a+(-b)}$$

It's the same thing, just watching the signs!

Just do the same thing for any other operation as it comes up.

Negative exponents also give you flexibility

If you come across an exponential term in a denominator, you can write it as a negative exponent and remove the fraction. Then you can manipulate the equation the way you want to.

For example:

$$5 + \frac{6}{x^2}(x^3) = 5 + 6x^{-2}(x^3)$$

These are exactly the same expression.

#2 Table of values for graphing

We mentioned them a couple of times, and even showed one, but what exactly is a table of values?

A ***table of values*** is a table that is set up with both of the variables that are in an equation and allows you to easily keep track of the results for different substitutions. It's another way to keep track of points for graphing. For lines, it's not typically necessary because you only need two points, but for other shapes...

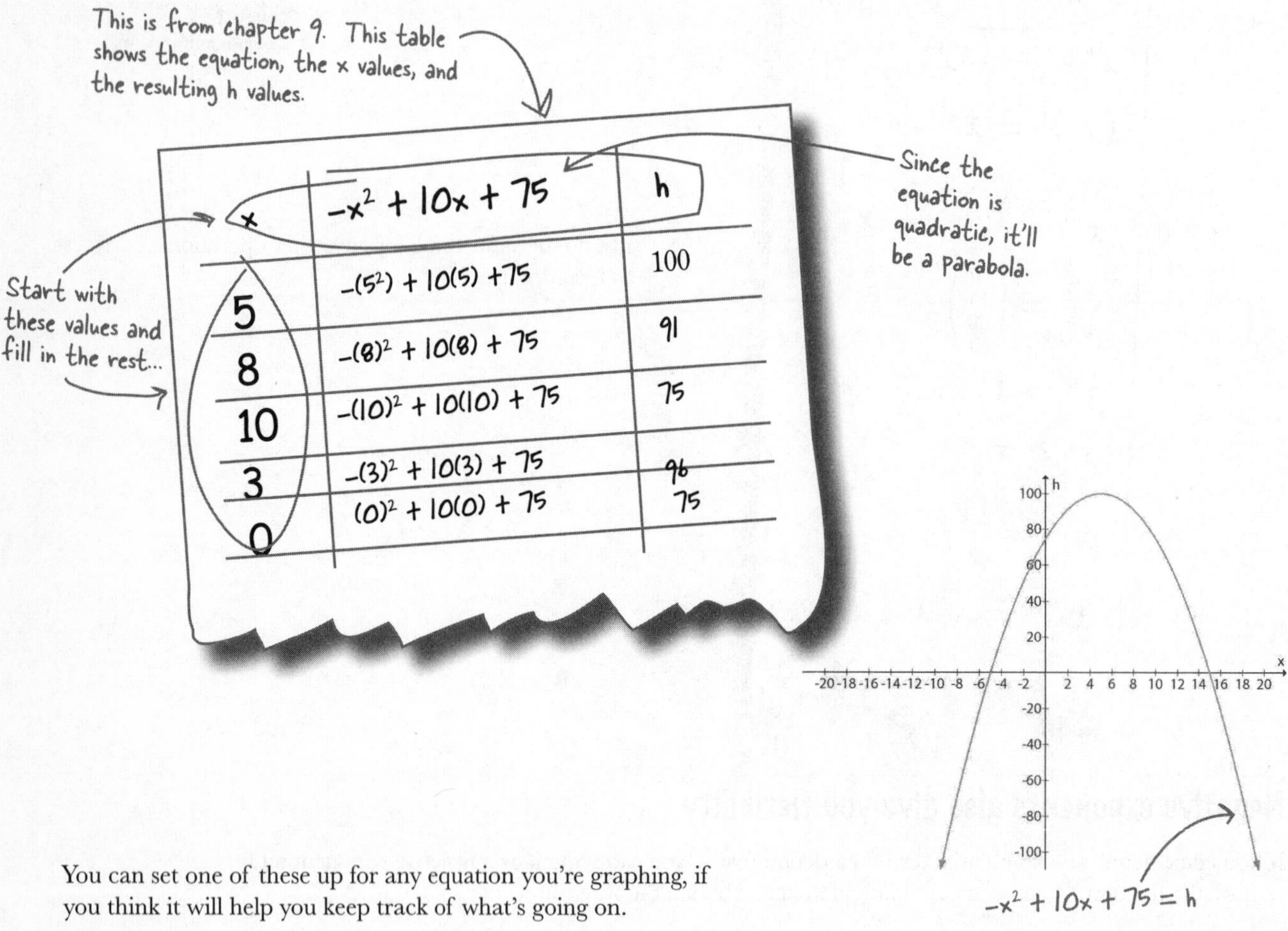

This is from chapter 9. This table shows the equation, the x values, and the resulting h values.
Since the equation is quadratic, it'll be a parabola.
Start with these values and fill in the rest...
x
-x² + 10x + 75
h
5
-(5²) + 10(5) +75
100
8
-(8)² + 10(8) + 75
91
10
-(10)² + 10(10) + 75
75
3
-(3)² + 10(3) + 75
96
0
(0)² + 10(0) + 75
75
h
x
-x² + 10x + 75 = h

You can set one of these up for any equation you're graphing, if you think it will help you keep track of what's going on.

#3 Absolute value equations

You've learned a lot about how to manipulate and solve equations, but we didn't get to cover how you handle an equation when there's an absolute value in there.

You know how to handle the absolute value of a number, but what if you have a variable inside the absolute value signs that you want to isolate? When you have an equation with an absolute value, even though there is only one unknown, there are two solutions.

$$\frac{77}{11} = \frac{11}{11} \cdot |x|$$

$$\frac{77}{11} = |x|$$

$$7 = |x|$$

$$x = 7 \text{ or } x = -7$$

This must be true because you strip the number of it's sign with the absolute value.

What happens if there's more than one term inside the absolute value signs? If that happens, you have to treat the whole quantity inside the absolute value as an unknown, and isolate it before you can do anything else.

You have to treat the "x + 3" part as an unknown and isolate it.

$$2 + |x+3| - 2 = 0 + 2$$

$$|x+3| = 2$$

Now this is where it gets weird. Remember the absolute value signs mean that the absolute value of whatever is in between them is equal to 2. That means whatever is inside those bars could be 2 or -2!

$$|x+3| = 2$$

$$x + 3 = 2$$

$$x + 3 = 2 - 3$$

$$x = -1$$

or

$$x + 3 = -2$$

$$x + 3 = -2 - 3$$

$$x = -5$$

To remove an absolute value, you need to isolate it, and then solve whatever's left using an option for both signs. It means you have to solve it twice.

#4 Calculators

Generally, the work in this book can be completed by hand (if you have enough paper). If you used a calculator, it didn't need to be anything past a basic calculator with exponents.

There are a lot of calculators out there that can graph, solve equations, and plug and chug the quadratic formula for you. For now:

Don't use a calculator to solve your equations!

The point of this book and the material in it is to learn how to manipulate the first level of equations while understanding what is going on. If you simply plug things into a calculator, what you've really understood is how to work your calculator!

As you progress in math, you'll need to use technology more - but not yet!

#5 More practice, especially for factoring

The best way to get good at all of this work is to practice it more. Working through the exercises in this book is a great start, but you should pull out your classroom textbook and work on those problems, too.

This book describes all of the principles you'll need to work in most Algebra I textbooks, and the more you work on them, the better you'll get.

Factoring, in particular, is a skill that you will be able to do much faster the more you do it. So keep practicing....

appendix ii: pre-Algebra review

Build on a solid foundation

I didn't start out as a prima ballerina—first there were little tutus.

Do you ever feel like you can't even get started?

Algebra is great, but if you want to learn it, you have to have a good understanding of number rules. Suppose you're rolling along and realize that you forgot how to multiply integers, add fractions, or divide a decimal? Well, you've come to the right place! Here we're going to cover all the pre-Algebra you need—*fast*.

Algebra starts with numbers

If the you hear the weatherman say it's "minus five," you know that it's **really** cold, colder than zero. Numbers sometimes need to indicate that they are less than 0, and they do that by adding a negative (or minus sign) in front of the number.

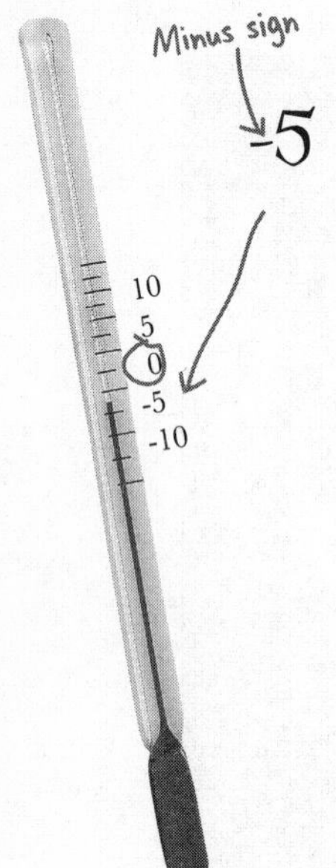

So, that's a negative number. What are the other numbers, the plain ones? Those are positive numbers. They are indicated by ***no sign or a positive sign***. If you're working with whole positive and negative numbers, it means that you're working with **integers**.

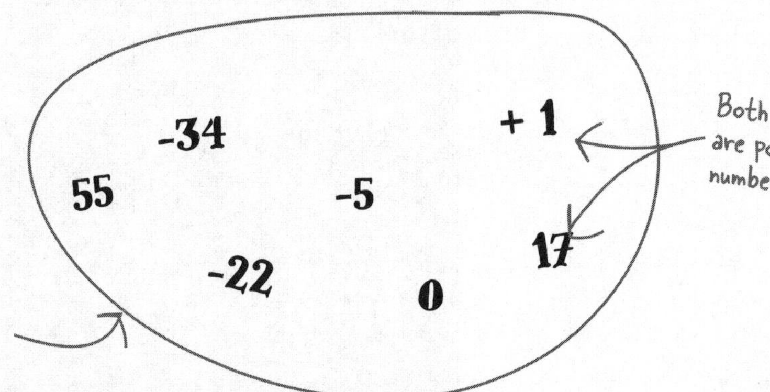

Sum it up

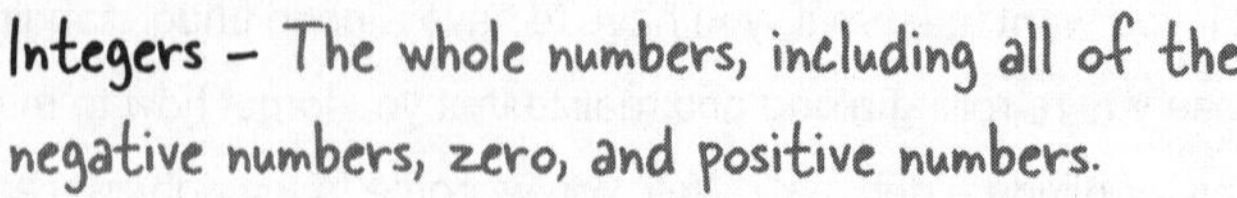

Integers – The whole numbers, including all of the negative numbers, zero, and positive numbers.

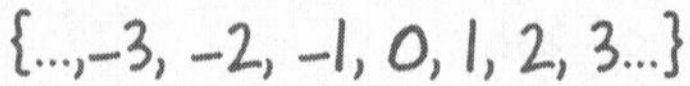

How do you work with negative numbers?

Operations with negative numbers are similar to operations with positive numbers; you just have to keep track of the sign of the numbers that you're dealing with. The first thing to understand is how the negative and positive numbers relate to each other. The number line will help with this.

The number line

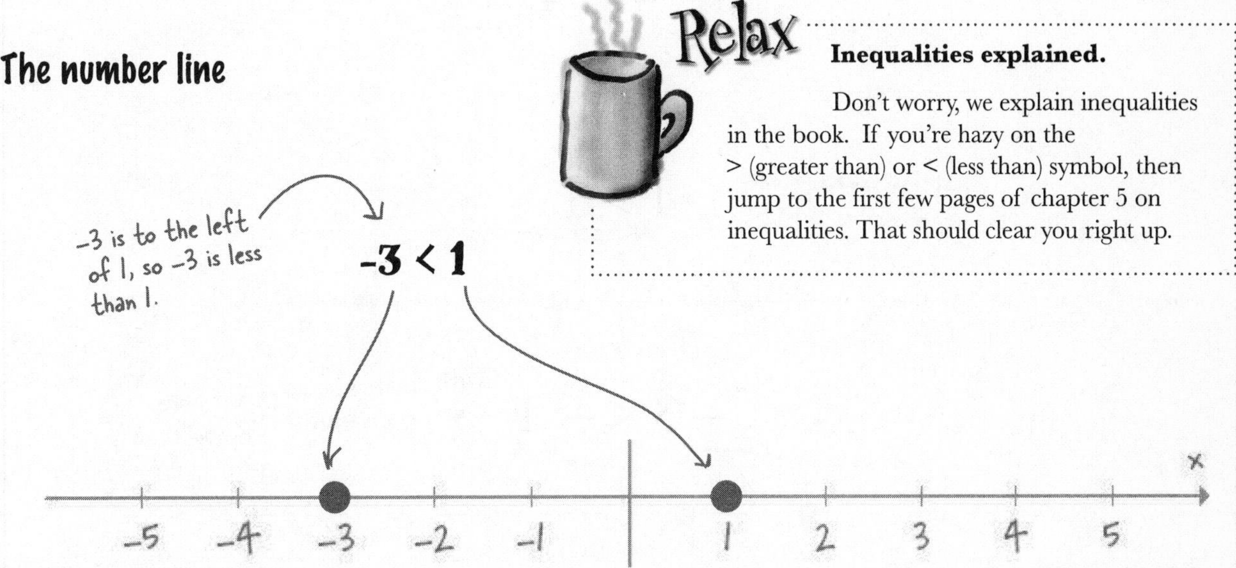

Relax
Inequalities explained.
Don't worry, we explain inequalities in the book. If you're hazy on the > (greater than) or < (less than) symbol, then jump to the first few pages of chapter 5 on inequalities. That should clear you right up.
-3 is to the left of 1, so -3 is less than 1.
-3 < 1
x
-5 -4 -3 -2 -1 1 2 3 4 5

To determine the relationship between integers, plot both numbers on a number line. The number farthest to the left is always less. That's because the left side of the number line is headed off to negative infinity, a very, very small number.

The number farthest to the right is always the larger number. Since as you go farther to the right on the number line, you are closer and closer to positive infinity—a huge number. Looking at -3 and 1 on the number line, it's clear that -3 is less than 1 since it's farther to the left.

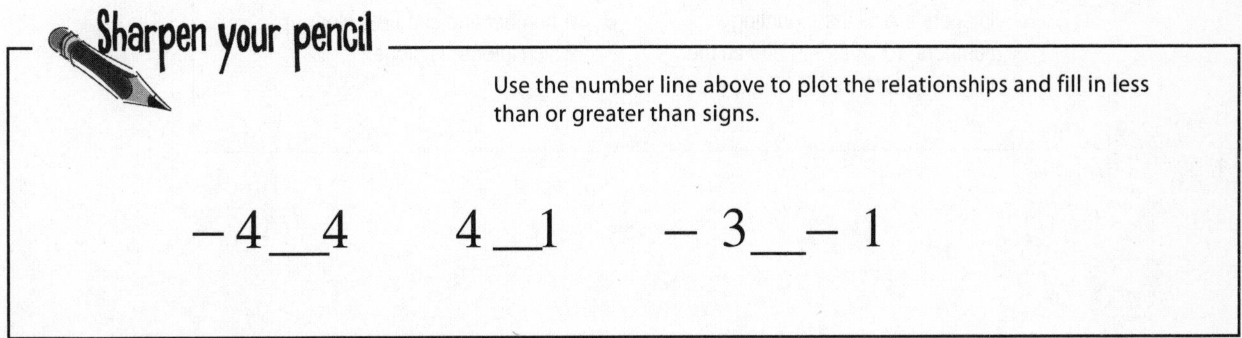

Sharpen your pencil Solution

Use the number line above to plot the relationships and fill in less than or greater than signs.

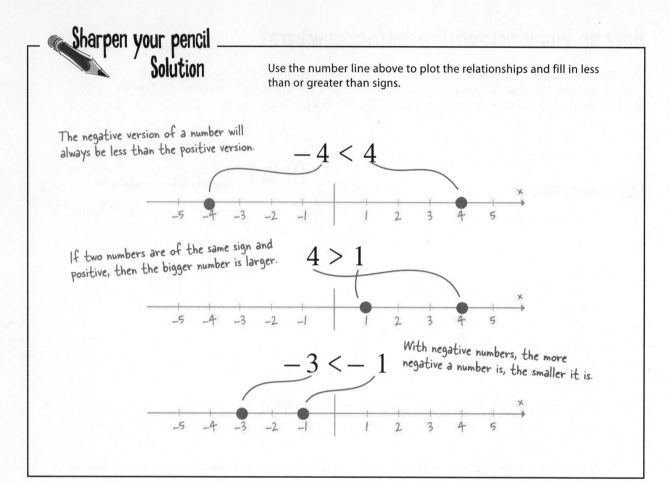

The negative version of a number will always be less than the positive version.

$$-4 < 4$$

-5 -4 -3 -2 -1 1 2 3 4 5 x

If two numbers are of the same sign and positive, then the bigger number is larger.

$$4 > 1$$

-5 -4 -3 -2 -1 1 2 3 4 5 x

With negative numbers, the more negative a number is, the smaller it is.

$$-3 < -1$$

-5 -4 -3 -2 -1 1 2 3 4 5 x

BULLET POINTS

- Integers are all the "counting numbers" (0, 1, 2, 3...) and all the negative numbers (-3, -2, -1, etc.).
- A number line can help clear up which integer is bigger.

Addition and subtraction of integers

You've probably been adding positive numbers for ages, so you have no problem with that. The good news is that when you're working with numbers of the same sign, there are some simple rules to follow.

1. **Adding two positive numbers is a positive number.**
   Nothing's changed here. They just keep going up!

2. **Adding two negative numbers is a negative number.**
   With two negative numbers, just add the number part first and put a negative sign in front of it; that's it.

We'll practice this in a second.

Working with mixed integers

Did you notice that there's no "subtracting positive numbers" on that list? Now that you're working with integers, the line between addition and subtraction gets a bit murky. Subtracting a number is the same thing as adding a negative number. For example, 2-3 is the same as 2 + (-3).

How do you know? To actually perform the operation, you can work with the number line. The rules are simple: for every negative number (or subtraction), move left on the number line, and for every positive number (or addition), go to the right.

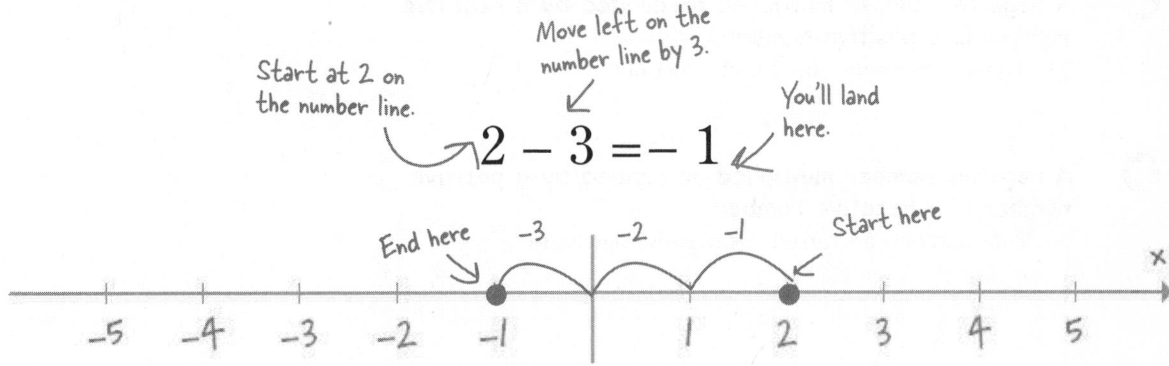
Start at 2 on the number line.
Move left on the number line by 3.
You'll land here.
2 − 3 = − 1
End here
-3
-2
-1
Start here
x
-5 -4 -3 -2 -1 1 2 3 4 5

3. **Rules for adding & subtracting mixed integers.**
   Move left on the number line for negative numbers, move right for positive numbers.

Multiplication and division of integers

To multiply and divide integers, you use the same operations as with regular positive numbers, with a few simple rules about how you work with the signs. When you come across an integer multiplication problem, first figure out the number piece:

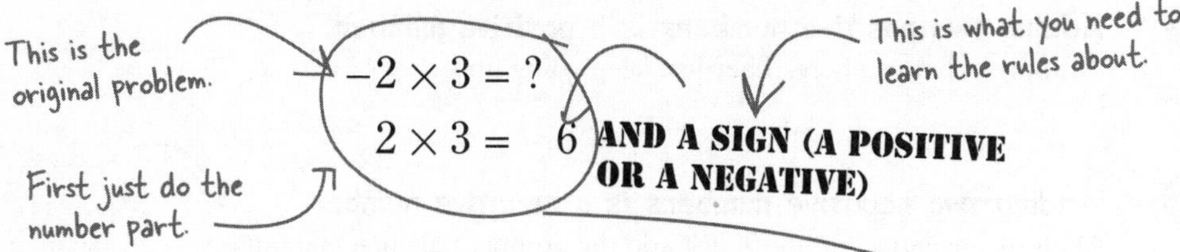

The rules for integer signs - multiplication and division

Now that you know where you need the rules, here they are. Remember, you just carry out the multiplication (or division) as you would for positive numbers, and then add the sign in front to get your answer.

1. **A positive number multiplied or divided by a positive number is a positive number.**
   So, you figure out the number part, and that's it.

2. **A negative number multiplied or divided by a negative number is a positive number.**
   The two negative signs cancel each other out.

3. **A negative number multiplied or divided by a positive number is a negative number.**
   So, if the numbers are mixed, the negative sign carries.

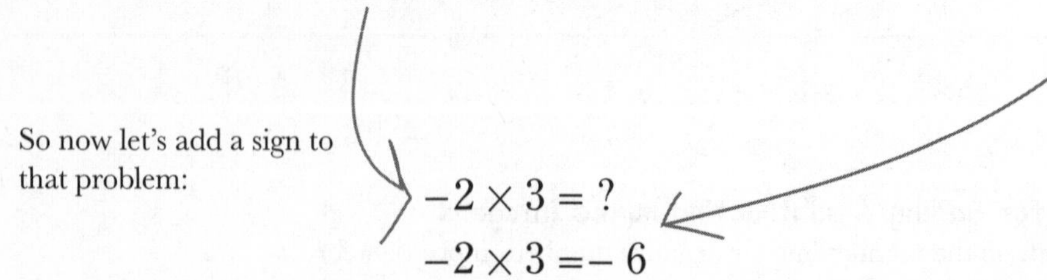

So now let's add a sign to that problem:
$-2 \times 3 = ?$
$-2 \times 3 = -6$

Let's try it...

BE the calculator

Your job is to play calculator and figure out these integer operations. Remember all the new rules you've learned! Work with the number lines - they help.

$7 + 4 =$ ___

$-3 - 8 =$ ___

You can use the number line, but since they're both negative, they'll stay negative.

Just watch the signs...

$-1 \times -32 =$ ___

$4 \times -5 =$ ___

Do the number part first, then figure out the sign.

$3 - 6 =$ ___

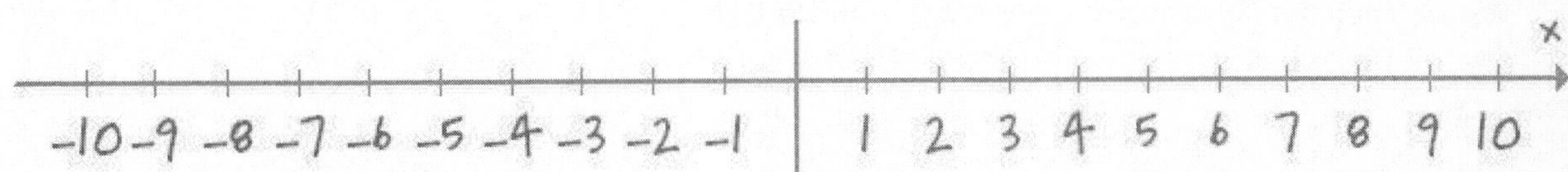

$-5 + (-4) =$ ___

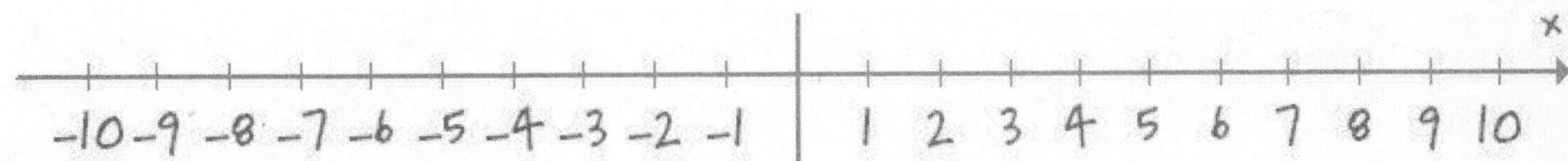

$-3 + 7 =$ ___

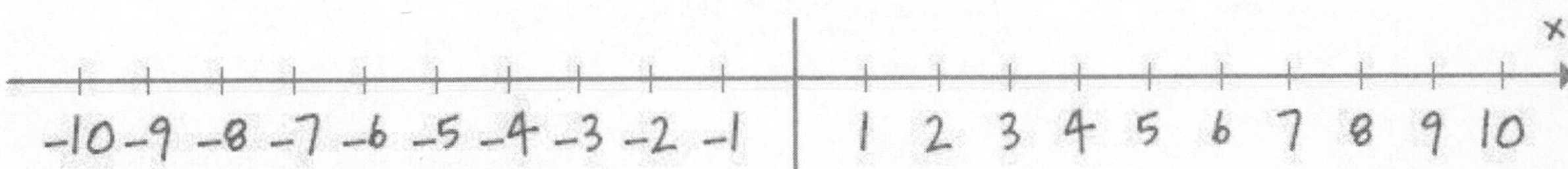

BE the calculator solution

Your job is to play calculator and figure out these integer operations. Remember all the new rules you've learned! Work with the number lines - they help.

$7 + 4 = 11$ ← This is a warm up.

$-3 - 8 = -11$ ← You can use the number line, too, but since they're both negative, you can just add them up and keep going.

$-1 \times -32 = 32$ ← The negatives cancel each other out, so it's positive.

$4 \times -5 = -20$ ← There's only one negative, so it stays negative.

Remember, the number part of the operation doesn't change at all!

$3 - 6 = -3$

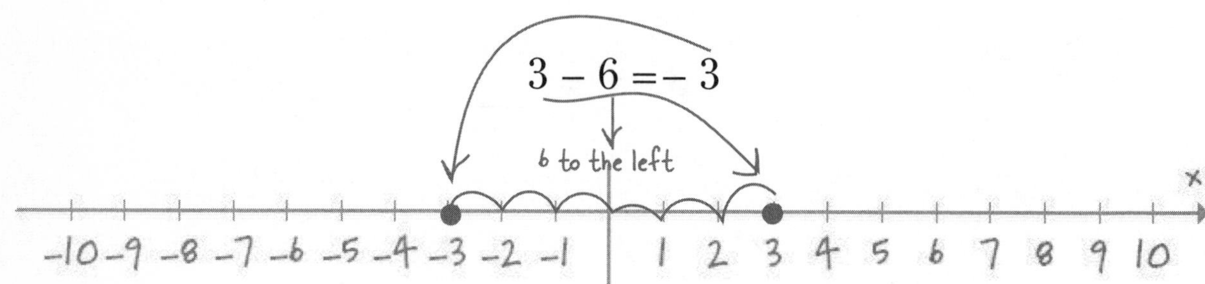

This is exactly the same thing as just subtracting 4.

$-5 + (-4) = -9$

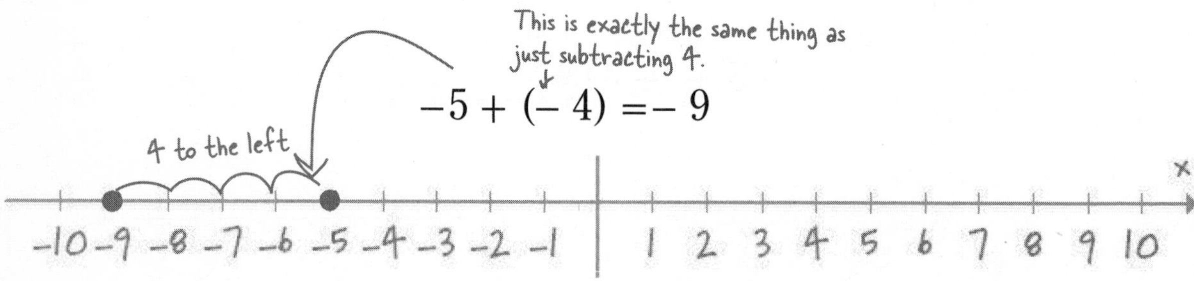

If you add a positive number, go to the right.

$-3 + 7 = 4$

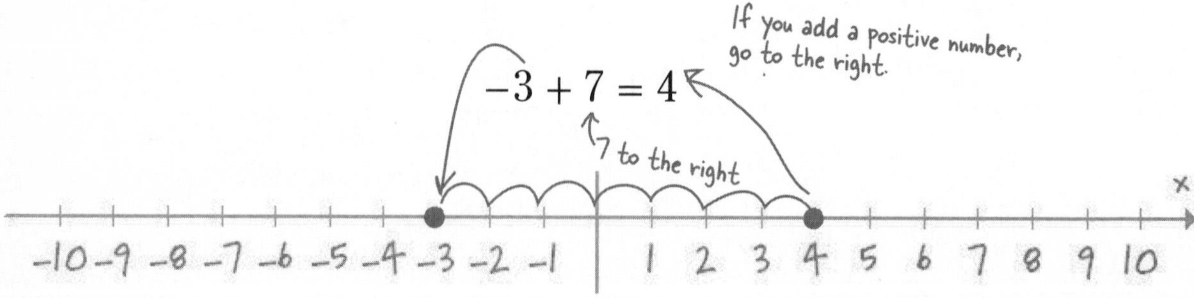

Absolute Value

Absolute value is the operation of stripping an integer of its sign. The upshot of that, of course, is that all of the numbers end up positive. Here's what it looks like:

The absolute value sign means "take the sign off of this number."

This is the absolute value sign.

$$|-6| = 6 \textit{ and } |6| = 6$$

How do you treat the absolute value sign and the numbers inside it? The absolute value sign acts like parentheses; you have to do what's inside it first and then take off the sign. So if it shows up in an expression you're working with, you have to handle it before moving on.

If you think of the absolute value sign as a brick wall, that might help.

This is the original expression.

$$|6 - 8| = |-2| = 2$$

You have to do this first.

Then, strip the sign.

$$6 - 8 \quad = \quad -2 \quad = 2$$

BULLET POINTS

- Absolute value means take the sign away from the number.
- The absolute value of a positive number is a positive number.
- The absolute value of a negative number is a positive number.
- An expression inside the absolute value sign needs to be simplified before the absolute value is taken.

What absolute value means

Absolute value really means the distance between a number and zero on the number line. It's important in cases where you're more interested in difference than direction. The number alone tells you how far you'll travel, and the sign tells you the direction you'll move in—to the left of zero, if the sign is negative, or to the right of zero, if the sign is positive.

Absolute value means caring about the distance, not the sign.

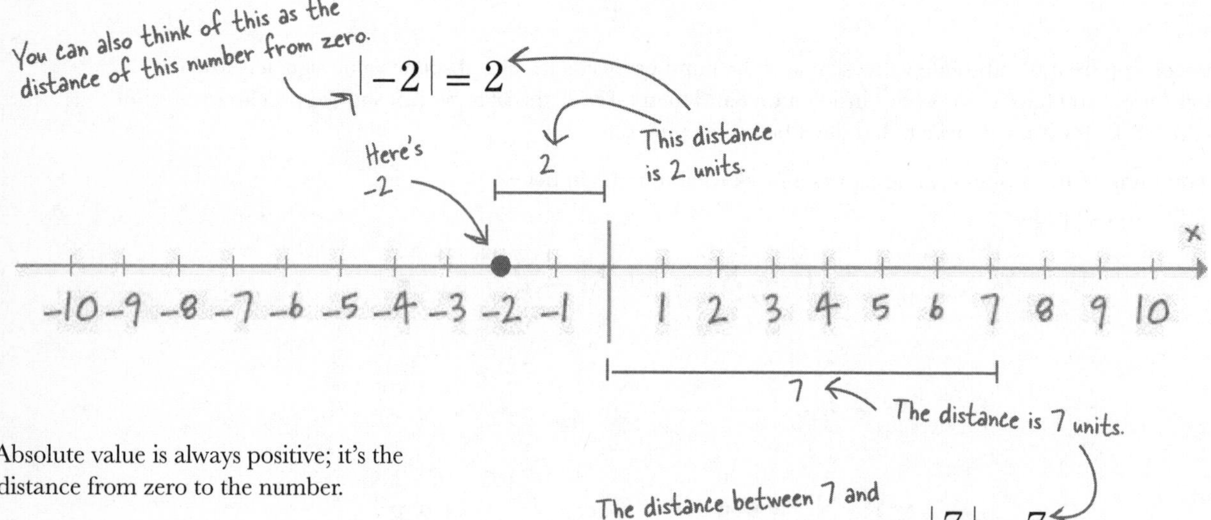
You can also think of this as the distance of this number from zero.
$|-2| = 2$
Here's -2
2
This distance is 2 units.
x
-10 -9 -8 -7 -6 -5 -4 -3 -2 -1 1 2 3 4 5 6 7 8 9 10
7
The distance is 7 units.
The distance between 7 and zero is...
$|7| = 7$

Absolute value is always positive; it's the distance from zero to the number.

Sum it up

Absolute value – The absolute value of a number is the value of the number without its sign. It is the distance of the number from zero on the number line.

y=mx+b

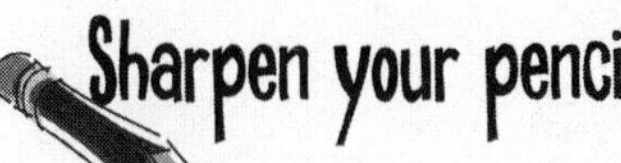

Simplify the absolute value problems below. Remember, you're looking for the distance between zero and the number.

It doesn't matter which sign is there; the absolute value means take it away.

$|-22| =$ .... $|+75| =$ .... $|172| =$ ....

$|10 + 3| =$ .... $|15 - 16| =$ ....

Simplify the absolute value first, then do the subtraction.

$25 - |13 + 4|$

....

....

....

$25 + |-13 - 4|$

....

....

....

Simplify the absolute value problems below. Remember, you're looking for the distance between zero and the number.

The positive sign gets dropped, although the absolute value is still 75.

$|-22| = 22$

$|+75| = 75$

$|172| = 172$

You have to do the operation inside the absolute value and then take the absolute value.

$|10 + 3| = |13| = 13$

Same deal here – inside first, then the absolute value.

$|15 - 16| = |-1| = 1$

Simplify the absolute value first, then do the subtraction

$$25 - |13 + 4|$$
$$25 - |17|$$
$$25 - 17$$
$$8$$

$$25 + |-13 - 4|$$
$$25 + |-17|$$
$$25 + 17$$
$$42$$

there are no Dumb Questions

Q: What is absolute value used for in the real world?

A: Distance is, at its core, just absolute value. It's the answer to "How far are you going?" It doesn't matter which way, so you're the distance you give will always be positive—an absolute value.

Another example is temperature change. Since temperatures are measured in terms of above and below zero, you may need to know how much the temperature needs to be raised. If you start at below zero and are going above zero (say, from -10 degrees to +32), you need the absolute value of those numbers added together, not just numbers themselves.

Q: Isn't the number line a little juvenile?

A: No, the number line is a great, easy way to keep track of what's going on when you're adding positive and negative numbers. Don't feel like you need to use something complicated just because you're getting ready for Algebra!

Q: What if we have three integers that are multiplied together? Which sign wins?

A: If you have a string of integers multiplied together, here's the deal: If it's all positive, the answer's positive. If you have an even number of negative signs, then the answer is positive. If you have an odd number of negative signs, the answer is negative.

Q: Zero is an integer, too?

A: It sure is. It's included because the integers are really whole numbers, so you need to keep the zero. Stay tuned, later in the chapter, we'll learn about zero and how it works.

Q: What if my integers get really big? Then the number line won't work as well?

A: Yes and no. You might not be able to count out your answer, that's true, but you could draw a number line with the tick marks meaning 10's, for example.

Q: Is subtraction done?

A: It's just a matter of perspective, really. Subtraction and addition of a negative are the same thing. That being said, it doesn't really matter what you call them; you do the same thing to deal with the operation.

The big take away from this is that you can put the negative inside the parentheses and work with it that way if it's easier in any given situation.

Number sets - all together

Number sets are a way of grouping number types, like integers. Knowing how to group numbers together can help you learn how to work with them when you go into Algebra.

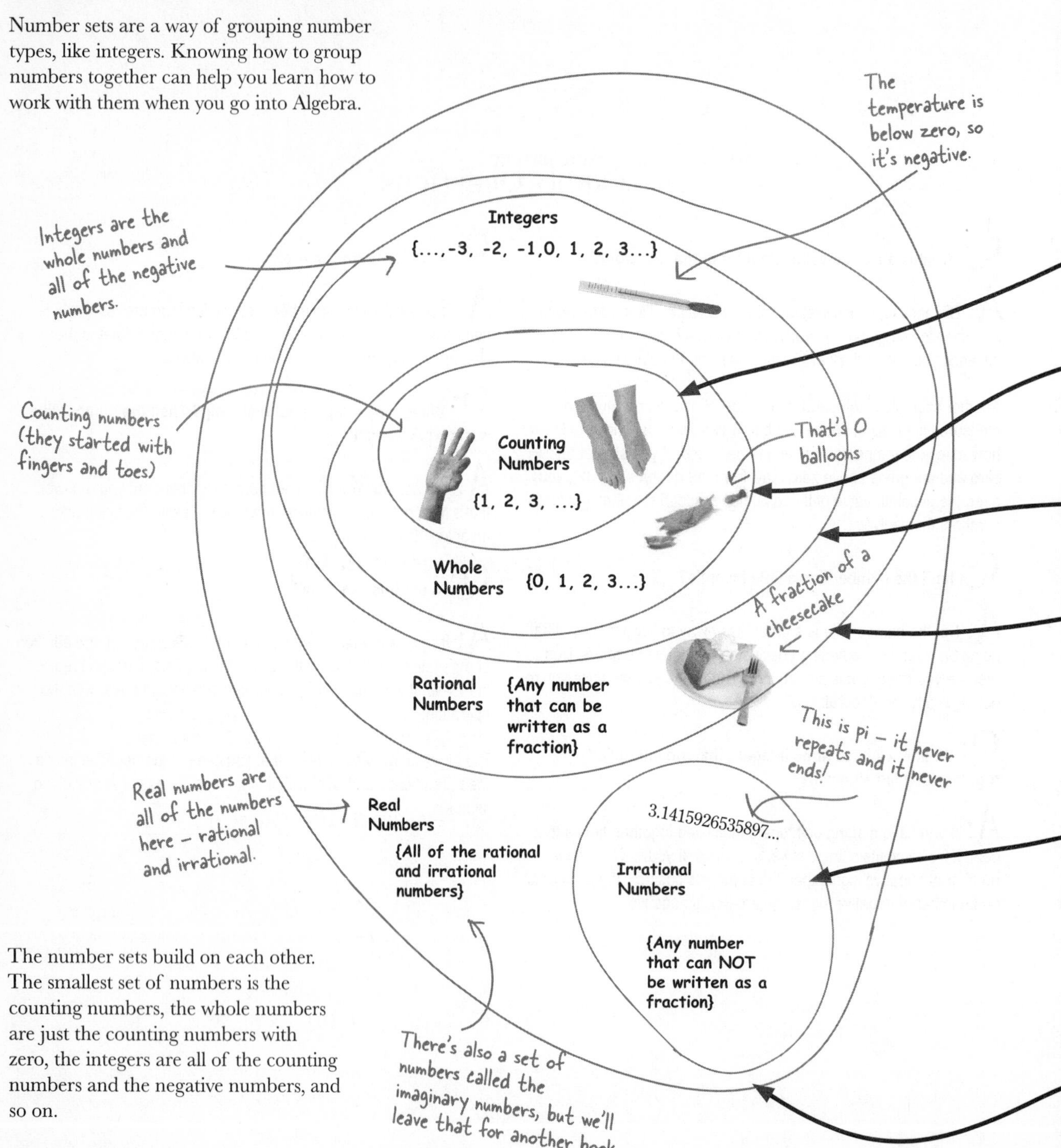

The temperature is below zero, so it's negative.
Integers are the whole numbers and all of the negative numbers.
Integers
{...,-3, -2, -1,0, 1, 2, 3...}
Counting numbers (they started with fingers and toes)
Counting Numbers
{1, 2, 3, ...}
That's 0 balloons
Whole Numbers
{0, 1, 2, 3...}
A fraction of a cheesecake
Rational Numbers
{Any number that can be written as a fraction}
This is pi – it never repeats and it never ends!
3.14159265358979...
Real numbers are all of the numbers here – rational and irrational.
Real Numbers
{All of the rational and irrational numbers}
Irrational Numbers
{Any number that can NOT be written as a fraction}
There's also a set of numbers called the imaginary numbers, but we'll leave that for another book.

The number sets build on each other. The smallest set of numbers is the counting numbers, the whole numbers are just the counting numbers with zero, the integers are all of the counting numbers and the negative numbers, and so on.

The number sets

In mathematics, there is a typical notation for sets of numbers. Here's what it looks like:

The curly braces mean "a mathematical set."

The numbers listed are included in the set and also establish the pattern for the rest of the set.

Counting Numbers: {1, 2, 3, ...}

This means "more in this direction like this" (4, 5, 6, etc.).

The counting numbers are the smallest set. They're just the numbers that you use to count things. It's also the first set of numbers that you work with when you're learning about math.

Whole Numbers: {0,1, 2, 3, ...}

Just the counting numbers with zero added

The whole numbers are just the counting numbers and zero. You'll need zero to indicate there's nothing of something.

Integers: {..., -3, -2, -1, 0, 1 , 2, 3, ...}

The whole numbers plus all of the negative numbers.

Rational Numbers: {Any number a/b}

Rational numbers are a bit more complicated. Any number that can be written as a fraction is a rational number. Since counting and whole numbers (like 2) can be written as a whole number divided by 1 (like 2/1), they're rational, too, just like all of the integers. This number set is helpful when we work with fractions. Decimals that can be converted to fractions are also rational.

Irrational Numbers: {Any number that can't be written as a fraction}

Irrational numbers are numbers that can't be fractions. You'll see them more and more as you get into geometry and the real world, but there are numbers that go on and on—non-repeating and non-terminating decimals. There are square roots that go on forever, and there's pi, the ratio between the circumference and diameter of a circle.

Real Numbers: {All of them so far}

Real numbers are just the set that encompasses all of the numbers, both rational and irrational. Since a rational number can't be irrational, but they're both numbers that can exist in the world, this set includes all of them.

Three ways to split things up

It's important to be able to communicate a piece of a whole thing, and there are three ways you should be familiar with: fractions, decimals, and percentages.

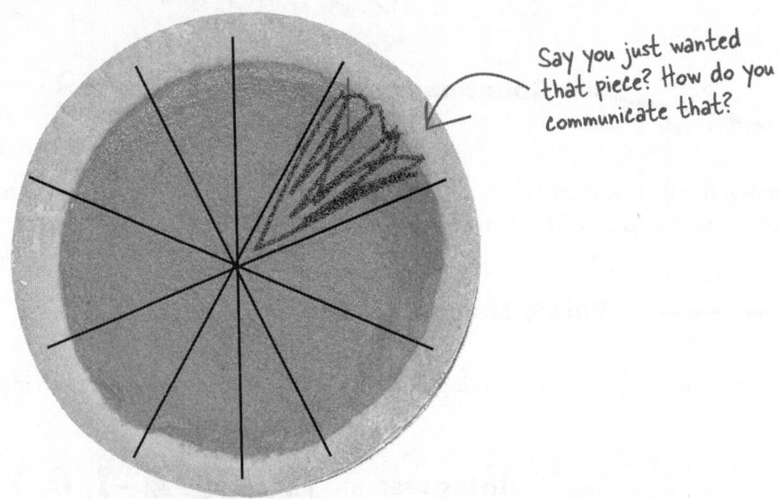

Say you just wanted that piece? How do you communicate that?

Fractions

Fractions are a rational way of expressing a piece.

This is the general form of a fraction

$$\frac{\text{Number of pieces that you have}}{\text{The total number of pieces in a whole}}$$

Our pie piece is: $\frac{1}{10}$ ← One piece out of 10 pieces

Decimals

This notation is the way calculators and computers work.

The decimal point

Number of whole things that you have ● Size of piece less than one

The form of this part is specific, we'll get into it later.

0.10

We're talking about less than one pie, so this is zero.

Percents

This notation is really just a tweak of the decimal. 100% means all of it

The percent sign

Whole Number %

10%

Do you see a relationship between the percent and the decimal?

Now we need to figure out how to work with all these different kinds of numbers. We'll start with decimals...

Decimal's Anatomy

Decimals are the easiest for calculators to work with, and sometimes the easiest for people, too. We'll start with the details of the format itself. Just like whole numbers have ones, tens, and hundreds, decimals have places, too.

Anything to the left of the decimal is the whole number.
tenths
hundredths
1.234
the decimal point
thousandths
The places go on, and on, and on...

How decimals communicate

Just by having the format of a decimal, you can tell two big things about the number.

1. **The number of whole things you're dealing with.**
   The number of whole things is the number to the left of the decimal point. If it's less than one thing, then it'll be zero.

2. **The size of the piece of the thing.**
   The numbers to the right of the decimal point communicate that size in pre-determined pieces. For example, if a number is in the tenths place, it is that many tenths.

If you need to deal with parts, and you need to deal with a calculator, you're probably going to need decimals.

Addition and subtraction with decimals

Addition and subtraction with decimals are almost exactly the same as addition and subtraction with whole numbers, but you must **line up the decimal points.** Just like with whole numbers, with decimals, you have to add the right place, so tenths must be added to tenths, hundredths to hundredths, etc. The way to do that is to make sure that the decimal points line up.

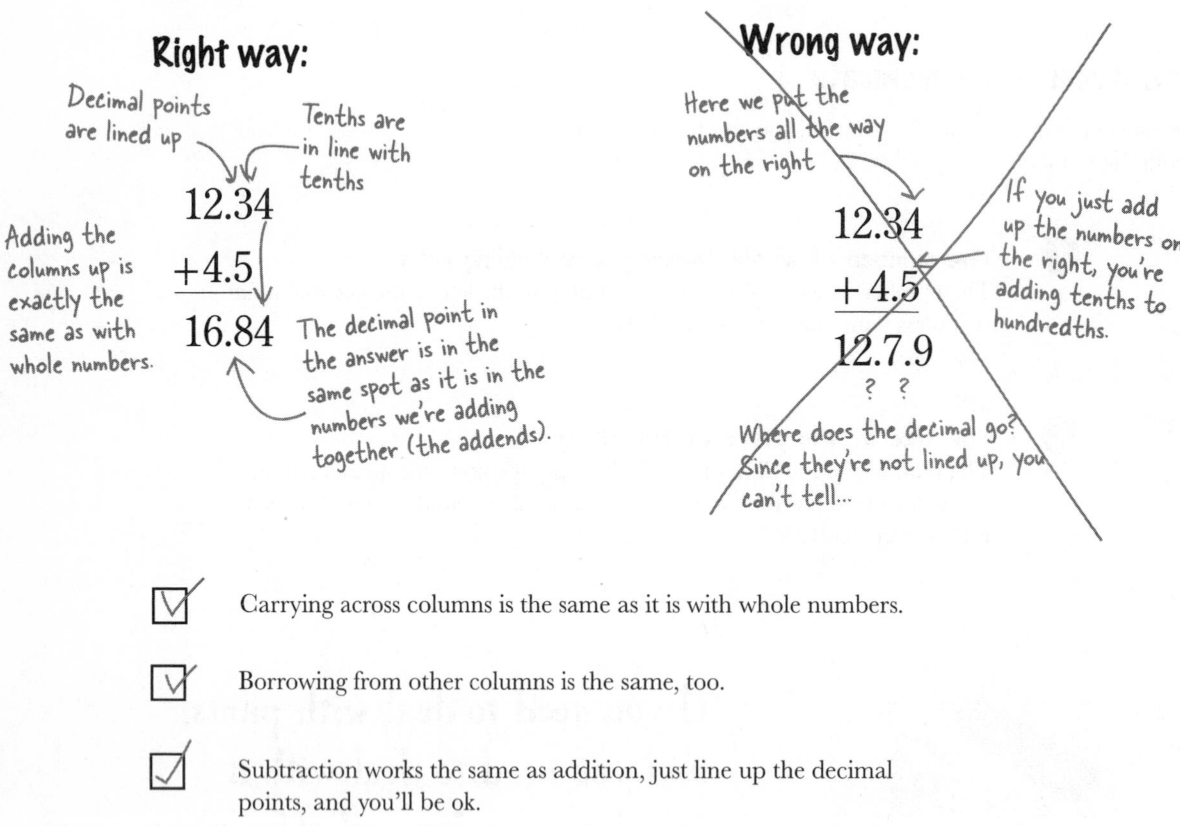
Try this: 12.34 + 5.6 = ?
Right way:
Decimal points are lined up
Tenths are in line with tenths
12.34
+4.5
16.84
Adding the columns up is exactly the same as with whole numbers.
The decimal point in the answer is in the same spot as it is in the numbers we're adding together (the addends).
Wrong way:
Here we put the numbers all the way on the right
12.34
+ 4.5
12.7.9
? ?
If you just add up the numbers on the right, you're adding tenths to hundredths.
Where does the decimal go? Since they're not lined up, you can't tell...
Carrying across columns is the same as it is with whole numbers.
Borrowing from other columns is the same, too.
Subtraction works the same as addition, just line up the decimal points, and you'll be ok.

When you add and subtract decimals you must line up the decimal points.

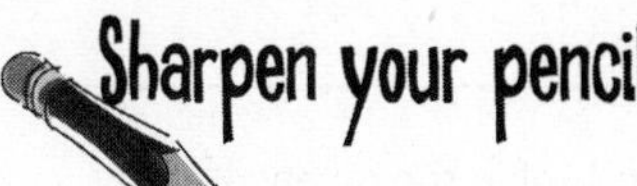

Complete the addition or subtraction and solve the given problems. Make sure to line up the decimals!

$6.9 + 12.41 = ?$

You have to line up the decimal points, remember.

$14.27 - 3.6 = ?$

$3 + 16.01 = ?$

$21.24 - 9.7 = ?$

Sam wants to buy the latest Y-box game—it's $49.99, plus $2.50 in sales tax and $13.65 in shipping (overnight!). How much is this game going to cost him?

Ella, Sam's sister, just got her birthday money, and she has $80 to spend. She wants to get a DVD for $13.35 and a new outfit for $42.35. How much will she have left?

Sharpen your pencil Solution

Complete the addition or subtraction and solve the given problems. Make sure to line up the decimal points!

6.9 + 12.41 = ?

```
  6.90
+12.41
------
 19.31
```

You have to line up the decimal points, remember.

14.27 – 3.6 = ?

```
 14.27
 -3.60
------
 10.67
```

Just borrow from the next column, like regular subtraction.

Add a zero to help keep track of the decimal places.

3 + 16.01 = ?

You have to add the decimal point and two zeros

```
  3.00
+16.01
------
 19.01
```

21.24 – 9.7 = ?

```
 21.24
 -9.70
------
 11.54
```

Lots of borrowing for this one – just do it like you do with whole numbers.

Sam wants to buy the latest Y-box game—it's $49.99, plus $2.50 in sales tax and $13.65 in shipping (overnight!). How much is this game going to cost him?

You can add up as many decimals as you want.

```
  49.99
  +2.50
  13.65
 ------
$ 66.14
```

Just line up the decimal points and add them up!

Ella, Sam's sister, just got her birthday money, and she has $80 to spend. She wants to get a DVD for $13.35 and a new outfit for $42.35. How much will she have left?

First, you have to add up her purchases

```
 13.35
+42.35
------
 55.70
```

Then you can subtract that value from her birthday money.

```
  80.00
 -55.70
 ------
$ 24.30
```

Decimal multiplication

Multiplying decimals is almost the same as multiplying whole numbers; however, there is a twist at the end. You can set up a decimal multiplication problem exactly the same way you would whole numbers: line the numbers up on the right, and start multiplying. After you've multiplied the first number by each digit of the second number, you can add them up.

The last step, the one that's different with decimals, is you need to count the number of places to the right of the decimal point in both of the factors, and then put that total number of decimal places in the answer.

Try this: $14.45 \times 1.5 = ?$

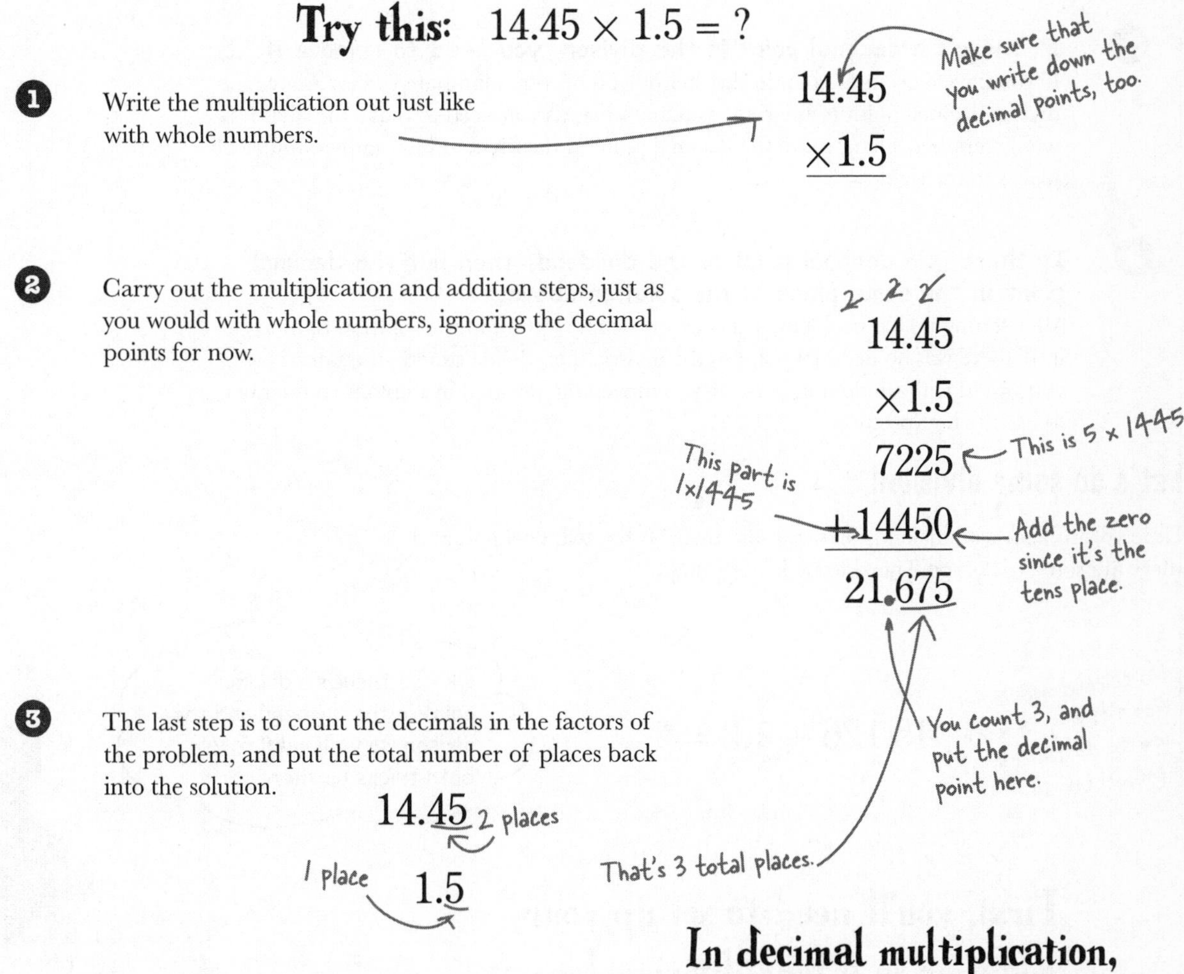

1 Write the multiplication out just like with whole numbers.
14.45
×1.5
Make sure that you write down the decimal points, too.
2 Carry out the multiplication and addition steps, just as you would with whole numbers, ignoring the decimal points for now.
2 2 2
14.45
×1.5
7225
This is 5 x 1445
This part is 1x1445
+14450
Add the zero since it's the tens place.
21.675
3 The last step is to count the decimals in the factors of the problem, and put the total number of places back into the solution.
You count 3, and put the decimal point here.
14.45
2 places
1 place
1.5
That's 3 total places.

In decimal multiplication, count the total number of decimal places in the factors.

Decimal division

Dividing numbers with decimals is similar to regular long division, but with some minor changes. Before we can talk about it, a quick refresher on some terms.

Dividend ÷ Divisor = Solution

These are the two ways you should be used to seeing division.

$$\begin{array}{r} \textit{Solution} \\ \textit{Divisor}\overline{)\textit{Dividend}} \end{array}$$

If you're dealing with decimals, take regular long division and make the following changes:

1. **If there's a decimal point in the divisor, you need to remove it.**
   If you have a decimal point in the divisor, you need to eliminate it. First, move the divisor decimal point to the right as many times as you need to make the divisor a whole number. Then, move the decimal point in the dividend the same number of places to the right.

2. **If there is a decimal point in the dividend, then put the decimal point in the same place in the solution above.**
   After setting up the problem, just stick the decimal point in the solution right above it. If there was no decimal point in the divisor, then divide exactly the same way you would with whole numbers. If you moved the decimal in step one, remember to use the new position.

Let's do some division!

Here's a problem with all kinds of twists and turns. It has two decimals and there are some tricks you'll need to make it through.

No calculators please! We're trying to learn how to do this by hand, after all.

$$15.126 \div 1.2 = ?$$

Ok - so there's a decimal point in the dividend and the divisor, so we'll need to use both tricks up there.

First, you'll need to set up your equation in a long division format...

Decimal division training

1. Write your problem out in long division form.

$1.2\overline{)15.126}$

2. Adjust the decimal points as needed. If the divisor has a decimal point, move it to the right until you have a whole number. Then move the same number of places in the dividend. After that's done, write the decimal point in the solution in the same place as in the dividend.

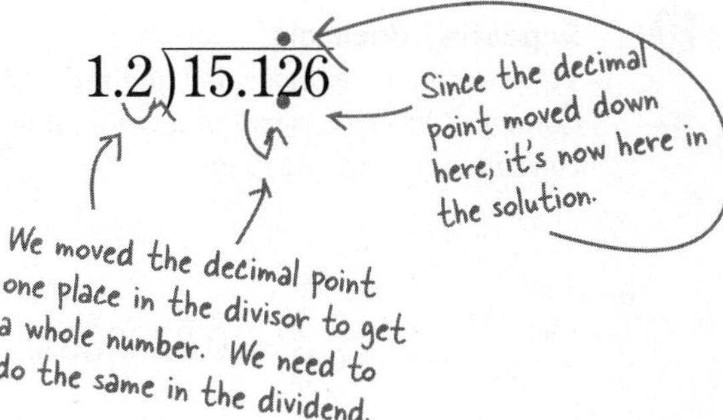
1.2)15.126
Since the decimal point moved down here, it's now here in the solution.
We moved the decimal point one place in the divisor to get a whole number. We need to do the same in the dividend.

3. Proceed with the long division. This is exactly the same as with whole numbers.

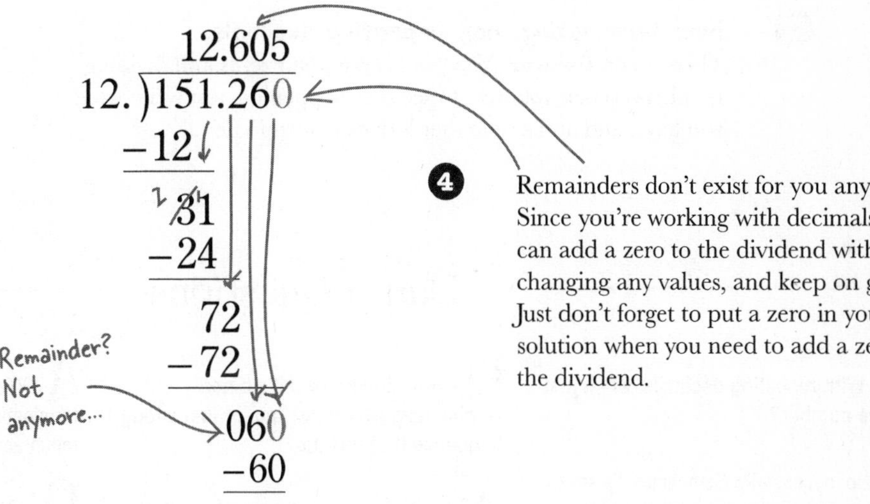
12.605
12.)151.260
−12
31
−24
72
−72
060
−60
0
Remainder? Not anymore...

4. Remainders don't exist for you anymore! Since you're working with decimals, you can add a zero to the dividend without changing any values, and keep on going. Just don't forget to put a zero in your solution when you need to add a zero to the dividend.

5. Keep going with your division. When you finish a division with no remainder, you're done!

There's a couple of exceptions to this – turn the page...

Special decimals

There are a couple of special cases that you need to be prepared for with decimals. Decimal division can actually end in three ways:

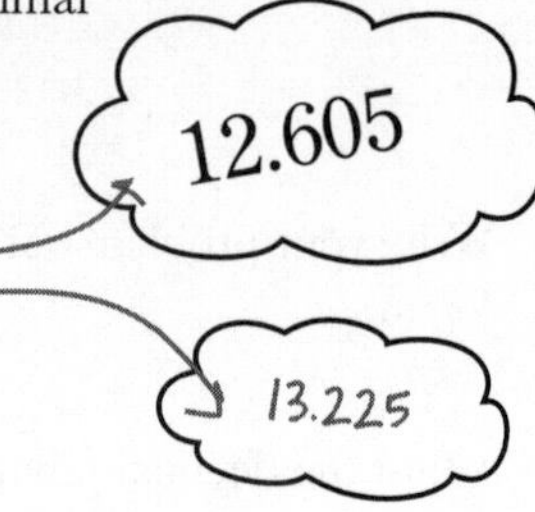

1. **Terminating decimals.**
   The one we just did is a terminating decimal. That means it ends.

2. **Repeating decimals.**
   These decimals get in a rut. The thirds (1/3, 2/3) are the easiest example. Once you start the division, it never ends. There's a convention for writing them:

Go ahead and try this. It's 1 divided by three.

$$\frac{1}{3} = 0.3333333... \ or \ 0.\overline{3}$$

This bar means the number after the decimal is repeating.

3. **Non-terminating, non-repeating decimals.**
   These go on **forever**. You just keep adding zeros and dividing numbers. When you start to see that happening, just write what you have, and add a note that it doesn't terminate.

there are no Dumb Questions

Q: With repeating decimals, will it just be one number?

A: Not necessarily. Sometimes it's several numbers that repeat, like 1.234234234.

These numbers are written with a bar on top of the entire sequence that repeats:

$$1.\overline{234}$$

Q: How do I know if it's a non-terminating, non-repeating, or just a long sequence that repeats?

A: Most of the time, the problem will tell you how many decimals to use before you decide. For example, it may say, "Only take the division out 5 places."

Q: How do you use repeating decimals? You can't add them to other numbers, right?

A: With repeating decimals and non-terminating, non-repeating decimals, it's much easier to work with them as fractions.

Q: Why can you just add zeros on the end when dividing by decimals?

A: Since the zero comes after the decimal, it's not changing the value of the dividend at all. 15.126 has the same value as 15.12600000. Adding the zero is a trick that makes it easier to finish long division.

Solve the following multiplication and division problems. You can add as many zeros as you need after the decimal, but don't forget about repeating and non-terminating, non-repeating decimals. If you go out 4 places and still don't have an answer, you can stop.

$15.1 \times 0.72 = ?$

$23.2 \div 5 = ?$

$\times$

Just because you don't start with decimals, doesn't mean you won't end up there.

$56 \div 3 = ?$

$10.6 \div 0.34 = ?$

Solve the following multiplication and division problems. You can add as many zeros as you need after the decimal, but don't forget about repeating and non-terminating, non-repeating decimals. If you go out 4 places and still don't have an answer, you can stop.

$15.1 \times 0.72 = ?$

There's no need to line up the decimals points, just keep track of them.

3 places in the solution because there is 1 in the first factor and 2 in the second.

```
   15.1
  ×0.72
    302
  10570
 10.872
```

The only new thing you have to do is put the decimal point in here.

$23.2 \div 5 = ?$

Just put the decimal point in the answer directly above the decimal point in the dividend.

```
   4.64
5)23.20
 -20
    32
   -30
    20
```

Just because you don't start with decimals, doesn't mean you won't end up there.

$56 \div 3 = ?$

This bar means repeating. The number it's over (six) is what repeats.

```
   18.66  (bar over final 6)
3)56.00
 -3
  26
 -24
   20
  -18
   20
   18
    2
```

Hmmm...this is going to go on forever. Looks like a repeating decimal.

$10.6 \div 0.34 = ?$

The answer decimal point needs to go right above the divisor decimal point.

You have to move this decimal two places

So this decimal point has to move the same two places.

```
           311764
0.34)10.600000
     -102
        40
       -34
        60
       -34
        260
       -238
        220
       -204
        160
       -136
         24
```

This is still going! Let's call it non-repeating, non-terminating.

If you want to check this, then go ahead and pull out the calculator. It'll fill up your screen!

You're 100% right!

Decimals and percentages are almost the same thing. Percentages are just a convention that we use to work with decimals, primarily those between zero and one. 0% is 0, and 100% is 1; everything else is some decimal in between. There are a lot of times when you need to talk about some consistent piece of something that's between zero and one—tax rates, contribution rates...

Dealing with percentages is easy. Just convert them back to decimals and work with them that way.

Saying 5% sounds way better than 5 hundredths.

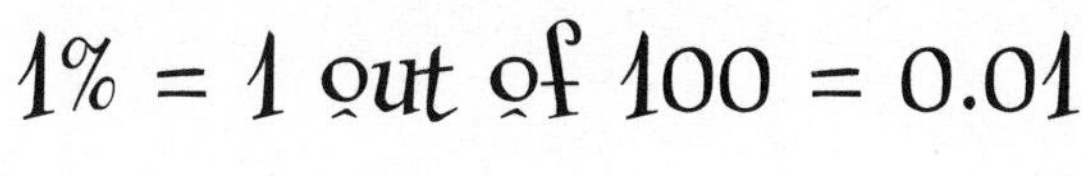

To convert a percent to a decimal, just move the decimal point two places to the left.

there are no Dumb Questions

Q: Does it work the other way—going from a decimal to a percent?

A: Yes. If you want to convert a decimal to a percent, just move the decimal point to the right two places.

Q: What if the percent is bigger than 100?

A: The process stays the same; you move the decimal two places to the left. The number that you come up with at the end will be over one, that's all.

Q: Can a percentage be written as a fraction?

A: It sure can, and we'll find out how to do it soon. For now, just remember that all of these things—decimals, percents and fractions— are all ways to work with ***parts*** of numbers.

Working with percents

With a typical percentage problem, you're looking for some percent of something, like sales tax. In chapter 1, we help Jo buy a game console, and when you buy something, you have to pay the taxes:

SUPER BUY!
KillerX 2.0
Gaming Syste
The brand new KillerX
2.0 includes full circle
entertainment value.
game controller inclu
(OPOD-112)
special value
$199

The game console is $199.

First let's deal with sales tax...

The base price is $199. After that, we need to think about taxes. In Mathopolis, the sales tax rate is 5%. Let's figure out how much we'll have to pay in taxes.

To find the percent of something, there a three steps.

1. Figure out the number you are taking the percentage of.

You need to find the percentage of the base price, $199.

199

2. Convert the percentage to a decimal. You do that my simply moving the decimal point in the percent two places to the ***left***.

Start with 5%

5%

Moving the decimal two places means you have to add a zero.

Take out the percent sign.

.0 5 % = 0.05

This decimal is there, even if it's not written.

3. Multiply the decimal to the first number. The value you'll get is the result you're looking for!

$$
\begin{array}{r}
199 \\
\times 0.05 \\
\hline
\\
+\quad\quad \\
\hline
\end{array}
$$

Do the math!

Whatever this value turns out to be is the value you're looking for.

Answer: 9.95 (printed upside down)

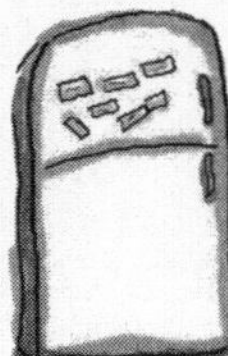

Percent Magnets

Use the magnets to complete the problems with percents.

Just move the decimal two places to the left.

Percents can still have decimals in them...

15% = ____

0.027 = ____

More than 100% is over one.

117% = ____

0.39 = ____

Really Big Mall is trying to decide if their customers would use Wi-Fi in the mall if it was installed. So, they took a survey of 618 shoppers and found that 61% said that they would use Wi-Fi if it was available. How many customers does that represent?

61% = ____

____ × 618 = ?

$$\begin{array}{r} 618 \\ \times \underline{\quad\quad} \\ 618 \\ + \underline{\quad\quad} \\ 37080 \end{array}$$

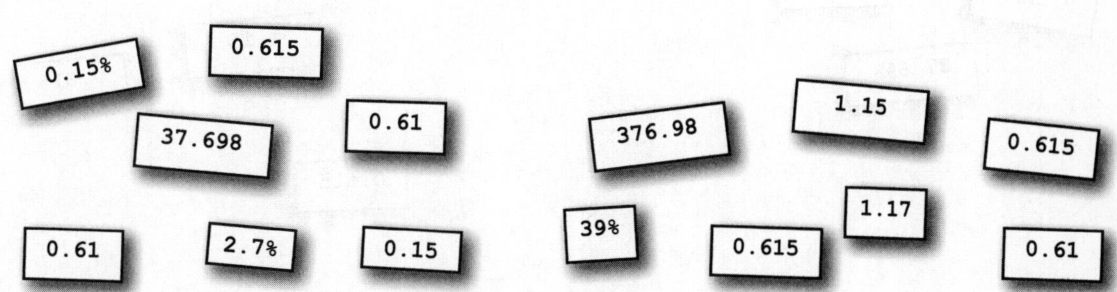

Percent Magnets Solution

Use the magnets to complete the problems with percents.

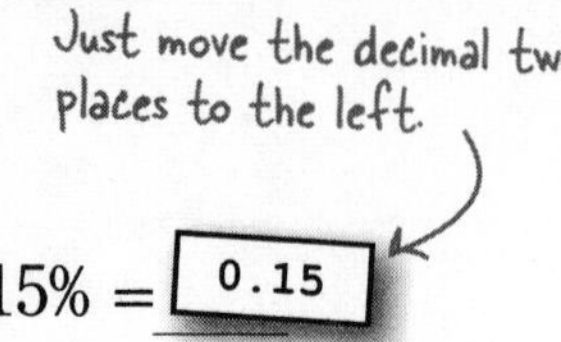
Just move the decimal two places to the left.
15% = 0.15

Percents can still have decimals in them...

Go the other way – two places to the right.

0.027 = 2.7%

More than 100% is over one.

117% = 1.17

0.39 = 39%

Really Big Mall is trying to decide if their customers would use Wi-Fi in the mall if it was installed. So, they took a survey of 618 shoppers and found that 61% said that they would use Wi-Fi if it was available. How many customers does that represent?

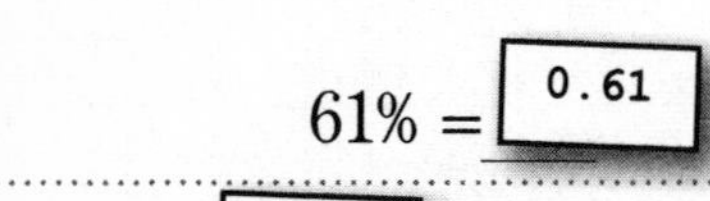
61% = 0.61

0.61 × 618 = ?

618
× 0.61

618
+ 37080

376.98

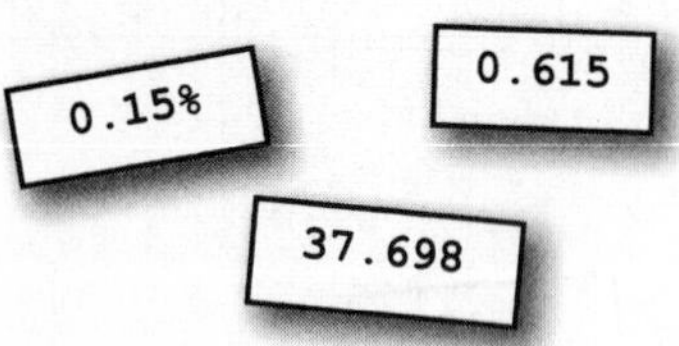
0.15%
0.615
37.698

1.15

0.615

0.615

Fractions

I hate fractions. They're hard and I don't see how they help anything. I'm just going to use my calculator. I'm so out of here...

Wait! They're not that bad.

Fractions are really helpful. Once you get the hang of them, they can be faster and more precise than decimals.

Remember repeating decimals?

They are a little tricky to get the hang of, so let's start with a review of exactly what a fraction means.

Fractions show parts of a whole

A fraction tells you how many pieces you have of something and how many pieces that thing is broken into.

This is the numerator → **Number of pieces that you have**

———

The total number of pieces in a whole ← This is the denominator

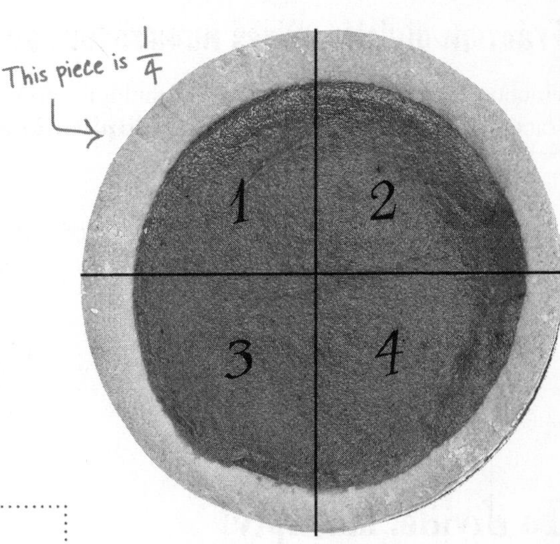

Relax

If you're worried about fractions, we'll fix you up.

Fractions are a fact of life, but once you get the hang of working with them, they're quite handy. Keep going, and you'll be fraction genius in no time.

Fraction multiplication

Fraction multiplication is the easiest operation to work with. You simply multiply the numerators together to get the answer's numerator, and then multiply the denominators together to get the answer's denominator.

Just multiply the numerators together.

$$\frac{1}{2} \times \frac{1}{3} = \frac{1 \times 1}{2 \times 3} = \frac{1}{6}$$

Write the new fraction – that's it!

Start with two fractions.

Then multiply the denominators together.

1. Multiply the numerators, and write the value as the numerator of the solution.
2. Multiply the denominators, and write the value of the denominator for the solution.

Fraction division mixes numerators and denominators

Fraction division is actually more like multiplication with a twist. To divide a fraction by a fraction, use a process called **cross multiplication**. Here's how to cross multiply:

To cross multiply, just multiply the first numerator by the second denominator to get the solution numerator.

Start with two fractions.

$$\frac{1}{3} \div \frac{1}{2} = \frac{1 \times 2}{3 \times 1} = \frac{2}{3}$$

Then multiply the first denominator by the second numerator to get the solution denominator.

To divide, multiply!

1. Multiply the ***first numerator*** by the ***second denominator***, and write the value as the **numerator** of the solution.
2. Multiply the ***first denominator*** by the ***second numerator***, and write the value as the **denominator** of the solution.

Now try it...

Match each problem to it's solution.

Problems	Solutions
$\frac{1}{20} \div \frac{6}{7}$	$\frac{5}{30}$
$\frac{3}{7} \times \frac{1}{11}$	$\frac{7}{120}$
$\frac{1}{12} \div \frac{2}{9}$	$\frac{3}{8}$
$\frac{1}{8} \div \frac{1}{3}$	$\frac{1}{24}$
$\frac{1}{2} \times \frac{3}{1}$	$\frac{9}{24}$
$\frac{1}{10} \div \frac{3}{5}$	$\frac{3}{77}$
$\frac{1}{8} \times \frac{1}{3}$	$\frac{3}{2}$

Solution

Match each problem to it's solution.

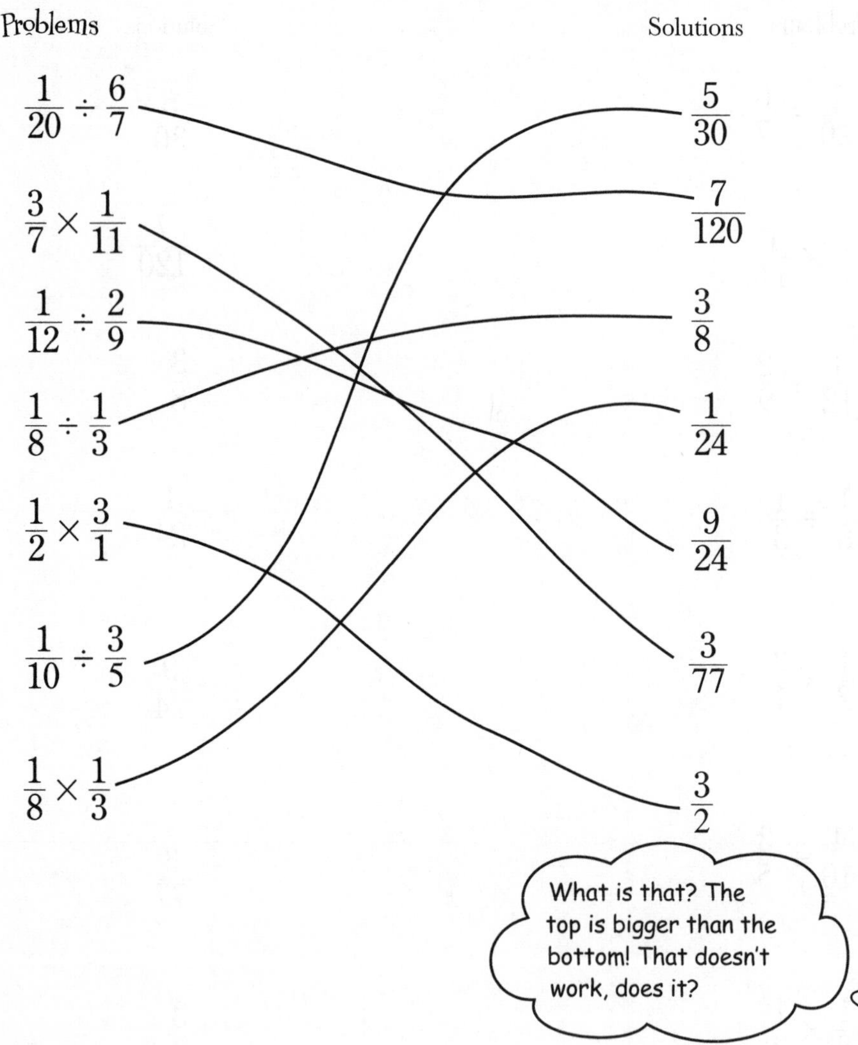

What is that? The top is bigger than the bottom! That doesn't work, does it?

Improper fractions

An improper fraction is one where the numerator is larger than the denominator. Since the numerator is larger, the fraction actually represents ***more than*** one.

For example, the last fraction in the previous exercise has a denominator of 2, which means that the whole is cut into 2 pieces. The numerator of three means that there are three pieces of it (more than one total pie).

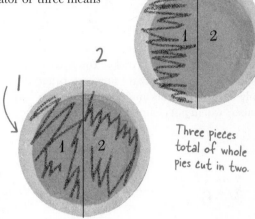

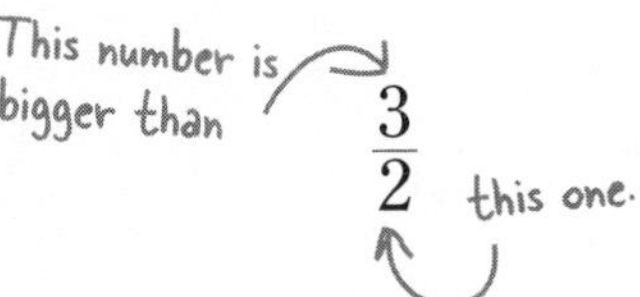

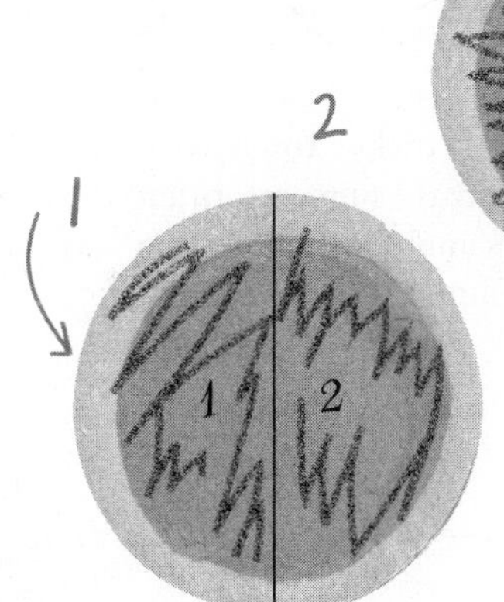

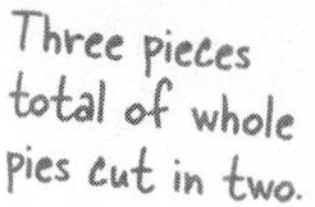

Three pieces total of whole pies cut in two.

Sometimes (like with multiplication and division) it makes more sense to keep working with the improper fraction. But what if you just wanted to know how many pies there are?

Divide to make an improper fraction proper

To convert an improper fraction to a proper fraction, you need to remember that the line in the fraction means division.

This line means division. $\frac{3}{2}$ Another way to read this would be "3 divided by 2."

So, do the division!

$$
\begin{array}{r}
1 \\
2\overline{)\,3} \\
-2 \\
\hline
1
\end{array}
$$

The answer is 1 remainder 1.

Any remainder of the division just goes back over your denominator. The whole number part of the division stays as a whole number.

$$1\frac{1}{2}$$

The one left over is still 1 over 2.

Here's your proper fraction (the official term is a mixed number).

If you look up at the pie picture again, it's one whole plus one half of a pie.

More about improper fractions

So what if you had to solve this problem?

$$2\frac{1}{6} \times 4\frac{1}{2} = ?$$

The mixed numbers make things tricky. You have to make these improper fractions before you can multiply the numerators and the denominators. That way you only have numerators and denominators and no whole numbers. Whole numbers would just complicate this problem.

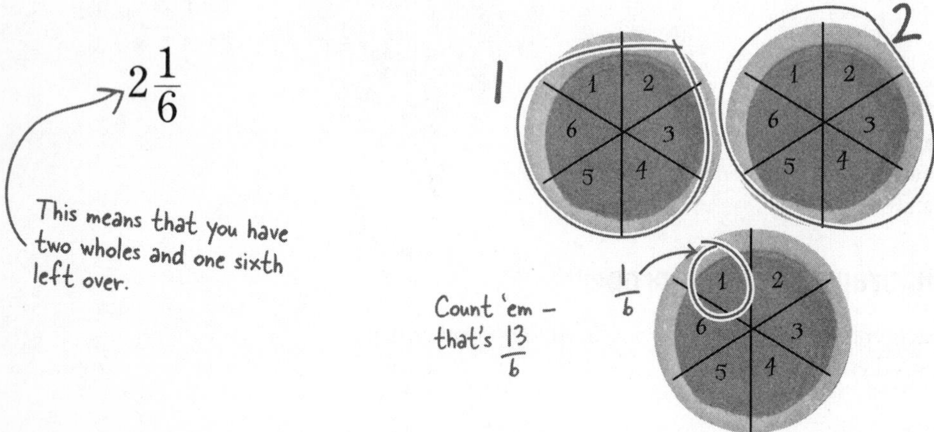

$2\frac{1}{6}$
This means that you have two wholes and one sixth left over.
1
2
1/6
Count 'em - that's 13/6

To get to a mixed number, you had to do division, so to get back to the improper number, you have to do the opposite—multiplication.

1. To convert an improper fraction, first write the denominator for the solution. It will be the same as the denominator for the fraction portion of the mixed number.

2. To find the **new numerator**, multiply the whole number times the denominator and add the **old** numerator. That's it!

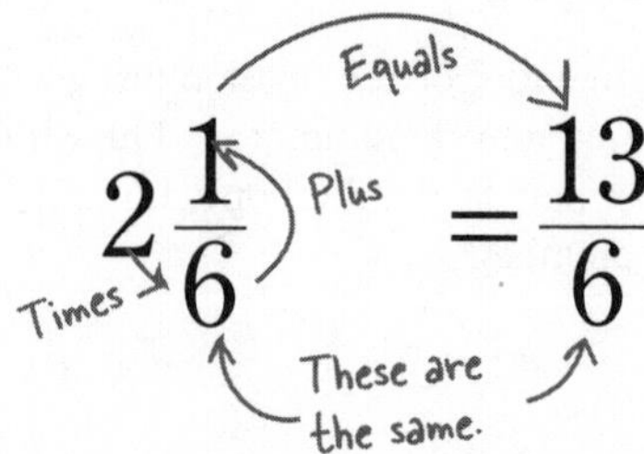

Equals
Plus
Times
$2\frac{1}{6} = \frac{13}{6}$
These are the same.

Convert the following fractions from proper to improper fractions to complete the operations. Reduce all of your answers to proper fractions..

times $1\frac{2}{9}$ plus = ______

$3\frac{3}{8}$ ______

times $2\frac{5}{7}$ plus ______

1 times 3 plus 1

$1\frac{1}{3} \times \frac{7}{8} =$ ....

....

....

$2\frac{6}{7} \times 1\frac{1}{4} =$ ....

....

....

Use this space for reducing back to a proper fraction.

Work here to reduce the solution to a proper fraction

Bake your fractions Solution

Convert the following fractions from proper to improper fractions to complete the operations. Reduce all of your answers to proper fractions..

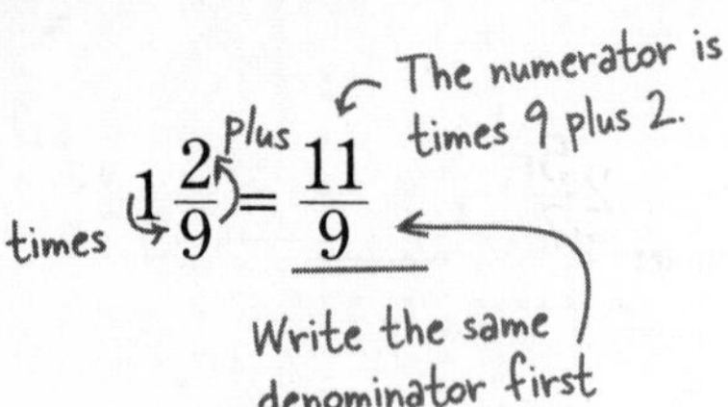

$1\frac{2}{9} = \frac{11}{9}$

The numerator is 1 times 9 plus 2.

Write the same denominator first

$3\frac{3}{8} = \frac{27}{8}$

3 times 8 plus 3

Denominator first

$2\frac{5}{7} = \frac{19}{7}$

2 times 7 plus 5

1 times 3 plus 1

$1\frac{1}{3} \times \frac{7}{8} = \frac{4}{3} \times \frac{7}{8}$

$= \frac{28}{24}$

This needs to be reduced back to a proper fraction again.

$24\overline{)28}$ = 1, -24, remainder 4: $1\frac{4}{24}$

To reduce this back to a proper fraction, divide by 24 and write the remainder as the new numerator.

$2\frac{6}{7} \times 1\frac{1}{4} = \frac{20}{7} \times \frac{5}{4}$

$= \frac{100}{28}$

To write these, you can just write the factors of both 20 and 6 multiplied together.

$28\overline{)100}$ = 3, -84, remainder 22: $3\frac{22}{28}$

To reduce this back to a proper fraction, divide by 28 and write the remainder as the new numerator.

BULLET POINTS

- To multiply fractions, multiply the numerators and multiply the denominators.
- To divide fractions, cross multiply.
- To convert an improper fraction to a mixed number, divide.
- To convert a mixed number to an improper fraction, multiply.

I learned how to do fraction division a different way. We used reciprocals.

It's true, there's a shortcut to divide fractions.

Cross multiplication is the method that is the most straightforward, but you can also use the reciprocal.

Invert a fraction to get its reciprocal

The reciprocal of a fraction is a fraction with the numerator and denominator switched. To divide two fractions using this method, you can multiply by the reciprocal instead of cross multiplying.

$$\frac{1}{2} \qquad \frac{2}{1}$$

The reciprocal of this is this.

See, we just flipped them over!

Fraction division - option #2

Here's the reciprocal way of doing things. It will give you *exactly* the same answer as cross multiplication:

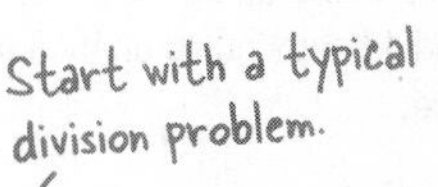
Start with a typical division problem.

$$\frac{1}{4} \div \frac{6}{7} =$$

1. Replace the division sign with multiplication, and replace the second factor with its reciprocal.

$$\frac{1}{4} \times \frac{7}{6} =$$

2. Treat it just like any other multiplication problem and multiply straight across.

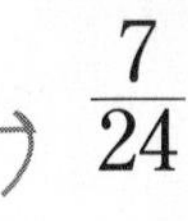

$$\frac{7}{24}$$

BRAIN
BARBELL

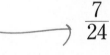

Try the same problem using cross multiplication - you'll get the same answer!

there are no Dumb Questions

Q: Why do we have to learn fractions? Don't most people use calculators?

A: It's true that most people use calculators. The problem is that when you get to funky decimals like repeating decimals, for instance, it's actually way easier to work with the fraction. If you need to carry that small number through a few steps in Algebra, it could get ugly.

Q: How do you know what each decimal place is? 10ths, 100ths?

A: That is basic number knowledge, and you just have to memorize it. The good news is that 0.1 is one tenth, 0.01 is one hundredth, and it keeps going up by a factor of 10 each time.

Knowing that, it's easy to convert a decimal to a fraction.

Q: Addition and subtraction of decimals are the same as whole numbers?

A: Yes if you LINE UP THE DECIMALS. Don't line them up to the right, because then you'll have 100ths added to 10ths, and that won't work.

Q: How long can I keep adding zeroes if I'm dividing?

A: Good question. As long as you want, really. If you have a repeating decimal, you'll figure it out pretty fast—probably by the hundredths place. Otherwise, keep going until you have no remainder or you have enough decimal places that you can answer your question. That will depend upon the context of the problem.

Q: Percents are really just decimals?

A: That's right! They were developed so we could easily talk about one hundredth of a thing. Since that's also how divide up our money, it's so convenient!

Q: For what do you use improper fractions for?

A: They are great for speed. If you have several steps to go through with multiplication and division of fractions, it's much easier to keep working with the improper fraction. If you convert improper fractions in the middle of a problem to a proper fraction (say, a whole number and a fraction), and then you need to multiply, you'll just end up going back again.

Q: Which is better, cross multiplication or using the reciprocal for fraction division?

A: It's a style thing, really. Both of them work fine, but for one, you need to rewrite the fraction, and for the other, you don't. It may bother you that cross multiplication doesn't involve writing exactly what you're doing. When you write out the reciprocal, the math you're doing is exactly the same thing, but the notation is different.

Adding and subtracting fractions

Addition and subtraction of fractions are a bit more complicated than multiplication and division. You can multiply and divide any two fractions without much trouble. To add and subtract, two fractions must have ***the same denominator.***

Why? Because the answer you are looking for will be in a given denominator, which says how many pieces your whole is cut up into.

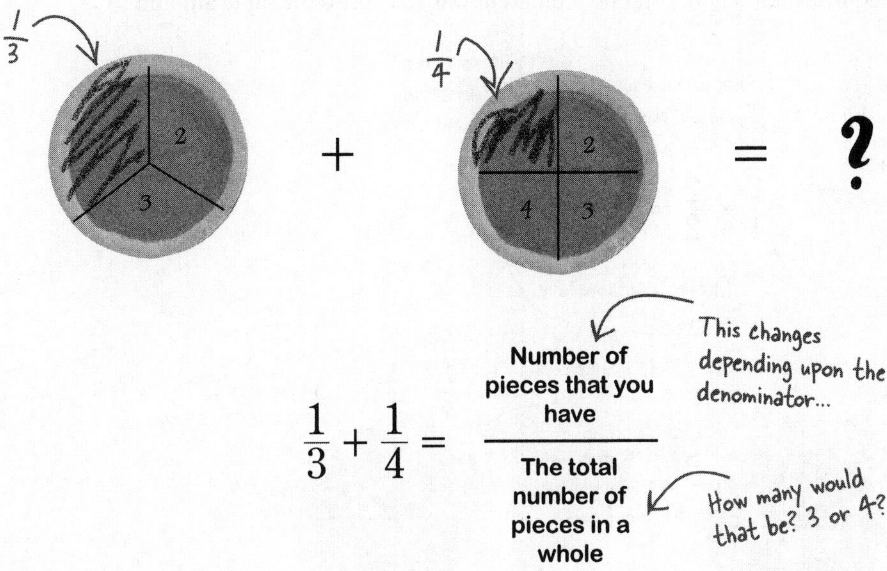

1/3
1/4
2
3
2
4
3
+
=
?
This changes depending upon the denominator...
$\frac{1}{3} + \frac{1}{4} = \frac{\text{Number of pieces that you have}}{\text{The total number of pieces in a whole}}$
How many would that be? 3 or 4?

You need a common denominator

The moral of the story is that we need to find a way to change the denominator of the fraction without changing it's value. But we've got a couple things we need to learn about before you can make that happen.

Equivalent fractions get you matching denominators

You may remember from your times tables that any number times one is itself. This one little fact makes it possible to change a denominator without changing the value of the fraction—the size of the pie piece that you're going to get.

To get an **equivalent fraction** (say, to add them, since they need the same denominator), you have to multiply the numerator and the denominator by the *same number* (like 2 over 2). You can do that because 2 over 2 is equal to one, so you're not changing the value of the fraction, just the way it's written.

When you do this to get an equivalent fraction, you find a different way to express the same amount of pie. Let's try it out.

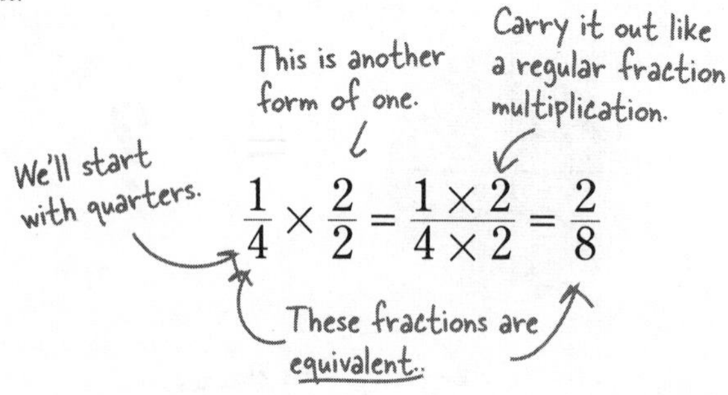
This is another form of one.
Carry it out like a regular fraction multiplication.
We'll start with quarters.
These fractions are equivalent.

$$\frac{1}{4} \times \frac{2}{2} = \frac{1 \times 2}{4 \times 2} = \frac{2}{8}$$

1/4
These are both the same amount of pie to eat.
2/8

Ok, so we can come up with equivalent fractions. But we still don't know how to add up two fractions with different denominators. How do you know what to multiply by?

We're trying to add these two together, remember?

$$\frac{1}{3} + \frac{1}{4} = \textbf{?}$$

Use the lowest common denominator for addition

What we're looking for is a common denominator for both fractions. To keep the math simple, it should be the **lowest common denominator**.

Lowest common denominator

The smallest number... that all the fractions involved *could* share... as a denominator.

To find the lowest common denominator, you need to find the **least common multiple** of the numbers that are the denominators of your fractions (in our case 3 and 4). Once you've done that, then you can figure out how to find the equivalent fractions of each addend with the right denominator.

A multiple of a number is the result of the original number times any other number.

These are the multiples that the denominators have in common.

Multiples of 3: 3, 6, 9, 12, 15, 18, 21, 24...

Multiples of 4: 4, 8, 12, 16, 20, 24...

We're looking for the least common multiple, so we're stopping here, but we could go on forever.

When you're presented with two fractions, and you want the lowest common denominator, just list the multiples for each number and choose the smallest one that they share.

Since we've found 12 is our least common multiple, now we can use it as the lowest common denominator.

Sum it up

Least common multiple (LCM) – The lowest multiple that numbers have in common.

Lowest common denominator (LCD) – The lowest common multiple of denominators.

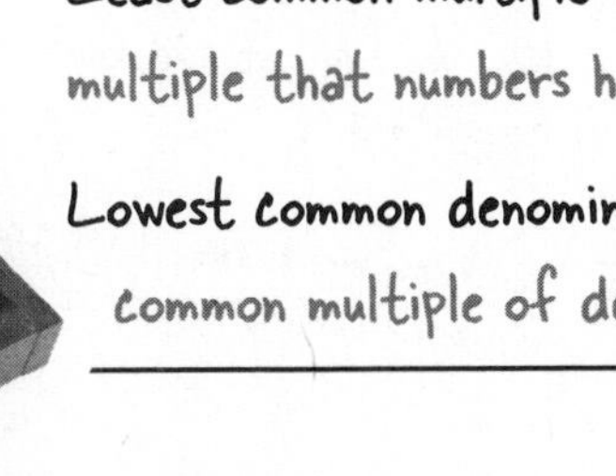

Fraction addition and subtraction training

Still feel like you can't add fractions? You're closer than you think—you just need to put it all together. Find the lowest common denominator (LCD) and multiply by 1 to get a common denominator.

1. **Write out the initial problem.** If you have a word problem, make sure that you know what you're working with by writing it out first.

This is that same problem, again. You're ready to do it this time!

$$\frac{1}{3} + \frac{1}{4} = ?$$

2. **Figure out the LCD.** Take the denominators you're working with and find the LCD, just like we did earlier.

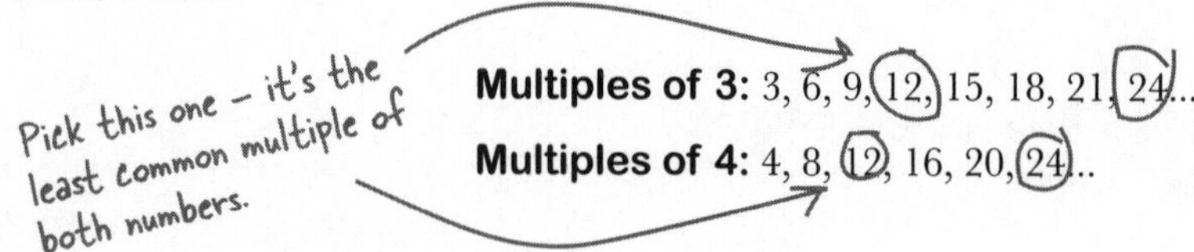

Pick this one - it's the least common multiple of both numbers.
Multiples of 3: 3, 6, 9, 12, 15, 18, 21, 24...
Multiples of 4: 4, 8, 12, 16, 20, 24...

3. **Determine what version of 1 to use to get both fractions with the LCD.** You know what denominator you need to get to (12). Now you have to figure out what fractional form of 1 to use to get there.

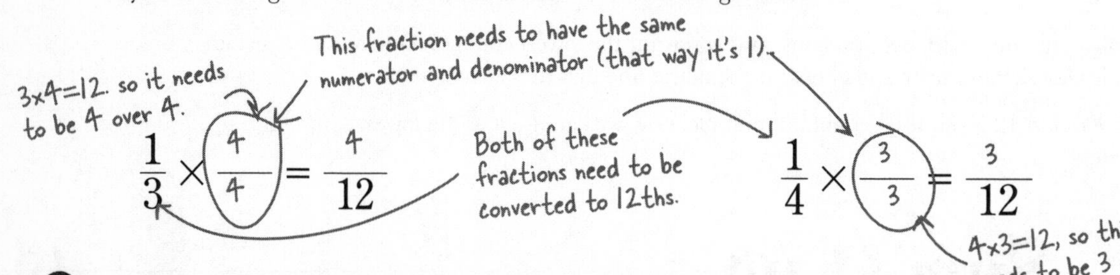

3x4=12, so it needs to be 4 over 4.
This fraction needs to have the same numerator and denominator (that way it's 1).
1/3 × 4/4 = 4/12
Both of these fractions need to be converted to 12ths.
1/4 × 3/3 = 3/12
4x3=12, so this fraction needs to be 3 over 3.

4. **Add (or subtract) the new equivalent fractions.** Replace the fractions from the original problem with their equivalent ones and then add them to find out the solution.

Replace the old fractions with the equivalent fractions you figured out here.

$$\frac{1}{3} + \frac{1}{4} =$$

$$\frac{4}{12} + \frac{3}{12} = \frac{7}{12}$$

Now you can just add the numerators up.

You know right away that the answer will be in 12ths since that's your LCD.

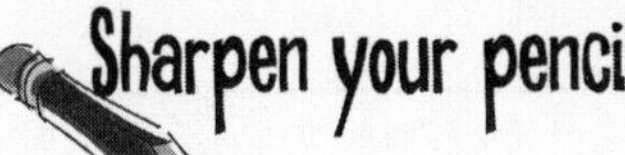

Use your new skills converting fractions to common denominators to add and subtract the fractions.

$\frac{2}{5} + \frac{3}{10} = ?$

Multiples of 5: ....

Multiples of 10: ....

Multiply the first fraction by 2 over 2 to get the LCD.

The second fraction you can multiply by one.

....

$\_ + \_ = \_$

....

Rewrite the original equation with the new common denominator

$\frac{5}{6} + \frac{2}{15} = ?$

Multiples of 6: ....

Multiples of 15: ....

....

$\_ + \_ = \_$

....

Rewrite the original equation with the new common denominator

$\frac{3}{4} + \frac{1}{6} = ?$

Multiples of 4: ....

Multiples of 6: ....

....

$\_ + \_ = \_$

....

Rewrite the original equation with the new common denominator

$\frac{16}{12} + \frac{1}{5} = ?$

Multiples of 12: ....

Multiples of 5: ....

....

....

$\_ + \_ = \_$

....

Rewrite the original equation with the new common denominator

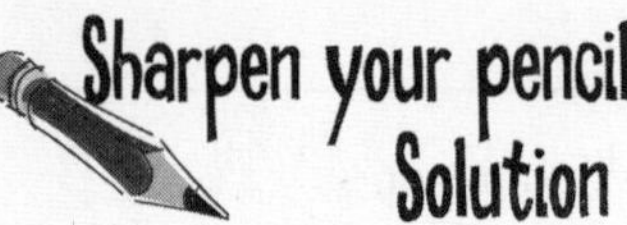

Use your new skills converting fractions to common denominators to add and subtract the fractions.

$\frac{2}{5} + \frac{3}{10} = ?$

This is the LCM

Multiples of 5: 5, 10, 15, 20, 25, 30...

Multiples of 10: 10, 20, 30, 40, 50...

Multiply the first fraction by 2 over 2 to get the LCD

$\frac{2}{5} \times \frac{2}{2} = \frac{4}{10}$ $\frac{3}{10} \times \frac{1}{1} = \frac{3}{10}$

$\frac{4}{10} + \frac{3}{10} = \frac{7}{10}$

Since this fraction is already in the LCD, you can multiply by one

$\frac{5}{6} + \frac{2}{15} = ?$

The LCM

Multiples of 6: 6, 12, 18, 24, 30, 36

Multiples of 15: 15, 30, 45, 60

5 over 5

$\frac{5}{6} \times \frac{5}{5} = \frac{25}{30}$

2 over 2 for the LCD

$\frac{2}{15} \times \frac{2}{2} = \frac{4}{30}$

$\frac{25}{30} + \frac{4}{30} = \frac{29}{30}$

$\frac{3}{4} + \frac{1}{6} = ?$

Multiples of 4: 4, 8, 12, 16, 20, 24

Multiples of 6: 6, 12, 18, 24, 30

$\frac{3}{4} \times \frac{3}{3} = \frac{9}{12}$ $\frac{1}{6} \times \frac{2}{2} = \frac{2}{12}$

$\frac{9}{12} + \frac{2}{12} = \frac{11}{12}$

$\frac{16}{12} + \frac{1}{5} = ?$

We had to go out a ways to find a common denominator

Multiples of 12: 12, 24, 36, 48, 60, 72

Multiples of 5: 5, 10, 15, 20, 25, 30, 35, 40, 45, 50, 55, 60

$\frac{16}{12} \times \frac{5}{5} = \frac{80}{60}$ $\frac{1}{5} \times \frac{12}{12} = \frac{12}{60}$

$\frac{80}{60} + \frac{12}{60} = \frac{92}{60}$

$\frac{92}{60}$, really? That's not very user friendly at all. I do not want to be finding LCD's that big.

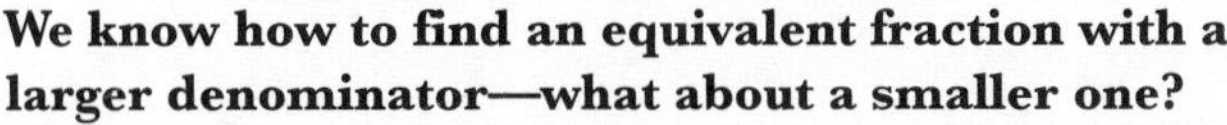

We know how to find an equivalent fraction with a larger denominator—what about a smaller one?

It happens pretty often that you get into a situation where your LCD is still pretty big, and then your answer is this big fraction. When that happens, you just need to divide by one in a way that gets you a smaller, more user friendly, but still equivalent, fraction. It's called ***reducing the fraction***.

But why is that ok?

Dividing by one doesn't change the value

Dividing by one doesn't change anything either, which means that you can divide a fraction by one (in any of its many forms) without changing the size of the piece of pie that you have.

All of this is just another way to express the identity of one. **Any number divided by one is itself**. Now the question is how do you figure out what to divide by?

Sum it up

Multiplicative Identity Property of One – Any number multiplied or divided by one is itself.

Reduce fractions by dividing by 1

To reduce a fraction, you need to find out the version (or versions) of 1 that you can divide into both the numerator and the denominator. You can do this division as many times as you want to, so to get started, you just need to find a common factor for both numbers—any one will do.

Reduce this to make it more user friendly.

These numbers are both even, so you can divide the top and the bottom by 2.

$$\frac{92}{60} = \frac{92 \div 2}{60 \div 2} = \frac{46}{30}$$

That's better, but if there's another common factor, you need to pull it out too – that way the fraction will be completely reduced.

To completely reduce a fraction, you need to remove all of the common factors from the fraction. That means you keep trying to divide out numbers until they're aren't any more in common. Since 46 and 30 are both even, you know that there's at least another 2 in there.

$$\frac{46}{30} = \frac{46 \div 2}{30 \div 2} = \frac{23}{15}$$

Ok, so that's it! 23 is prime, so there aren't any more factors left. The fraction can't be reduced any further.

You can't stop reducing the fraction until there aren't any more common factors. That means that either the numerator or the denominator must be prime, or there aren't any more factors in common.

You know, doing this over and over is boring. Is there a way to speed this up?

Factor trees can eliminate lots of little steps

There's an easy way to come up with factors of a number; it's called a **factor tree**. A factor tree is a table that lists all of the factors, so it's easier to come up with your factor list and then reduce your fractions.

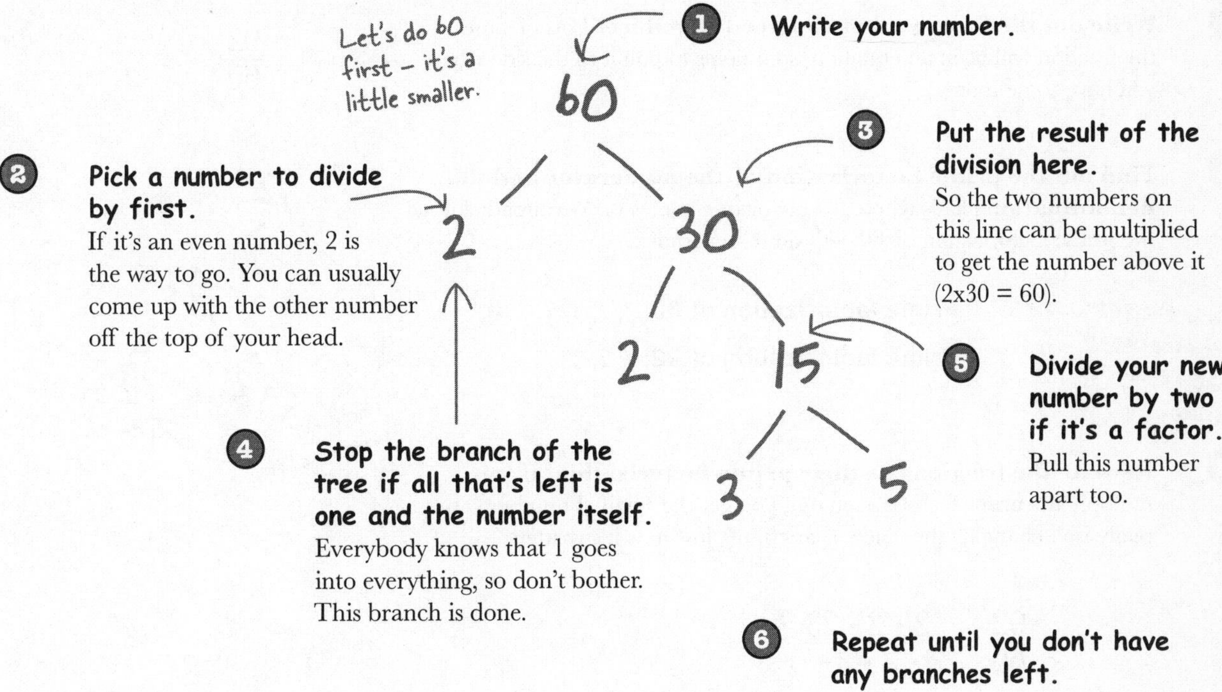

Let's do 60 first - it's a little smaller.
1 Write your number.
60
2 Pick a number to divide by first.
If it's an even number, 2 is the way to go. You can usually come up with the other number off the top of your head.
3 Put the result of the division here.
So the two numbers on this line can be multiplied to get the number above it (2x30 = 60).
2
30
2
15
3
5
4 Stop the branch of the tree if all that's left is one and the number itself.
Everybody knows that 1 goes into everything, so don't bother. This branch is done.
5 Divide your new number by two if it's a factor.
Pull this number apart too.
6 Repeat until you don't have any branches left.

Pick out the prime factors

You may not know how to read it yet, but you now have a list of all of the factors that make up your number. The form of the tree tells you one important thing, too. The end of each of the branches are the **prime factors** of the number.

If you list all of the prime factors, it's called **prime factorization**, and those are the smallest pieces that can be multiplied together to get your big number. Not only that, but multiplying them in different combinations will give you all of the factors of the numbers.

In this form, you can use the list of factors as a check list to figure out all of the factors.

Prime factorization of 60: 2, 2, 3, 5

You write both 2's out because it's the prime factorization, not just the prime factors.

Reduce fractions with the factor tree

Now that you know how to use the factor tree, you can reduce fractions quickly. Let's try our fraction again.

1. **Write out the fraction that you need to reduce.** Lots of times, the fraction will be in an equation, so it helps to pull it to the side where you have some room.

$$\frac{92}{60}$$

2. **Find out the prime factorization of the numerator and the denominator.** That's where the factor tree comes in. We already know the prime factorization of 60; let's do 92 real quick.

This is the prime factorization of 60 that we came up with earlier.

Prime factorization of 60: 2, 2, 3, 5

Prime factorization of 92: 2, 2, 23

92
2
46
2
23

These are the prime factors.

3. **Rewrite the fractions as their prime factorization.** If you multiply the prime factorization out, you get the original number. This really isn't changing the value of anything, just how it's written.

$$\frac{92}{60} = \frac{2 \cdot 2 \cdot 23}{2 \cdot 2 \cdot 3 \cdot 5}$$

4. **Divide out all the common factors.** Every factor that's in both the numerator and denominator can be removed.

$$\frac{92}{60} = \frac{\cancel{2} \cdot \cancel{2} \cdot 23}{\cancel{2} \cdot \cancel{2} \cdot 3 \cdot 5}$$

What's left here is the fully reduced fraction.

5. **Simplify the remaining factors.** If there are factors left that need to be combined again, do that before you write the final fraction.

$$\frac{92}{60} = \frac{23}{3 \cdot 5} = \frac{23}{15}$$

Prime factorizations make reducing fractions fast.

there are no Dumb Questions

Q: What's the GCF? I've heard it before, but I'm not sure what it's used for.

A: The GCF stands for greatest common factor. That is the largest number that can be pulled out of two larger numbers.

Q: I've heard of using the GCF for reducing fractions. How does that work?

A: The GCF is actually used in much the same way for reducing fractions as the LCM is used for coming up with the LCD. If you can list the factors that go into both your numerator and denominator, it's easy to find the biggest one.

Once you do that, you can simply divide the numerator and denominator by the GCF. The problem is, it's tricky to come up with the factor list. The factor tree and the prime factorization always work the first time.

Q: What's the difference between the LCD and the LCM?

A: The LCM (lowest common multiple) is a numeric property of any two numbers. You use the LCM of two numbers as the LCD (least common denominator). It's basically the LCM applied to a specific situation—fractions.

Q: Working with fractions seems to be much harder than working with decimals. Is it worth it?

A: Getting started with fractions is a bit trickier than starting with decimals. The thing is, once you get the hang of it, fractions can actually be faster and easier because they tend to stay neater than decimals.

Think about it, working with non-terminating, non-repeating decimals would not be pretty.

Q: When do you need to reduce fractions?

A: Sometimes you'll want to do it just to make the fractions easier to deal with. Working with large denominators gets pretty ugly if you need to come up with a LCD and add or subtract.

Other times, your problem will ask you to reduce the fraction, or you will get an answer that just doesn't make much sense if it's not reduced.

Q: Why is it called a factor tree?

A: Not really sure, but did you ever think that they kinda look like evergreen trees?

Q: What if I can do the factorization in my head? Do I still need to do the factor tree?

A: Nope, you can skip it (but the tree tends to keep you from messing up).

Q: Apparently dividing by one and multiplying by one are both important?

A: Very. Working with the identity property of one makes all of this fraction manipulation possible.

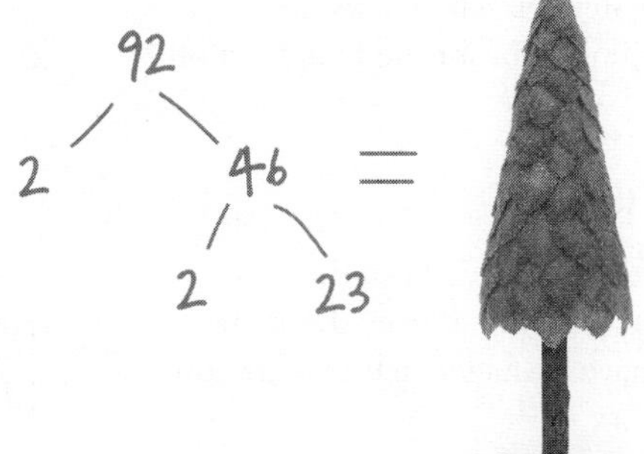

Getting started with fractions is trickier than decimals, but it's easier in the end.

Putting it all together - fractions

$\times$ **To multiply fractions**, just multiply the numerators together to get the new numerator; multiply the denominators to get the new denominator.

$$\frac{1}{2} \times \frac{1}{3} = \frac{1 \times 1}{2 \times 3} = \frac{1}{6}$$

$\div$ **To divide fractions**, just cross multiply. So multiply the first numerator by the second denominator to get the numerator solution. Then multiply the first denominator by the second numerator to get the solution for the denominator.

$$\frac{1}{3} \div \frac{1}{2} = \frac{1 \times 2}{3 \times 1} = \frac{2}{3}$$

$+$ **To add fractions**, you first need to make sure the fractions have a common denominator. After that, just add the numerators and keep the common denominator.

$$\frac{1}{3} + \frac{1}{4} =$$

$$\frac{4}{12} + \frac{3}{12} = \frac{7}{12}$$

$-$ **To subtract fractions**, you first need to make sure the fractions have a common denominator. After that, just subtract the numerators and keep the common denominator.

$$\frac{1}{3} - \frac{1}{4} =$$

$$\frac{4}{12} - \frac{3}{12} = \frac{1}{12}$$

To convert improper fractions, do the division. Just divide the numerator by the denominator. Whatever is left over as the remainder is the numerator for the final fraction.

$$\frac{3}{2} \rightarrow 2\overline{)3} = 1,\ 3 - 2 = 1 \rightarrow 1\frac{1}{2}$$

The one left over is still 1 over 2.

To reduce fractions, divide both the numerator and the denominator by the same factor until they have no factors left in common.

$$\frac{92}{60} = \frac{2 \cdot 2 \cdot 23}{2 \cdot 2 \cdot 3 \cdot 5} = \frac{23}{15}$$

Work through the different type of fraction problems, and reduce all of your answers!

$1\frac{1}{3} + 1\frac{5}{7} = 2 + \frac{1}{3} + \frac{5}{7}$

For this operation, you can add up the whole numbers, and then the fractions.

$2\frac{3}{7} \div \frac{1}{7} = ?$

$5\frac{1}{4} - \frac{1}{3} = ?$

$7\frac{5}{7} \div \frac{8}{7} = ?$

Work through the different type of fraction problems, and reduce all of your answers!

$1\frac{1}{3} + 1\frac{5}{7} = 2 + \frac{1}{3} + \frac{5}{7}$

For this operation, you can add up the whole numbers, and then the fractions.

Multiples of 3: 3, 6, 9, 12, 15, 18, 21, 24

Multiples of 7: 7, 14, 21, 28, 35, 42

$\frac{1}{3} \times \frac{7}{7} = \frac{7}{21}$ $\frac{5}{7} \times \frac{3}{3} = \frac{15}{21}$

$2 + \frac{7}{21} + \frac{15}{21} = 2\frac{22}{21}$

The fraction needs to be moved to a proper fraction.

$2 + 1\frac{1}{21} = 3\frac{1}{21}$

You can do the division to reduce this fraction, but since it's so close the denominator, you can see that it's just one and one left over.

$5\frac{1}{4} - \frac{1}{3} = ?$

$\frac{21}{4} - \frac{1}{3} = ?$

For this operation, you need to convert the first fraction to an improper, then subtract.

Multiples of 4: 4, 8, (12), 16, 20, 24

Multiples of 3: 3, 6, 9, (12), 15, 18, 21

$\frac{21}{4} \times \frac{3}{3} = \frac{63}{12}$ $\frac{1}{3} \times \frac{4}{4} = \frac{4}{12}$

$\frac{63}{12} - \frac{4}{12} = \frac{59}{12}$

This fraction needs to be converted to a proper fraction.

$$\begin{array}{r} 4 \\ 12\overline{)59} \\ -48 \\ \hline 11 \end{array} \rightarrow 4\frac{11}{12}$$

$2\frac{3}{7} \div \frac{1}{7} = ?$

First, convert the fraction.

$2\frac{3}{7} = \frac{17}{7}$

Then, cross multiply.

$\frac{17}{7} \div \frac{1}{7} = \frac{119}{7}$

The same!

Or, you can use the reciprocal and multiply..

$\frac{17}{7} \times \frac{7}{1} = \frac{119}{7}$

Do the division to reduce the fraction.

$$\begin{array}{r} 17 \\ 7\overline{)119} \\ -7 \\ \hline 49 \\ -49 \\ \hline 0 \end{array} \rightarrow 17$$

No remainder because the answer is a whole number.

$7\frac{5}{7} \div \frac{8}{7} = ?$

Convert the fraction

$7\frac{5}{7} = \frac{54}{7}$

Cross multiply or...

$\frac{54}{7} \div \frac{8}{7} = \frac{378}{56}$

Clearly this needs to be reduced.

$\frac{378}{56} = \frac{3 \cdot 3 \cdot 3 \cdot 2 \cdot 7}{2 \cdot 2 \cdot 2 \cdot 7}$

$\frac{27}{4}$

Do the division to make proper.

$$\begin{array}{r} 8 \\ 4\overline{)27} \\ -24 \\ \hline 3 \end{array}$$

$8\frac{3}{4}$

Can I just say, I am so over fractions. Is there a way to just convert them to decimals? Then I can just deal with them that way.

YES! You can convert fractions to decimals.

Now that you have the skills to reduce fractions and change their denominators, you're ready to find out how to convert decimals to fractions and back again. And since decimals and percentages are pretty much the same thing, you can convert to percentages, too.

Converting from a fraction to a decimal

Converting from a fraction to a decimal is actually very simple. **You just do the division.** We know from converting improper fractions that the line in the middle of the fraction just means division. We also know, from decimal division, that you can add as many zeros as you need after the decimal to finish the division. If you put those together, you can convert fractions to decimals.

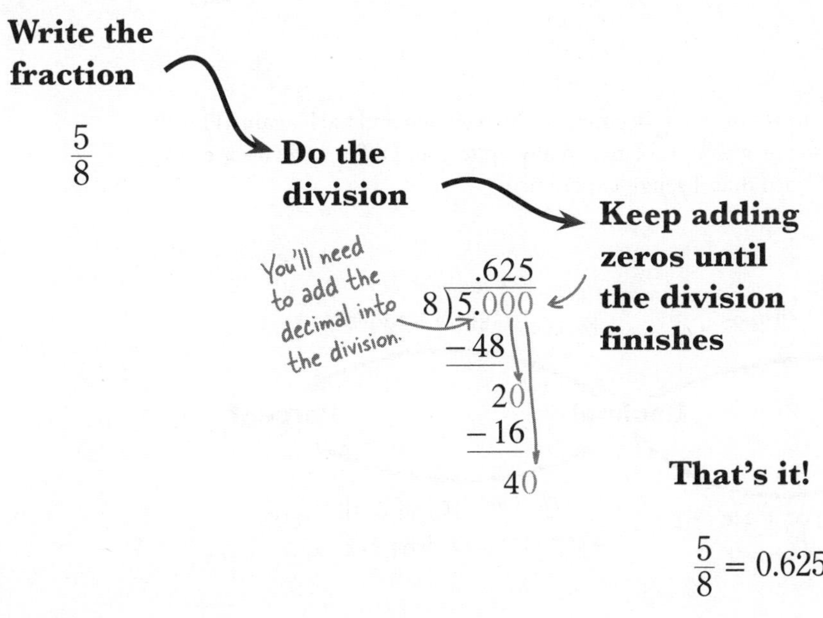

That's it!

$$\frac{5}{8} = 0.625$$

Converting decimals to fractions

The key to converting decimals to fractions is the decimal places themselves. Remember that 0.1 is one tenth? As in, one over ten. So, to convert a decimal, just drop the decimal, put the numerals of the decimal in the numerator, and then put a one and the number of zeros for the decimal (like, 1000 for thousandths) in the denominator.

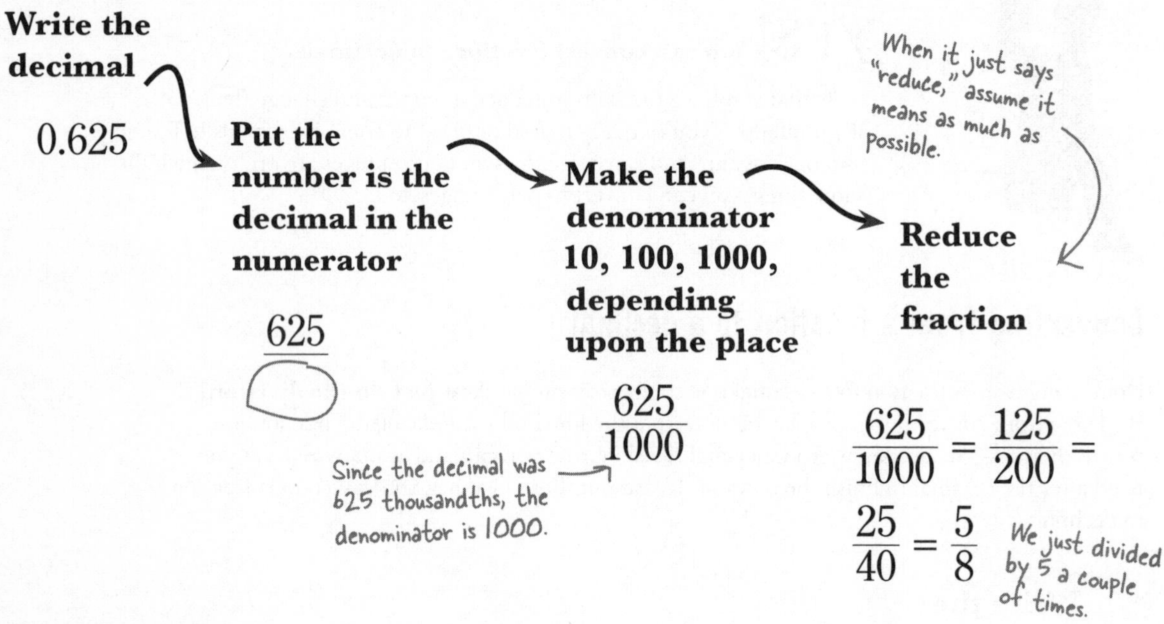

Write the decimal
0.625
Put the number is the decimal in the numerator
625
Make the denominator 10, 100, 1000, depending upon the place
625/1000
Since the decimal was 625 thousandths, the denominator is 1000.
Reduce the fraction
When it just says "reduce," assume it means as much as possible.
625/1000 = 125/200 = 25/40 = 5/8
We just divided by 5 a couple of times.

Conversions everywhere

Just to sum up—you also know how to get from the decimal to a percentage and back again. This means that which form you use to do any given problem is really up to you. Different forms are good for different things, and you'll learn that through experience.

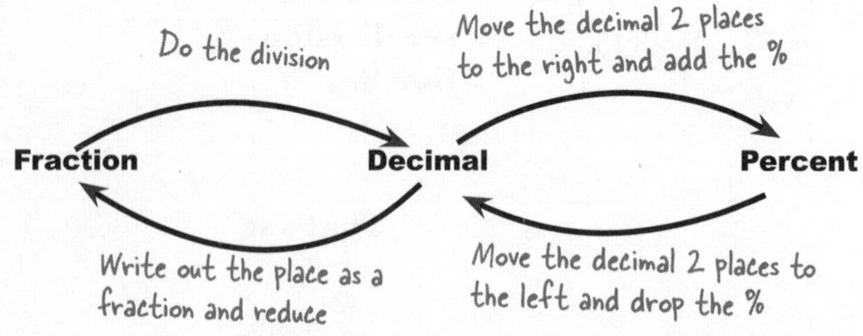

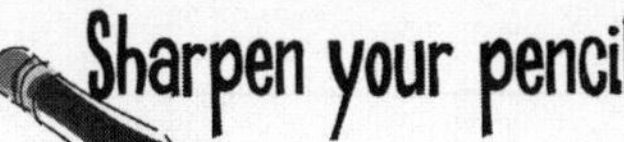

Practice converting fractions to decimals.

$\frac{13}{16} = \quad 16\overline{)13}$

$\frac{13}{16} =$ ....

$\frac{33}{250} = \quad 250\overline{)33}$

$\frac{33}{250} =$ ....

$\frac{3}{8} = \quad 8\overline{)3}$

$\frac{3}{8} =$ ....

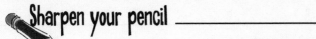

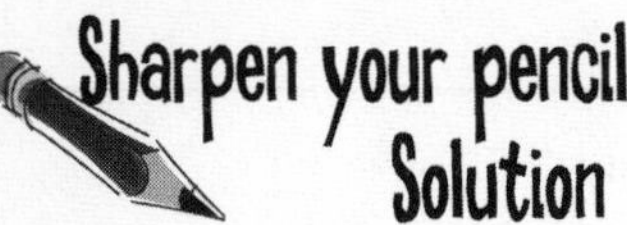

Practice converting fractions to decimals.

$$\frac{13}{16} = \quad 16\overline{)13.0000}$$

0.8125
-128
20
- 16
40
- 32
80
- 80
0

Just divide the numerator by the denominator

Carry down the zeroes

$\frac{13}{16} = 0.8125$

$$\frac{33}{250} = \quad 250\overline{)33000}$$

0.132
-250
800
-750
500
-500
0

$\frac{33}{250} = 0.132$

$$\frac{3}{8} = \quad 8\overline{)3000}$$

0.375
-24
60
-56
40
-40
0

$\frac{3}{8} = 0.375$

Division by Zero doesn't work

There's a special case with division: division by zero doesn't work (and since fractions are just division, it's a fraction problem, too). Mathematically speaking, ***division by zero is undefined***. To understand division by zero, it's best to start off with division by a few numbers close to zero. If you move to smaller and smaller numbers, a disturbing trend starts appear.

This is just re-written using regular fraction rules.

This is pretty small.

$$\frac{5}{\frac{1}{2}} = 5 \cdot \frac{2}{1} = 10$$

A pile of 10 coconuts.

This is smaller.

$$\frac{5}{\frac{1}{4}} = 5 \cdot \frac{4}{1} = 20$$

Really small

$$\frac{5}{\frac{1}{100}} = 5 \cdot \frac{100}{1} = 500$$

The closer you get to dividing by zero, the bigger the answer gets. If 1/100 gets you 500, imagine 1/1000 or 1/1,000,000! As you get closer and closer to zero, you get closer and closer to infinity. You just can't divide by zero—there's no answer.

Division by 0 is undefined. That means that there is no answer.

Sometimes multiplication takes forever!

What if you wanted to write out an expansion of multiplying the same number over and over again? Say you send a chain letter to two people, who then also send it to two people. How many people would get it by the third day?

That sounds like multiplication...

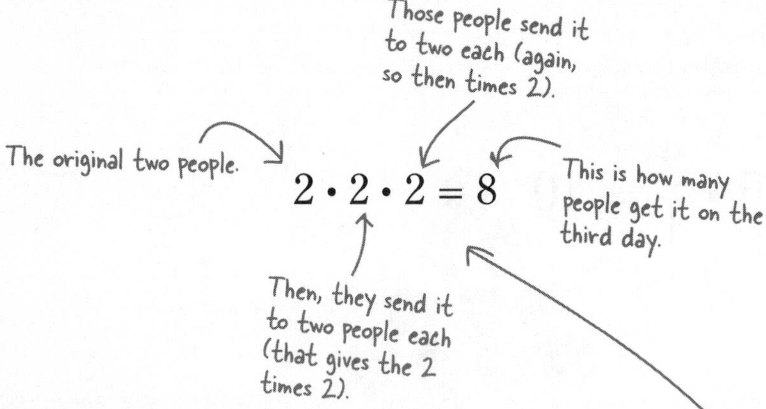
Those people send it to two each (again, so then times 2).
The original two people.
$2 \cdot 2 \cdot 2 = 8$
This is how many people get it on the third day.
Then, they send it to two people each (that gives the 2 times 2).

Is there a shorter way?

Seems like there should be an easier way to write this thing out. It's pretty easy to mess up how many 2's get written, and you're just writing the same thing over and over again. That's why we have **exponents.** An exponent is a notation above a number that means "multiply the base by itself this many times."

If we rewrite the equation we wrote before with an exponent, it would look like this:

This is the exponent, 3.

This number is called the "base." It's the number that gets multiplied.

$$2^3 = 8$$

These equations are exactly the same thing, just with exponent notation.

BRAIN POWER

Now that you have an idea of how exponents work, how many people will have the chain letter on Day 4? How about on Day 5?

How quickly things spread...

We've figured out what happens to the chain letter on the third day. What if we wanted to know about the 4th day, or the 10th day? It'd be helpful to generalize the equation.

We know that each person who gets the letter will send it on to two people—that won't change. What does change is how many days have passed, which is the exponent.

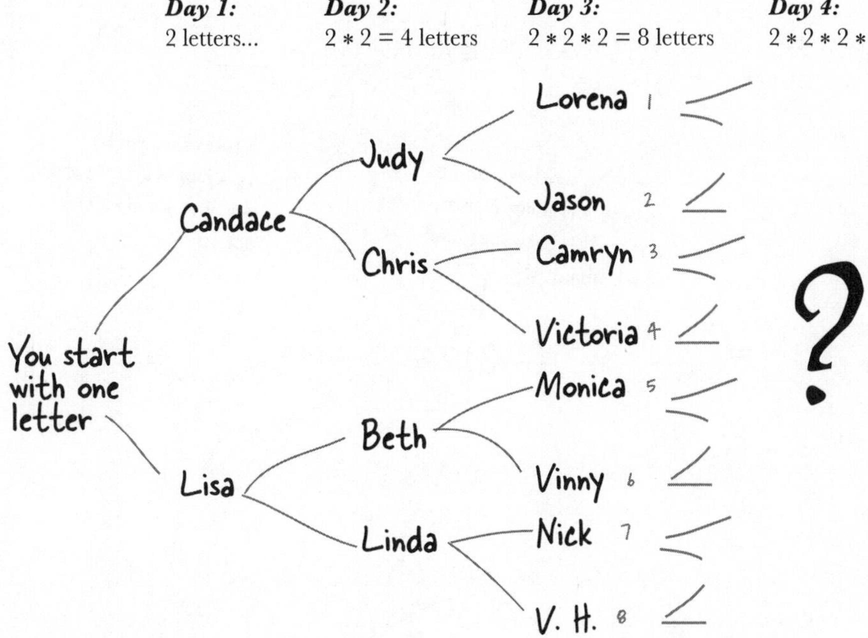

Since each person sends the letter to two other people, our base is two. Each day is basically how many 2's we have—that's our exponent.

The exponent is 4 (# of days).

$$2^4 = 2 \cdot 2 \cdot 2 \cdot 2 = 16$$

16 people get the video on day 4...

The base is 2.

Sum it up

Exponent – A superscript notation that indicates multiplication of the base that many times.

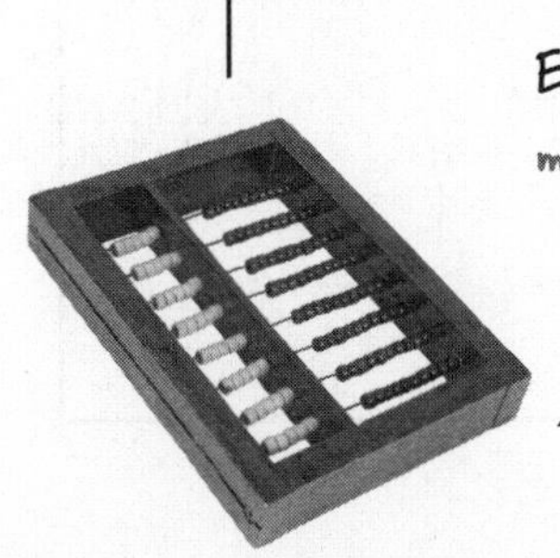

Sharpen your pencil

Simplify some exponents. Make sure to write out the whole expanded expression first. You're probably going to need a calculator for some of these.

The 5 is multiplied 5 times.

$5^5 = \dots\dots\dots = \dots\dots$

Just because this is a fraction, don't worry – just write it out like you do for whole numbers.

$\left(\frac{1}{2}\right)^2 = \dots\dots\dots = \dots\dots$

Remember, with fraction multiplication, you just go straight across.

Negative numbers don't change anything either, just write them out and think about it!

$1^6 = \dots\dots\dots = \dots\dots$

$(-3)^3 = \dots\dots\dots = \dots\dots$

Arts and crafts:

If you want to see a real life example of exponents in action, go grab a piece of paper. Fold it in half. The stack is now two sheets thick, right? Nothing exciting yet...

Then fold it in half again, and now it's four sheets thick. One more time - now it's eight sheets thick.

How would you express this as a base and an exponent so that if someone told you how many times they folded it you'd know how many sheets thick the stack was?

Why does all this matter?

Now you're ready to tackle Algebra. And Algebra is the beginning of the good stuff. You wouldn't believe the problems we've got to solve.

Kate
Jack and Kate duke it out over math to get to Australia.
Jack
Help Jo find out all about gaming system and how to pay for them.
Now, let's learn Algebra.
Paul and Amanda use Algebra to figure out the details of a road trip.
Kathleen masters the art of fantasy football, while staying under her salary cap.

Sharpen your pencil Solution

Simplify some exponents. Make sure to write out the whole expanded expression first. You're probably going to need a calculator for some of these

The 5 is multiplied 5 times.

$$5^5 = \underline{5 \cdot 5 \cdot 5 \cdot 5 \cdot 5} = \underline{3215}$$

You need to use the calculator, or you can do it by hand: 5 x 5 = 25, 25 x 5 = 125, etc.

$$\left(\frac{1}{2}\right)^2 = \underline{\frac{1}{2} \cdot \frac{1}{2}} = \frac{1}{4}$$

With integer multiplication, an odd number of negatives keeps it negative.

$$1^6 = \underline{1 \cdot 1 \cdot 1 \cdot 1 \cdot 1 \cdot 1} = \underline{1}$$

Just another expression of the identity property of one = one to any exponent is still one.

$$(-3)^3 = \underline{-3 \cdot -3 \cdot -3} = -27$$

Paper folding is an example of exponents in action.

Each time we fold it, we double the number of sheets, so the base is two.

$$2^a$$

The exponent is the number of times we fold the paper.

BULLET POINTS

- Exponents are shorthand for repetitive multiplication.
- The **base** is the number that gets multiplied.
- The **exponent** is how many times the base is multiplied.

Index

Symbols

A

B

C

D

E

G

H

I

K

L

M

N

Q

R

S

Z